Lecture Notes in Earth Sciences

91

Editors:
S. Bhattacharji, Brooklyn
G. M. Friedman, Brooklyn and Troy
H. J. Neugebauer, Bonn
A. Seilacher, Tuebingen and Yale

W0107495

Springer-Verlag Berlin Heidelberg GmbH

Iain Gilmour   Christian Koeberl

Impacts
and the Early Earth

With 164 Figures and 28 Tables

 Springer

Editors:

Dr. Iain Gilmour
Planetary Sciences Institute
The Open University
MK7 6AA Milton Keynes
United Kingdom

E-mail: i.gilmour@open.ac.uk

Dr. Christian Koeberl
University of Vienna
Institute of Geochemistry
Althansstrasse 14
1090 Vienna
Austria

E-mail: christian.koeberl@univie.ac.at

Cataloging-in-Publication data applied for

Die Deutsche Bibliothek - CIP-Einheitsaufnahme

Impacts and the early earth: proceedings of the first workshop of the
European Science Foundation Scientific Program on the Response of the
Earth System to Impact Processes/ ESF IMPACT. Ed. by Iain Gilmour and
Christian Koeberl.
(Lecture notes in earth sciences; 91)
ISBN 978-3-540-67092-6   ISBN 978-3-540-46578-2 (eBook)
DOI 10.1007/978-3-540-46578-2

"For all Lecture Notes in Earth Sciences published till now please see final pages of
the book"

ISSN 0930-0317
ISBN 978-3-540-67092-6

Typesetting: Camera ready by author
Printed on acid-free paper   SPIN: 10837475   32/3111-54321

Preface

This volume is the first one to result from the activities of the new scientific program on "The Response of the Earth System to Impact Processes" (IMPACT) of the European Science Foundation (ESF). The ESF is the European association of national funding organizations of fundamental research, with more than 60 member organizations from more than 20 countries. One of the main goals of the ESF is to bring European scientists together to work on topics of common concern. The ESF IMPACT program deals with all aspects of impact research, mainly through the organization of workshops, exchange programs, short courses, and related activities. An important aspect of the program is to bring people together to help stimulate interdisciplinary and international research.

Impacts of asteroids or comets on the Earth surface have played an important role in the evolution of the planet. The ESF IMPACT program is an interdisciplinary program aimed at understanding impact processes and their effects on the Earth System, including environmental, biological, and geological changes, and consequences for the biodiversity of ecosystems. The program is geared towards understanding of the linkage between impact processes and the Earth System, i.e., defining and studying the effects of impact events on the environment, including atmospheric, climatic, biologic, and geologic effects and interactions between these subsystems. Important aspects of future research regard also the consequences of the high-energy impact events for the biodiversity of ecosystems. Comprehending the processes that are responsible for these interactions is the key goal of the program. Currently 12 countries sponsor the program, which was launched in 1998, but an expansion to 14 countries is anticipated.

The first workshop to be organized as part of the ESF-IMPACT program was held at Robinson College, Cambridge, United Kingdom, from December 13th to 15th, 1998, in. A total of 56 scientists from 10 European countries, the United States, Russia and South Africa attended.

Contributions at the meeting covered the following themes:

- Impacts and the Origin of Life on Earth and Mars
- Late heavy bombardment and impact processing of the Earth
- The Earth's Early Cratering Record
- Micrometeorites and Spherule layers
- Impacts through geological time

A selection of 17 papers resulting from this meeting is included in the present volume. Manuscripts were reviewed by two referees and were considered for publication on the basis of originality and the themes discussed at the workshop.

There is currently some discussion on the present state of scientific understanding of the role impacts may have played in the biological and geological evolution of the Early Earth. There is good evidence that simple single celled life forms were already present at the start of the Archean, which means that chemical evolution must have occurred during the Hadean and life had probably originated at around 4 Ga. Since there are no Hadean rocks, our knowledge of the Earth's history during this period is necessarily based on indirect evidence. The Earth acquired an oceanic-type crust formed shortly after the process of core-mantle separation occurred some 4.5 Ga, although we have no record of it in the rocks. By 3.8 Ga light granitic continental crust had formed from the mantle. Proof of this comes from the 3.8 Ga granites and associated metamorphosed volcanic and sedimentary rocks at Isua in western Greenland, which represent the oldest piece of primitive crust known to geologists. Conditions at the Earth's surface and in the atmosphere during the Hadean must have had a powerful influence on chemical evolution. Estimates for the growth of continents on the early Earth cover a wide range of parameters save that there were no continents at the beginning. This results in most scientists assuming that a global ocean existed in which the pre-biotic chemistry that led to the origin of life to place.

The possibility of continued intense bombardment by meteorites long after the Earth had formed has led some scientists to conclude that the emergence of life could have been deterred for hundreds of millions of years. Many of the projectiles would have been much large enough that they would have generated enough heat to boil the surface of the oceans as well as throwing large clouds of dust and molten rock into the atmosphere. The implication of this hypothesis is that the impacts would have deterred the emergence of life anywhere near the surface of the oceans until perhaps as late as 3.8 billion years ago. On the other hand, comets and meteorites may have delivered important organic molecules to the Earth. The question of how this delivery could have influenced the development of life on the early Earth, and if bacterial life forms could survive the forces of an impact, are discussed in the paper by Burchell et al. The macromolecular organic materials present in meteorites and their role in the origin of life on Earth are subsequently discussed by Sephton and Gilmour, and in a related paper, Wright et al. discuss the effects of atmospheric heating on carbon compounds in meteorites and micrometeorites.

Only 10% of the 150 or so known impact craters on Earth date from the Precambrian, an Era spanning some 88% of the Earth's history. The Precambrian encompasses fundamental events in the origin and evolution of our planet from the origin of life itself, the development of continents, to the development of eukaryotic organisms and the vast diversification of life that occurred at the end of this Era. We know from the lunar cratering record that the Earth was subject to intense bombardment by asteroids and comets early in its history on a scale far greater than anything it has experienced since. Koeberl et al. discuss their search for evidence of such a heavy bombardment on Earth in about 3.8 to 3.9 million-year-old rocks from Greenland.

Another hot topic is the geological evidence for Early Archean impact events, in particular the occurrence of spherule layers in the Precambrian Barberton formation (South Africa), which may represent the remnants of large-scale impact events. This is a controversial area, because many of the normally accepted criteria for the identification of impact structures, such as shock metamorphism in rock-forming minerals, are not preserved. Shukolyukov et al. present new data based on chromium isotopic ratios that they argue proves that the large platinum group element enrichments associated with these spherule layers are extraterrestrial in origin. In contrast, Reimold et al. argue that the enrichments are too great to be of primary origin and suggest that the mineralogical and geochemical features of the spherule layers are the result of (secondary) mineralization.

The importance of the recognition of an impact record in the Precambrian sedimentary succession is also emphasized by Simonson et al. Impact structures per se, as well as their proximal ejecta blankets, are limited to areas on the order of a few hundred kilometers in diameter, whereas distal ejecta from the largest impacts can be distributed globally. These authors examined various late Archean to Paleoproterozoic formations from Australia and South Africa that contain preserved spherule layers. They interpreted these layers as altered microtektites and microkrystites and therefore inferred that they have an impact origin.

A different type of extraterrestrial deposit is described by Kettrup et al., who discuss occurrences of fossil micrometeorites from about 18 to 1.9 billion year old sediments in Finland. It is astonishing that these materials have survived with only little alteration.

A major theme of the workshop was the need for the better recognition of ancient impact events in the geological record. Vishnevsky and Raitala describe the use of impact-produced diamonds as indicators of shock metamorphism in Precambrian rocks. Gibson and Reimold provide a useful review of the Vredefort impact structures in South Africa (with an age of 2.02 billion years the oldest, and by chance also the largest known, impact structure on Earth) as a case study for old, deeply eroded impact structures. Kenkman and co-workers cover techniques of structural geology in their contribution that might be applied in future from a better understanding of the geology of large ancient impact structures. In a related paper, Abels et al. describe the use of remote sensing techniques in the study of old and deeply eroded impact structures.

More general topics follow. Hughes provides a review of cratering rate investigations, and Jones et al. discuss rarely studied features of carbonate rocks that result of from impact-induced melting processes. At the end is a trio of papers by Lilljequist, Suuroja and Suuroja, and Puura et al., describing various Scandinavian structures as case studies of more recent impact events.

From these contributions, and from the discussions at the workshop, it is clear that impact cratering events have played a major role in the very early history of our planet. We are only beginning to try and decipher the evidence of such early events, as we currently do not even have good criteria for the recognition of Early Archean impact events, and a lot remains to be done. We hope that the current volume is a first step in the right direction.

Iain Gilmour   Christian Koeberl
Open University   University of Vienna
Milton Keynes, UK   Vienna, Austria

December 1999

Acknowledgments

The editors are grateful to the ESF IMPACT program for organizing the workshop and for providing financial support for the preparation of this proceedings volume. We appreciate the help and cooperation of all reviewers and authors in trying to produce this book in a timely fashion. We furthermore thank Springer Verlag, especially W. Engel, A. Weber-Knapp, and A. Bernauer-Budimann for their support of this project.

Contents

List of Contributors

A. Abels
Institute for Planetologie
D-48149 Münster
Germany

L. Bischoff
Institute for Planetologie
D-48149 Münster
Germany

A. Bunch
BioSciences Laboratory
University of Kent
Canterbury
Kent CT2 7NR
United Kingdom

M.J. Burchell
Unit for Space Sciences and Astrophysics
Physics Laboratory
University of Kent
Canterbury, Kent CT2 7NR
United Kingdom

G.R. Byerly
Department of Geology and Geophysics
Louisiana State University
Baton Rouge, LA 70803
USA

P. Claeys
Museum für Naturkunde
Institut für Mineralogie
Humboldt-Universität Berlin
Invalidenstraße 43
D-10115 Berlin
Germany

A. Deutsch
Institute for Planetologie
D-48149 Münster
Germany

R.L. Gibson
Department of Geology
University of the Witwatersrand
Johannesburg 2050
South Africa

I. Gilmour
Planetary Sciences Research Institute
The Open University
Milton Keynes, MK7 6AA
United Kingdom

S. Hassler
Department of Geological Sciences
California State University
Hayward, CA 94542
U.S.A

S. Heuschkel
Museum für Naturkunde
Institut für Mineralogie
Humboldt-Universität Berlin
Invalidenstraße 43
D-10115 Berlin
Germany

M. Hornstein
Department of Geology
Oberlin College
Oberlin, OH 44074-1044
U.S.A

D. Hughes
Department of Physics and Astronomy
The University
Sheffield S3 7RH
United Kingdom

B.A. Ivanov
Institute for Dynamics of Geospheres
Russian Academy of Science
Moscow 117939
Russia

S. Johnson
Impact Cratering Research Group
Department of Geology
University of the Witwatersrand
Johannesburg 2050
South Africa

A.P. Jones
Department of Geological Sciences
University College London
Gower Street
London, WC1E 6BT
United Kingdom

A. Kärki
Department of Geosciences
University of Oulu
P.O. Box 333
FIN-90571 Oulu,
Finland

T. Kenkmann
Museum für Naturkunde
Institut für Mineralogie
Humboldt-Universität Berlin
Invalidenstraße 43
D-10115 Berlin
Germany

D. Kettrup
Institute for Planetologie
D-48149 Münster
Germany

K. Kirsimäa
Institute of Geology
University of Tartu
Vanemuise 46, 51014
Tartu
Estonia

J. Kirs
Institute of Geology
University of Tartu
Vanemuise 46, 51014
Tartu
Estonia

C. Koeberl
Institute of Geochemistry
University of Vienna
Althanstrasse 14
A-1090 Vienna
Austria

M. Konsa
Institute of Geology
Tallinn Technical University
Estonia Ave 7
10143 Tallinn
Estonia

F.T. Kyte
Center for Astrobiology
Institute of Geophysics and Planetary Physics
University of California
Los Angeles, CA 90095-1567
USA

R. Lilljequist
North Atlantic Natural Resources
Kungsgatan 62, S-752 18
Uppsala
Sweden

D.R. Lowe
Department of Geological and Environmental Sciences
Stanford University
Stanford, CA 94305
USA

G.W. Lügmair
Max-Planck- Institute for Chemistry, Cosmochemistry
PO 3060
55020 Mainz
Germany

I. McDonald
School of Earth & Environmental Sciences
University of Greenwich
Chatham Maritime
Kent, ME4 4AW
United Kingdom

M. Niin
Geological Survey of Estonia
Kadaka tee 80/82
12618 Tallinn
Estonia

L.J. Pesonen
Geological Survey of Finland
FIN-02150 Espoo
Finland

P. Pihlaja
Geological Survey of Finland
FIN-02150 Espoo
Finland

C.T. Pillinger
Planetary Sciences Research Institute
The Open University
Milton Keynes, MK7 6AA
United Kingdom

J. Plado
Institute of Geology
University of Tartu
Vanemuise 46, 51014
Tartu
Estonia

V. Puura
Institute of Geology
University of Tartu
Vanemuise 46, 51014
Tartu
Estonia

J. Raitala
University of Oulu
P.O Box 3000 Fin-90401
Oulu
Finland

W.U. Reimold
Impact Cratering Research Group
Department of Geology
University of the Witwatersrand
Johannesburg 2050
South Africa

M. Rosing
Geologisk Museum
Oster Voldgade 5-7
DK-1350 Copenhagen K
Denmark

M.A. Sephton
Planetary Sciences Research Institute
The Open University
Milton Keynes, MK7 6AA
United Kingdom

N.R.G. Shrine
Unit for Space Sciences and Astrophysics
Physics Laboratory
University of Kent
Canterbury, Kent CT2 7NR
United Kingdom

A. Shukolyukov
Scripps Institute of Oceanography
University of California San Diego
La Jolla, CA 92093-0212
USA

B.M. Simonson
Department of Geology
Oberlin College
Oberlin, OH 44074-1044
U.S.A

D. Stöffler
Museum für Naturkunde
Institut für Mineralogie
Humboldt-Universität Berlin
Invalidenstraße 43
D-10115 Berlin
Germany

K. Suuroja
Geological Survey of Estonia
Kadaka tee 80/82
EE12618
Tallinn
Estonia

S. Suuroja
Geological Survey of Estonia
Kadaka tee 80/82
EE12618
Tallinn
Estonia

S. Vishnevsky
Institute of Mineralogy and Petrology
Russian Academy of Sciences
Novosibirsk
Russia

I.P. Wright
Planetary Sciences Research Institute
The Open University
Milton Keynes, MK7 6AA
United Kingdom

P.D. Yates
Planetary Sciences Research Institute
The Open University
Milton Keynes, MK7 6AA
United Kingdom

J.C. Zarnecki
Unit for Space Sciences and Astrophysics
Physics Laboratory
University of Kent
Canterbury, Kent CT2 7NR
United Kingdom

H. Zumsprekel
Institute for Planetologie
D-48149 Münster
Germany

1 Exobiology: Laboratory Tests of the Impact Related Aspects of Panspermia

M.J. Burchell[1], N.R.G. Shrine[1], A. Bunch[2], and J.C. Zarnecki[1]

[1]Unit for Space Sciences and Astrophysics, Physics Laboratory, University of Kent, Canterbury, Kent CT2 7NR, United Kingdom.
[2]BioSciences Laboratory, University of Kent, Canterbury, Kent CT2 7NR, United Kingdom.

Abstract. The idea that life began elsewhere and then naturally migrated to the Earth is known as Panspermia. One such possibility is that life is carried on objects (meteorites, comets and dust) that arrive at the Earth. The life (bacteria) is then presumed to survive the sudden deceleration and impact, and then subsequently develop here on Earth. This step, the survivability of bacteria during the deceleration typical of an object arriving at Earth from space, is studied in this paper. To this end a two-stage light gas gun was used to fire projectiles coated with bacteria into a variety of targets at impact speeds of 3.8 to 4.9 km s^{-1}. Targets used were rock, glass, metal and aerogel (density 100 kg m^{-3}). Various techniques were used to search for bacteria that had transferred to the target material during the impact. These included taking cultures from the target crater and ejecta, and use of fluorescent dyes to mark sites of live bacteria. So far only one sample has shown a signal for bacteria surviving an impact. This was for bacteria cultured from the ejecta spalled from a rock surface during an impact. However, this result needs to be repeated before any firm claims can be made for bacteria surviving a hypervelocity impact event.

Introduction

The Question of the origin of Life

During the 19th century, what were previously certainties in Western thought came under increasing attack. One of these concerned the origin of life. Certainties offered by religious explanations for the origin of life were challenged, and a new source for the origin of life had to be found. Many lines of thought emerged, but two stand out, spontaneous creation (with the drawback that this had never been observed) and Panspermia, the origin of life away from the Earth and its natural migration to here (again, never observed and which, in some senses, merely serves to push the problem to one side, i.e., life still had to originate somewhere). As indicated, both of these ideas had their drawbacks.

One consequence of spontaneous creation is that it might happen anywhere. So by the end of the 19th century, literature had populated the other planets of the Solar System with a variety of life-forms. It seems to have been widely accepted that however it arose, life was not unique to the Earth. This was of course mostly based upon ignorance of the true nature of the environments of other planets. However, without a firm knowledge (true or false) of how life originated and what conditions are necessary for its survival, speculation and fantasy are not unreasonable. Since then, advances in astronomy and exploration by spacecraft, have revealed the true (non-Earth like) nature of many of the possible habitats in the Solar System, and have also found a lack of evidence for life elsewhere than the Earth.

As a field, exobiology (the study of life away from the Earth) would therefore seem to be a problematic discipline. But there are many significant reasons why this is not so. For example, the understanding of life on Earth has increased enormously in the 20th century. The role of viruses, DNA, etc., has opened up a whole new world of activity. Equally, material from only two non-terrestrial bodies has been studied by experiments searching for life (ignoring studies of meteorites that may have undergone substantial modification on their journey to Earth). One such body is the Moon, which has been studied in-situ and from which some 382 kg of rocks were returned by the Apollo missions in the early 1970s. Automatic Russian craft also returned 321 g of lunar samples to the Earth (many references to this are in Russian, but a readily accessible English source is Surkov 1997, pp 98-110). The other place where explicit searches for life have been carried out is Mars (studied in-situ by the bio-science instruments of the two Viking landers in 1976, see Synder 1979 for a summary).

In both cases one should ask if the lack of definitive signals for life could be taken as indicative of a lack of life on the whole body in question. On the Moon the landing sites were relatively few in number and sample collection was limited. However, the Moon appears to be a fairly uniformly inert body, with no history of a significant atmosphere or evidence for extensive periods with liquid water. It is presumed to have cooled relatively early in its history. Given this, the lack of biological activity in the returned samples is widely held to be significant.

On Mars, the situation is different. The Viking landers carried a set of three biological experiments (Klein 1978). The instruments used were designed over 25 years ago and were deployed in just two places, using samples that were picked up off the surface. The experiments were designed to look for evidence of metabolic activity by placing soil samples in incubators and monitoring the evolved gases. One experiment looked for the release of gas (oxygen) after the soil was moistened and again after the moistened soil was added to a nutrient broth. Sudden releases of oxygen were found, even after just the moistening step. The other two experiments used radioactive carbon as a tracer. In one, the carbon was in the nutrient added to the soil, and the atmosphere was sampled to see if it was being processed and released as a gas (e.g., carbon dioxide). The result was positive, and when the soil was heated (hopefully to kill any bacteria) the process stopped. Taken by itself this seemed a very positive result. In the last experiment, the radioactive carbon was in the atmosphere (in the experimental chamber), which was monitored to see if the soil samples took up the carbon. Again a positive result was obtained, and when heated the rate of take-up decreased (although not to zero). However, viewed as a whole, the consensus of most of the scientists involved was that the soil was highly reactive and was releasing or processing the various materials involved via purely chemical (non-biological) processes which had not initially been expected. These conclusions were undoubtedly influenced by the non-observation of organic material in the same soil by other experiments on the Viking landers (Bieman 1977). This is commented on, for example, in Horneck (1995) pp 199-200, and also in McKay et al. (1998), and serves to illustrate the difficulty of carrying out definitive studies by remote control where the general physical properties of the target body are not already well understood. A more complete summary of how to carry out searches for life in the Solar System using spacecraft is given in Fitton (1997).

However, for Mars the lack of clear cut results from the Viking landers is not considered by all scientists to be the totally definitive result. One can imagine many more interesting sites to investigate on the Martian surface (e.g., near the polar caps, in the permafrost believed to be sub surface on Mars, beneath former river beds, etc.) than just the two sampled. Also, whilst life may not be present now, it may have been in the past and left fossil remains which can be searched for.

Given the restricted nature of these searches for life elsewhere in the Solar System, the lack of a positive result should not be considered definitive. Meanwhile, studies of life here on Earth have advanced even in the last 25 years. The list of possible habitats where life is found here on Earth seems to expand continually, as does the range of conditions under which bacteria can live (see

Madigan and Marrs 1997 for a recent review, or Davies 1998 Chap. 7). Currently there is evidence that terrestrial life manages to survive in a wide range of conditions; e.g., temperatures up to 113 °C (with the lower limit below 0 °C, but how low is unclear), salinity up to 37%, pH values in the range 0.7 to 11.5 and pressures up to 500 atmospheres. Indeed, the meaning of the word 'life' itself undergoes change, as questions of whether spores are alive, whether viruses are alive, and so on, undergo debate.

Thus, given that searches for life in the Solar System away from the Earth have been relatively rare and that the limits on the possible environments for life are undergoing continual expansion, then exobiology should still be considered a vital discipline in science even though there is no answer as to how life first appeared on Earth or knowledge as to whether it exists (or existed) on other planets. Even if exobiology fails to answer the question of how life started, if evidence were found for life elsewhere in space, it would at the very least revolutionize mankind's view of the Universe and the role we play in it. An example of how great an issue this is, was the impact of the recent reports concerning possible fossil remains in the Martian meteorite ALH84001 (McKay et al. 1996). Many scientists subsequently doubted the interpretation of the structures observed (e.g., Harvey and McSween 1996 or Grady et al. 1997), but the enormous publicity surrounding the matter indicates the importance of the search for evidence of life elsewhere (see, for example, Jakosky 1998 Chap. 9, or Davies 1998 pp 170-177 for fuller discussions of ALH84001).

Panspermia

Interestingly, part of the search for life away from the Earth can be carried out on the Earth itself. Ancient literature and pictorial records indicate strong evidence for observations of meteor showers throughout history. Many summaries are readily available (e.g. Heide and Wlotzka 1995 and Norton 1994) and it is clear that on countless occasions meteorites have been recovered and used for a variety of purposes (for worship, as talismans, as sources of metals, etc.). However, the realization that meteorites were objects which had come from space has only gained widespread acceptance in our Western civilization in comparatively recent times. As recently as the 18th century, claims that objects fell from the skies were often denounced as nonsense by scientific investigators. It was only as the 18th century ended and the 19th began, that this situation changed, with reputable scientists accepting the evidence that rocks and metal objects were indeed falling from the skies. Debate still raged as to whether these were rocks somehow condensed from the air, or ejected from volcanoes etc. But there was soon general acceptance that they did indeed come from space.

Once the idea that "messengers from space" were arriving at the Earth gained ground, other speculations followed. If life existed elsewhere, and if the meteorites came from elsewhere, maybe they carried life? Early in the 19th century, chemical analysis of a French meteorite showed that it contained carbon compounds. By the late 19th century people were claiming to have found fossils in meteorites (see Davies 1998 p 186 and following). Although derided, such claims fuelled speculation about life raining down on the Earth from space. A well-known account of such ideas is by Arrhenius (1908). Although life itself has not been discovered, much evidence has since been accumulated for the existence of organic molecules in space. By the middle of the 20th century astronomers had shown that organic molecules were present in interstellar clouds. By the 1960s analytic techniques had advanced sufficiently that studies of an old meteorite (the Orgueil meteorite) revealed organic material and hydrocarbons. Criticisms of these results rapidly emerged, including seemingly unanswerable questions about possible terrestrial contamination of the meteorite during its time on Earth. However, in 1969 in Australia a new meteorite (the Murchison meteorite) fell to Earth and was rapidly recovered. Investigations revealed organic material, including many types of amino acids, which could not have been terrestrial contaminants. In parallel to this ample evidence has accumulated from other sources that there is organic material widely distributed throughout the Solar System. For example the mass spectrometer onboard the Giotto spacecraft that flew past Halley's comet in 1986 found that some of the material ejected from the comet was organic (Kissel and Krueger 1987). Equally, spectroscopic measurements of Titan (a moon of Saturn) have revealed a wealth of organic chemistry occurring in its atmosphere (Samuelson et al. 1983).

Whilst not demonstrating the existence of life elsewhere, such results do show that the building blocks of life (organic molecules, amino acids etc) are widely distributed throughout the Solar System. If so, then perhaps life did evolve somewhere else, and perhaps meteorites could bring examples of it to the Earth. A mechanism needed to be found, however, that would remove life from its original habitat and place it in a meteorite. Several such possibilities have been proposed.

For example, one might imagine life evolving in space itself. This idea has waxed and waned in popularity over the years. In the United Kingdom perhaps the leading proponents of this theory for many years have been Sir Fred Hoyle and Professor N.C.Wickramasinghe. In essence they call upon many different ideas to suggest that bacteria are present in space as individual objects independent of a planet to live on. For a brief summary of their ideas see Hoyle (1998) or Hoyle and Wickramasinghe (1981). They note that bacteria have properties of radiation hardness and survivability in vacuum to a degree that is quite unnecessary to survive in a terrestrial environment. They then assume a usually unnamed principle of evolution, to wit that if a life form has properties that are not needed in its present mode of life, then it needed them in the past at an earlier stage in its evolution. They thus suggest that since these characteristics are not needed on Earth, the habitat where they would have been needed is in space. This implies an extraterrestrial origin for bacteria. Taking this further, they have shown that light extinction curves found by astronomers, can be explained by absorption of light by bacteria in space (Hoyle 1998 pp 16-18), although other explanations also exist.

Hoyle and co-workers also note that life appears to have been present on the Earth very soon after the end of the period of intense bombardment from space that occurred early in the Earth's history. The exact dates for the end of the period of intense bombardment and the origin of life are still slightly uncertain. It is thought the Earth formed some 4.55 billion years ago, with the period of intense bombardment ending some 4.0 to 3.8 billion years ago. Rocks which date from up to 3.85 billion years ago contain carbon isotope ratios compatible with those found in living organisms (the slight mass difference between carbon-12 and carbon-13 causes the former to be favoured in chemical processes used by living organisms). However, this may not indicate the presence of life as non-biological chemical processes can show the same bias. The first single cell fossils appear in rocks dated from 3.6 to 3.5 billion years ago. Whichever one uses as the date for the first appearance of life (the carbon isotope ratio rock data or the appearance of fossils) life emerged relatively quickly once the heavy bombardment from space ended. During the period of heavy bombardment some impacts would have generated a heat pulse sufficient to vaporize the Earth's oceans and heat the planetary surface sufficiently to kill all life. As the period of intense bombardment ended, smaller impacts would still have occurred with decreasing frequency. What Hoyle and Wickramasinghe (and others) note is that there is a continuous chain of life on Earth stretching back to almost when the life-killer impacts stopped, indicating that life appeared almost immediately after stable conditions emerged. They then suggest that if bacteria exist between stars and are carried to Earth on meteorites or comets, then life will have continually seeded the Earth (and been wiped out by killer impacts), hence its apparent origin on Earth as soon as stability occurred.

A mechanism favoured by Hoyle and Wickramasinghe for delivering the bacteria to Earth is via comets arriving from the outer Solar System. The inner Solar System was heated during its formation to a temperature that probably killed any bacteria, but icy comets from the outer Solar System would have preserved the bacteria and brought it direct to the Earth's surface in an impact. Equally the bacteria could have arrived on grains from the tail of a comet which passed by the Earth. The use of dust grains to carry the bacteria would deliver the bacteria (or virus) over a large part of the Earth almost instantaneously. They then claim this allows the rapid spread of pandemics (i.e., illness that infect large numbers over wide areas, spreading faster than can be allowed for by normal transmission methods). This claim was again widely reported at the time of the recent Leonid meteor shower (Wickramasinghe 1998).

Such claims for bacteria evolving in space and travelling to the Earth represent Panspermia in perhaps its purest form. A variant does exist, which some find more plausible. This is the so-called Rocky Panspermia (see Melosh 1988). In this idea life evolved on a planet (or minor body) which was subsequently struck by a large meteorite (at least 10 or 20 km across) at a high impact speed. Due to orbital mechanics, collisions of bodies in the Solar System usually occur anywhere from a few to over 50 km s^{-1}. Such a collision is called a hypervelocity impact. These collisions are very violent, resulting in a crater much larger than the dimensions of the impacting object, which is itself mostly vaporized in the impact. During the later stages of the crater formation, material, which was originally part of the target surface, is thrown off at high velocity. This can exceed the local escape velocity, resulting in ejection of material off the planetary surface into space. Life forms (or fossils) could be trapped in this ejected material.

Dependent on its velocity, the ejecta can travel into interplanetary space and a variety of outcomes are possible. It may collide with a natural satellite (moon) of the parent planet, it may enter a relatively stable orbit around the Sun, or eventually it may collide with a second planet. The requirements for rocky Panspermia are thus fulfilled: A possibly life containing object may migrate perfectly naturally from one planet to another.

Rocky Panspermia

The ingredients of rocky Panspermia, advanced above, are all assertions. To place the idea onto a sounder footing requires evidence for the various steps in the argument. One important ingredient is that large impacts should have occurred on planetary surfaces. In the early part of the Solar System's history, these were frequent. Evidence is readily visible on the lunar surface in the form of large craters. The cratering record on the Earth itself is however subject to loss due to erosion by wind and rain and by geological processes, yet craters are found. At the start of this century Meteor Crater in Arizona (USA) was extensively surveyed and explored (it is about 1.2 km wide and nearly 200 m deep). Claims for a meteoritic origin were advanced and disputed. Searches for the buried bulk of the meteorite failed (although small fragments were readily found), weakening the claim for a meteoritic origin for the crater. However, it was subsequently realized that due to the hypervelocity nature of the impact, the shock waves would have deposited so much energy into the meteorite that most of it would have vaporized on impact.

Despite such doubts, other impact craters have subsequently been identified on the Earth and the meteoritic origin of 12 large craters is now accepted. In total, there are approximately 150 craters held to be probable meteorite impact sites (see Grieve 1998 for a discussion). These have diameters than range up to 250-300 km and their ages vary from an estimated two billion years to a few thousand years. Thus occasional large impacts still occurred after the end of the period of intense bombardment. The violent impacts that caused these large craters may well have ejected material back into space, an important pre-requisite for rocky Panspermia. And of course, not only will this have occurred on Earth, but also on other planets.

As well as large impactors (yielding large craters), there are also smaller objects arriving at the Earth's surface. The shock process related to an impact is not restricted to the moment of impact on the planetary surface. As soon as an object encounters the atmosphere it is subject to heating of its surface and shock waves are generated in its interior. For some meteorites the resulting internal stresses can be sufficient to exceed the strength of the object, causing it to explode during its descent through the atmosphere. The result is that the arrival at the top of the atmosphere of a large object can result in a fall of a multitude of smaller objects onto the planetary surface. A mid-air explosion may accompany this, as if a large bomb has gone off. The most famous example of this in recent times is probably the Sikhote-Alin meteorite, which fell in Siberia in 1947 (see Norton 1994 pp 100-109). In this event a meteor was visually observed to explode in the atmosphere, and atmospheric shock waves were noted up to 150 km away. At the main impact site over 120 craters were subsequently found (in which iron meteorite fragments were found with masses ranging up to 2000 kg), the largest crater being just some 30 metres across. Siberia of course also witnessed the Tunguska event in 1908, when an object (now widely believed to be a comet) effectively detonated in the air causing surface damage over an area 50 km across, although without any large craters being formed.

There are also small meteorites as well as large ones arriving at the Earth, and these can descend relatively intact to the surface. Whether the small meteorites are part of a shower (from the disruption of a parent body during its entry) or are an individual fall, they do not leave craters on the Earth's surface. They are decelerated during their descent through the atmosphere and embed themselves into the Earth's surface in shallow pits or trenches. Some may even end up just sitting on the surface. Either way they survive relatively intact (unlike their larger counterparts which vaporize on impact and leave a large impact crater) and are fairly accessible. Thus far, several thousand such meteorites have been collected and sampled.

As can be seen from the above, objects arriving from space do not necessarily completely vaporize on impact. Thus bulk material can be transferred to the Earth's surface in an impact, another pre-requisite for rocky Panspermia. However, it should be noted that the majority of objects impacting the Earth do not originate on another planetary surface. Indeed it is only in the last 20 years that evidence has emerged that any meteorites did indeed originate from another planetary body. Key to this discovery are two classes of meteorites that have been found here on Earth. These are the lunar meteorites and the SNC meteorites. The former are identified because some meteorites have been shown to have the same composition and mineralogy as samples of Lunar rocks retrieved by the US and USSR space programs (Delaney 1989, Eugster 1989). The first of these to be clearly identified was ALH81005 (whose name identifies it as the 5th sample classified of those recovered in the Allan Hills in Antarctica in 1981) and several more have been similarly categorized. The existence of Lunar meteorites here on Earth, shows that the necessary transfer mechanism for materials to move from one Solar System body to another does indeed exist.

The SNC meteorites are held to come from Mars (McSween 1994, Vickery and Melosh 1987). Rocky meteorites come in two main classes, chondrites and achondrites (for an extensive discussion of these see Taylor 1992 Chap. 3). The former contain chondrules, small spheres (mostly olivine or pyroxene) typically 1 mm in diameter, which comprise upwards of 40% of the mass of the meteorite. The chondrites are believed not to have undergone extensive heating during their formation and have thus not been resident in a larger body. The achondrites do not contain chondrules, and are differentiated materials showing evidence of formation from a melt. The achondrites are sub-divided into several sub-classes, one of which (SNC) is so-distinct that it is often considered separate from the rest of the achondrites. The mineral content of the SNCs labels them as shergottities, nakhlites or chassignites (named after standard meteorites of each type and hence SNC). What is significant is that their ages are found to range from 180 million to about 1.3 billion years. If they had formed in space during the formation of the planets themselves, they should all be at least 4 billions years old. It is thus inferred that they were formed in a planetary body, which continued to have a volcanic/magmatic activity long after its formation. This suggests their origin as being from a large terrestrial planet; Mars being the favoured point of origin for these meteorites. This latter conclusion was reached after isotopic ratios of gases found trapped in pockets in the rocks (Bogard 1982), were found to match the results from similar isotope ratio measurements made of the Martian atmosphere by the Viking landers (for a discussion see for example Pillinger 1988, or McSween 1994, and references therein). The implication is clear, the SNC meteorites arrived on Earth after being ejected somehow from Mars. Much modelling of the transfer process has now being carried out following early work by Wetherill (1984). Recent work (e.g., Gladman et al. 1996 and Gladman 1997) has focussed on establishing possible timescales for transfers, most likely destinations for material ejected from Mars and so on. Whilst timescales for transfers can be long, in Gladman (1997) it was found that, as a minimum, it may take only a few years for material to move from Mars to the Earth. Of further interest is the finding of Gladman (1997) that meteorites ejected from Mars are more likely to reach the Earth than vice versa. Thus, the transfer mechanism for Panspermia from Mars to Earth may be more likely than that from Earth to Mars.

In exobiology terms the transfer mechanism for moving materials around the Solar System is thus established, but where is the evidence for life being transferred? There are several stumbling blocks to supposing that viable life-forms can be transferred via the meteorite route. Perhaps the most worrying objection is that in impacts on atmosphere-less surfaces, the meteorite can be expected to vaporize. For impacts onto bodies that possess an atmosphere, heating can be expected during passage through the atmosphere that may well sterilize the meteorites. Indeed the shock involved in accelerating the rock off a planetary surface at the start of its journey may be sufficient to cause a degree of heating compatible with sterilization. Thus even if life was present on the other planetary body, the mechanisms involved in the transfer and capture of material between planets may be such as to kill the life form, preventing Panspermia occurring.

An interesting coda to this is that any fossil remains in the rock may survive intact. Thus searches in meteorites for fossilized micro-organisms are a sensible approach to searches for life on other planets, although of course they do not represent evidence for Panspermia. The controversy over claims of the discovery of possible nano-scale fossils in meteorite ALH84001, which were advanced in 1996 (McKay et al. 1996), indicates how significant a verified such find would be. That the claims for the nano-fossils are now subject to revision and possible re-interpretation by other authors (e.g., see Harvey and McSween 1996, or Grady et al. 1997, and references therein), does not disguise the need for continual investigation in this field.

The field of exobiology can perhaps be summarized in the cartoon of Fig. 1. Any of the possibilities (a) to (e) shown in the figure may be the correct one, the true situation is simply unknown. If it occurs, Panspermia is only present in (b) or (c), and is not a necessary consequence of the existence of life elsewhere in the Solar System. Thus whilst the transfer of material between planets can now be taken as demonstrated, what is lacking is a demonstration that any of this material can be a viable life form.

Survivability of Bacteria Under Impact Conditions

One of the key steps for Panspermia (rocky or otherwise) is to show that a life form can survive a hypervelocity impact. For example, it is estimated in Heide and Wlotzka (1995) p 44, that an impact on a solid target at 15 km s^{-1} will generate in the target a pressure of order 100 GPa. More detailed calculations regarding impacts are summarized in Melosh (1989) p. 63, where peak pressures are given vs. depth below the surface for an iron meteorite striking an anorthosite target over a range of velocities from 5 to 45 km s^{-1}. One consequence of the great pressures is melting and vaporization of the projectile and some of the target material. Detailed descriptions of the physics of shock compression are readily available in texts such as Asay and Shahinpoor (1993), and specific applications of shock physics to hypervelocity impacts relevant to meteorites and impact cratering on planetary bodies are summarized in Melosh (1989).

However, since meteorites are recovered here on Earth, these extreme conditions are not always reached (as outlined above). Therefore, if laboratory tests can determine the range of temperatures and pressures that bacteria can tolerate, it would be possible to hypothesize conditions in which Panspermia might be possible. This has been the approach of past work, where for example static loading of pressure has been applied to bacteria and it has been shown after being subject to pressures of up to 42.5 GPa, bacteria are still viable (Horneck 1995). However, all such demonstrations are indirect. A slightly more relevant approach is to fire organic material into targets at hypervelocities. Results of such tests are reported by Bunch et al. (1992) who fired phthalic acid crystals and polycyclic aromatic hydrocarbons into various targets. They reported successful transfer of the organic material on the targets. However, what would be more convincing (in the absence of finding non-terrestrial bacteria in a meteorite) would be a laboratory demonstration that a life form, e.g. bacteria, could survive a transfer from projectile to target in a hypervelocity impact. A preliminary experimental program investigating this is described in the rest of this paper.

Experimental Program

Light Gas Gun

Hypervelocity impact experiments required a laboratory means of accelerating an object to a velocity of greater than a few km s^{-1} and directing it onto a target. This is limited to small objects, and indeed the size is inversely correlated to the achievable velocity (i.e. the highest velocities are only achievable for the smallest particles). For macroscopic objects (i.e., mm or cm sized) this is best achieved with a two-stage light gas gun (Crozier and Hume 1957). Normal guns cannot exceed a muzzle velocity far in excess of 1 km s^{-1}, but two-stage guns can reach higher velocities, i.e., into the hypervelocity regime (although a practical limit is approximately 10 or 11 km s^{-1} even for the most efficient guns).

A two-stage gun (see Fig. 2 for an example) works by using a normal powder charge to drive a piston along the first stage gun barrel (called a pump tube). This piston compresses a light gas (usually hydrogen), which is confined to the pump tube by a disk of material which bursts when the gas pressure is sufficient. The light gas is then released into the second stage barrel (called the launch tube), where it accelerates the projectile. The projectile is thus accelerated by a lighter gas than that obtained from the burning of a powder charge. The expansion of the light gas achieves a higher velocity than that of the heavy gas, driving the projectile above the approximately 1 km s^{-1} limit of a single stage gun.

The two-stage gun used in this work (shown schematically in Fig. 2) is at the University of Kent at Canterbury (Burchell et al. 1999a). It is a small gun, designed for frequent use rather than for achieving the highest possible velocities. The maximum velocity obtained with the gun has been 6.8 km s^{-1}, but in the present work lower velocities are used. This is because the current study is aimed at investigating a concept involving hypervelocity impacts, rather than at investigating behaviour at a specific velocity. One important feature of this gun is that the launch tube is rifled, spinning the object being fired. The gun can thus fire a sabot, which if it has been pre-cut (here into 4 segments), can be discarded in flight (impacting on the stop plate shown in Fig. 2), leaving just the load carried centrally in the sabot to proceed to the target (see Burchell et al. 1999a for details).

In the experiments reported here the projectile loaded for each shot (in a discardable sabot) consisted of pieces of a porous ceramic. The ceramic was crushed to produce fragments typically a few hundred micrometres to a millimetre across. The pores in the ceramic were approximately 1 μm across. The presence of the pores, plus the rough surface due to the crushing, meant that each piece of ceramic had an irregular shape and surface. This is illustrated in Fig. 3, with an image from a scanning electron microscope of some of the projectile material used in this work. In a shot several pieces of ceramic were loaded into the sabot (see Fig. 4). The load spreads laterally slightly during firing, with only some of it passing through the aperture in the stop plate and hitting the target. The rest intercepts the stop plate. In a typical shot between one third and one half of the load reaches the target.

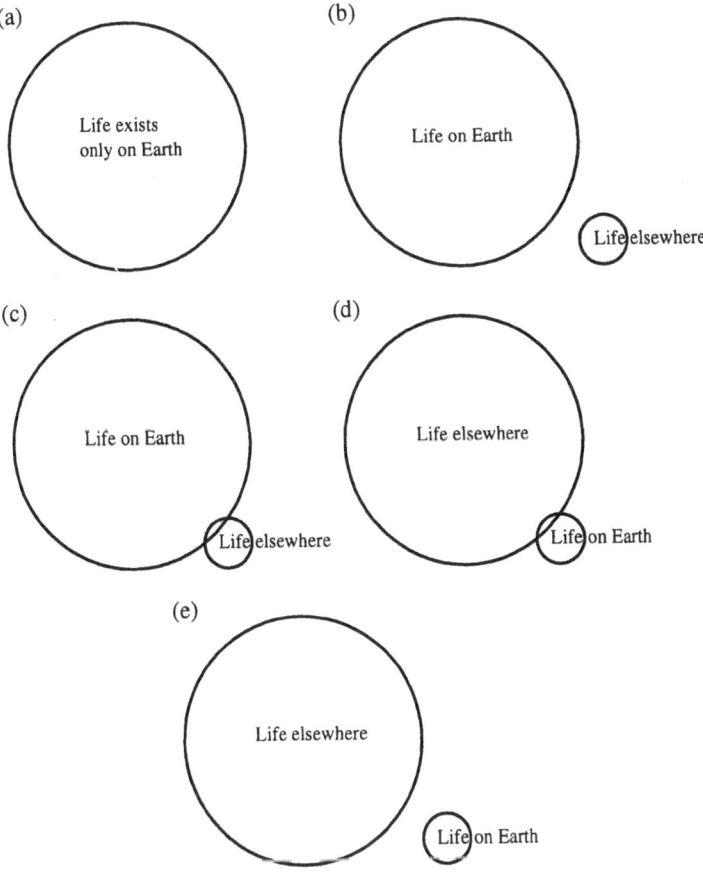

Fig. 1. Cartoon showing possibilities for distribution of life.

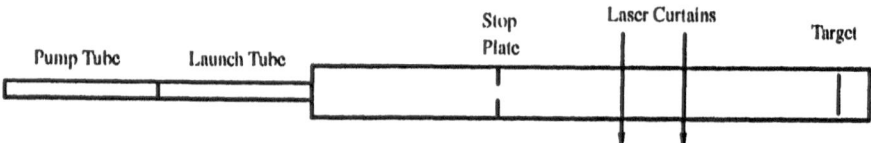

Fig. 2. Schematic of the light gas gun. (Projectile is fired from left to right.)

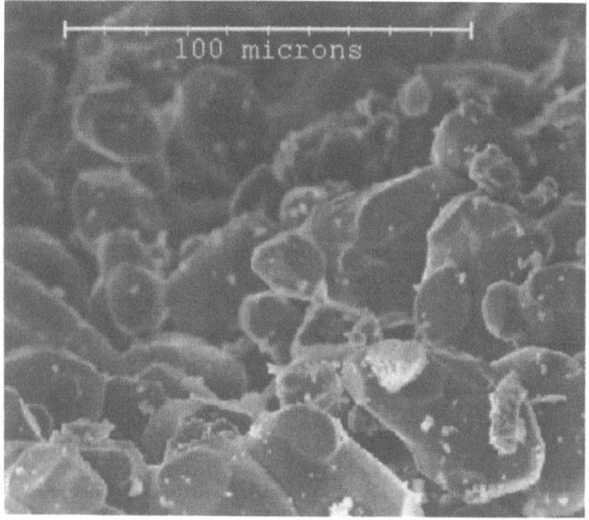

Fig. 3. SEM micrograph of projectile material.

The velocity of a shot was found from a combination of two methods. When it crossed the laser curtains shown in Fig. 2, the projectile interrupted the light, changing the signal from the photodiode on which the laser was focussed. The two laser curtains had a separation of 49.9 cm, which, combined with the time between the signals, provided the velocity. The photodiode signals are shown in Fig. 5 (for one of the shots). Note that in this shot two pieces of projectile travelled to the target, and hence two signals were present on the photodiode output. If the projectile was smaller than 200 μm in size it may have been too small to give a signal on the photodiode. In such cases the velocity was obtained from timing information from two piezo-electric transducers (pzt). One was mounted on the stop plate, the other to the target. An impact on the stop plate (target) would shake the pzt, and the resulting compression of the pzt gave rise to an electrical signal. Once corrected for the transit time of the compression wave through the stop plate (target) from the impact site to the pzt, the relative timing of the signals from the two pzt's was used to obtain the velocity. This system was used in every shot, but was less accurate than the laser curtain method, the results of which supersede it whenever available.

The target chamber in this work was maintained at a pressure of between 0.1 to 0.3 mbar (it varied slightly from shot to shot). This was sufficient to prevent significant deceleration of the projectiles in flight. A variety of targets were used. Metal and glass targets simulated impacts onto solid surfaces in the absence of an atmosphere. A porous limestone rock was used to see if use of a weaker target with modest porosity affected the results. Finally a very porous target was used, this was aerogel (a low density dried silica gel). Its porous nature (typical pore sizes of 40-80 nm for the aerogel used here, see Burchell et al. 1999b) means that its apparent density (in this case 100 kg m^{-3}) is much lower than the normal density of solid materials. It has been shown previously (see Tsou et al. 1988 or Burchell et al. 1999b and references therein) that aerogel of this density can capture relatively intact, particles hitting it at impact speeds in the hypervelocity regime. The particles impact the surface of the aerogel and tunnel in, before coming to rest (relatively intact) inside the aerogel sample. This crudely simulates impacts on a planetary atmosphere, where the meteorite is decelerated and lands fairly intact.

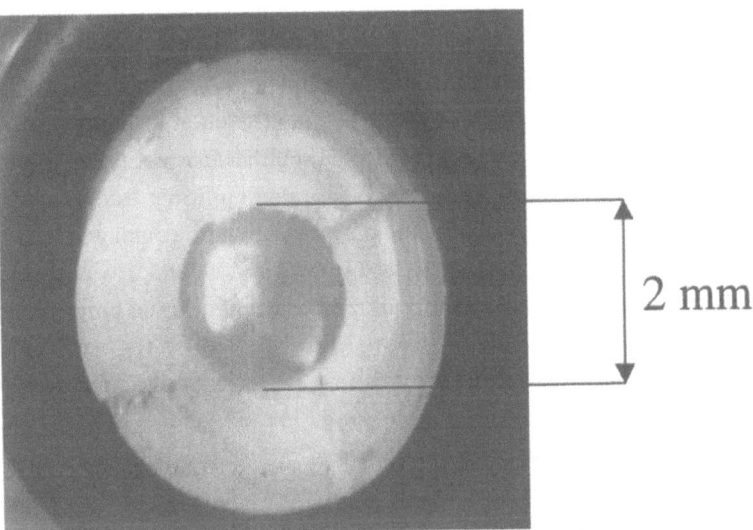

Fig. 4. Projectiles loaded in a sabot. (The projectile pieces are in the central hole.)

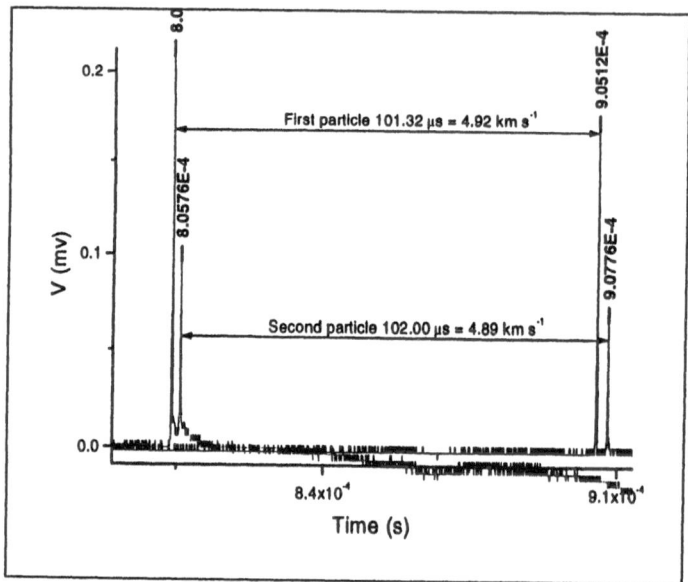

Fig. 5. Photodiode signals used for velocity measurement (shot 2).

Preparation of Bacteria

The bacteria used were rhodococcus. These have the characteristics of being hardy (resistant to damage) and are typically rod shaped (size 1 – 4 µm). Other reasons for choosing this as the bacteria include that although occurring naturally in nature it is not usually airborne in large quantities in a research laboratory, and can be eliminated with simple swabbing by alcohol. Further it is not harmful to humans and once cultured develops a pinkish colour, marking it out from other common bacteria. The rhodococci were grown on a rich nutrient medium and harvested by centrifugation. The projectile material (a powdered ceramic) was cleaned with alcohol before use. A small quantity of the powdered ceramic projectile material was then poured into a solution of the bacteria. This was left for two days to allow the bacteria to adhere to the rough surface of the projectiles. The projectiles were then placed in a sealed vial, which was stored in a refrigerator at 4° C. For a shot requiring bacteria, between 5 and 7 pieces of projectile would be removed from the vial and loaded into the sabot. At all stages of handling sterile latex gloves were used and face masks worn to prevent contamination.

Shot Program

To simulate impacts onto atmosphere-less surfaces a variety of targets were used: rock, glass and metal. The rock targets were Cotswold stone (a typical limestone), approximately 15 x 15 x 5 cm in size. The glass targets were disk shaped float glass blanks, of 15 cm diameter and 3 cm thickness. The metal targets were the stop plates used in the experiments; since some of the projectile pieces would impact them in each shot, it was possible to obtain craters on the metal. In some shots these were distinct from the craters caused by the impacts of the sabot parts and could thus be analyzed. A new target and stop plate were used for each shot. As well as craters in the targets, ejecta from the targets were studied in the experiments. These were collected by placing a steel tray along the base of the target chamber below the target.

The sabot, target, target holder and ejecta capture tray were cleaned before each shot, as were tools used in handling of the materials. This was done by immersing small parts (e.g., sabots, tips of tweezers, etc.) in alcohol. For larger surfaces alcohol was sprayed onto the surface which was then wiped with clean cotton swabs. The rock targets were also baked in an oven for 10 hours at 80° C. This was to drive any moisture out of the stone (otherwise it out-gases during evacuation of the target chamber, limiting the pressure to 1 to 2 mbar). During handling, all staff wore latex gloves to minimize contamination of thesurfaces. This was done for each shot of the gun, independent of whether or not bacteria were to be fired in that shot.

The shot program is shown in Table 1. The first two shots were at rock targets. Shot 1 featured clean projectiles, i.e., pieces of ceramic that had not been exposed to the bacteria. Shot 2 used projectiles contaminated by the bacteria. This allowed a control, permitting a contrast between shots when bacteria were and were not present on the projectile. Shot 3 used a glass target and clean projectiles and shot 4 used a glass target and bacteria contaminated projectiles.

Shots 5 and 6 used aerogel targets. Due to reactions between alcohol and the aerogel, it was not possible to clean the surfaces of the aerogel with alcohol. In Shot 5, clean projectiles were used with no bacteria present. In shot 6, bacteria impregnated projectiles were used. Note that in shots 4 and 5, the stop plates were recovered for analysis, providing impacts of projectiles on metal targets.

Examples of rock and glass targets, after impact, are shown in Fig. 6 and Fig. 7. In Fig. 8 is a stop plate (shot 4), where four larger craters are present (from the four sabot pieces) and some smaller craters are visible from the projectile pieces. In Fig. 9a sample of aerogel is shown mounted in the end of the light gas gun. Whilst it was possible with minimal handling (and hence minimal risk of contamination) to image the other target materials after a shot, this was not the case for the aerogel. Therefore in Fig. 10, an image of aerogel with tracks caused by impacts of particles is shown, but it is not from an impact in this work. It is from a shot where 50 µm diameter Al_2O_3 spheres were fired at the aerogel at 5.2 km s^{-1}, and is only shown here for illustrative purposes.

Analysis

To determine if the doping of the projectile bacteria had been successful, unused projectile samples were taken after the shot program was completed and were cultured. After two days the sample with no bacteria doping revealed little bacteria content (Fig. 11), indicating that problems of accidental contamination were minimal. The bacteria impregnated sample however showed a massive growth of bacteria (saturating the nutrient medium), which had turned pink (the signature of rhodococcus). Thus at the time of shooting it can be assumed that the projectiles which were supposed to carry rhodococcus bacteria, did indeed do so.

Table 1. Shot program. Where more than one velocity is shown, it means that velocities were obtained for individual particles travelling to the target in the shot. The uncertainty in the velocity is ± 1%, except for shot 6 where velocity was found by the pzt method and the uncertainty is ± 5%

Shot	Velocity (km s^{-1})	Target	Bacteria on projectile?
1	3.80	Rock	No
2	4.89 and 4.92	Rock	Yes
3	4.69	Glass	No
4	3.48	Glass	Yes
5	3.88	Aerogel	No
6	4.1	Aerogel	yes

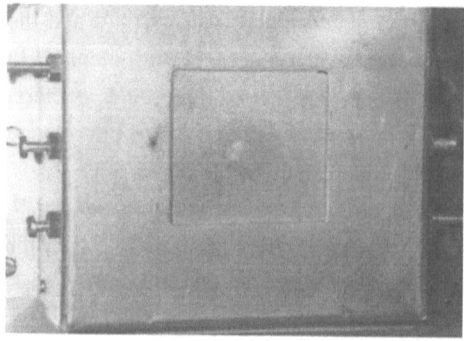 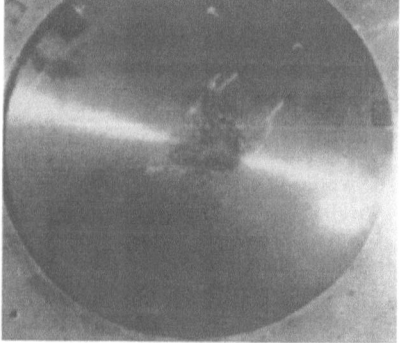

Fig. 6. Rock target after impact by 2 projectiles.   **Fig. 7.** Glass target after impact

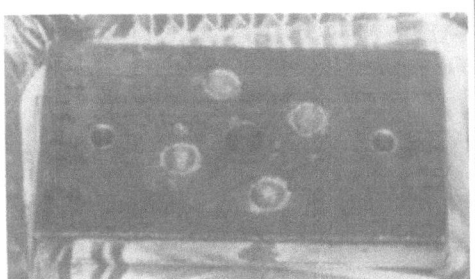

Fig. 8. Metal stop plate after a shot.　　　**Fig. 9.** Aerogel target (before a shot).

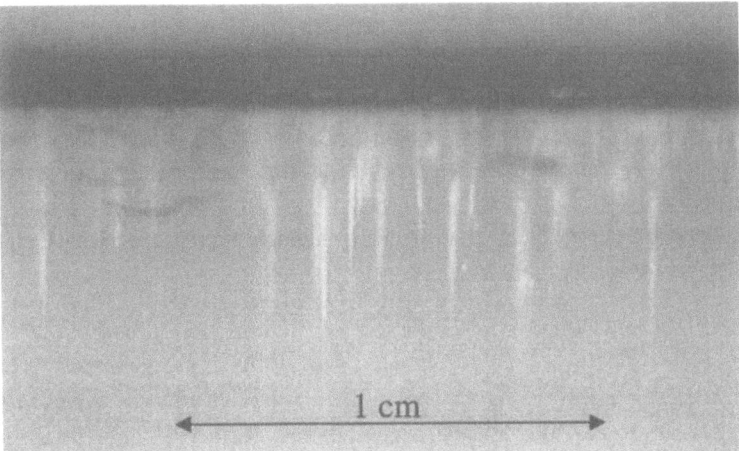

Fig. 10. Tracks in aerogel. (The impact was from above.)

After a shot the targets and stop plates were handled with sterile latex gloves and placed in clean containers (swabbed out with alcohol). They were then stored in a refrigerator at 4 °C until transfer to the Bio-Sciences Laboratory at the University of Kent. There they were subject to several types of examination. The most simple was to use swabs moistened with distilled water. These were wiped on crater surfaces. The water in these swabs was then cultured to see if bacteria could be grown from it. The method of culturing was to suspend the recovered material from the swabs in a liquid nutrient broth. The liquid broth was then spread onto a solidified form of the broth (made solid using agar). Individual viable cells (those able to divide) form visible colonies after incubation under the same conditions as they were originally grown. Although a simple technique this approach had several difficulties. It was made difficult for the rock samples by the porous and indeed absorbent nature of the rock, which prevented a good pickup of material from the crater surfaces by the swabs. For the glass samples, it was made difficult by the way pieces of shattered glass were picked from the target by the swab, preventing uniform sampling of the surface.

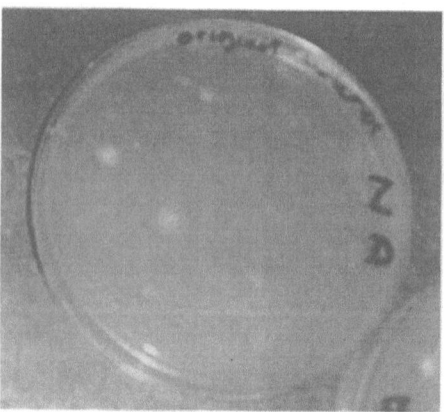

Fig. 11. Bacteria sample cultured from an undoped sample of projectile material left unused at the end of the program.

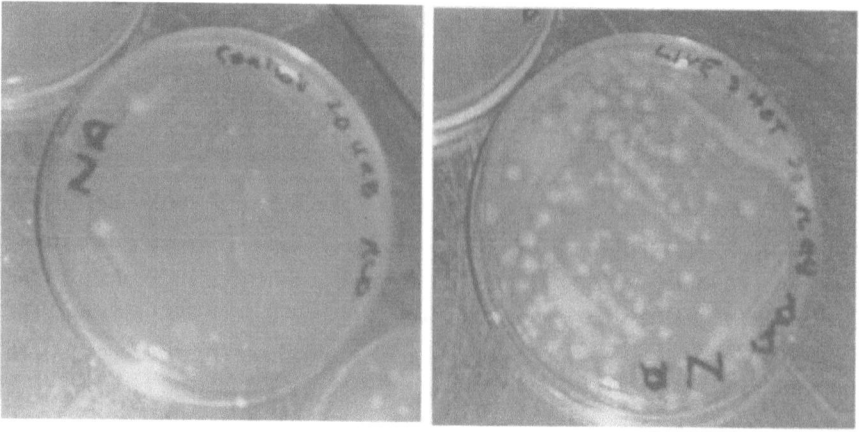

Fig. 12. Bacteria samples cultured on petri dishes from rock ejecta. Left is shot 1 (no bacteria on projectile), right is shot 2 (where bacteria were present on the projectile).

The aerogel was not suitable for swabbing, as the trapped particles were inside the aerogel. Thus whilst superficially an attractive approach to analysis, difficulties were encountered which may have prevented any definitive results being obtained from this technique.

The second technique involved study of the ejecta. This was collected after each shot and placed in clean glass containers. The samples were subsequently cultured in a broth, which was spread onto petri dishes. After two days growth was visible on some of the dishes. The dishes containing the ejecta from the two rock shots are shown in Fig. 12. A clear difference in degree of bacteria content is visible. The shot where the projectile had no bacteria, has minimal bacteria growth on the relevant dish (This low level of bacteria growth was common to all shots where no bacteria were used, and indicates that the general handling techniques were sufficient to minimise the risk of contamination. If observed in any sample, this level of bacteria was thus taken as a null result.) By contrast, the dish for the shot where the projectile had been doped with bacteria, shows substantial numbers of bacteria. Initially, the bacteria on this latter dish did not initially develop the pinkish hue expected for rhodococcus. However, after four more days the pinkish hue had developed (but was still absent on the equivalent petri dish for the undoped projectile). The suggestion is thus straightforward: In the ejecta from the rock target (shot at with bacteria on the projectile), bacteria were found at a level above the general background contamination level, and also had the expected pinkish hue for rhodococcus. This can be taken as indicative of the successful transfer of bacteria from the projectile to the rock target. However, no such bacteria could be cultivated from the glass target ejecta.

The third analysis technique was applied to the aerogel samples. A fluorescent dye (bactolight) was injected into the tracks in the aerogel. The dye can be hydrolyzed by live cells only, i.e., if it encountered live bacteria it fluoresces. The aerogel was then examined under a confocal scanning microscope, searching for this fluorescence. After preliminary observations, no traces of fluorescence were determined from either aerogel sample.

Discussion of Results

A Positive Result?

Some of the results of this initial program are quite promising. The method of doping the projectile with bacteria does indeed work (an essential pre-requisite for this investigation). Also, the methods used to prepare clean samples were sufficient to reduce the normal bacteria concentrations to levels which makes feasible the search for bacteria transferred in the impacts. However, beyond that the pattern of results is not conclusive. Some of the analysis techniques were not as applicable as originally thought (e.g., swabbing of crater surfaces). Equally other techniques, e.g. the methods for examination of the ejecta and the study of the aerogel, do appear to be viable methods of analysis.

Proceeding to the interpretation of the results poses several interesting questions. First it has to be acknowledged that attempting an experiment once is not sufficient to produce a conclusive result. Accidental contamination of the samples was always possible. This is why rhodococcus were chose with their pinkish colour, it being unlikely that bacteria present in the laboratory anyway would mimic this colour tag. However, cross-contamination from one sample to another is always a risk. The only satisfactory way to avoid this possible explanation for a positive result is to repeat the experiment (this is planned).

If taken at face value as a positive result, it is interesting that the ejecta from the rock target carried bacteria transferred from the projectile. A possible interpretation is that during the impact the bacteria were transferred to the rocky target surface, and this surface (around the impact site) was then thrown off the target in a later stage of crater formation. Equally one might imagine part of the projectile being ejected during the impact (as a result of the shock waves which propagate through not just the target but also the projectile during an impact). The lack of an observation of bacteria on the crater after an impact does not distinguish between these possibilities. Further, bacteria might have transferred to the target at the impact site, but then been killed due to elevated temperatures on the target surface during the impact.

What is interesting is that the ejecta from the glass target did not show any bacteria. This suggests that the transfer mechanism requires a rock rather than glass target, or perhaps a porous target is required. This is an area where more experiments are required.

The lack of a positive result from aerogel is disappointing. That projectiles can be captured relatively intact in aerogel is well known. However, possibly the heating during the capture sterilizes the projectile surface. Again, this is a topic worthy of further investigation.

Surviving Fraction

If the presence of the bacteria in rock ejecta in Fig. 12 is taken as evidence for successful transfer of bacteria to a target surface during a hypervelocity impact, one can calculate a surviving fraction. This is defined as the number of captured bacteria divided by the number of incidence bacteria. At present, this calculation is primitive, since no accurate knowledge of the number of bacteria in a shot exists. Accordingly, this number has to be estimated. The first step in the estimate is to assume that the bacteria cover the entire surface of a piece of projectile in a layer one bacterium thick. If we assume that the bacteria are rods of 1 by 4 μm length, and that the projectile pieces are spheres 500 μm across, then an estimated 196,250 bacteria are present on each piece of projectile. If typically two pieces of projectile hit each target, this gives an incidence flux of approximately 400,000 bacteria. From Fig. 11, the number of bacteria found on the rocky ejecta when a bacteria doped projectile was used can be estimated by assuming one colony equals one bacteria. This yields an estimate of approximately 50 bacteria. Thus, the surviving fraction is 1.25×10^{-4}, equivalent to one survivor per 8,000 incident. This number is however only a crude estimate.

Conclusions

The experiments described in this paper have simulated in the laboratory the mechanisms that can occur during hypervelocity impacts of bacteria loaded objects onto a variety of atmosphere-less surfaces (rock, metal, glass), and also impacts into an atmospheric analogue (aerogel). The level of initial sterility of the targets is shown to be controllable, in that none of the test shots (i.e., shots of projectiles without bacteria) produced large-scale signals for bacteria after the shots were carried out. Thus this experimental approach is a viable one in terms of testing the feasibility of bacteria transferring from projectile to target in a hypervelocity impact.

However, the results of the current round of experiments are ambiguous. One shot did give a positive result, i.e., bacteria were initially present on the projectile, and bacteria were subsequently found on the ejecta from a rocky target. But, none of the other experiments produced such results. The need to reproduce this result is thus paramount. Thus at the moment, whilst the present work has shown that such experiments are possible and can readily be carried out, the single positive result needs to be confirmed (i.e., reproduced), or will be left in isolation without a convincing explanation. If we interpret the result as coming from bacteria launched from the gun (and not accidental contamination) it is interesting that the only positive result came not from the crater itself, but from material ejected from the crater during the impact. Thus the bacteria might in effect be scrapped off the projectile during impact and carried by ejecta to the surrounding area. Or perhaps part of the projectile flies apart during the impact, carrying bacteria away from the impact site. Either way, it is not necessarily an impact crater itself, but the surrounding region which may show contamination from bacteria after an impact. An alternate explanation is that the bacteria never reached the target, but were somehow scattered from the projectile as it passed over the collection tray before it reached the target. Even this would be remarkable, as it would imply the bacteria survived the acceleration to 4.9 km s^{-1} in the gun.

The relevance to possible Panspermia is self-evident. A convincing demonstration that bacteria can survive hypervelocity impacts would put Panspermia onto a much sounder base. Of course, the final proof would be direct evidence itself. Meanwhile, there are still questions. As pointed out by many observers, life on Earth seems to have a common origin (the so-called Tree of Life, see, for example, Davies 1998, p 49). If Panspermia does provide the origin of the tree of life (i.e. a common ancestor to all Earth life), why did it only happen once? Meanwhile, the bigger question would still remain open: Where did life first originate?

Acknowledgements

We thank PPARC for a Rolling Grant which is used to fund operation of the light gas gun at the Univ. of Kent. We thank R.Newsam for carrying out the optical microscopy of the aerogel targets. Finally we thank the referees for helpful comments particularly on the first half of the manuscript.

References

Arrhenius S (1908) Worlds in the making: the evolution of the universe. Harper Bros, London New York, 230 pp

Asay JR, Shahinpoor M (1993) High pressure shock compression of solids. Springer, Berlin Heidelberg New York, 465 pp

Biemann K, Oro J, Toulmin P, Orgel LE, Nier AO, Anderson DM, Simmonds PG, Flory D, Diaz AV, Rushneck DR, Biller JE, Lafleur AL (1977) The search for organic substances and inorganic volatile compounds in the surface of Mars. J Geophys Res 82: 4641-4658

Bogard DD (1982) Trapped noble gas in the EETA79001 shergottite. Meteoritics 17: 185-186

Bunch TE, Becker L, Bada J, Macklin J, di Brozolo FR, Flemming RH, Erlichman J (1992) Hypervelocity impact survivability experiments for carbonaceous impactors. In: Levine AS (ed) LDEF-69 months in space: 2nd LDEF post-retrieval symposium (NASA CP 3194), pp 453-477

Burchell MJ, Cole M, McDonnell J, Zarnecki J (1999a) Hypervelocity impact studies using the 2 MV van de graaff accelerator and two-stage light gas gun of the University of Kent at Canterbury. Meas Sci Technol 10: 41-51

Burchell MJ, Thomson RT, Yano HY (1999b) Capture of hypervelocity particles in aerogel: In ground laboratory and low Earth orbit. Planet Space Sci 47: 189-204

Crozier WD, Hume W (1957) High velocity, light gas gun. J Applied Physics 28: 892-898

Davies P (1998) The fifth miracle—the search for the origin of life. Simon and Schuster, New York, 400 pp

Delaney JS (1989) Lunar basalt breccia identified among Antarctic meteorites. Nature 342: 889

Eugster O (1989) History of meteorites from the Moon collected in Antarctica. Science 245: 1197

Fitton B (1997) European Space Agency exobiology science team study on the search for Life in the solar system — Final Report, 155 pp

Gladman J, Burns J, Duncan M, Lee P, Levison H (1996) The exchange of impact ejecta between terrestrial planets. Science 271: 1387-1392

Gladman J (1997) Destination: Earth. Martian meteorite delivery. Icarus 130: 228-246.

Grady MM, Wright IP, Pillinger CT (1997) Microfossils from Mars: a question of faith? Astronomy Geophysics 38: 26-29

Grieve RAF (1998) Extraterrestrial impact on the earth; the evidence and the consequences. In: Grady MM, Hutchison R, McCall GJH, Rothery DA (eds) Meteorites: Flux with time and impact effects. Geological Society London Special Publication 140: 105-131

Harvey RP, McSween HY (1996) A possible high temperature origin for the carbonates in the Martian meteorite ALH84001. Nature 382: 49-51

Heide F, Wlotzka F (1995) Meteorites—messengers from space. Springer, Berlin Heidelberg New York. 231 pp

Horneck G (1995) Exobiology, the study of the origin, evolution and distribution of life within the context of cosmic evolution: a review. Planet Space Sci 43: 189-217

Hoyle F (1998) Comets: a matter of life and death. In: Bondi H, Weston-Smith M (eds) The universe unfolding. Oxford University Press, Oxford, pp 3-22

Hoyle F, Wickramasinghe C (1981) Comets – a vehicle for Panspermia. In: Ponnamperuma C (ed), Comets and the origin of life. D Reidel Publishing Company, Dordecht, pp 227-239

Jakosky B (1998) The search for life on other planets. Cambridge University Press

Kissel J, Krueger FR (1987) The organic component of dust from comet Halley as measured by the Puma mass spectrometer on board Vega 1. Nature 326: 755-760

Klein HP (1978) The Viking biological experiments on Mars. Icarus 34: 666-674

Madigan M, Marrs B (1997) Extremophiles. Scientific American 276 (4): 66

McKay D, Gibson EK, Thomas-Keptra KL, Vali H, Romanek CS, Clemett SJ, Chillier XDF, Maechling CR, Zare RN (1996) Search for past life on Mars: possible relic biogenic activity in Martian meteorite ALH84001. Science 273: 924-930

McKay CP, Grunthaner FJ, Lane AL, Herring M, Bartman RK, Ksendzov A, Manning CM, Lamb JL, Williams RM, Ricco AJ, Butler MA, Murray BC, Quinn RC, Zent AP, Klein HP, Levin GV (1998) The Mars oxidant experiment (MOx) for Mars '96, Planet Space Sci 46: 769-777

McSween HY (1994) What have we learnt about Mars from SNC meteorites. Meteoritics 29: 757-779

Melosh HJ (1988) The rocky road to Panspermia. Nature 332: 687

Melosh HJ (1989) Impact cratering; a geologic process. Oxford University Press, Oxford New York, 245 pp

Norton OR (1994) Rocks from space. Mountain Press Publishing Company, Missoula, Montana, 467 pp

Pillinger CT (1988) Stable isotope genealogy of meteorites. Phil Trans Roy Soc Lond A 325: 525-533

Samuelson RE, Maguire WC, Hanel RA, Kunde VG, Jennings DE, Yung-Yuk L, Aikin AC (1983) CO_2 on Titan. J Geophys Res 88: 8709-8715

Snyder CW (1979) The planet Mars as seen at the end of the Viking mission. J Geophys Res 84: 8478-8519

Surkov Y (1997) Exploration of terrestrial planets from spacecraft. Wiley, New York, 446 pp

Taylor SR (1992) Solar system evolution. Cambridge University Press, Cambridge, 307 pp

Tsou P, Brownlee DE, Laurance MR, Hrubesh L, Albee AL (1988) Intact capture of hypervelocity micrometeoroid analogs. Lunar Planet Sci 19: 1132-1133

Vickery AM, Melosh HJ (1987) The large crater origin of SNC meteorites. Science 237: 738-743

Wetherill G (1984) Orbital Evolution of impact ejecta from Mars. Meteoritics 19: 1-12

Wickramasinghe C (1998) The Times (London newspaper), p. 1, 20th Nov

2 Macromolecular Organic Materials in Carbonaceous Chondrites: A Review of their Sources and their Role in the Origin of Life on the Early Earth

Mark Sephton and Iain Gilmour

Planetary Sciences Research Institute, The Open University, Milton Keynes, MK7 6AA, United Kingdom. (M.A.Sephton@open.ac.uk, I.Gilmour@open.ac.uk)

Abstract. In the early solar system, the impact of comets, asteroids and micrometeorites on the primitive Earth resulted in extraterrestrial organic matter reaching the planet surface intact. This organic matter contributed to the organic inventory available for chemical reactions on the early Earth and may have played a role in the origin of life.

Carbonaceous chondrites contain up to 5% by weight of organic matter, including important prebiotic molecules. The majority of the organic matter in carbonaceous chondrites (some 70-90%) occurs as complex, cross-linked macromolecular materials. These macromolecular materials may have had a role in the prebiotic chemistry of the primitive Earth. We have undertaken a series of experiments involving the hydrous pyrolysis of macromolecular materials from the Murchison (CM2) meteorite aimed at investigating the fate of extraterrestrial macromolecules on 'wet' planets such as the Earth.

Our experiments solubilised 55% of the carbon in the macromolecular materials. Thus, an order of magnitude more free organic compounds can be generated by hydrothermal processing of macromolecular materials than are available as free compounds within the bulk meteorite. The major products were alkyl-substituted benzenes, naphthalenes and indanes, alkyl-substituted thiophenes and benzothiophenes, and alkyl-substituted phenols. The pyrolysis products are comparable to those compounds present as free entities in Murchison, though with substantially increased abundances of phenols and thiophenes. Similar macromolecular material-derived compounds may have contributed to the organic inventory available for prebiotic chemistry on the early Earth.

Introduction

The impact of comets, asteroids and micrometeorites on the primitive Earth may have been both beneficial and detrimental to the origin of life. Extraterrestrial organic matter may have reached the prebiotic Earth intact and could have played a significant role in the origin of life. However, large impacts are likely to have frustrated life's development. The energy associated with giant impacts may have repeatedly destroyed the early biosphere. Inferences, drawn from the lunar cratering record and the sparse terrestrial geological record, appear to constrain the time during which life on Earth originated to the final stages of the late heavy bombardment (e.g., Chyba and Sagan 1992). At this time, the planets of the inner solar system were subjected to an intense barrage by asteroids and comets. There were few viable environments for life on the Earth during the first several hundred million years after the formation of the planet.

Geological approaches to studies of the origin of life are limited by the scarcity of the Archean geological record and the inability of organic matter to survive geological processing over long periods of time. Nevertheless, by 3.5 Ga ago microfossil evidence from the Warrawoona volcanic sequence in Western Australia suggests that life had established itself (Schopf 1993). More controversially, carbon isotopic data from the oldest identified sedimentary rocks on Earth, the 3.8 Ga old Isua metasediments in Greenland, suggests that autotrophic photosynthesis may have left its signature (Schidlowski 1979; Mojzsis et al. 1996; Nutman et al. 1997; Rosing et al. 1999). This evidence of ancient life in the Isua metasediments implies that there is no geological record of the Earth's early environment prior to the existence of a biosphere.

Attempts to explore the geological record in the Archean are fraught with contamination hazards and, therefore, interpreting the significance of organic compounds in Archean rocks is difficult (Imbus and McKirdy 1993). A better record of early planetary evolution may exist on Mars, but that awaits further planetary exploration (Brack and Pillinger 1998). Faced with these problems, biochemical approaches to understanding the origin of life attempt to extrapolate backwards in time from extant biological structures and mechanisms to more primitive systems (Morowitz 1992). On the other hand, chemical evolution experiments address the problem in the opposite direction, by attempting to model the production of important biological molecules from simple precursors in the laboratory (e.g., see Miller 1993 for a review). These forward-looking studies have been successful in identifying a variety of synthetic processes that could have occurred under primitive Earth conditions to yield simple biologically useful organic building blocks (biomonomers), such as amino acids (Hennet et al. 1992).

Meteorite studies also employ a forward-looking approach. The cosmic history of the biogenic elements can be traced from the formation of organic molecules in interstellar clouds to their presence in biological systems on the Earth. Organic molecules are ubiquitous throughout the galaxy and a number of extraterrestrial environments may have contributed organic matter to meteorites and comets. These environments include carbon star atmospheres, interstellar clouds, the solar nebula, giant-planet sub-nebulae, and the asteroidal meteorite parent body (Anders 1991). The record preserved in meteorites is also partly the result of secondary thermal and aqueous processing on the meteorite parent bodies (Sephton et al. 1999). Hence, carbonaceous chondrites provide us with organic molecules that are the direct products of abiotic chemical reactions.

It now seems likely that the primitive Earth accumulated exogenous organic matter from impacting objects ranging in magnitude from the micrometer-sized interplanetary dust particles to the kilometer-sized asteroids and comets (Chyba et al. 1990; Chyba and Sagan 1992). Some fraction of this total inventory undoubtedly arrived under conditions that would have allowed the survival of organic compounds (Anders 1989; Chyba and Sagan 1997).

Comets are thought to be extremely primitive objects and may have preserved some of their interstellar molecular constituents (e.g., Cottin et al. 1999). Among the molecules identified in cometary dust grains by the Giotto mission to comet Halley were hydrogen cyanide (HCN) and formaldehyde (HCHO) (Geiss et al. 1991). Other molecules such as methane, acetylene, acetonitrile, formic acid, hydrogen isocyanide, cyanoactylene, and thioformaldehyde have been identified in comets Hyakutake and Hale-Bopp (Lis et al. 1997) and there is some evidence for polymeric organic matter (Huebner and Boice 1997).

The CI and CM carbonaceous chondrites comprise a unique subset of meteorites, containing up to 5% by weight of organic matter. The Murchison meteorite, in particular, has received detailed examination. Over 70 amino acids have been identified in Murchison, consisting of many non-protein amino acids and relatively high concentrations of eight protein amino acids (glycine, alanine, valine, leucine, isoleucine, proline, aspartic acid, and glutamic acid) (Cronin et al. 1981; Cronin et al. 1985; Cronin and Pizzarello 1986; Cronin and Chang 1993). The presence of numerous non-protein amino acids, i.e., those not utilised by terrestrial biology, in this meteorite is an important indicator of the extraterrestrial provenance of these compounds.

In this paper, we review the nature and origin of organic macromolecular materials in meteorites and examine the possible response of these entities to heat and water on the early Earth, using laboratory hydrous pyrolysis experiments.

Table 1. Previously identified moieties in the Murchison macromolecular materials

Moiety	Example	References
aromatic hydrocarbons	benzene	Studier et al. 1972 Levy et al. 1973 Biemann 1974 Holtzer and Oró 1977 Hayatsu et al. 1977 Cronin et al. 1987 Sephton et al. 1998
aliphatic hydrocarbons	cyclohexane	Hayatsu et al. 1977 Cronin et al. 1987
phenols	phenol	Studier et al. 1972 Hayatsu et al. 1977 Hayatsu et al. 1980 Sephton et al. 1998
ketones	benzophenone	Biemann 1974 Hayatsu et al. 1977
carboxylic acids	benzene carboxylic acid	Studier et al. 1972

Table 1. (continued)

Moiety	Example	References
ethers	methylfuran	Holtzer and Oró 1977 Hayatsu et al. 1977 Hayatsu et al. 1980
thiophenes	benzothiophene	Levy et al. 1973 Biemann 1974 Holtzer and Oró 1977 Hayatsu et al. 1977 Sephton et al. 1998
nitriles	benzonitrile	Levy et al. 1973 Holtzer and Oró 1977
nitrogen heterocycles	pyridine	Studier et al. 1972 Hayatsu et al. 1977

Macromolecular organic materials in carbonaceous chondrites

Elemental composition

The elemental composition of the Murchison macromolecular materials has been determined as $C_{100}H_{71}N_3O_{12}S_2$ based on elemental analysis (Hayatsu et al. 1977) and revised to $C_{100}H_{48}N_{1.8}O_{12}S_2$ based on pyrolytic release studies (Zinner 1988). Such large macromolecules are difficult to study and analytical approaches primarily attempt to break the structure down into fragments that are easier to deal with, using techniques such as pyrolysis or chemical degradation.

Aromatic and aliphatic hydrocarbons

The majority of the carbon in the Murchison macromolecular materials is present within aromatic ring systems (Table 1). This aromatic nature was revealed by a series of early pyrolysis studies in which the macromolecular materials were thermally fragmented to produce benzene, toluene, alkylbenzenes, naphthalene, alkylnaphthalenes indene, acenaphthene, fluorene, phenanthrene, and biphenyl (Studier et al. 1972; Levy et al. 1973; Biemann 1974; Holtzer and Oró 1977).

Further identification of the aromatic units in the Murchison macromolecular materials was achieved by Hayatsu et al. (1977), who used sodium dichromate oxidation to selectively remove aliphatic side chains in the macromolecular materials and, thereby, isolate and release the aromatic cores present. These studies have revealed a dominance of single-ring aromatic entities, but also indicated a significant amount of two- to four-ring aromatic cores bound to the macromolecular materials by a number of aliphatic linkages.

Nuclear magnetic resonance (NMR) spectroscopy was used by Cronin et al. (1987) to detect aromatic carbon within the Murchison macromolecular materials. Cronin et al. (1987) reported the presence of extensive polycyclic aromatic sheets significantly larger than the two- to four-ring aromatic entities isolated by Hayatsu et al. (1977).

Recently, Sephton et al. (1998) investigated the composition of the Murchison macromolecular materials using hydrous pyrolysis followed by gas chromatography-isotope ratio-mass spectrometry. The structural and isotopic data led these authors to suggest that the macromolecular materials may be present as at least two organic phases: a reactive open organic network, containing mainly one- or two-ring aromatic cores and a more unreactive condensed aromatic component.

Aliphatic hydrocarbon moieties (Table 1) are present in significant amounts within the Murchison macromolecular materials. Several pyrolysis studies have indicated that these moieties exist within or around the aromatic network as hydroaromatic rings and short alkyl substituents or bridging groups (Levy et al. 1973; Biemann 1974; Hayatsu et al. 1977; Holtzer and Oró 1977). NMR studies have also revealed abundant aliphatic units in the form of short, branched groups that link and decorate aromatic centers (Cronin et al. 1987).

Oxygen-, sulphur- and nitrogen-containing moieties

Several pyrolysis experiments have released oxygen-containing moieties (Table 1) such as phenols (Studier et al. 1972; Hayatsu et al. 1977; Sephton et al. 1998), benzene carboxylic acids (Studier et al. 1972), propanone (Levy et al. 1973; Biemann 1974) and methylfuran (Holtzer and Oró 1977) from the Murchison macromolecular materials. Sodium dichromate oxidation of the macromolecular materials liberated aromatic ethers such as dibenzofuran and aromatic ketones such as fluorenone, benzophenone and anthraquinone. Each of these organic units appears to be bound into the macromolecular network by two to four aliphatic linkages (Hayatsu et al. 1977). Hayatsu et al. (1980) used alkaline cupric oxide to selectively cleave organic moieties incorporated into the Murchison macromolecular materials by ether groups. These authors established the presence of a significant amount of phenols bound to the macromolecular materials by ether linkages.

Thiophenes (Table 1) are common pyrolysis products of the Murchison macromolecular materials and thiophene, methylthiophene, dimethylthiophene, and benzothiophene have been detected using pyrolysis (Levy et al. 1973; Biemann 1974; Holtzer and Oró 1977; Sephton et al. 1998). Sodium dichromate oxidation has revealed the presence of substituted benzothiophene and dibenzothiophene moieties within the macromolecular materials (Hayatsu et al. 1980).

Pyrolysis has also led to the tentative identifications of nitrogen heterocyclics (Table 1). Cyanuric acid (Studier et al. 1972) and alkylpyridines (Hayatsu et al. 1977) have been detected in the Murchison macromolecular materials. Similar types of analyses have revealed acetonitrile and benzonitrile (Levy et al. 1973; Holtzer and Oró 1977). Substituted pyridine, quinoline, and carbazole were observed in sodium dichromate oxidation of the macromolecular materials (Hayatsu et al. 1977).

Environments of formation

Interstellar space

The macromolecular materials in carbonaceous chondrites are deuterium-rich and it has been suggested that this enrichment is a result of the formation of the macromolecular materials from interstellar molecules (Kolodny et al. 1980; Robert and Epstein 1982). Interstellar clouds contain simple gaseous compounds, such as CH_4, CH_2O, H_2O, N_2, and NH_3. Some of these molecules are ionized and fragmented by cosmic radiation. These ions may then combine with neutral molecules resulting in deuterium-enriched reaction products. Such ion-molecule interactions produce increasingly deuterium-enriched and complex organic matter that condenses on to dust grains (Robert and Epstein 1982). Within interstellar clouds simple gas-phase molecules condense as icy mantles around silicate dust grains (Greenberg 1984). Increased temperatures or periods of ultraviolet radiation lead to thermally- or photolytically-driven polymerization reactions in these mantles. This process results in complex organic products that may also be deuterium-enriched (Sandford 1996). Nitrogen isotopes also suggest an interstellar origin for much of the macromolecular material (Alexander et al. 1998). Once formed, by reactions in the gas phase, solid phase or combinations of both, these isotopically-enriched complex organic residues would become incorporated into carbonaceous chondrites following the collapse of the interstellar cloud.

Solar nebula

Hayatsu et al. (1977) proposed that the macromolecular materials are the result of the thermally-induced polymerization and aromatization of aliphatic hydrocarbons produced by the Fischer-Tropsch reaction in the solar nebula. The Fischer-Tropsch reaction involves the production of n-alkanes from CO and H_2 on the surfaces of mineral catalysts. Supporting evidence for this theory includes an apparently similar carbon isotopic fractionation between the macromolecular materials and coexisting carbonates in the meteorite to that seen in Fischer-Tropsch syntheses (Kvenvolden et al. 1970; Lancet and Anders 1970), the occurrence of the macromolecular materials as coatings on the surfaces of mineral nuclei (Alpern and Benkeiri 1973) and the presence of aliphatic moieties within the macromolecular materials as revealed by photochemical oxidation (Hayatsu et al. 1977).

However, as a means of producing macromolecular materials, the Fischer-Tropsch theory presents significant problems. Carbon isotopic measurements of the carbon-containing phases of the Murchison meteorite reveal that CO is substantially lighter than the macromolecular materials (Kvenvolden et al. 1970). If this CO represents residual source gas for the Fischer-Tropsch-produced organic matter, then this measurement is inconsistent with the laboratory-produced results of Lancet and Anders (1970). Petrographic evidence indicates that the catalysts necessary to trigger the reaction in the primitive solar nebula were formed much later on the meteorite parent body (Kerridge et al. 1979; Bunch and Chang 1980). The aliphatic species present in the macromolecular materials can be accounted for by alkyl substitution, hydroaromatic rings, and bridging groups between aromatic units, none of which are characteristic of Fischer-Tropsch reactions.

It has also been proposed that the macromolecular materials were produced in the solar nebula by the gas-phase pyrolysis of simple aliphatic hydrocarbons such as acetylene (C_2H_2) and methane (CH_4) at temperatures of 900 to 1100 K (Morgan et al. 1991). The addition of acetylene molecules by a free radical mechanism leads to polymerization and the formation of high molecular weight aromatic structures. Acetylene and methane are believed to be present in interstellar clouds and would have been inherited by the solar nebula following cloud collapse (Morgan et al. 1991). In fact, it is suggested that 1 to 10% of the initial gaseous carbon in the solar nebula was present as small aliphatic hydrocarbons (Morgan et al. 1991).

Meteorite parent body

Peltzer et al. (1984) proposed that following the formation of the Murchison meteorite parent body, electric discharges and ultraviolet light may have acted on transient atmospheres leading to the production of organic matter by Miller-Urey reactions. Miller-Urey reactions are known to be efficient processes for the production of complex organic matter (Sagan and Khare 1979).

The internal heating of the Murchison parent body is thought to have initiated periods of aqueous alteration, as liquid water was present at temperatures of less than 20 °C (Clayton and Mayeda 1984). Organic synthesis could have occurred as hot fluids reacted on the surfaces of mineral catalysts during transport from the interior of the parent body. Similar processes have been proposed for the Earth (Shock et al. 1998) and Mars (Zolotov and Shock 1999). In addition, pre-existing organic matter in cometary ices may have been transformed by hydrolytic reactions (Bunch and Chang 1980).

Constraints on possible environments of formation

Theories that rely on *de novo* synthesis of the macromolecular materials in the solar nebula or on the meteorite parent body suffer from a major flaw, i.e., the inability to produce the deuterium enrichment seen in the macromolecular materials. This suggests that events in the solar nebula and on the meteorite parent body are more likely to amount to the secondary processing of pre-existing interstellar organic matter rather than primary synthesis.

Experimental

Hydrous pyrolysis

The insoluble organic materials in the Murchison meteorite were isolated by digesting the inorganic matrix with cycles of HF and HCl acids – Murchison K procedure until post HF/HCl step (Amari et al. 1994) – and by removing any free organic compounds with solvent extraction. To produce fragments of the macromolecular materials, this residue was subjected to hydrous pyrolysis. 136 mg of the HF/HCl residue was placed in a 1 ml stainless steel insert and 0.4 ml of high purity water was added. The insert was then purged with nitrogen gas, sealed and placed into a 71 ml stainless steel high pressure reactor (series 4740, Parr Instrument Co.), which was filled with 20 ml water to minimize pressure differentials. The whole arrangement was heated to 320 °C in a muffle furnace for 72 hours. These conditions were chosen to achieve the higher yields possible without compromising the safe operation of the high-pressure equipment. When cooled, the pyrolysis products were extracted and analyzed.

Supercritical fluid extraction

Free organic matter and fragments of macromolecular materials were extracted from the bulk meteorite by supercritical fluid extraction (SFE) (Gilmour and Pillinger 1993). Two extractions, one using 1.5 g of crushed, whole unextracted meteorite, and one using 136 mg of hydrously pyrolysed hydrofluoric/ hydrochloric (HF/HCl) acid residue (see below), were performed. Each consisted of an initial static extraction with pure CO_2 (99.9995%, 276 bar) for 90 mins, followed by a dynamic extraction (276 bar, 1 ml/min) for 45 mins. The extracts were collected by immersing the fused silica capillary SFE vent in diethyl ether cooled to approximately 0 °C. These conditions allowed the selective extraction of non-polar organic compounds and resulted in a final extract that was dissolved in a few hundred microliters of solvent, ready for immediate analysis by gas chromatographic techniques.

Gas chromatography-mass spectrometry

Compound detection and identification was performed by gas chromatography-mass spectrometry (GC-MS) using a Hewlett Packard 5890 gas chromatograph interfaced to a 5971 mass selective detector. Analyses were performed by on-column injection onto a HP5 capillary column (50 m x 0.32 mm x 0.17 μm). Following a 10 min holding period at 25 °C the GC oven was programmed from 25 °C to 220 °C at 5 °C min^{-1} and then from 220 to 300 °C at 10 °C min^{-1}. The final temperature was held for 12 min.

Results and discussion

Alkylbenzenes, alkylnaphthalenes and alkylindanes

The distribution patterns for alkylbenzenes, alkylnaphthalenes and alkylindanes in the Murchison hydrous pyrolysate are shown in Figures 1, 2 and 3. The alkylbenzenes are present as C_1 to C_4 compounds (Table 2). The volatility of benzene precludes its detection even with the use of SFE. The distribution of alkylbenzenes is dominated by toluene with substantially lower abundances of C_2 and C_3 alkylbenzenes and minor amounts of C_4 alkylbenzenes. The alkylnaphthalenes are less alkylated with significant amounts of C_0 to C_2 compounds present, but only trace amounts of C_3 species (Table 2). Alkylindanes in the hydrous pyrolysate are present as C_0-C_2 compounds. Although Komiya et al. (1993) reported the presence of indene and indane in pyrolysis products of the several Antarctic carbonaceous chondrites, only fully hydrogenated species were observed in these experiments. Together, these results indicate that the macromolecular materials, sampled by the hydrous pyrolysis experiment, are dominated by cores of one or two aromatic rings linked by aliphatic branches containing up to four carbon atoms.

The primary structure of the aromatic hydrocarbon species in the macromolecular materials may have been determined by the pre-terrestrial aqueous event to which they have been subjected. During hydrous pyrolysis, water acts as a source of exogenous hydrogen that is incorporated into the organic matter. This process inhibits aromatization. The water present during the aqueous event on the meteorite parent body may have performed a similar function. If this were the case then the extent of aromatization in the macromolecular materials may be a direct function of the level of aqueous alteration they have experienced.

In this context, the aliphatic branches attached to the aromatic cores may represent relics of hydroaromatic structures either produced, or protected, by the water-derived hydrogen supplied by the pre-terrestrial aqueous event. This explanation would be consistent with the dominance of short aliphatic chain lengths observed in the pyrolysis products of the macromolecular materials.

Sulfur-containing hydrous pyrolysis products

The distribution patterns of benzothiophenes and thiophenes in the hydrous pyrolysate are shown in Figures 4 and 5. Thiophenes are present in C_1 to C_4 alkylated forms, whereas the benzothiophenes are present as C_0 to C_2 derivatives (Table 2). Thiophenes in hydrous pyrolysates may represent primary structures released from the macromolecular materials or secondary products synthesized during the pyrolysis procedure itself. During hydrous pyrolysis, labile organic sulfur is released as H_2S which, together with any elemental sulfur present, is

Table 2. Aromatic moieties in the hydrous pyrolysate of the Murchison macromolecular materials (this paper).

Moiety	Structure	R (Alkylation)
alkylbenzenes		C_0, C_1, C_2, C_3, C_4
alkylnaphthalenes		C_0, C_1, C_2, C_3
alkylthiophenes		C_0, C_1, C_2, C_3, C_4
alkylbenzothiophenes		C_0, C_1, C_2
alkylphenols		C_0, C_1, C_2

incorporated into the organic matter. The chemical composition of the new sulfur-containing species is determined by the temperature at the time of incorporation. At lower temperatures (ca. 200 °C) more alkylthiophenes form, while at higher temperatures (ca. 330 °C) the more stable alkylbenzothiophenes are favored.

It is possible that reactions of this type occurred during the aqueous event on the meteorite parent body. If this was the case then primary sulfur-containing moieties in meteoritic organic matter may be useful indicators of the temperatures reached during the aqueous event. The presence of the thermodynamically stable benzothiophenes in the hydrous pyrolysate may reflect sulfur incorporation during a relatively high temperature aqueous event. These benzothiophenes must be at least partly primary sulfur-containing moieties as they are also present as free compounds in unheated extracts of Murchison (Sephton et al. 1998).

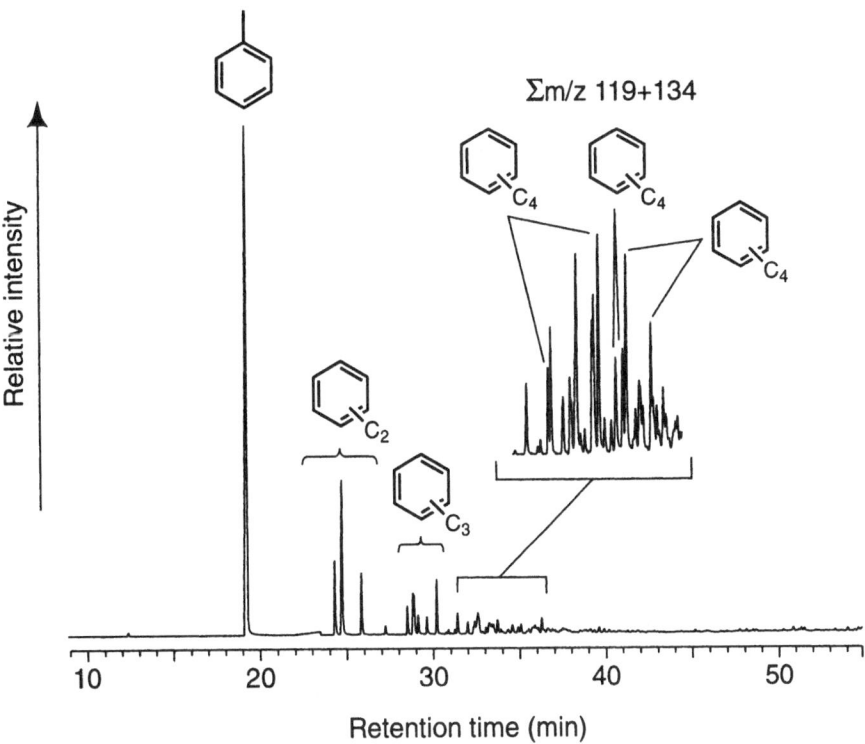

Fig. 1. Mass chromatogram (m/z 78, 91, 92, 105, 106, 119, 120), characteristic of C_0-C_3 alkylbenzenes, from a supercritical fluid extract of the Murchison hydrous pyrolysate. Inset is a partial mass chromatogram (m/z 133, 134) for C_4 alkylbenzenes from the same extract.

Oxygen-containing hydrous pyrolysis products

Figure 6 shows the distribution pattern of phenol and its C_1 to C_2 derivatives (Table 2). As with the thiophenes, oxygen-containing species in hydrous pyrolysates may represent primary or secondary products. During the hydrous pyrolysis experiment, water-derived oxygen can be added to organic matter to produce phenols or other oxygenated species (Stalker et al. 1994).

Oxygen-containing components have been detected in the macromolecular materials by previous authors using other methods. Hayatsu et al. (1977) used selective chemical degradation to establish the presence of phenolic ethers in the organic network. Aryl ethers are easily cleaved to produce two phenol moieties under hydrous pyrolysis conditions (Siskin and Katritzky 1991). Therefore, the phenols in the hydrous pyrolysate of the Murchison macromolecular materials may be pyrolysis products of the ethers detected by Hayatsu et al. (1977). Furthermore, it is conceivable that hydrous alteration on the meteorite parent body may have led to analogous ether cleavages. If this was the case, then phenols, both free in the Murchison meteorite and bound to the macromolecular materials, are partly the products of aqueous processing.

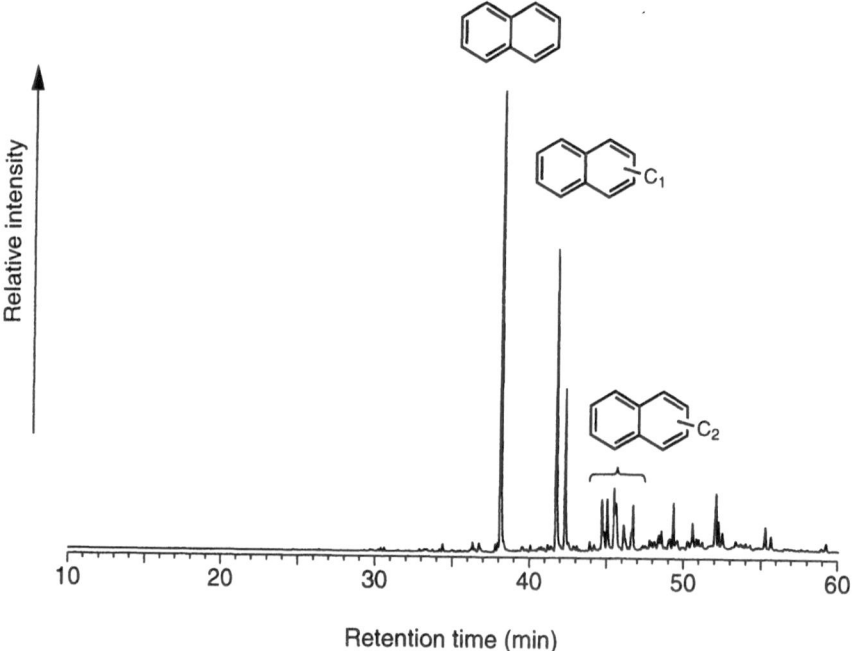

Fig. 2. Mass chromatogram (m/z 128, 141, 142, 155, 156), showing C_0-C_3 alkyl-naphthalenes.

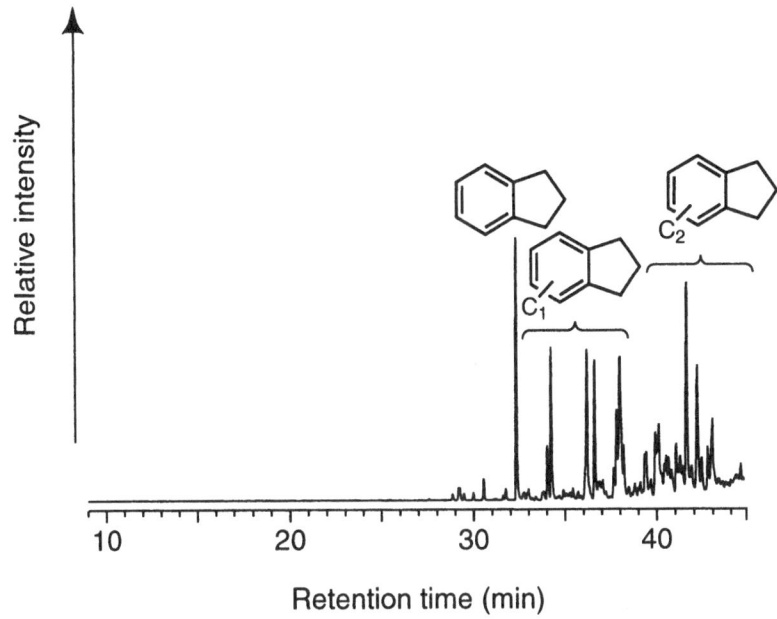

Fig. 3. Mass chromatogram (m/z 117, 118, 131, 132, 145, 146), showing C_0-C_2 alkylindanes.

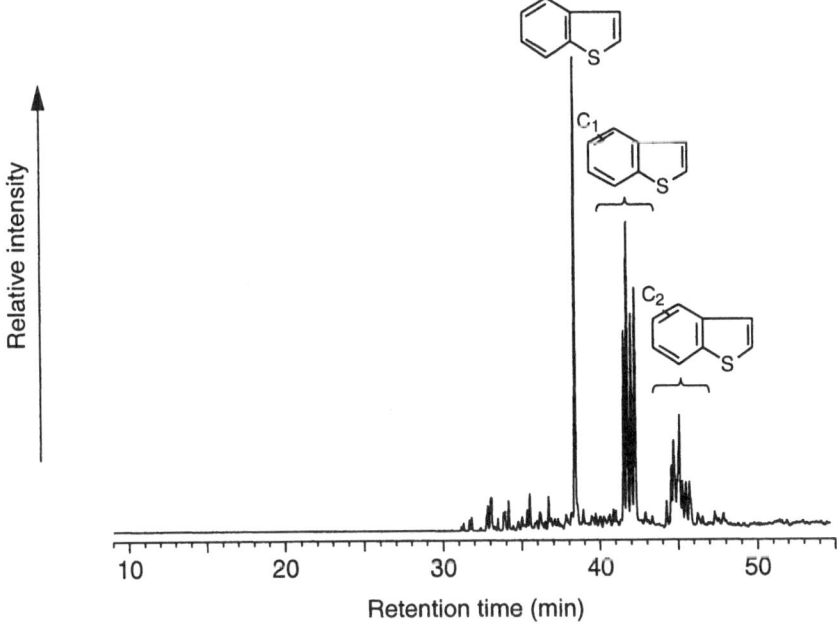

Fig. 4. Mass chromatogram (m/z 134, 147, 148, 161, 162), showing C_0-C_2 alkylbenzo[b]thiophenes

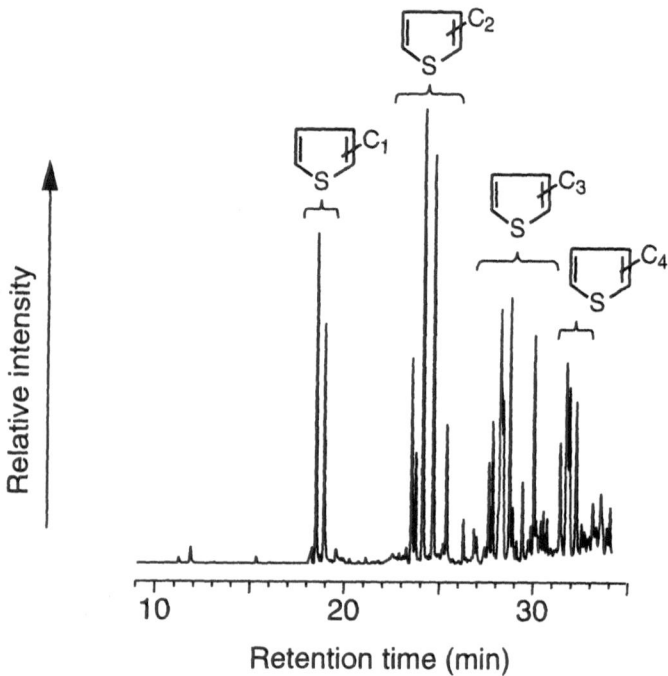

Fig. 5. Mass chromatogram (m/z 84, 97, 98, 111, 112, 125, 126, 139, 140) showing C_0-C_4 alkylthiophenes.

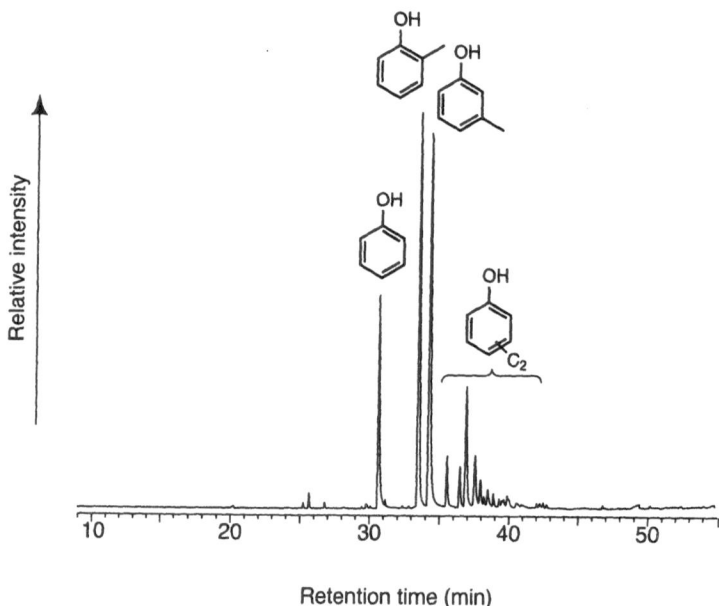

Fig. 6. Mass chromatogram (m/z 94, 107, 108, 121, 122) showing C_0-C_2 alkylphenols.

Meteorite macromolecular materials and the origin of life

Accretion of organic matter on the early Earth

The majority of organic matter delivered to the early Earth arrived in the form of micrometeorites. A study of the entry conditions under which micrometeorites survive atmospheric heating indicates that most particles greater than 400 μm (or larger for non-spherical particles) melt during entry (Hunten 1997). Anders (1989) has calculated an accretion rate of organic carbon from meteors and micrometeorites at the present time. This rate is based on a size range from 10^{-12} g, below which organic matter would have been destroyed by photolysis, to 10^{-6} g, above which most of the organic matter would have been destroyed by ablation or on impact. The calculation gave an accretion rate of 3.2×10^8 g yr^{-1}, which corresponds to a mean organic carbon concentration in the oceans of 2×10^{-5} g l^{-1} over 10^8 yr. The value during the early history of the Earth would have been higher, because meteorite infall rates were higher. Anders (1989) estimated a value for late accretional material ~10^3 times greater than the present day influx, although, he points out that the fraction of organic matter that survived may have been lower because of recurrent vaporization of the oceans by large-scale impacts.

The relationship between organic carbon in micrometeorites and carbonaceous chondrites is unclear. The carbon isotopic composition of macromolecular carbon in carbonaceous chondrites varies from −10 to −20 ‰ (Kerridge 1985). Aromatic molecules occurring as free compounds or produced by pyrolytic degradation of the macromolecule have a wider range in $\delta^{13}C$ values from -5 to −25 ‰ (Gilmour and Pillinger 1994; Sephton et al. 1998). Carbonaceous components released from Greenland and Antarctic micrometeorites have $\delta^{13}C$ values in the range −18 to −22 ‰ and −22 to −25 ‰ respectively (Wright et al. 1997) suggesting some similarities between micrometeorites and carbonaceous chondrites, although it should be noted that these values are also similar to those of potential petroleum-derived hydrocarbon contaminants (Gilmour et al. 1984). Clemett et al. (1993) analyzed stratospherically collected interplanetary dust particles (IDPs) with a microprobe two-step laser mass spectrometer. Several polycyclic aromatic hydrocarbons (PAHs) and their alkylated derivatives were identified and the spectra were similar to those observed in the carbonaceous chondrite, Allende (CV3). However, the IDP mass spectra were generally more complex and indicated the presence of nitrogen-containing functional groups (Clemett et al. 1993). The detection of the non-protein amino acid α-amino-isobutyric acid in micrometeorites provides supporting evidence for the existence of extraterrestrial organic matter in these objects (Brinton et al. 1998). Differences have been noted between the amino acid ratios in micrometeorites and carbonaceous chondrites, which have led to the suggestion of a cometary source for micrometeorite organic matter (Brinton et al. 1998). However, even if cometary in origin, micrometeorites may still have delivered macromolecular organic matter to the early Earth (Huebner and Boice 1997).

Hydrothermal processing of accreted organic matter

There is a growing awareness that organic compounds can undergo transformations at elevated temperatures and pressures (Siskin and Katritzky 1991). Changes in the physics and chemistry of water, e.g., a decrease in its dielectric constant and an increase in its ionic product (Siskin and Katritzky 1991), at elevated temperatures produce solvent properties similar to those of polar organic solvents at room temperature. The systematic study of the reactivity of various organic compounds in water, over temperature ranges of 200 to 350 °C, indicates that condensation, bond cleavage, hydrolysis and autocatalytic reactions occur (Siskin et al. 1990). It might be expected that the highly aromatic nature of Murchison macromolecular materials (Cronin et al. 1987; Sephton et al. 1998) would render them unreactive. However, our results suggest that this is not the case. For instance, the abundance of phenols in the hydrous pyrolysate may be attributed to the cleavage of aryl ethers within the macromolecular materials (Siskin and Katritzky 1991). Furthermore, a study by Mautner et al. (1995) subjected a sample of bulk Murchison meteorite to typical terrestrial hydrothermal conditions of 350 °C and 25 MPa; compounds including nonanoic acid, glycine, and pyrene were released and Mautner et al. (1995) suggested that some of these molecules may have been derived from the Murchison macromolecular materials.

An assessment of both the stability and reactivity of meteoritic macromolecular materials in the presence of water on the early Earth is essential in order to evaluate its potential as a prebiotic source of organic matter. Interpreting the results of hydrothermal experiments can be difficult (Shock et al. 1998). Yet, hydrous pyrolysis indicates that aqueous processing of meteoritic macromolecular materials will facilitate the transfer of biogenic elements between organic and inorganic matter. It may also release molecules that could be involved in the synthesis of biologically interesting compounds.

Some of the molecules released by the hydrous pyrolysis of Murchison macromolecular materials are amphipathic, that is molecules with hydrophobic and hydrophyllic parts. Deamer (1985) noted that amphiphatic molecules, isolated by solvent extraction of Murchison, could provide the starting materials for membrane formation. These molecules can organize themselves so that their hydrophobic parts are clustered towards the center of a globule or bilayer. These globules or bilayers can act as primitive membranes that enclose areas in which molecular species can interact in a microenvironment insulated from their surroundings.

Hydrothermal environments were apparently widespread in the early solar system (Carr 1996; Jakosky and Shock 1998) and their importance as locations for pre-biotic organic synthesis has been recognized for some time (Miller and Lazcano 1995). The possibility that comet- or meteorite-derived organic matter might enter these environments through accretion has also been recognized as a means of processing cometary volatiles in the outer solar system (Shock and McKinnon 1993), or as a means of releasing prebiotic molecules from meteorites (Mautner et al. 1995). In general, the breakdown of Murchison macromolecular materials by hydrous pyrolysis leads to the formation of thermodynamically stable condensed aromatic products. However, these products can undergo further reactions. The phenolic group is activating and nitration, halogenation, acetylation, and alkylation electrophilic substitution reactions in the nucleus are relatively easy. Furthermore, Shock and Schulte (1990) have proposed that under conditions in which aqueous organic compounds are in metastable equilibrium with polycyclic aromatic hydrocarbons the formation of amino and carboxylic acids is energetically favorable at moderate temperatures and pressures. This possibility is intriguing, as it would greatly increase the potential of meteoritic macromolecular materials to take part in pre-biotic chemistry.

Acknowledgements

This work was supported by PPARC and the Royal Society. Reviews by Sara Russell and an anonymous reviewer helped improve the manuscript. The IMPACT program of the European Science Foundation is thanked for logistical support.

References

Alexander CMO, Russell SS, Arden JW, Ash RD, Grady MM, Pillinger CT (1998) The origin of chondritic macromolecular organic matter: A carbon and nitrogen isotope study. Meteoritics Planet Sci 33: 603-622

Alpern B, Benkeiri Y (1973) Distribution de la matière organique de la météorite d'Orgueil par microscopie en fluorescence. Earth Planet Sci Lett 19: 422-428

Amari S, Lewis RS, Anders E (1994) Interstellar grains in meteorites; I, Isolation of SiC, graphite, and diamond; size distributions of SiC and graphite. Geochim Cosmochim Acta 58: 459-470

Anders E (1989) Pre-biotic organic matter from comets and asteroids. Nature 342: 255-257

Anders E (1991) Organic matter in meteorites and comets - possible origins. Space Sci Rev 56: 157-166

Biemann K (1974) Test result on the Viking gas chromatograph-mass spectrometer experiment. Origins Life 5: 417-430

Brack A, Pillinger CT (1998) Life on Mars: chemical arguments and clues from Martian meteorites. Extremophiles 2: 313-319

Brinton KLF, Engrand C, Glavin DP, Bada JL, Maurette M (1998) A search for extraterrestrial amino acids in carbonaceous Antarctic micrometeorites. Origins Life 28: 413-424

Bunch TE, Chang S (1980) Carbonaceous chondrites II: Carbonaceous chondrite phyllosilicates and light element geochemistry as indicators of parent body processes and surface conditions. Geochim Cosmochim Acta 44: 1543-1577

Carr MH (1996) Water on early Mars. In: Bock G, Goode JA, Walker M (eds) Evolution of hydrothermal ecosystems on Earth (and Mars?) Ciba Foundation Symposia 202. Wiley, New York, pp 249-267

Chyba C, Sagan C (1992) Endogenous production, exogenous delivery and impact-shock synthesis of organic molecules - an inventory for the origins of life. Nature 355: 125-132

Chyba CF, Sagan C (1997) Comets as a source of prebiotic organic molecules for the early Earth. In: Thomas PJ, Chyba CF, McKay CP (eds) Comets and the origin and evolution of life. Springer Verlag, Berlin, pp 147-174

Chyba CF, Thomas PJ, Brookshaw L, Sagan C (1990) Cometary delivery of organic molecules to the early Earth. Science 249: 366-373

Clayton RN, Mayeda TK (1984) The oxygen isotope record in Murchison and other carbonaceous chondrites. Earth Planet Sci Lett 67: 151-161

Clemett SJ, Maechling CR, Zare RN, Swan PD, Walker RM (1993) Identification of complex aromatic molecules in individual interplanetary dust particles. Science 262: 721-725

Cottin H, Gazeau MC, Raulin F (1999) Cometary organic chemistry: a review from observations, numerical and experimental simulations. Planet Space Science 47: 114-162

Cronin JR, Chang S (1993) Organic matter in meteorites: molecular and isotopic analysis of the Murchison meteorite. In: Greenberg JM (ed) The Chemistry of Life's Origins. Kluwer Academic Publishers, Dordrecht, pp 209-258

Cronin JR, Gandy WE, Pizzarello S (1981) Amino acids of the Murchison meteorite: 1. 6 carbon acyclic primary alpha-amino alkanoic acids. J Molec Evol 17: 265-272

Cronin JR, Pizzarello S (1986) Amino acids of the Murchison meteorite: 3. 7 carbon acyclic primary alpha-amino alkanoic acids. Geochim Cosmochim Acta 50: 2419-2427.

Cronin JR, Pizzarello S, Fyre JS (1987) ^{13}C NMR spectroscopy of the insoluble carbon of carbonaceous chondrites. Geochim Cosmochim Acta 51: 229-303

Cronin JR, Pizzarello S, Yuen GU (1985) Amino acids of the Murchison meteorite: 2. 5 carbon acyclic primary beta-amino, gamma-amino and delta-amino alkanoic acids. Geochim Cosmochim Acta 49: 2259-2265

Deamer DW (1985) Boundary structures are formed by organic components of the Murchison carbonaceous chondrite. Nature 317: 792-794

Geiss J, Altwegg K, Anders E, Balsiger H, Ip WH, Meier A, Neugebauer M, Rosenbauer H, Shelley EG (1991) Interpretation of the ion mass spectra in the mass per charge range 25-35 amu/e obtained in the inner coma of Halley comet by the HIS-sensor of the Giotto IMS experiment. Astron Astrophys 247: 226-234

Gilmour I, Pillinger CT (1993) Extraction and isotopic analysis of medium molecular weight hydrocarbons from Murchison using supercritical carbon dioxide. Lunar Planet Sci 24: 535-536

Gilmour I, Pillinger CT (1994) Isotopic compositions of individual polycyclic aromatic hydrocarbons from the Murchison meteorite. Mon Not Roy Astron Soc 269: 235-240.

Gilmour I, Swart PK, Pillinger CT (1984) The carbon isotopic composition of individual petroleum lipids. Org Geochem 6: 665-670

Greenberg JM (1984) Chemical evolution in space. Origins Life 14: 25-36

Hayatsu R, Matsuoka S, Scott RG, Studier MH, Anders E (1977) Origin of organic matter in the early solar system-VII. The organic polymer in carbonaceous chondrites. Geochim Cosmochim Acta 41: 1325-1339

Hayatsu R, Scott RG, Studier MH, Lewis RS, Anders E (1980) Carbynes in meteorites: detection, low temperature origin and implications for interstellar molecules. Science 209: 1515-1518

Hennet RJ-C, Holm NG, Engel MH (1992) Abiotic synthesis of amino acids under hydrothermal conditions and the origin of life: A perpetual phenomenon? Naturwissenschaften 79: 361-365

Holtzer G, Oró J (1977) Pyrolysis of organic compounds in the presence of ammonia: The Viking Mars lander site alteration experiment. Org Geochem 1: 37-52

Huebner WF, Boice DC (1997) Polymers and other macromolecules in comets. In Thomas PJ, Chyba CF, McKay CP (eds) Comets and the origin and evolution of life. Springer Verlag, Berlin, pp 111-130

Hunten DM (1997) Soft entry of micrometeorites at grazing incidence or by aerocapture. Icarus 129: 127-133

Imbus SW, McKirdy DM (1993) Organic geochemistry of Precambrian sedimentary rocks. In: Engel MH, Macko SA (eds) Organic Geochemistry. Plenum Press, New York, pp. 657-684

Jakosky BM, Shock EL (1998) The biological potential of Mars, the early Earth, and Europa. J Geophys Res (Planets) 103: 19359-19364

Kerridge JF (1985) Carbon, hydrogen and nitrogen in carbonaceous chondrites - abundances and isotopic compositions in bulk samples. Geochim Cosmochim Acta 49: 1707-1714

Kerridge JF, Mackay AL, Boynton WV (1979) Magnetite in CI carbonaceous chondrites: Origin by aqueous activity on a planetesimal surface. Science 205: 395-397

Kolodny Y, Kerridge JF, Kaplan IR (1980) Deuterium in carbonaceous chondrites. Earth Planet Sci Lett 46: 149-158

Komiya M, Shimoyama A, Harada K (1993) Examination of organic compounds from insoluble organic matter isolated from some Antarctic carbonaceous chondrites by heating experiments. Geochim Cosmochim Acta 57: 907-914

Kvenvolden K, Lawless J, Peterson E, Flors J, Ponnamperuma C, Kaplan IR, Moore C (1970) Evidence for extraterrestrial amino acids and hydrocarbons in the Murchison meteorite. Nature 228: 923-936

Lancet MS, Anders E (1970) Carbon isotope fractionation in the Fischer-Tropsch synthesis and in meteorites. Science 170: 980-982

Levy RL, Grayson MA, Wolf CJ (1973) The organic analysis of the Murchison meteorite. Geochim Cosmochim Acta 37: 467-483

Lis DC, Keene J, Young K, Phillips TG, Bockelee-Morvan D, Crovisier J, Schilke P, Goldsmith PF, Bergin EA (1997) Spectroscopic observations of Comet C 1996 B2 (Hyakutake) with the Caltech submillimeter observatory. Icarus 130: 355-372

Mautner MN, Leonard RL, Deamer DW (1995) Meteorite organics in planetary environments - hydrothermal release, surface-activity, and microbial utilization. Planet Space Sci 43: 139-147

Miller SL (1993) The prebiotic synthesis of organic compounds on the early Earth. In Engel MH, Macko SA (eds) Organic Geochemistry. Plenum Press, New York, pp 625-637

Miller SL, Lazcano A (1995) The origin of life - Did it occur at high temperatures? J Molec Evol 41: 689-692

Mojzsis SJ, Arrhenius G, McKeegan KD, Harrison TM, Nutman AP, Friend CRL (1996) Evidence for life on Earth before 3,800 million years ago. Nature 384: 55-59

Morgan WA, Feigelson ED, Wang H, Frenklach M (1991) A new mechanism for the formation of meteoritic kerogen-like material. Science 252: 109-112

Morowitz HJ (1992) Beginnings of cellular life: Metabolism recapitulates biogenesis. Yale University Press, New Haven 195 pp

Nutman AP, Mojzsis SJ, Friend CRL (1997) Recognition of >=3850 Ma water-lain sediments in West Greenland and their significance for the early Archaean Earth. Geochim Cosmochim Acta 61: 2475-2484

Peltzer ET, Bada JL, Schlesinger G, Miller SL (1984) The chemical conditions on the parent body of the Murchison meteorite: some conclusions based on amino, hydroxy and dicarboxylic acids. Adv Space Res 4: 69-74

Robert F, Epstein S (1982) The concentration and isotopic composition of hydrogen, carbon and nitrogen in carbonaceous meteorites. Geochim Cosmochim Acta 46: 81-95

Rosing MT (1999) C-13-depleted carbon microparticles in >3700-Ma sea-floor sedimentary rocks from west Greenland. Science 283: 674-676

Sagan C, Khare BN (1979) Tholins: Organic chemistry of interstellar grains and gas. Nature 277: 102-107

Sandford SA (1996) The inventory of interstellar materials available for the formation of the solar-system. Meteoritics Planet Sci 31: 449-476

Schidlowski M (1988) A 3,800-million-year isotopic record of life from carbon in sedimentary-rocks. Nature 333: 313-318

Schopf JW (1993) Microfossils of the Early Archean Apex Chert: new evidence of the antiquity of life. Science 260: 640-646

Sephton MA., Pillinger CT, Gilmour I (1998) $\delta^{13}C$ of free and macromolecular aromatic structures in the Murchison meteorite. Geochim Cosmochim Acta 62: 1821-1828

Sephton MA, Pillinger CT, Gilmour I (1999) Small-scale hydrous pyrolysis of macromolecular material in meteorites. Planet Space Sci 47: 181-187

Shock EL, Schulte MD (1990) Amino acid synthesis in carbonaceous meteorites by aqueous alteration of polycyclic aromatic hydrocarbons. Nature 343: 728-731

Shock EL, McKinnon WB (1993) Hydrothermal processing of cometary volatiles - applications to Triton. Icarus 106: 464-477

Shock EL, McCollom T, Schulte MD (1998) The emergence of metabolism from within hydrothermal systems. In: Weigel J, Adams MWW (eds) The keys to molecular evolution and the origin of life. Taylor and Francis, Washington, pp 59-76

Siskin M, Brons G, Katritzky AR, Balasubramanian M (1990) Aqueous organic chemistry: 1. Aquathermolysis - comparison with thermolysis in the reactivity of aliphatic compounds. Energ Fuel 4: 475-482

Siskin M, Katritzky AR (1991) Reactivity of organic compounds in hot water: geochemical and technological implications. Science 254: 231-237

Stalker L, Farrimond P, Larter SR, Telnaes N, van Graas G, Oygard K (1994) Water as an oxygen source for the production of oxygenated compounds (including CO_2 precursors) during kerogen maturation. Org Geochem 22: 477-486

Studier MH, Hayatsu R, Anders E (1972) Origin of organic matter in early solar system-V: Further studies of meteoritic hydrocarbons and discussion of their origin. Geochim Cosmochim Acta 36: 189-215

Wright IP, Yates P, Hutchison R, Pillinger CT (1997) The content and stable isotopic composition of carbon in individual micrometeorites from Greenland and Antarctica. Meteoritics Planet Sci 32: 79-89

Zinner E (1988) Interstellar cloud material in meteorites. In: Kerridge JF, Matthews MS (eds) Meteorites and the early solar system. University of Arizona Press, Tucson, pp 956-983

Zolotov M, Shock E (1999) Abiotic synthesis of polycyclic aromatic hydrocarbons on Mars. J Geophys Res (Planets) 104: 14033-14049

3 Effects of Atmospheric Heating on Infalling Meteorites and Micrometeorites: Relevance to Conditions on the Early Earth

I.P. Wright, P.D. Yates and C.T. Pillinger

Planetary Sciences Research Institute, The Open University, Milton Keynes MK7 6AA, United Kingdom. (I.P.Wright@open.ac.uk)

Abstract. Since micrometeorites may have played an important role in the development of conditions on the early Earth, bringing vital supplies of the biogenically important elements, it is apposite to appraise the effects that atmospheric heating exerts on such particles as they head towards the planet's surface. An expedient way to approach this problem is to undertake an investigation of the effects of pulse-heating on the carbon inventories of micrometeorites arriving at the present time. Unfortunately analytical techniques for carbon have not yet advanced to the point where levels of sensitivity are compatible with a study of this type. However, it is still possible to contemplate the effects of atmospheric heating by utilizing samples of relatively well documented meteorites as analogs of micrometeorites. To this end small samples of three meteorites (Allende, Weston and Goalpara) have been subjected to pulse-heating in order to investigate the effects of atmospheric infall on carbon chemistry. In this preliminary study the samples were heated in air thereby simulating present day atmospheric conditions. The results show, not surprisingly, that heating removes carbon. However, the extent of this effect is dependent upon combustion kinetics, which in turn is presumably related to the nature of individual components and their location within the samples. For instance, the results show that refractory carbon-bearing components are not only able to survive pulse-heating to 1500°C, but may actually become enriched in samples as a result of a heating episode (i.e., as a consequence of preferential removal of other more labile materials). Although carbon is removed by heating, this does not appear to affect the carbon isotopic composition of the material that remains. This means that analyses of micrometeorites falling to Earth at the present time, although recording artificially low carbon contents, will be to some extent valid in terms of carbon isotopic measurements (especially where stepped combustion is used to acquire the relevant data above a certain threshold temperature, which is itself related to the extent of atmospheric heating). Furthermore, a detailed calibration of the extent of carbon removal may assist the development of atmospheric heating models, thereby constraining entry velocities, and ultimately sources. To simulate effectively the nature of infall heating in former times, more sophisticated experiments would need to be carried out involving the use of synthetic "atmospheres" prepared from various gas mixtures.

Introduction

There are a number of pertinent reasons to study the carbon contained in micrometeorites which fall to Earth at the present time, not least of all because they sample a wide-spread Solar System reserve of material (from cometary dust, to collisional debris from the asteroid belt, to planetary materials, and perhaps even interstellar grains). Likewise, it is desirable to assess the effects that infalling micrometeorites could have on the development of conditions on the early Earth; for instance, the primitive surface having been sterilized by relatively large impacts early in the history of the planet may have subsequently become gently seeded with biogenically important compounds from micrometeorites. This may have been an important step in the origin and development of life on Earth. The exact relevance of micrometeorites in this regard requires the resolution of one important issue: to what extent were the original pre-terrestrial carbon-bearing constituents modified during the process of deceleration through the Earth's atmosphere? In principle it should be possible to assess the magnitude of this effect through the study of micrometeorites falling to Earth at the present time. However, there is an additional problem here, namely that micrometeorites falling today suffer from problems of contamination (in an environment that is far more diverse than that of the primitive Earth).

Studies of the atmospheric trails and fireballs associated with the 1998 Leonid "meteor storm", along with attempts to collect particles from a high altitude balloon, demonstrate one aspect of trying to comprehend the nature of comets (Tempel-Tuttle in this case). A far more sophisticated project aimed at obtaining cometary materials involves the sending of a spacecraft (*Stardust*) to collect dust from comet p/Wild and return it to Earth in 2006. In the mean time extraterrestrial dust that is incident upon space hardware in low Earth orbit, which is subsequently returned to Earth (e.g., Hubble Space Telescope solar array panels), is a valuable resource for study (Graham et al. 1997), but in this case the dust is extensively melted by the high velocity impact process. As such the best immediate hope for studying cometary (and/or asteroidal) dust is the acquisition of random particles which enter the atmosphere and survive to be collected either at ground level, or from within the stratosphere. A potential impediment to the successful prosecution of studies of terrestrially captured dust is the detrimental influence of atmospheric heating; the concern for investigations of carbon is that during infall, original constituents may succumb to volatilization, or oxidation, followed by subsequent loss. Genge and Grady (1998) describe a melted micrometeorite that contains features compatible with extensive reduction during atmospheric heating, most probably because the sample was originally carbon-rich. There are two important consequences here: firstly, if we want to study present-day micrometeorites it is necessary to understand the effects of atmospheric heating in order to be able to document the primary constituents of interest. Secondly, the effects of atmospheric heating may be an important parameter when considering the role of micrometeorites during the development of the early Earth; in particular, it is desirable to assess to what extent biologically important elements and compounds could survive to collect at the surface since this input could clearly influence the development of life on the planet.

There are, of course, many hypotheses concerning the origin of life on Earth (see, for instance, Horgan 1991; Miller 1993; Schidlowski 1993; etc.). To some extent atmospheric chemistry may have played a part (e.g., Walker 1977; Schopf 1983; Kasting 1993; Summers and Chang 1993), but equally, hydrothermal vent systems in deep oceans (e.g., Miller and Bada 1988; Deming and Baross 1993) may also have been important. Even once life had got started, giant impacts early in the history of the Earth may have been responsible for various episodes of extermination, frustrating the development of primitive life (e.g., Maher and Stevenson 1988; Chyba 1993). An open question is where the various chemicals and elements necessary for life came from originally; they may simply have accreted with the planet, although various exotic sources have been proposed, including the input of organic materials from interstellar grains (Goldanskii 1977; Greenberg 1981), comets (Oro et al. 1980) or micrometeorites (e.g., Anders 1989). Whilst very large bodies from space are discounted because the temperatures and pressures on impact are sufficiently high to destroy organic compounds it is clear that such materials in bodies of typical meteorite size (say 1-1000 kg) survive intact (e.g., carbonaceous chondrites, which contain a rich variety of organic compounds; see, for instance, Cronin et al. 1988). Furthermore, Anders (1989) has shown that particles in the size range of ca. 1-100 μm are gently decelerated within the atmosphere and consequently may only be subjected briefly to temperatures of a few hundred °C. As such they have the potential to contribute significant amounts of unaltered organic materials to Earth.

As the struggle continues to understand the chemistry and stable isotopic composition of carbon in micrometeorites (Wright et al. 1988, 1996; Yates et al. 1994) a number of common features are emerging from the various types of sample that have been analyzed, viz. particles of >100 μm size retrieved from the deep-sea environment, Greenland cryoconite and Antarctic ice (note that herein the term "micrometeorite" is used to describe particles collected at the Earth's surface and does not include materials collected from the stratosphere, more usually referred to as "interplanetary dust particles", a.k.a. IDPs). Particles that are obviously melted contain very little, if any, indigenous carbon. Unmelted, or partially melted samples on the other hand appear to have certain affinities with primitive carbonaceous chondrites (CI/CM, and possibly C3); for instance, the results from stepped combustion experiments can be interpreted to show carbon in the form of macromolecular organic materials, microdiamonds and silicon carbide (Wright et al. 1988, 1996). Notwithstanding these observations and that, known from other studies, of localized concentrations of carbon in Antarctic micrometeorites (up to 11 wt%, i.e., 3.13 x CI1, Perreau et al. 1993), perhaps the most notable feature is the relatively low *overall* concentration of carbon in the samples (compared to normal carbonaceous chondrites). This suggests either one of two possibilities: the samples are fundamentally different from meteorites like carbonaceous chondrites, or the major carbon-bearing constituents have largely been removed in the terrestrial environment. While the exact nature and origin of micrometeorites remains a subject of research and debate it seems likely that at least some of the particles will be related to conventional meteorite types. As such, the low carbon contents are considered to be the result of secondary processes.

Now, upon arrival at Earth an extraterrestrial sample that is originally devoid of carbon will undoubtedly begin to acquire the element simply through the processes of contamination and weathering. The former involves the addition of components such as water-soluble organic components, airborne dust and aerosol particles, atmospheric hydrocarbons etc. Weathering, on the other hand, involves chemical reactions such as that between slightly acidic groundwater (containing carbonic acid) and silicates, producing bicarbonate and carbonate minerals. In addition, there is a further aspect of weathering that is relevant here and that is the effect of terrestrial fluids on macromolecular organic materials. It is already known from studies of terrestrial coal samples that weathering affects the overall nature and distribution of organic compounds (Martinez and Escobar 1995). Furthermore, the long-term exposure of meteorites with well-known pre-terrestrial organic signatures to fluid-related processes results in a breakdown of macromolecular materials (Sephton and Gilmour 1998); the implication here is that weathering processes can also act to remove indigenous organic compounds.

The apparent deficit of carbon in micrometeorites could therefore be a consequence of weathering; however, herein we explore the extent to which atmospheric entry heating was responsible for removal of original carbon-bearing components.

The effect of atmospheric heating on IDPs (typical size 5-15 μm, up to a few 10s of micrometers) has been considered previously; studies have been either theoretical/computational (e.g., Flynn 1989; Love and Brownlee 1991), or experimental. In the detailed theoretical approach of Love and Brownlee (1991), various heating profiles were developed for particles of different mass, entry angle and velocity. A typical result shows particles experiencing a very brief period of heating, followed by rapid cooling; the entire episode takes only a few seconds, with peak temperatures in the range 800-1700°C. In the experimental studies, samples of IDPs have been subjected to pulse-heating under vacuum to assess the level of de-gassing experienced by atmospheric heating, as witnessed by measurements of He and Ne (Brownlee et al. 1993; Nier and Schlutter 1993) and Zn (Klöck et al. 1994). The holy grail of experimental work is the determination of the peak temperatures encountered during infall, thereby constraining the entry velocity, which should distinguish between asteroidal and cometary sources (e.g., Nier and Schlutter 1993). In fact this philosophy could now be extended to the recognition of interstellar particles of the type observed as radar meteoroids (Taylor et al. 1996).

An equivalent experimental study of the behavior of carbon during atmospheric heating is desirable to establish the degree to which the element can be removed and also the extent of any accompanying isotopic fractionation. The latter is important since one of the goals of studying carbon is to determine its *original* $^{13}C/^{12}C$ ratio, thereby delimiting sources and providing background information on the nature of the element throughout the Solar System. Because an episode of heating has the potential to modify an original carbon isotopic composition, assessing the magnitude of this phenomenon is very important. However, the difficulty of the task is made plain by the following observation - assuming a nominal carbon concentration of 5 wt% (i.e., CI/CM levels) an IDP of 10 μm may only have an *original* carbon content of 0.25 ng. Even with carbon contents of 47 wt% (currently the record for an IDP, Thomas et al. 1995) this only translates to 2.4 ng of carbon, although larger samples would, of course, contain more. Although isotopic studies of carbon by static vacuum mass spectrometry has advanced to picomole levels of detection, i.e., 10s of pg, (Prosser et al. 1990), extraction procedures are constrained by ng blanks. Furthermore, while determination of the carbon isotopic composition may be possible by ion probe (witness the spectacular results from analyses of individual SiC and graphite grains extracted from meteorites, e.g., Zinner 1998) the errors remain large and the state-of-the-art as far as IDPs is concerned is still the 3 analyses by McKeegan et al. (1985). The technique of two-step laser desorption-mass spectrometry may be applicable in assessing the nature of some of the carbon (e.g., Hahn et al. 1988; Clemmett et al. 1993), but this has not yet been used for isotope ratio measurements.

Rather than await the arrival of a technique capable of analyzing precise stable isotope ratios on the quantities of carbon typically encountered in IDPs, we decided to investigate the effects of pulse-heating on carbon chemistry at a level that was more compatible with micrometeorites. Instead of using micrometeorites themselves, we chose to work with samples that are available in abundance and which are already known to contain manageable quantities of carbon - namely meteorites. Samples were analyzed for their carbon contents and stable isotopic compositions using stepped combustion (Swart et al. 1983a; Wright and Pillinger 1989) and static mass spectrometry (Wright 1984, 1995). The philosophy of the study was to use small pieces of meteorites (i.e., samples in the size range 300-500 µm) as analogs of micrometeorites. In this way it was perceived that it would be possible to compare results obtained by pulse-heating with those from the "unheated" samples, i.e., those taken from interiors. Such a comparative study is not possible using micrometeorites alone since all samples have been heated to a significant degree during atmospheric entry.

As well as providing insights into the effects of atmospheric heating on the carbon budget of incoming bodies, the research also provides a useful framework with which to tackle the problem of understanding what carbon there is in micrometeorites. This latter possibility arises because of the vast body of knowledge that has accrued in recent years concerning meteoritic carbon constituents (e.g., Swart et al. 1983b; Pillinger and Russell 1993; Anders and Zinner 1993). This information has, in most cases, been obtained from the use of relatively large samples (~10-50 g), which have been subjected to various forms of chemical and physical treatments to separate the components of interest. Such an approach is clearly impossible in the case of micrometeorites. However, since the separated carbon-bearing components of interest have been analyzed by stepped combustion, we can use this information to interpret stepped combustion data obtained during equivalent analyses of bulk (untreated) meteorites. In other words, we can gain an appreciation of how the known carbon-bearing components of meteorites "appear" during analyses that only use 50-500 µg of bulk samples. In this way it should be possible to contemplate results from bulk analyses of micrometeorites in the light of the calibrations made between bulk meteorites and their known constituents.

The present work is fairly limited in scope. Only three meteorites have been studied; Allende (CV3), Weston (H4), and Goalpara (ureilite). These were chosen to represent a range of different types of extraterrestrial materials, from chondrites to achondrites, the individuals comprising carbon in a variety of forms. The samples were subjected to pulse-heating in air (i.e., laboratory air at atmospheric pressure), rather than vacuum, since carbon (unlike He and Ne, for instance) is likely to undergo chemical reactions with oxygen (or oxygen-bearing species) during atmospheric heating. Two temperature regimes were used: 800 and 1500°C (to represent the lower and upper peak temperatures typically encountered during infall). After pulse-heating (carried out for 10 seconds) the samples were analyzed for their carbon content and stable isotopic composition using stepped combustion-mass spectrometry (see Wright and Pillinger 1989). In addition to the artificially heated samples, the bulk (unheated) meteorites were also analyzed to provide appropriate reference points, as well as samples of the fusion crusts of each meteorite. The results of the investigation are considered alongside data from previous analyses of micrometeorites.

Experimental

Working in a class-100 clean room, a small quantity of sample chips was broken from each parent meteorite using stainless steel implements (which had themselves been pre-cleaned by ultrasonication in a 50:50 mixture of toluene and methanol, and then air-dried at 100°C overnight). These collections of chips constituted the reservoirs from which samples were taken for all experiments, except the analyses of fusion crusts, which were obtained separately from the outer surfaces of the meteorites. For each experiment an individual chip was washed in ca. 50 ml of dichloromethane (Analytical grade) and then loaded into a pre-cleaned and pre-weighed receptacle made of platinum foil. This was then weighed to give the mass of each particle. The sample, contained in its sealed envelope, was then placed in a glass petri-dish (pre-cleaned by combustion in air at ca. 600°C for 1 hour) and removed from the clean room.

The pulse-heating apparatus consisted of a high-temperature furnace, constructed from a silicon carbide element, mounted vertically and including a platinum-rhodium thermocouple. The furnace temperature was set to either 800 or 1500°C by appropriate control of the current to the element. Into this was inserted a 15 cm length of quartz glass tube (3 mm internal diameter; pre-cleaned by combustion in air at 1000°C for 10 hours); this was left to stabilize for 5 minutes. A sample envelope was then dropped into the glass tube and after 10 seconds the tube was removed and the envelope tipped into another pre-cleaned quartz glass tube (5 mm internal diameter). This second tube was held at room temperature to facilitate rapid cooling of the sample.

Individual samples were then loaded into the gas extraction system of the MS 86 static vacuum mass spectrometer (Prosser et al. 1990). After evacuation and manipulation of the sample into the combustion furnace, each sample was then analyzed by heating in an atmosphere of pure oxygen gas (at ~20 torr) in the following increments: 200, 300, 400, 500, 600, 800 and 1200°C. A résumé of the technique is given in Carr et al. (1986) and Yates et al. (1992).

As well as the analyses of the meteorites, three blank experiments were carried out. The first was a full procedural blank, involving all the handling steps of the main sample program, in which an empty platinum foil packet was analyzed for carbon using the standard stepped combustion procedure. In two further experiments, empty platinum foil packets were subjected to pulse-heating at 800 and 1500°C, respectively, according to the same conditions as the meteorite samples.

Results

Since the experiments herein have been conducted on quantities of materials that would be considered small compared to traditional analyses of meteorites, it was necessary firstly to evaluate the effects of working with such samples. Thus, before discussing the results from the pulse-heating experiments it is instructive to consider data from the unheated samples. In Figure 1 the overall $\delta^{13}C$ and carbon contents for the small samples are displayed along with literature values for the same meteorites. Note that for each meteorite the previously acquired data cluster in fairly tight groups, with spreads in $\delta^{13}C$ and carbon concentrations of a few permil and percent respectively. In every case the small samples analyzed herein fall outside these fields, by relatively large amounts in terms of carbon contents; $\delta^{13}C$ values, on the other hand, are within the published ranges. This demonstrates that whilst Allende, Weston and Goalpara are heterogeneous with respect to their overall distribution of carbon (at the scale of sampling used, i.e., 50-500 µg, say around 400 µm in size), within any particular location of an individual the relative abundance of carbon-bearing components probably remains the same. This is reflected by the fact that the summed $\delta^{13}C$ values for the three small samples analyzed herein (i.e., $\Sigma\delta^{13}C$ obtained by summing and weighting all the $\delta^{13}C$ values for each individual step of the 3 separate analyses), reflect those of the bulk meteorites recorded from samples that were orders of magnitude larger.

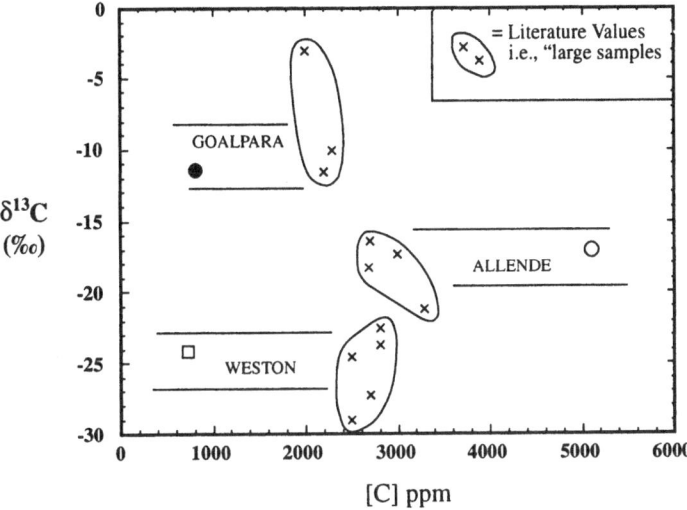

Fig. 1. Plot of carbon stable isotopic composition ($\delta^{13}C$, in ‰) versus carbon contents ([C], in ppm) obtained from three different meteorite samples – Goalpara (filled circles), Allende (open circles) and Weston (filled squares). Literature values are plotted as crosses and grouped in fields. Note that there are large disparities in [C] between the small samples of the three meteorites analyzed herein and literature data; however, $\delta^{13}C$ in each case are equivalent. This shows that meteorite samples are heterogeneously distributed with respect to carbon, although overall bulk $\delta^{13}C$ remains the same.

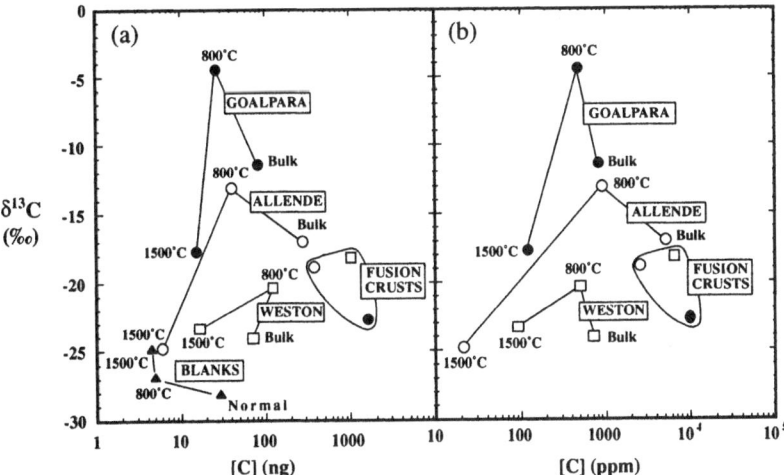

Fig. 2. (a) Plot of $\delta^{13}C$ (‰) versus carbon contents ([C], in ng) for the meteorite samples analyzed herein along with data for blank experiments (the latter being either unheated, i.e., "normal", or pulse-heating experiments at 800°C and 1500°C). (b) Equivalent relationship of $\delta^{13}C$ versus [C], this time in ppm (note: the blank data cannot be included in such a plot). In both plots are shown data for meteorite fusion crusts, unheated ("bulk") samples, and the results of pulse-heating at 800°C and 1500°C.

In Figure 2a the overall $\delta^{13}C$ and carbon contents of all the experiments are plotted along with data for the blanks. The sample results are shown *without* making a correction for the blanks. Dealing with the bulk samples first, comparisons should be made to the "normal blank", as labelled on the figure ($\Sigma C = 28.5$ ng, $\Sigma\delta^{13}C = -28.0‰$). In the case of Weston, applying a correction for blank exerts a small influence in the measured carbon isotopic composition (changing $\delta^{13}C$ from -24.1 to $-21.6‰$), while that for Goalpara is somewhat larger ($\delta^{13}C$ from -11.4 to $-2.9‰$); Allende is not substantially affected because of the relatively larger amounts of carbon contained in this sample. Note that none of these corrections would move the $\delta^{13}C$ values outside the ranges of published data (Figure 1); as such we will continue to consider the uncorrected data. In the case of the 800°C pulse-heated samples, a correction for the equivalent blank ($\Sigma C = 4.8$ ng, $\Sigma\delta^{13}C = -26.7‰$) would make no significant changes to the data. For those samples heated to 1500°C it can be seen that the data tend towards the high-temperature blank ($\Sigma C = 4.3$ ng, $\Sigma\delta^{13}C = -24.7‰$), which is perhaps not surprising since intuitively we would expect almost all indigenous carbon to have been removed by heating to such high temperatures.

The same data from Figure 2a are reproduced in 2b, but in this case the carbon contents are plotted as concentrations rather than absolute yields. Considering the results in this format it is readily apparent that the carbon concentration decreases as a function of the peak-temperature attained during pulse-heating. Note also that $\delta^{13}C$ values increase between the unheated samples and the 800°C experiments. This reflects the removal of low-temperature organic carbon in each sample. For Allende this carbon is probably indigenous to some extent; in the case of Weston and Goalpara it is more likely to be of terrestrial biogenic origin. This poses an immediate problem for the interpretation and applicability of the pulse-heating results - clearly, as analogs for micrometeorites, normal meteorites would not be entirely suitable if heavily contaminated. All extraterrestrial samples (micrometeorites and meteorites) become contaminated with carbon when exposed to the terrestrial environment, but this is after any heating episode that they may have experienced. Thus, a pulse-heating experiment that merely acts to remove the contaminant carbon is of no real value in assessing the effects of atmospheric heating on the removal of *indigenous* carbon. It will be necessary to bear this is mind when considering the data further.

In a similar vein, caution should be expressed with the results from the fusion crusts. Originally it was anticipated that fusion crust data might approximate some of the results from the pulse-heating experiments. Indeed, for Allende it could be argued that this is the case since the fusion crust sample shows a depletion in carbon relative to bulk. However, it is necessary to recall the observations from

Figure 1, which demonstrate the heterogeneity of meteorites at the sample levels employed in the present study. Since fusion crust samples were necessarily not taken from the same reservoirs as the other experiments, it is impossible to attach any significance to their carbon contents in relation to the bulks. When we consider the data from all three fusion crust experiments it is noteworthy that they form tight groupings on Figures 2a and b. This seems to suggest a common process to explain the results - namely heating and subsequent loss of any indigenous low-temperature carbon, followed by an input of terrestrial biogenic materials. This could explain why all samples have similar $\delta^{13}C$ values. However, it should be pointed out that $\delta^{13}C$ values of −18.2‰ (Weston) and −18.8‰ (Allende) are not typical of terrestrial contamination ($\delta^{13}C = -25 \pm 5$‰). We will return to this below. An observation that should be made at this point is that while the interior samples of the meteorites can be considered to have experienced minimum levels of exposure to contamination, we have no such confidence over the fusion crusts. After all, the meteorite samples in question were at one point in their history picked up and collected - they may have succumbed to all manner of handling (one of us [IPW] is reminded that his tutor in meteorite recognition, Dr. S.O.Agrell, now sadly deceased, was occasionally observed to spit on the surfaces of hand-specimens as an aid to identification).

Figures 3a-c display the amounts of carbon released from the three meteorites by the individual stepped combustion experiments. Note that for the 800°C pulse-heated samples of all three meteorites there is a decline in the relative quantity of low-temperature carbon (that released below 600°C) compared to the corresponding bulks. In contrast, all three of the 800°C heated samples contain amounts of high-temperature carbon that are roughly equivalent with those in the respective bulks - in fact, in Weston and Goalpara the levels are enhanced. In comparing the 1500°C- with the 800°C-heated experiments the amount of low-temperature carbon in Weston and Goalpara does not show significant depletions, but does in the case of Allende. All three meteorites show relative depletions in high-temperature carbon in the 1500°C experiments, as would be expected. For the fusion crust experiments the overall pattern of carbon released from Allende is similar to that of the bulk, whereas in Goalpara, carbon is enhanced at low temperatures. In the case of Weston there is a relatively "flat" release of carbon from the fusion crust, which is actually similar to that from the 800°C experiment, albeit at much higher levels.

Turning now to the carbon isotopic data from the stepped combustion experiments (Figures 4a-c) it can be seen that for Allende the low temperature parts of each experiment are the same. At higher temperatures a ^{13}C-rich "component" identified as a release at 700-800°C in the bulk sample has been lost from the pulse-heating experiments and the fusion crust. The 1500°C-heated experiment, with $\delta^{13}C$ values around $-25‰$, appears to be merely system blank as concluded before. Note that in terms of carbon isotopic composition at least, the 800°C-heated experiment and the fusion crust sample are very similar, which is somewhat at odds with the carbon yield data (Figure 3a). For Weston (Figure 4b), we can again assume that the data from the 1500°C-heated sample represents system blank (with $\delta^{13}C$ values in the $-25 \pm 5‰$ window). In terms of carbon isotopic composition the bulk sample of Weston is indistinguishable from blank, although the yield of carbon is clearly well in excess of it (Figures 1 and 3b). Interestingly, the 800°C-heated and fusion crust samples, which appeared to be similar in respect of carbon yields (see above), show progressive evidence for a high-temperature, ^{13}C-rich, component ($\delta^{13}C$ up to $-13.1‰$ at 800-1200°C from the fusion crust). This phenomenon may have parallels with laboratory investigations of meteorite samples where the technique of "preparative pre-combustion" is implemented (e.g., Ash et al. 1990). In this technique, successive stepped combustion experiments on samples pre-combusted at progressively higher temperatures, reveal the presence of carbon-bearing constituents that were otherwise obscured in the previous extractions. An accompanying phenomenon involves chemical reactions (oxidation, for instance) acting on mineral constituents of the samples, causing physical degradation and ultimately exposure of previously protected forms of carbon. On this basis we may suppose that Weston contains a refractory ^{13}C-rich component - if true, it may be particularly prevalent within the fusion crust (Figure 3b).

In the 1500°C-heated experiment of Goalpara, unlike that of Weston, there is clearly some survival of indigenous carbon. Indeed, carbon released from the 800-1200°C combustion step of the 1500°C-heated experiment has a $\delta^{13}C$ of about 0‰, which is clearly not from blank; rather this is from the same high temperature component that is observed in both the unheated and 800°C-heated samples. Comparing the data from the 800°C-heated sample with that of the bulk extraction there are obvious similarities with just a slight hint that the pulse-heated sample may have lost some of its indigenous carbon. In contrast, the fusion crust of Goalpara is dominated by contamination (and at fairly high levels - i.e., Fig. 3c).

Discussion

A factor that is of immediate relevance to studies of carbon in micrometeorites is that, even without atmospheric heating, the heterogeneous distribution of carbon in the meteorite samples results in measured abundances of the element that can vary substantially. In the case of Goalpara and Weston the carbon contents in the small samples are less than half of that expected; in the sample of Allende the carbon content is more than doubled. This should not be surprising - at the scale employed herein, meteorites of the type studied *are* heterogeneous. Furthermore, it is known from detailed studies of micrometeorites that C/O ratios, as measured by electron energy loss spectrometry on sub-µm spots, can vary tremendously (Perreau et al. 1993). Thus, overall carbon contents of micrometeorites may not be of any relevance. However, this makes the assumption that the relationship between micrometeorites and their original hosts is analogous to that between small and large samples of the same meteorite. In the case of micrometeorites produced by collisions in the asteroid belt this is entirely reasonable. For samples of cometary dust the issue is more to do with the grain size of individual particles and whether they are sufficiently fine-scale to allow a truly homogenous distribution of carbon at the sample sizes typically employed. Intuitively, cometary grains are likely to be very heterogeneous, so again overall carbon contents of individual micrometeorites would appear to be inconsequential.

As has been pointed out previously, the carbon contents of micrometeorites are low compared to carbonaceous chondrites. Part of the reason for this could simply be due to the heterogeneity that probably exists between samples. But this is too much of a simplification; it would be reasonable to assume that natural heterogeneity should have produced samples that were carbon-rich as well as carbon-poor. So, perhaps there is a reason for the dearth of samples with high carbon contents. For instance, they might not survive to reach the Earth's surface on account of atmospheric heating. Consider that high carbon samples (i.e., enriched in volatiles) may be particularly susceptible to destruction during heating, and thus have lower chances of survival. Alternatively, high carbon samples may produce melted/heated samples that are more porous and in consequence more easily weathered at the Earth's surface; or, they may have a fundamentally different appearance physically and have not, as yet, been recognized. Notwithstanding these caveats it should be noted that the $\delta^{13}C$ values recorded from the small meteorite samples were the same as those from larger (bulk) samples. In principle therefore, carbon isotopic analyses of micrometeorites that were not heated significantly during atmospheric infall should be representative of their parent bodies. The question is thus, to what extent does infall heating affect the carbon inventory of micrometeorites.

Next we consider the results of the pulse-heating experiments in concert with the bulk analyses of each individual set of experiments (and here internal comparisons of yields are possible on account of how the starting materials were obtained). From the results obtained herein it is shown that, in general, heating

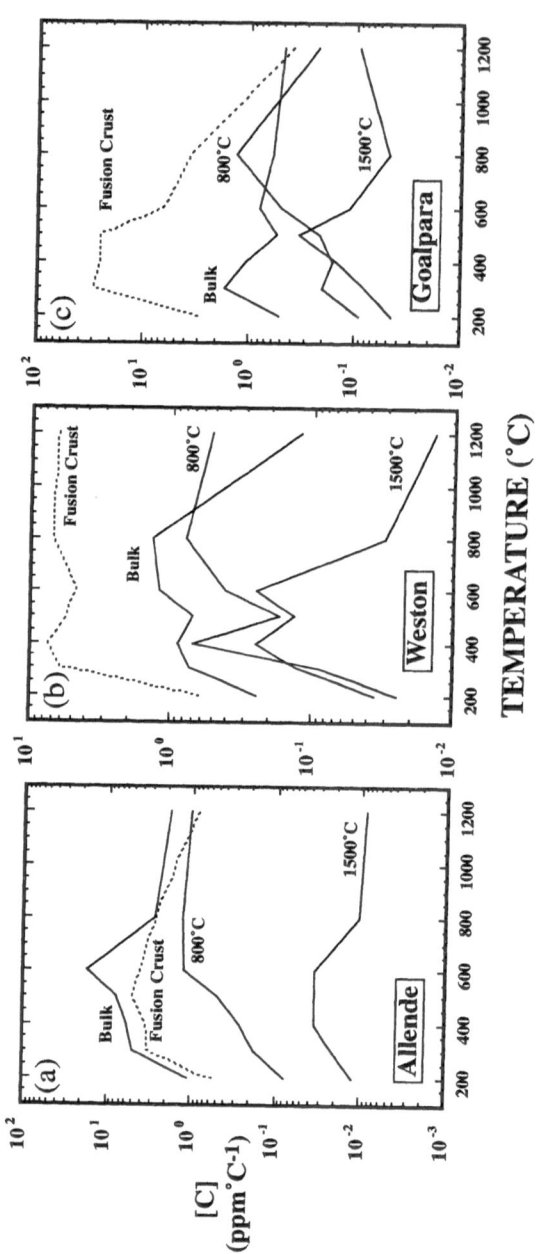

Fig. 3. Stepped combustion data from (a) Allende, (b) Weston and (c) Goalpara. Each plot shows the yield of carbon (as ppm °C⁻¹) versus temperature (°C) – note the changes of scale on the ordinate axes. On each plot is shown stepped combustion data from samples of whole-rock ("bulk"), fusion crust, and materials subjected to pulse-heating at 800°C and 1500°C.

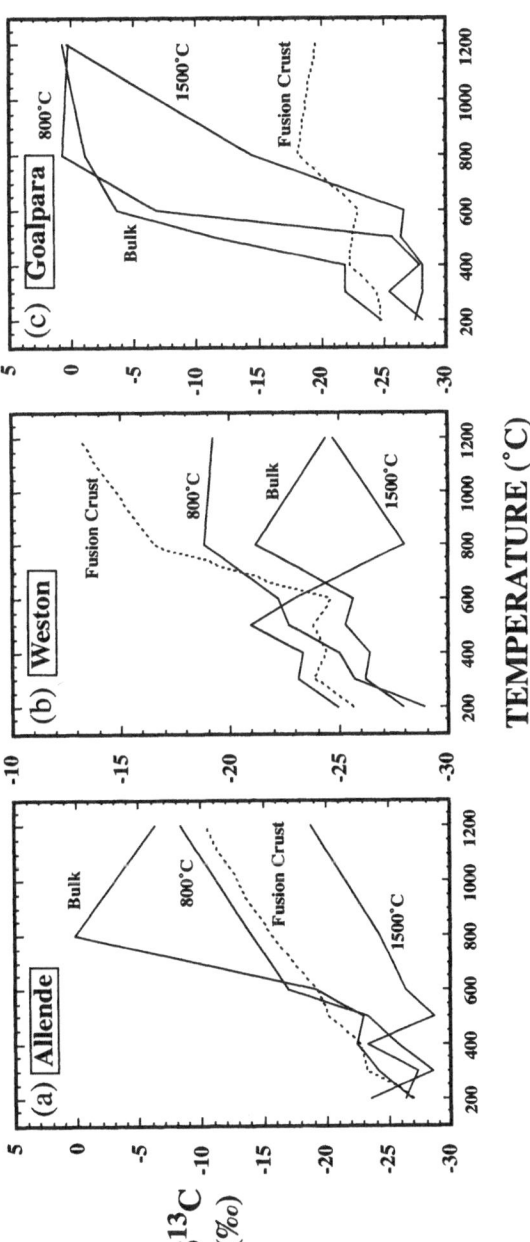

Fig. 4. Stepped combustion data from (a) Allende, (b) Weston and (c) Goalpara. Each plot shows the carbon stable isotopic composition ($\delta^{13}C$) versus temperature (°C) – note the changes of scale on the ordinate axes. On each plot is shown stepped combustion data from samples of whole-rock ("bulk"), fusion crust, and materials subjected to pulse-heating at 800°C and 1500°C.

meteorites to higher temperatures progressively removes the carbon-bearing constituents. While this observation could have been gleaned from the results of any stepped combustion analysis of a bulk sample, such an experiment is not a good simulation of atmospheric heating because step durations are typically 30 minutes, and the overall heating rate is relatively slow (and monotonic). As an illustration of how results from stepped combustion on their own could be misleading, it is seen that in Allende the major carbonaceous material, poorly ordered graphite, is removed by 800°C when using this technique (this is an observation made from many analyses of bulk Allende in our laboratory, including the sample analyzed herein). In contrast during pulse-heating to 800°C, the major carbonaceous component survives to some extent (witness the fact that in Figure 3a there is still a release of carbon from 600-800°C during the stepped combustion analysis of the sample that had already been heated to 800°C). This comparison exemplifies how carbon removal is dependent upon the kinetics of combustion, which are not instantaneous.

Continuing with a description of the Allende results, note that in both the bulk sample and that which was pulse-heated to 800°C, there is evidence for refractory ^{13}C-rich carbon, i.e., as observed in the 800-1200°C steps. From Figure 4a it can be seen that this "component" has a $\delta^{13}C$ of, at most, –6 to –8‰. Carbon of this nature has been observed before in analyses of bulk Allende (e.g., Swart et al. 1983b), where it is found preferentially located in fine-grained size fractions, i.e., matrix (Ash 1990). The complicated nature of the isotopically anomalous forms of carbon in Allende has been documented by Ash et al. (1988, 1990); the meteorite is known to contain both ^{13}C-rich and ^{13}C-poor components, with $\delta^{13}C$ values of +527 and –261‰ respectively, Ash et al. (1990). It is a sobering thought that analyses of the bulk sample (i.e., which approximates all that is possible for studies of micrometeorites) produces only mere indications of what is actually present.

Weston was selected for analysis on the basis that it is an ordinary chondrite (H4), with a previously documented carbon chemistry, that could be taken as somewhat representative of the majority of infalling meteorite-sized extraterrestrial materials. Being asteroidal in origin it may be considered as having at least some relationships with micrometeorites from an equivalent source. A previously published stepped combustion determination of the meteorite (Grady et al. 1989) showed Weston to have, in addition to terrestrial organic contamination, a component of "graphitic" carbon with a $\delta^{13}C$ of around –20‰. The "graphitic" component, which is demonstrably indigenous to the sample and probably quite disordered in nature is located in C-rich aggregates (Grady et al. 1989). It is considered that this carbon is probably very similar to the major component in

Allende and also survives to large extent the effect of pulse-heating to 800°C. The sample of fusion crust from Weston appears to be substantially enriched (by a factor of 10) in this "graphitic" component. It is possible that this is simply the result of sample heterogeneity although it might also demonstrate the resilience of such materials to brief periods of oxidation during atmospheric heating. Certainly graphite itself does not "melt" below several 1000s of °C, which makes it far more refractory than some of the other constituents of the meteorite (including silicates). Of relevance here is the $\delta^{13}C$ measured for the "graphitic" material in the fusion crust, which at $-13.1‰$ is considerably higher than that determined in unheated bulk samples (ca. $-20‰$, Grady et al. 1989). Because the "graphitic" component is easily resolved from organic materials during extraction (Grady et al. 1989), it is clear that the measured $\delta^{13}C$ values in the unheated sample are not artificially low as a result of component-masking. Rather, the results from the fusion crust show a genuine elevation in $\delta^{13}C$. This could be due to one of three processes: (1) atmospheric heating, causing the destruction of the graphite and revealing another refractory component (Russian Doll syndrome), (2) isotopic fractionation arising from distillation of the graphite, or (3) the addition of a terrestrial graphitic component. We cannot rule out (3), but it seems highly unlikely. No matter which of the other two processes are responsible, the inescapable conclusion is that there has been an overall concentration of carbon in the fusion crust. In other words, as material ablated from the surface of the meteorite during atmospheric infall, carbon, for some reason, was retained. It is impossible to explore this further here but perhaps fusion crusts may be of more value in studying the carbon chemistry of meteorites than realized previously. This phenomenon could have some relevance for the study of micrometeorites since it implies that melting processes may actually act to concentrate carbon. Indeed, carbon could become enriched towards the surfaces of particles although, intuitively, this seems unlikely. In any event it would be difficult to study this possible effect on account of contamination issues; in addition, at present, we can only contemplate analyzing whole-particles because of the limits of detection imposed by the analytical system.

The main carbon-bearing component in Goalpara is graphite, with accessory diamond (e.g., Vdovykin 1970). Upon stepped combustion this mixture of elemental carbon in Goalpara has been shown to burn above 500°C, with a peak at around 600-700°C, giving $\delta^{13}C$ values of about $-2‰$ (Grady et al. 1985). The bulk sample analyzed herein showed the same general features, although the level of organic contamination was relatively high. The 800°C pulse-heated experiment showed a much clearer release of graphite/diamond, a consequence of the contamination having been removed. Note that the graphite/diamond still survived to some extent even after pulse-heating to 1500°C. In contrast, the fusion crust appeared to be devoid of this component, showing instead only the effects of contamination. In light of the graphite/diamond survivability in the 800 and 1500°C experiments it seems likely that the absence in the fusion crust was due to sample selection rather than loss by heating. The graphite in Goalpara is likely to be more ordered (i.e., more crystalline) than that in either Allende, or Weston (certainly in the case of Allende the major carbon constituent is amorphous, or poorly crystalline, e.g., Smith and Buseck 1981, whilst that in Goalpara, although fine-grained, is graphite, e.g., Vdovykin 1970). In other words, if certain micrometeorites were to contain crystalline graphite then this type of carbon is likely to survive atmospheric heating. However, the exact degree to which such material could be preserved is probably a reflection of its grain size, in the same way that diamonds of nanometre dimensions are less resilient to gas phase oxidation than larger samples (e.g., Wright and Pillinger 1989).

Conclusions

Three samples of meteorites (Allende, Weston, and Goalpara) have been subjected to pulse-heating (10s in air) to simulate atmospheric heating under conditions pertinent to the present day; subsequent analyses, using stepped combustion, have documented the effects of this process on the indigenous carbon components. Analyses of fusion crusts from the same three meteorites have also been undertaken in attempts to discover the actual effects of atmospheric heating. Unfortunately, but not unexpectedly, fusion crusts were found to suffer from the additional complication of carbon contamination, which makes interpretation of the results rather ambiguous. It should be noted however, that the meteorites chosen for study have perhaps not been collected and curated in the most ideal ways. A study of fusion crusts from meteorites collected in Antarctica would seem a more useful approach - certainly in the case of an SNC meteorite collected from Antarctica (LEW 88516), the amount of carbon in the fusion crust is the same as in the interior, suggesting low levels of surficial contamination (Wright et al. 1993).

Ignoring the fusion crust data the effects of pulse-heating on meteorite samples can be summarized as follows:
1. It acts to remove terrestrial organic contaminants.
2. An experiment performed at a particular peak temperature does not fully remove components that would otherwise have been lost during a stepped combustion extraction conducted up to an equivalent temperature (which shows that the removal of carbon is dependent upon combustion kinetics).

3. Although carbon-bearing components may be lost as a consequence of volatilization or oxidation, where remnants of the equivalent materials are left in the sample they appear not to have suffered from isotopic fractionation.

4. Pulse-heating may induce effects in samples similar to those experienced during preparative pre-combustion, whereby certain components are removed revealing the presence of materials that are otherwise masked.

5. Some refractory carbon-bearing components may survive, to some extent, brief periods of pulse-heating to 1500°C; this is certainly the case for graphite/diamond intergrowths, of the type found in ureilites.

If the fusion crust data are included (while ignoring the problems of contamination) then it appears that pulse-heating may actually act to concentrate refractory forms of carbon, as more volatile components (such as silicates) are removed.

A major ramification of the experiments is that while carbon contents determined for micrometeorites may be low, as a result of the loss of carbonaceous components during atmospheric entry, carbon isotopic measurements of any material remaining will not be compromised and can therefore be used in the quest to assess the sources of the particles. Considering the wider issue of atmospheric heating, it is suspected that peak temperatures of 800°C will probably remove any indigenous organic components. Unfortunately the results obtained herein do not constrain this problem unequivocally. Regardless, the apparent survival of organic materials in certain micrometeorites (e.g., Wright et al. 1996) must surely constrain peak temperatures in some way. On the basis of stepped combustion data acquired over many years it is anticipated that a refractory carbon-bearing component such as silicon carbide will survive atmospheric heating, provided peak temperatures are below a few hundred °C. Furthermore, if organic constituents can survive then so too must the more refractory carbon phases, diamond and graphite. Indeed, well crystalline diamond/graphite could survive a brief episode of heating to 1500°C.

The experiments conducted herein were relatively simple and represent a first attempt at constraining the effects of atmospheric heating on the carbon budget of micrometeorites. More sophisticated studies immediately present themselves - perhaps the most useful would be a series of pulse-heating experiments in the region 400-800°C, and for varying lengths of time. In addition it would be desirable to extend the range of analog types investigated by repeating the type of experiments performed herein but with samples of CI and CM meteorites. The next logical stage of the investigation is to perform the same types of experiments, but under different atmospheric conditions. These could range from highly reduced (H_2, NH_3, N_2, CH_4, or mixtures thereof) to more oxidized (CO_2, H_2O, etc.). In this way it would be possible to investigate the nature of atmospheric heating on micrometeorites during conditions pertinent to the early Earth, and thereby constrain the inputs of biogenically important elements and compounds to the primitive surface.

Acknowledgements

Financial support was from PPARC. The authors acknowledge the helpful reviews of Mark Burchell and Mark Sephton.

References

Anders E (1989) Pre-biotic organic matter from comets and asteroids. Nature 342: 255-257

Anders E, Zinner E (1993) Interstellar grains in primitive meteorites: diamond, silicon carbide, and graphite. Meteoritics 28: 490-514

Ash RD (1990) Interstellar dust from primitive meteorites: a carbon and nitrogen isotope study. PhDThesis, Open University, 220 pp

Ash RD, Arden JW, Grady MM, Wright IP, Pillinger CT (1988) An interstellar dust component rich in ^{12}C. Nature 336: 228-230

Ash RD, Arden JW, Grady MM, Wright IP, Pillinger CT (1990) Recondite interstellar carbon revealed by preparative precombustion. Geochim Cosmochim Acta 54: 455-468

Brownlee DE, Joswiak DJ, Love SG, Nier AO, Schlutter DJ, Bradley JP (1993) Identification of cometary and asteroidal particles in stratospheric IDP collections. Lunar Planet Sci 24: 205-206

Carr RH, Wright IP, Joines AW, Pillinger CT (1986) Measurement of carbon stable isotopes at the nanomole level: a static mass spectrometer and sample preparation technique. J Phys E: Sci Instrum 19: 798-808

Chyba CF (1993) The violent environment of the origin of life: progress and uncertainties. Geochim Cosmochim Acta 57: 3351-3358

Clemett SJ, Maechling CR, Zare RN, Swan PD, Walker RM (1993) Identification of complex aromatic molecules in individual interplanetary dust particles. Science 262: 721-725

Cronin JR, Pizzarello S, Cruikshank DP (1988) Organic matter in carbonaceous chondrites, planetary satellites, asteroids and comets. In: Kerridge JF, Matthews MS (eds) Meteorites and the Early Solar System. University of Arizona Press, Arizona, pp 819-857

Deming JW, Baross JA (1993) Deep-sea smokers: windows to a subsurface biosphere? Geochim Cosmochim Acta 57: 3219-3230

Flynn GJ (1989) Atmospheric entry heating: a criterion to distinguish between asteroidal and cometary sources of interplanetary dust. Icarus: 77, 287-310

Genge, MJ, Grady, MM (1998) Melted micrometeorites from Antarctic ice with evidence for the separation of immiscible Fe-Ni-S liquids during entry heating. Meteoritics Planet Sci 33: 425-434

Goldanskii VI (1977) Interstellar grains as possible cold seeds of life. Nature 269: 583-584

Grady MM, Wright IP, Swart PK, Pillinger CT (1985) The carbon and nitrogen isotopic composition of ureilites: implications for their genesis. Geochim Cosmochim Acta 49: 903-915

Grady MM, Wright IP, Pillinger CT (1989) A preliminary investigation into the nature of carbonaceous material in ordinary chondrites. Meteoritics 24: 147-154

Graham GA, Sexton, A, Grady MM, Wright, IP (1997) Further attempts to constrain the nature of impact residues in the HST solar array panels. Adv Space Res 20: 1461-1465

Greenberg JM (1981) Chemical evolution of interstellar dust - a source of prebiotic material? In: Ponnamperuma C (ed) Comets and the origin of life. Reidel, Dordecht, pp 111-127

Hahn JH, Zenobi R, Bada JL, Zare RN (1988) Application of two-step laser mass spectrometry to cosmogeochemistry: direct analysis of meteorites. Science 239: 1523-1525

Horgan J (1991) In the beginning. Sci Am 264 (2): 100-109

Kasting JF (1993) Evolution of the Earth's atmosphere Hadean to recent. In: Engel MH, Macko SA (eds) Organic Geochemistry. Plenum Press, New York, pp 611-623

Klöck W, Flynn GJ, Sutton SR, Bajt S, Neuking N (1994) Heating experiments simulating atmospheric entry of micrometeorites. Lunar Planet Sci XXV: 713-714

Love SG, Brownlee DE (1991) Heating and thermal transformation of micro-meteoroids entering the Earth's atmosphere. Icarus 89: 26-43

McKeegan KD, Walker RM, Zinner E (1985) Ion-microprobe isotopic measurements of individual interplanetary dust particles. Geochim Cosmochim Acta 49: 1971-1987

Maher KA, Stevenson DJ (1988) Impact frustration of the origin of life. Nature 331: 612-614

Martinez M, Escobar M (1995) Effect of coal weathering on some geochemical parameters. Org Geochem 23: 253-261

Miller SL (1993) The prebiotic synthesis of organic compounds on the early Earth. In: Engel MH, Macko SA (eds) Organic Geochemistry. Plenum Press, New York, pp 625-637

Miller SL, Bada JL (1988) Submarine hot springs and the origin of life. Nature 334: 609-611

Nier AO, Schlutter DJ (1993) The thermal history of interplanetary dust particles collected in the Earth's stratosphere. Meteoritics 28: 675-681

Oro J, Holzer G, Lazcano-Araujo A (1980) The contribution of cometary volatiles to the primitive Earth. In: Holmquist R (ed) COSPAR - Life Sciences and Space Research. XVIII, pp 67-82

Perreau M, Engrand C, Maurette M, Kurat G, Presper Th (1993) C/O atomic ratios in micrometer-size crushed grains from Antarctic micrometeorites and two carbonaceous meteorites. Lunar Planet Sci 24: 1125-1126

Pillinger CT, Russell SS (1993) Interstellar SiC grains in meteorites. J Chem Soc Faraday Trans 89: 2297-2304

Prosser SJ, Wright IP, Pillinger CT (1990) A preliminary investigation into the isotopic measurement of carbon at the picomole level using static vacuum mass spectrometry. Chem Geol 83: 234-246

Schidlowski M (1993) The initiation of biological processes on Earth. In: Engel MH, Macko SA (eds) Organic Geochemistry. Plenum Press, New York, pp 639-655

Schopf JW (1983) The Earth's earliest biosphere: its origin and evolution. Princeton University Press, New Jersey

Sephton M, Gilmour I (1998) A "unique" distribution of PAH in ALH 84001 or a selective attack in meteorites from Mars. Meteoritics Planet Sci 33: A142-143

Smith PPK, Buseck PR (1981) Graphitic carbon in the Allende meteorite: a microstructural study. Science 212: 322-324

Summers DP, Chang S (1993) Prebiotic ammonia from reduction of nitrite by iron(II) on the early Earth. Nature 365: 630-633

Swart PK, Grady MM, Pillinger CT (1983a) A method for the identification and elimination of contamination during carbon isotopic analyses of extraterrestrial samples. Meteoritics 18: 137-154

Swart PK, Grady MM, Pillinger CT, Lewis RS, Anders E (1983b) Interstellar carbon in meteorites. Science 220: 406-410

Taylor AD, Baggaley WJ, Steel DI (1996) Discovery of interstellar dust entering the Earth's atmosphere. Nature 380: 323-325

Thomas KL, Keller LP, McKay DS (1995) A comprehensive study of major-, minor-, and light- element abundances in over 100 interplanetary dust particles. Meteoritics 30: 587-588

Vdovykin GP (1970) Ureilites. Space Sci Rev 10: 483-510

Walker JCG (1977) Evolution of the atmosphere. Macmillan, New York

Wright IP (1984) $\delta^{13}C$ measurements of smaller samples. Trends Anal Chem 3: 210-215

Wright IP (1995) Stable isotope ratio mass spectrometry. In: Encyclopedia of Analytical Science, Academic Press Ltd, pp 2869-2877

Wright IP, Pillinger CT (1989) Carbon isotopic analysis of small samples by the use of stepped-heating extraction and static mass spectrometer. In: Shanks WC, Criss RE (eds) New frontiers in stable isotopic research: Laser probes, ion probes, and small-sample analysis. US Geological Survey Bulletin, 1890, pp 9-34

Wright IP, Carr RH, Pillinger CT (1988) Carbon stable isotope analyses of individual deep-sea spherules. Meteoritics 23: 339-349

Wright IP, Douglas C, Pillinger CT (1993) Further carbon isotope measurements of LEW 88516. Lunar Planet Sci 24: 1541-1542

Wright IP, Yates PD, Hutchison R, Pillinger CT (1997) The content and stable isotopic composition of carbon in individual micrometeorites from Greenland and Antarctica. Meteoritics and Planet Sci 32: 79-89

Yates PD, Wright IP, Pillinger CT (1992) Application of high-sensitivity carbon isotope techniques - a question of blanks. Chem Geol (Isotop Geosci Sect) 101: 81-91

Yates PD, Wright IP, Hutchison R, Pillinger CT (1994) The carbon stable isotopic compositions of separated collections of type I and type S deep-sea spherules. Trends in Chem Geol 1: 215-225

Zinner E (1998) Stellar nucleosynthesis and isotopic composition of presolar grains from primitive meteorites. Ann Rev Earth Planet Sci 26: 147-188

4 Search for Petrographic and Geochemical Evidence for the Late Heavy Bombardment on Earth in Early Archean Rocks from Isua, Greenland

Christian Koeberl[1], Wolf Uwe Reimold[2], Iain McDonald[3], and Minik Rosing[4]

[1]Institute of Geochemistry, University of Vienna, Althanstrasse 14, A-1090 Vienna, Austria. (christian.koeberl@univie.ac.at)
[2]Impact Cratering Research Group, Department of Geology, University of the Witwatersrand, Johannesburg 2050, South Africa.
[3]School of Earth & Environmental Sciences, University of Greenwich, Chatham Maritime, Kent ME4 4AW, U.K.
[4]Geologisk Museum, Oster Voldgade 5-7, DK-1350 Kobenhavn K, Denmark.

Abstract. The Moon was subjected to intense post-accretionary bombardment between about 4.5 and 3.9 billion years ago, and there is evidence for a short and intense late heavy bombardment period, around 3.85 ± 0.05 Ga. If a late heavy bombardment occurred on the Moon, the Earth must have been subjected to an impact flux at least as intense. The consequences for the Earth must have been devastating. In an attempt to investigate if any record of such a late heavy bombardment period on the Earth has been preserved, we performed a petrographic and geochemical study of some of the oldest rocks on Earth, from Isua in Greenland. We attempted to identify any remnant evidence of shock metamorphism in these rocks by petrographic studies, and used geochemical methods to detect the possible presence of an extraterrestrial component in these rocks. For the shock metamorphic study, we studied zircon, a highly refractive mineral that is resistant to alteration and metamorphism. Zircon crystals from old and eroded impact structures were found earlier to contain a range of shock-induced features at the optical and electron microscope level. Many of the studied zircon grains from Isua are strongly fractured, and single planar fractures do occur, but never as part of sets; none of the crystals studied shows any evidence of optically visible shock deformation. Several samples of Isua rocks were analyzed for their chemical composition, including the platinum group element (PGE) abundances, by neutron activation analysis and ICP-MS. Three samples showed somewhat elevated Ir contents (up to 0.2 ppb) compared to the detection limit, which is similar to the present-day crustal background content

(≤ 0.03 ppb), but the chondrite-normalized siderophile element abundance patterns are non-chondritic, which could be a sign of either a small extraterrestrial component (if an indigenous component is subtracted), or terrestrial (re)mobilization mechanisms. In absence of any evidence for shock metamorphism, and with ambiguous geochemical signals, no unequivocal conclusions regarding the presence of extraterrestrial matter (as a result of possible late heavy bombardment) in these Isua rocks can be reached.

Introduction

The most probable theory of the origin of the Earth and other planets explains their formation by accretion of smaller objects (planetesimals). Towards the end of the accretion of the Earth, at about 4.55 Ga, it may have been impacted by a Mars-sized body, which is the currently accepted hypothesis for the origin of the Moon (e.g., Cameron and Benz 1991; Taylor 1993). The consequences of such an impact event for the proto-Earth would have been severe; they would have included almost complete re-melting of the Earth, loss of any primary atmosphere (Vickery 1990), and admixture of material from the impactor. The material remaining in orbit after accretion of the Moon would have continued to impact onto the Earth for quite some time. Due to later geological activity, no record of this very early bombardment remains on the surface of the Earth. On the other hand, there is abundant evidence from Apollo rocks that indicates that the Moon was subjected to intense post-accretionary bombardment between about 4.5 and 3.9 billion years ago (e.g., Ryder 1989, 1990). In addition, the lunar highlands show much evidence for isotopic resetting, consistent with a short and intense late heavy bombardment period, around 3.85 ± 0.05 Ga (e.g., Tera et al. 1974; Ryder 1990). However, these data were disputed by some workers (e.g., Baldwin 1974; Hartmann 1975), who interpreted the resetting as having been caused by the tail-end of the decaying late accretionary impact flux. There is also some evidence that meteorite parent bodies were subjected to a late cataclysmic bombardment (e.g., Bogard and Garrison 1991), indicating that this event may not have been restricted to the Moon (or the Earth-Moon system). During this period (and later in its history), the Earth would have been subjected to a significantly larger number of impact events, as it has a larger diameter and a much stronger gravitational attraction than the Moon.

It seems realistic to assume that if a late heavy bombardment occurred on the Moon, the Earth (and other terrestrial planets – cf. Wetherill 1975) must have been subjected to an impact flux at least as intense as that onto the Moon. The consequences for the Earth must have been devastating. Mantle temperatures shortly after the end of the accretion of the Earth were probably much higher than today. It has been estimated that about half of all heat produced by ^{235}U decay to ^{207}Pb was released before 3.9 Ga, which, together with thermal energy released during the impact of late accretionary bodies, added several hundred degrees to the internal temperature of the Earth (e.g., Smith 1981; Davies 1985; Taylor 1993, 1999). Atmospheres could have been generated (Matsui and Abe 1986), chemically altered (Fegley et al. 1986), and eroded (Vickery 1990) by large-scale impacts. There is some geochemical evidence (e.g., Nd isotope data) indicating that the Earth's upper mantle had already undergone some differentiation at the time of formation of the oldest rocks preserved on the Earth's surface (Greenland, Canada, Australia) (e.g., Harper and Jacobsen 1992; McCulloch and Bennett 1993; Bowring and Housh 1995; but see also Gruau et al. 1996, for a different viewpoint). The differentiated Nd isotope data (and the presence of some relict zircons with ages >3.8 Ga; e.g., Compston and Pidgeon 1986) could be interpreted to suggest that small amounts of crust had formed prior to 4 billion years ago, but had later been mixed back into the upper mantle. This view has been confirmed by Hf istope studies on single zircons (Amelin et al. 1999). However, it is unlikely that large amounts of crust were present at that time (although a few authors – e.g., Kröner and Layer 1992 – argue otherwise). It is evident that any large-scale early impacts had some influence on the development of the continental crust (Goodwin 1976), and it is conceivable that the late heavy bombardment has been responsible for the rehomogenization of the minor amounts of pre-3.85 Ga crust with the upper mantle.

Such very early impact events must have occurred, no matter if there was a significant peak in the impact flux at about 3.8 to 3.9 Ga, or if this was just the time of the cessation of the enhanced post-accretionary impact flux. The search for any evidence of these impacts has important implications, as the presently documented terrestrial impact record basically covers less than half of the Earth's history. So far the oldest known terrestrial impact structures are the Proterozoic Vredefort and Sudbury Structures of 2023±4 and 1850±3 Ma age, respectively (e.g., Deutsch and Schärer 1994; Gibson and Reimold, this volume). These structures also represent two of the three largest impact structures known on Earth, with respective original diameters of 250-300 and 200 km, and they represent the complete, current, pre-1.85 Ga terrestrial impact crater record. It has been estimated (e.g., Grieve 1980; Frey 1980) that about 200 impact structures >1000 km in diameter formed on Earth between 4.6 and 3.8 billion years ago, which would have covered about 40% of the surface of the Earth. Using the minimum estimate for the cratering frequency, Grieve (1980) calculated that these impact events would have added a cumulative energy of about 10^{29} J to the Hadean Earth.

So far, no firm evidence for this "early bombardment" - or even of early Archean impact events - has been found, except for some unusual spherule horizons in South Africa and Australia, whose origin is controversial and which are discussed elsewhere (cf. Lowe et al. 1989; Koeberl and Reimold 1995; Koeberl et al. 1999; Reimold et al., this volume). The suggestion that terrestrial greenstone belts, which formed early in the geologic history, are equivalents of the lunar mare (Green 1972), has not been confirmed.

Motivation and Background of Study

The time period preceding the available rock record on the Earth (i.e., pre-3.85 Ga) is called the Hadean Eon (see, e.g., Taylor, 1999). So far the available evidence for the pre-2 Ga impact record of the Earth-Moon system comes from lunar data. Whereas the records for the Hadean Eon on the Earth are very sparse, the pre-Nectarian and Nectarian periods cover this time interval on the Moon (cf. Ryder et al. 1999). Soon after the formation of the Moon from a giant impact, a feldspathic crust formed the lunar highlands around 4.44 ± 0.02 Ga (e.g., Carlson and Lugmair 1988; Premo and Tatsumoto 1992), which represents the oldest dated event in the Earth-Moon system. The currently observable morphology of the lunar highlands is almost exclusively the result of numerous impacts that occurred prior to the extrusion of the volcanic flows that form the visible mare plains (e.g., Wilhelms 1984, 1987). These ancient impact structures include giant multi-ring basins (cf. Spudis 1993), as well as a wide range of craters with smaller diameters. Hartmann (1965, 1966) suggested that most of the cratering recorded in the lunar highlands occurred early in lunar history and inferred a cratering rate that was about 200-times more intense during the first few hundred million years of the age of the Moon than today (although there are some indications of an increase of the cratering rate in the Phanerozoic; cf. McEwen 1998).

As discussed in detail by Ryder (1990), all impact melt rock samples from the lunar highlands cluster in age around 3.8 to 3.9 Ga; there are basically no such rocks with ages that are older than 3.9 Ga. These data were interpreted to either represent the tail end of a continuous heavy bombardment, which was declining at that time and comprises the remnant of the accretion of the Moon (e.g, Hartmann 1975; Neukum et al. 1975; Baldwin 1971, 1974, 1981, 1987; Taylor 1982; Wilhelms 1987), or the signature of a "terminal cataclysm" - a significant but short-time increase in impact flux at that time (e.g., Tera et al. 1974; Ryder 1990). For the purposes of the present paper, we use the term "late heavy bombardment" (LHB) to refer to the terminal cataclysm (i.e., a sudden, drastic, and short-lived increase in the impact flux around 3.85 billion years ago). Despite critical comments by Hartmann, Baldwin, and others, regarding the absence of a convincing mechanism for a LHB, Zappala et al. (1998) recently found a mechanism that would be in agreement with the requirements of celestial mechanics.

In an attempt to investigate if any record of such a late heavy bombardment period on the Earth has been preserved, we started a petrographic and geochemical study of some of the oldest rocks on Earth. The available geological record of the Earth dates back to only about 3900 Ma, with a few rare detrital zircons providing evidence for the existence of felsic melts around 4.27 Ga (e.g., Compston and Pidgeon 1986). The question arises if there are appropriate rocks on Earth that would record evidence of any early bombardment. The nature of the earliest crust on Earth, and the amount of crust present, has been debated, but it seems likely, from comparison with other planets, that the earliest crust on Earth was of basaltic composition (e.g., Taylor 1989, 1992, 1993; Arndt and Chauvel 1991).

The characteristics of the relict zircons mentioned above indicate a granitoid (continental) source and, thus, provide evidence for at least minor amounts of felsic igneous rocks in the Hadean. The existence of large continental areas is unlikely, because granitic crust is derived from primitive mantle by recycling of subducted basaltic crust through a "wet" mantle, which will - over the history of the Earth - slowly produce increasing amounts of granitic crust (e.g., Taylor 1992, 1993). It is possible that small amounts of felsic rocks may have formed from remelting of basaltic crust that sank back into the mantle (e.g., Taylor 1989, 1999). Taylor (1999) speculates that the Hadean Earth most probably had a thick basaltic crust, covered by an ocean, with little dry land and minor amounts of felsic rocks (granitoids). Any sedimentological record, which would host information specific to surface environments and processes, such as the rate and violence of meteorite impact and the presence of life, has been lost from Hadean times. Earliest evidence for such effects appears only much later and has been dated at around 3900 Ma (see below).

The present study is a first attempt to search the limited available pre-$\approx$3.8 Ga rock record on Earth for any indications of early bombardment (LHB or other), and includes petrographic and geochemical approaches to the problem. On the one hand, we are attempting to identify any remnant evidence of shock metamorphism by petrographic studies, and, on the other hand, we are trying to chemically identify the possible presence of an extraterrestrial component in these rocks.

Samples

The Itsaq Gneiss Complex of West Greenland covers an area of about 3000 km^2, and preserves some of the oldest rocks known on Earth. The rocks of the Itsaq area are dominated by orthogneisses, often associated with packages of sediments and volcanic rocks (supracrustals), including the Isua supracrustal belt and the Akilia association. The Isua belt is the oldest known complex of volcanic and sedimentary rocks on Earth, and has a range of ages from 3.7 to 3.85 Ga. The latter comprise massive amphibolites and complex metasomatic carbonates (possibly metamorphosed remnants of early Archean oceanic crust), abundant banded iron-formation (BIF), and some rare graywacke and metapelites. Most rocks in the belt are strongly deformed, and all were recrystallized under upper greenschist to amphibolite facies conditions. Nutman (1986) divided the belt into nine 'formations.' The metabasaltic (amphibolite facies) Garbenschiefer Formation (Rosing et al. 1996), derived mainly from tholeiitic volcanic protoliths with turbiditic and pelagic sedimentary screens, is a tenth major unit. Primary depositional features (e.g., conglomerates, felsic agglomerates, and some graded bedding in felsic volcanic rocks) were described by Allaart (1976), Dimroth (1982), Nutman et al. (1984, 1996), and Nutman (1986, 1997). However, the interpretation of some of these features as being of primary origin was questioned by Rose et al. (1996) and Rosing et al. (1996); these authors suggested - based on metasomatic alteration spatially associated with ultrabasic bodies - that the rocks underwent pervasive metasomatism, which obscured their origin to some degree.

The supracrustals at Isua were probably deposited in a sediment-poor arc or back-arc basin in relatively deep water (e.g., Nutman et al. 1984). Studies of early Archean sediments from the Isua Supracrustal belt and the Akilia association in the Godthbsfjord region of southern West Greenland suggest, based on zircon geochronology, that they are the oldest sediments yet identified on Earth with ages >3.8 billion years (e.g., Nutman et al. 1997). The isotopic composition of graphitic microinclusions in apatites from Isua banded iron formations (BIF) was interpreted by Mojzsis et al. (1996) to represent traces of pre-3.85 billion year old life forms. However, it has recently been suggested that the Isua rocks studied by Mojzsis et al. (1996) and Nutman et al. (1997) may be younger than previously assumed, with an age of maybe only 3.65 Ga, and also the age and origin of the apatite has been debated (Moorbath et al. 1997; Kamber et al. 1997; Moorbath and Kamber 1998; Kamber and Moorbath 1998; Rosing 1999; Frei et al. 1999). However, Moorbath and co-workers derive age estimates for these rocks of 3650 Ma, some which are based on whole-rock Pb errorchrons rather than zircon geochronology, and may, thus, be the product of open isotope system behavior.

On the other hand, the age of a zircon may not represent the final depositional age. It is further interesting to note that Nutman et al. (1997) measured ages ranging from 3.55 to 3.87 billion years for zircons from quartz-dolerite sheets from Akilia Island, which are assumed to be somewhat younger than the Akilia association; they interpreted the oldest measured age as the true age. While it is not possible to resolve the age issue here, it should be pointed out that this question has consequences for the search of traces of a LHB on the Earth, as it is these rocks that provide the only samples for a search for extraterrestrial components.

For the purposes of our study, we selected a suite of pre-3.7 billion year old rocks from the Isuasupracrustal belt. The identification of suitable rocks for such a study is difficult. Appel (1979) found chromite grains of possibly extraterrestrial origin in Isua rocks, but no follow-up studies have been done since. Sedimentary rocks would be best suited, but there is some controversy as to whether terrigenous clastic sediments occur at Isua (e.g., Rosing et al. 1996; Nutman et al. 1997). Most units that have been interpreted as metasediments turn out to be strongly metasomatized intrusive gneisses. There are a few graded turbidite units, but they also appear to be dominated by volcanic units. Rocks that preserve sedimentary structures have been described from a low-strain domain within the Garbenschiefer Formation (Rosing 1999). These rocks can be traced about 100 m along strike and form a 50 m thick unit that is dominated by 10 to 70 cm thick normally graded beds of clastic sediments (containing mainly quartz, mica, chlorite, and graphite), which have distinct bases and which are separated by up to 10 cm thick black slaty units (Rosing 1999). We focused on samples that have not been positively proven to be plutonic, and which have mixed zircon populations - either because they represent multi-phase intrusives, had an extended metamorphic history, or because they were formed from several sources.

We studied a variety of samples, including turbitites, felsic rocks, pelagic shales, and BIF. The turbidite sediments (samples 810194, 810196, 810205, 810215) are from well preserved Bouma sequences within a package of mafic volcanics. The complex was metamorphosed to amphibolite facies, and some units may have undergone sulfide mineralization, as reflected by late Archean to Proterozoic Pb-Pb step leaching ages (Frei et al. 1999). The sediments from the Bouma sequences and enclosing amphibolites give a Sm-Nd whole rock isochron age of 3779 ± 81 Ma (Rosing 1999). The turbidites most probably represent immature volcanogenic detritus mixed with pelagic mud. Sample 810213 represents pelagic mud probably accumulated over several thousands of years in an oceanic setting, i.e., it could contain integrated (cosmic? aeolian?) dust.

We analyzed four "felsic" samples: 810348 and 800494 (graywacke/felsic gneiss), and 810436 and 810437, which represent conglomeratic rock. These three samples are from the so-called A6 or felsic formation; samples from this unit yielded ages of 3.81 billion years (Nutman et al. 1996). Most authors have ascribed a sedimentary origin to this formation, but Rosing et al. (1996) questioned this interpretation. The "conglomerate" consists of granitoid nodules up to tens of centimeters across in a carbonate-rich matrix (Nutman et al. 1984). Some workers have suggested that this is a felsic tuff, whereas Rosing et al. (1996) argued that it is a boudinaged and carbonated orthogneiss package within the supracrustals. We have included some samples from these units in our study, even though they may have been uncritically assumed to be of sedimentary origin, just because they occur in a supracrustal belt.

Eight pelagic shale (or related) samples were analyzed: 810193 and 810216 (from a gravity flow part of Bouma sequence), 810207 (a phyllite from graded unit in pelagic shales), 810208 (similar to 810207, but rich in white mica), 810217 (pelagic shale that forms part of a Bouma sequence), 460512 (from a chlorite-rich part of the pelagic shale sequence), and 460528 and 460534 (slaty part of pelagic shale sequence). All samples are from within the Garbenschiefer Formation of the western segment of Isua; the location is described by Rosing (1999).

The BIF samples (see also Dymek and Klein 1988), which represent a chemical sedimentary precipitate dominated by quartz and magnetite, include samples 810401 and 810403, as well as 940081 from the Bouma sequence. They are representative of quartz-magnetite BIF, and are from two different localities and, possibly, two different stratigraphic units. 940081 is rich in amphibole, either due to the presence of volcanogenic clastic components in the protolith, or due to later carbonate metasomatism. Pb isotopic data indicate open system behavior during the late Archean (Frei et al. 1999), suggesting that carbonate impregnation at that time could be the case. The two other BIF samples are from the so-called major BIF, and contain minor diopside, tremolite, and garnet.

Analytical Methods

Major element analyses were done on powdered samples by standard X-ray fluorescence (XRF) procedures (see Reimold et al. 1994, for details on procedures, precision, and accuracy). The concentrations of V, Cu, Y, and Nb were also determined by XRF analysis. All other trace elements were determined on 200-mg-aliquots by instrumental neutron activation analysis (INAA); for details on the procedures (instrumentation, standards, accuracy, precision, etc.), see Koeberl (1993). For most samples, Sr and Zr concentrations were determined by both XRF and INAA, and for some of the low abundance samples, Ni data were also obtained by XRF. International standard reference materials (Govindaraju 1994) were used to check accuracy and precision of the analyses.

Noble metal concentrations were determined using a Ni sulfide fire assay with Te co-precipitation and ICP-MS procedure. This was modified from Jackson et al. (1990) in order to enhance detection limits and minimise the reagent blank. Typically, 6 g of Na-carbonate, 12 g borax, 0.7 g sulphur, 2.10 g Ni carbonate, and 3.5 g of silica were used for fusion of a 10 g sample aliquot. Samples containing high concentrations of Fe sulphides were pre-roasted at 700°C for 60 minutes in order to reduce the sulphur content as a high FeS content can lead to physicomechanical problems during slag-matte separation (Lenahan and Murray-Smith 1986). The reagents were thoroughly mixed and transferred into a fire clay crucible before being fired for 90 minutes at 1060°C in a VECSTAR electric furnace equipped with a EUROTHERM 91E temperature controller. After removal and cooling, the crucibles were broken open and the Ni sulphide buttons were separated from the silicate slag, weighed and fully described.

Following fire assay, each button was sealed inside a plastic bag, cracked using a sharp tap from a hammer, and the fragments transferred to a 500 ml beaker containing 120 ml of concentrated HCl. Moderate heat was applied using a hotplate and the buttons were dissolved in the HCl. After the sulphide had dissolved, any noble metals which had entered solution were co-precipitated with Te, using $SnCl_2$ as a reductant (Shazali et al. 1987). After co-precipitation, the insoluble noble metal-bearing residue was filtered under vacuum through a Whatman® 0.45-μm nitrocellulose filter paper and washed with 10% HCl and warm distilled water. Each filter paper was then placed inside a 100ml screw-top teflon vial. 8-12 crystals of salt (NaCl) were placed in the vial along with the filter paper, then 4.0 ml of concentrated Hcl and 3 ml of concentrated HNO_3 were added. Each vial was tightly sealed and placed in a waterbath at 40°C overnight. The addition of NaCl promotes the formation of soluble PGE chloro-complexes (Lenahan and Murray-Smith 1986). The vial cooled in an icebath for 20-30 minutes, after which it was opened and the solution containing the dissolved noble metals was made up to 100.0 g by dilution with cold distilled water. Despite attempts to minimise losses of Os as volatile OsO_4 during dissolution of the filter paper, Os recoveries were erratic with variations of up to 20% (well in excess of previously observed sample heterogeneity) between some standard replicates. For this reason, Os data are not reported in this study.

Finally, the solutions were spiked with internal standard monitors for instrumental drift and noble metal concentrations were determined by external calibration on a VG Elemental PQ2+ Plasmaquad ICP-MS using synthetic noble metal standard solutions and procedures similar to those outlined by Jackson et al. (1990). In order to accurately determine Au in samples containing very high concentrations of the metal, solutions were diluted appropriately to correspond with the range of noble metal standards. Reagent blanks were prepared using silica in place of the sample powder and processed as outlined above. Method detection and quantitation limits (defined as 3 and 10 standard deviations of the blank respectively) are given in Table 1. The accuracy of the analysis was determined by analysis of the certified reference materials Peridotite-WPR1 (Govindaraju 1994) and Komatiite-Wits1 (Tredoux and McDonald 1996; McDonald et al. in prep.). Method detection and quantitation limts as well as blank concentrations are given in Table 1.

Due to the limited amount of material available for some samples, the PGE and Au concentrations obtained for the present study generally represent analysis of a single powder aliquot. About half of the samples provided sufficient material for duplicate analysis. Precisions (at the 1σ level) for PGE and Au are given for these samples with the results. The lack of sufficient replicates precludes a measure of precision on noble metal concentrations in the other samples (McDonald 1998), but in each case it is unlikely to be larger than 10 rel%.

Shock Petrographic Study

For the shock metamorphic study, it is important to choose a mineral that would not undergo easy deformation, alteration, metamorphism, or recrystallization. One of the best suited (i.e., most resistant) minerals for this purpose is the refractory mineral zircon, which is present in trace amounts in most of these rocks. Zircon has been shown to contain a range of shock-induced features at the optical and electron microscope level (e.g., Bohor et al. 1993; Reimold et al. 1998; Leroux et al. 1999). While planar deformation features (PDFs) in quartz may have long been annealed or altered, those in zircon have a better chance to survive for several billion years, as is indicated by the preservation of shocked zircons in rocks from the about 2 Ga old Vredefort and Sudbury impact structures (see, e.g., Gibson and Reimold; this volume).

Five samples that were most likely of sedimentary origin (Bouma sequence turbidite) yielded, unfortunately, no zircons. Zircons could be separated, however, from felsic schists, which may be of sedimentary origin, but the origin of which is less well constrained than that of the turbidites. Grain mounts were made of hundreds of zircon crystals, which have a wide variety of shapes and zonation patterns. Shapes range from well-rounded to very high aspect ratios. Grain sizes are also variable, but mostly lie between 60 and 150 µm. Where extensive zonation is present, the outer zones of the crystals are generally of euhedral shape with well-developed pyramidal faces. Inner zones/cores may be well- rounded or variably euhedral to subhedral in shape. Many crystals contain inclusions (e.g., apatite needles). While many grains are strongly fractured, most fractures are of irregular shape or even of curved appearance. Single planar fractures do occur, but never as part of sets. In fact, sets of irregular fractures were not observed either. In none of the crystals studied any evidence of optically visible shock deformation was found. However, we hope to be able to study more zircons that are extracted from sedimentary rocks and from different locations (maybe even pre-4 Ga detrital zircons).

Table 1. Analytical data for noble metals determined by the ICP-MS method

	Ir	Ru	Rh	Pt	Pd	Au
Detection Limit	0.02	0.05	0.02	0.14	0.10	0.02
Quantitation Limit	0.06	0.16	0.06	0.46	0.33	0.06
Blank	0.12	<0.15	<0.02	<0.29	<0.22	0.39
WPR1-1	13.9	23.1	14.3	265	232	38
WPR1-2	13.4	21.8	13.5	256	237	185
Recommended	13.5	21.3	13.4	285	235	42
Wits1-1	1.4	4.2	0.98	6.1	4.6	4.4
Wits1-2	1.2	3.5	0.99	4.9	4.7	7.9
Wits1-3	1.2	3.3	1.1	5.2	4.2	4.9
Recommended	1.4±0.2	3.5±0.4	1.1±0.2	5.2±0.5	4.5±0.8	4.9±1.5

Measurements were done on blanks and standard reference materials. All data in ppb. Recommended values for standards WPR1 from Govindaraju (1994) and for Wits1 from Tredoux and McDonald (1996).

Table 2. Major and trace element composition of all Isua samples

	Turbidite				Greywacke		Conglomerate			BIF		Pelagic Shale							
	810194	810196	810205	810115	800494	810348	810436	810437	940081	810401	810403	810193	810207	810208	810216	810217	460512	460528	460534
SiO$_2$	60.05	56.48	52.9	53.57	68.4	69.22	76.54	54.53	41.5	54.46	54.04	59.98	71.77	69.81	53.78	78.88	68.52	59.59	65.47
TiO$_2$	0.36	0.45	0.24	0.49	0.315	0.25	0.15	0.02	0.08	0.22	0.2	0.64	0.35	0.38	0.72	0.39	0.39	0.58	0.50
Al$_2$O$_3$	16.43	18.12	17.64	21.49	11.47	10.3	8.02	8.95	2.49	4.57	0.07	16.89	14.56	15.72	20.85	11.12	16.05	15.74	16.23
Fe$_2$O$_3$	7.67	7.66	7.05	8.82	2.97	4.35	2.36	4.54	53.74	36.09	38.96	7.36	2.72	3.40	8.60	2.16	2.86	7.71	6.13
MnO	0.24	0.22	0.29	0.16	0.15	0.15	0.13	0.38	0.13	0.05	0.19	0.20	0.09	0.10	0.12	0.05	0.09	0.20	0.10
MgO	4.78	4.02	3.62	3.3	2.2	3.84	2.04	5.1	2.34	1.52	2.6	4.58	1.22	1.28	1.85	0.69	1.18	4.36	1.63
CaO	3.03	3.99	8.88	3.45	4.50	5.23	4.44	11.28	1.66	3.48	4.36	3.33	3.04	3.57	5.82	1.23	3.13	4.14	2.23
Na$_2$O	2.24	2.11	0.29	4.51	0.21	0.99	0.17	0.083	0.22	0.05	0	2.76	0.58	0.57	2.59	4.50	0.54	0.71	4.70
K$_2$O	1.72	3.15	2.84	2.93	5.29	1.86	2.21	3.79	0.14	0.29	0.042	1.86	3.18	3.25	0.12	0.87	3.52	3.41	1.48
P$_2$O$_5$	0.17	0.17	0.09	0.19	0.18	0.15	0.15	0.14	0.16	0.25	0.1	0.11	0.05	0.06	1.32	0.07	0.07	0.10	0.10
LOI	3.37	3.33	6.13	1.09	3.72	3.11	3.84	9.68	0.11	0.11	0.02	2.76	2.2	2.2		0.55	2.36	3.91	1.71
Total	100.06	99.70	99.97	100.00	99.40	99.45	100.05	98.49	102.57	101.19	100.58	100.47	100.16	100.34	100.34	100.51	98.71	100.45	100.28
Sc	12.5	13.3	39.9	11.9	4.81	11.7	3.11	3.09	2.11	3.67	0.27	14.2	8.02	7.75	12.3	5.21	8.11	12.9	8.81
V	59	62	127	54	29	55	18	17	5	3	<15	90	30	36	89	29	31	n.d.	62
Cr	137	137	170	74.9	18.9	51.2	15.9	13.1	25.1	55.9	3.4	171	28.4	35.2	71.4	31.1	27.9	162	61.1
Co	31.7	27.9	24.7	24.4	3.18	24	23.5	12.7		6.24	3.25	17.2	8.19	32.8	13.5	6.89	7.43	12.1	20.1
Ni	70	59	99	49	<2	8	10	4	40	32	7	90	35	31	53	28	31	25	70
Cu	<2	<2	<2	<2	27	<2	<2	<2	<3	18	7	<2	<2	<2	<2	<2	<2	<2	44
Zn	67	60	28	84	7	29	45	20	96	90	68	73	12	13	95	7	13	30	38
Ga	35	70	20	30		30	3		120	55	110	15	9	7	9		15	19	10
As	0.1	0.88	0.21	0.2	3.18	0.1	0.65	0.79	0.52	0.08	1.92	0.24	0.28	3.34	0.43	0.91	0.39	0.33	0.3
Se	0.42	0.11	0.26	0.15	0.17	0.15	0.15	0.26	0.09	0.08	0.29	0.1	0.05	0.27	0.14	0.12	0.27	0.12	0.15
Br	0.1	0.2	0.2	0.1	0.6	0.4	0.2	0.3	0.08	0.8	0.3	0.41	0.15	0.29	0.27	0.2	0.1	0.06	0.36
Rb	76.2	148	126	119	93.5	41.2	46.6	74.1	2.3	5.8	1.9	105	90.5	90.7	108	30.5	100	153	59.7
Sr	91	88	45	137	25	48	21	25	1	3	3	117	34	39	136	67	33	32	149
Y	12	9	12	10	104	6	5			10	12	10	24	26	157	79	29	n.d.	12
Zr	118	107	44	129	8.5	68	86	83	27	47	6	129	221	224		79	244	99	138
Nb	9	8	7	8		7	7	7	5	6	6	9	13	12	10	8	14	n.d.	9
Sb	0.11	0.25	0.08	0.08	0.64	0.18	0.24	0.31	0.11	0.45	3.12	0.05	0.09	0.11	0.25	0.25	0.14	0.16	0.096
Cs	1.04	2.05	1.76	3.56	1.68	2.61	1.79	1.51	0.095	0.25	0.034	1.92	0.83	1.38	2.21	0.85	1.23	1.69	1.09
Ba	68	145	166	366	1230	740	1090	724	12	25	20	85	230	255	350	166	275	95	98
La	15.2	18.3	4.61	14.1	19.9	13.1	15.9	18.6	3.84	2.34	1.42	13.1	24.6	27.8	15.6	6.07	24.6	11.5	8.62
Ce	31.2	37.2	9.63	27.5	39.5	25.7	31.5	36.8	6.32	4.85	2.21	28.5	51.7	51.3	31.5	12.6	51.6	24.3	19.4
Nd	16.1	18.6	5.69	13.4	19.1	11.9	15.8	17.4	3.44	3.37	1.18	14.3	26.1	25.5	16.3	6.1	26.7	14.5	10.2
Sm	3.09	3.59	1.31	2.57	3.59	2.14	2.27	2.52	0.68	0.51	0.32	2.77	4.81	4.41	2.69	1.05	4.46	2.37	2.11
Eu	0.72	1.09	0.27	0.77	0.76	0.59	0.61	0.68	0.29	0.95	0.28	0.76	0.86	0.92	0.57	0.28	0.85	0.67	0.55
Gd	2.85	2.7	1.7		2.2	1.7	1.7	1.8	0.82	0.22	0.39	2.4	4.5	4.7	1.8	0.9	4.78	1.9	1.8
Tb	0.41	0.45	0.28	0.37	0.27	0.24	0.24	0.27	0.14	0.13	0.089	0.41	0.71	0.71	0.31	0.16	0.79	0.32	0.31
Tm	0.24	0.22	0.19	0.21	0.11	0.12	0.084	0.086	0.08	0.78	0.065	0.22	0.41	0.36	0.18	0.09	0.44	0.18	0.16
Yb	1.61	1.37	1.51	1.34	0.61	0.83	0.51	0.52	0.49	1.11	0.47	1.21	3.04	2.51	1.31	0.66	3.11	1.18	1.01
Lu	0.22	0.21	0.23	0.19	0.084	0.13	0.071	0.081	0.065	0.11	0.075	0.18	0.41	0.33	0.19	0.10	0.38	0.13	0.21
Hf	3.32	3.76	1.02	3.69	2.49	1.98	1.99	2.19	0.56	0.8	0.037	3.42	6.45	6.24	3.94	2.06	6.61	3.09	3.71
Ta	0.44	0.38	0.1	0.37	0.32	0.28	0.37	0.26	0.047	1.11	0.013	0.28	0.68	0.78	0.31	0.15	0.69	0.22	0.29
Ir (ppb)	<0.5	<0.5	<0.2	0.1	0.2	0.2	0.3	<0.5	<0.2	0.11	0.2	<1.1	<0.5	<0.7	<0.7	<0.5	0.2	<1	<0.8
Au (ppb)	<3	0.2	12	1.2	0.5		<3	0.2	0.9	0.8	1.6	0.3	0.5	0.8	<3	0.8	<2	<1	0.5
Th	3.29	3.72	0.62	3.32	3.05	2.12	2.43	2.78	0.47	1.75	0.053	3.42	7.32	6.52	3.52	1.93	6.96	3.03	3.79
U	0.86	0.94	0.34	0.95	0.41	0.37	0.65	2.37	0.11	1.82	0.18	0.59	1.22	0.77	0.62	0.51	1.05	0.62	0.61
K/U	16667	27926	69608	25702	107520	41892	28333	80983	10606	7796	1944	26271	21721	35173	34812	14216	27937	45833	20219
La/Th	4.62	4.92	7.44	4.25	6.52	6.18	6.54	6.69	8.17	1.75	26.79	3.83	3.36	4.26	4.43	3.15	3.53	3.80	2.27
Hf/U	7.55	9.89	10.20	9.97	7.78	7.07	5.38	8.42	11.91	10.09	2.85	12.21	9.49	8.00	12.71	13.73	9.58	14.05	12.79
Th/U	3.83	3.96	1.82	3.49	7.44	5.73	3.74	7.13	4.27	4.32	0.29	5.80	6.00	8.47	5.68	3.78	6.63	4.89	6.21
La/Yb$_N$	6.38	9.03	2.06	7.11	22.04	10.67	21.07	24.17	5.30	2.03	2.04	7.32	5.47	7.48	5.68	6.21	5.35	6.59	6.21
Eu/Eu*	0.76	1.10	0.44	0.95	0.41	0.97	0.97	1.00	1.21	1.82	2.48	0.92	1.22	0.63	0.81	0.90	0.58	0.99	0.88

Major element data in wt%, trace element data in ppm, except as noted. All Fe as Fe$_2$O$_3$. n.d. = not determined.

Table 3. Noble metal contents of Isua samples, determined by ICP-MS.

Sample	Ir	Ru	Rh	Pt	Pd	Au
Turbidite						
810194	b.l.d.	b.l.d.	b.l.d	0.71	0.45	0.12
810205	b.l.d.	<0.10	0.18	3.7	7.9	8.1
Graywacke						
800494	0.09	0.19	0.21	0.57	0.40	0.97
810348	b.l.d.	b.l.d.	b.l.d.	1.4	1.5	0.33
Conglomerate						
810436	b.l.d.	b.l.d.	b.l.d.	b.l.d.	b.l.d.	b.l.d.
810437	b.l.d.	b.l.d.	b.l.d.	b.l.d.	b.l.d.	0.25±0.10
BIF						
940081	<0.07	0.23±0.10	0.06±0.01	0.49±0.16	0.57±0.04	0.32±0.08
810401	0.18	0.17	0.12	1.2	1.1	0.57
810403	b.l.d.	b.l.d.	b.l.d.	b.l.d.	0.60	1.2
Pelagic Shale etc.						
810193	b.l.d.	<0.16	b.l.d.	<0.16	0.23±0.05	<0.10
810207	0.06±0.01	<0.12	b.l.d.	<0.28	0.27±0.12	<0.08
810208	b.l.d.	<0.14	b.l.d.	0.54±0.09	0.24±0.08	<0.11
810216	0.08±0.02	<0.15	0.30±0.06	14.8±4.0	0.22±0.03	<0.03
810217	b.l.d.	<0.15	b.l.d.	0.54±0.10	0.18±0.03	0.52±0.21
460512	b.l.d.	<0.10	b.l.d.	<0.16	<0.12	1.4±0.6
460528	<0.04	0.23±0.02	b.l.d.	<0.42	0.32±0.02	1.6±0.1
460534	<0.03	<0.15	b.l.d.	0.61±0.40	0.33±0.02	3.7±1.4

All data in ppb. Values between the detection limit and the quantitation limit are reported as <x, where x = next greatest integer. b.l.d. = below limit of detection. Samples that show ± data ave been analyzed in duplicate.

Search for Meteoritic Component Geochemical Markers

Several samples of Isua rocks, which included turbidite and pelagic sediments and BIF samples from a well-preserved Bouma sequence, were analyzed for their chemical composition, including the abundances of siderophile elements, including the PGEs, to search for the possible presence of any unusual enrichment in siderophile element abundances, which could represent evidence for a meteoritic component. The same rocks were also analyzed for their complete major and trace element contents, to properly document the compositions of these samples (and to obtain supplementary siderophile element data on, e.g., Fe, Cr, Co, and Ni). The results are given in Table 2. The BIF samples, which show Fe_2O_3

contents of 36 to 54 wt% (Table 2), are, as mentioned above, from two different localities and possibly represent two different stratigraphic units. The compositions of the turbitite, pelagic shale, and graywacke samples are mostly similar to those major and trace element contents measured before on comparable rocks (e.g., Nutman et al. 1984; Dymek and Klein 1988; Rosing et al. 1996). Figure 1 shows chondrite-normalized rare earth element (REE) patterns for BIF and some other selected samples. The BIF samples have positive Eu anomalies, whereas the other samples show relatively steep patterns from La to Lu, with only the turbidite showing a slight negative Eu anomaly.

Of 17 samples analyzed for PGE contents by ICP-MS (Table 3), four yielded measurable amounts of Ir above the detection limit of this study (0.02 ppb). This detection limit is of the same order of magnitude as the present-day Ir content of average upper crustal rocks (about 0.02 ppb; Taylor and McLennan 1985). Contents of Ir determined by INAA (Table 2) seem to indicate some elevated contents as well, but these values are near the detection limit for this method and are obviously not reliable compared to the much more precise ICP-MS data. One of the samples (810401) is from the BIF sequence and has 0.18 ppb Ir; another (810494) is a graywacke with 0.09 ppb Ir. Samples 810207 (a phyllite from the graded unit in pelagic shales) and 810217 (the gravity flow part of the Bouma sequence) have 0.06 and 0.08 ppb, respectively (Table 3). The latter sample has a relatively high Pt content (14.8 ppb), possibly indicating the presence of an isoferroplatinum nugget. The BIF and graywacke samples also have Ru and Rh above detection limit; 810216 has only detectable Rh but not Ru. Most of the 17 samples analyzed for noble metal contents have contents of Pt, Pd, and Au that are above the detection limit, but the abundances of these elements are somewhat variable.

Figure 2 shows the abundances of the measured PGEs, together with the Cr, Co, and Ni abundances, normalized to carbonaceous chondrite abundances. To allow comparison of normalization for different chondrite types, Fig. 3 shows an example of one sample (BIF 810401) normalized to the abundances for the four major chondrite groups: carbonaceous and ordinary chondrites of the H-, L-, and E-type. This diagram shows that the differences between the four normalizations are not significant, compared to the overall variations between the individual elemental contents.

A meteoritic component could, if found to be present in some of these rocks, have two possible sources. One would be impact-derived, i.e., from meteoritic (asteroidal, cometary) material mixed in with terrestrial material during an impact event and distributed either as distal ejecta (similar to, e.g., the enrichment in PGEs found in the clay marking the Cretaceous-Teriary boundary; cf. Alvarez et al. 1980), or as locally produced and deposited crater fill and proximal ejecta (suevite, impact melt rock or breccia). The other possible source is cosmic dust, which accretes onto the Earth and is incorporated into BIF or pelagic sediments during sedimentation. These latter rocks have relatively slow sedimentation rates and are thought to accumulate trace elements from the ocean water. Studies of impact ejecta and crater rocks with known meteoritic components (see Koeberl 1998, for a review) indicate that it does not seem possible to easily distinguish between the two possible sources based on chemical composition alone.

The siderophile element abundance patterns show variations by a factor of 3 to 4 of the individual abundances. For example, normalized abundances of Cr, Co, Ni, Pd, Pt, and Au are much higher in almost all samples than the Ru, Rh, and Ir abundances. Whereas there seems to be a difference in the normalized patterns between BIF rocks and other sample types (Fig. 1), it remains unclear at this time how much the indigeneous (background) concentration of Pd and Pt (which are enriched compared to Ir) contributes to the total. Further studies should help to define a possible indigenous component, which could be subtracted from the contents of the rocks with higher siderophile element abundances, to yield a possibly "added" component. An attempt was made to quantify and remove an indigeneous component by assuming that the background concentrations are similar to those found in oceanic crust (considering that 4 billion years ago there was most likely very little continental crust, and impacts would not have penetrated into the mantle). The concentrations of Al and/or Sc in BIF were taken as proxies for the possible amount of oceanic crust, and corrections to the concentrations of the siderophile elements in the BIF samples were calculated. The results are plotted in Fig. 4. Whereas these corrections resulted in somewhat lower values of Cr, Co, Ni, Ru, and Pt, the resulting normalized patterns are still not flat and near-chondritic, indicating that any indigeneous component (if present) was not similar in composition to oceanic crust.

The observed PGE (and siderophile element) abundance patterns of the BIF samples are non-chondritic (because of elevated Pd and Pt and low Ru, Rh, and, in a few cases, Ir), but the elevated Ir content in the two BIF samples may still indicate a remnant meteoritic phase. On the other hand, it could also be the result of terrestrial concentration reactions (cf. Ryder et al., 1999). In the absence of any evidence for shock metamorphism, and with ambiguous geochemical signals, no unambiguous conclusions regarding the presence of extraterrestrial matter (as a result of the late heavy bombardment) in the Isua rocks studied here can be made.

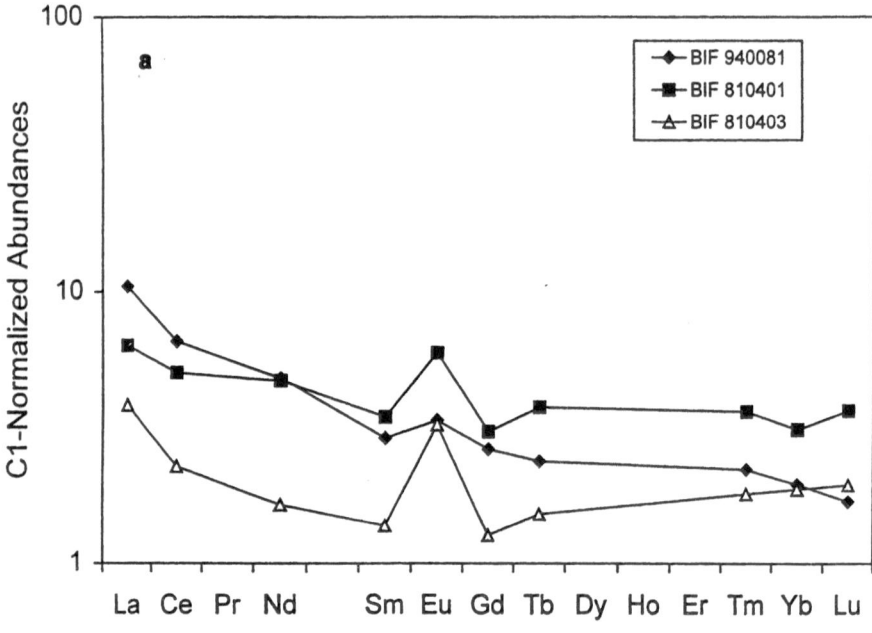

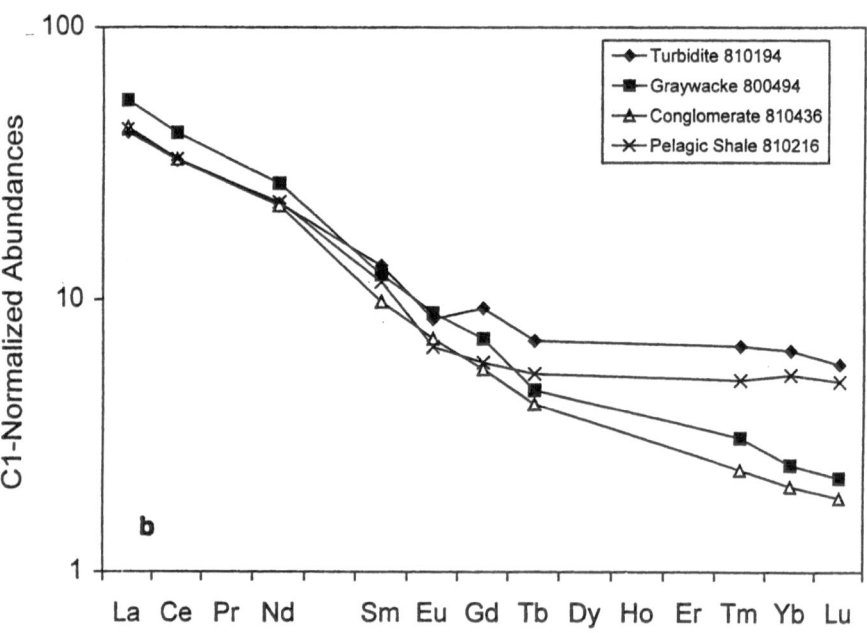

Fig. 1. Chondrite (C1)-normalized abundances of the rare earth elements in selected samples from Isua. a) BIF samples; b) representative samples of other groups (e.g., graywacke, turbidite). Normalization factors from Taylor and McLennan (1985).

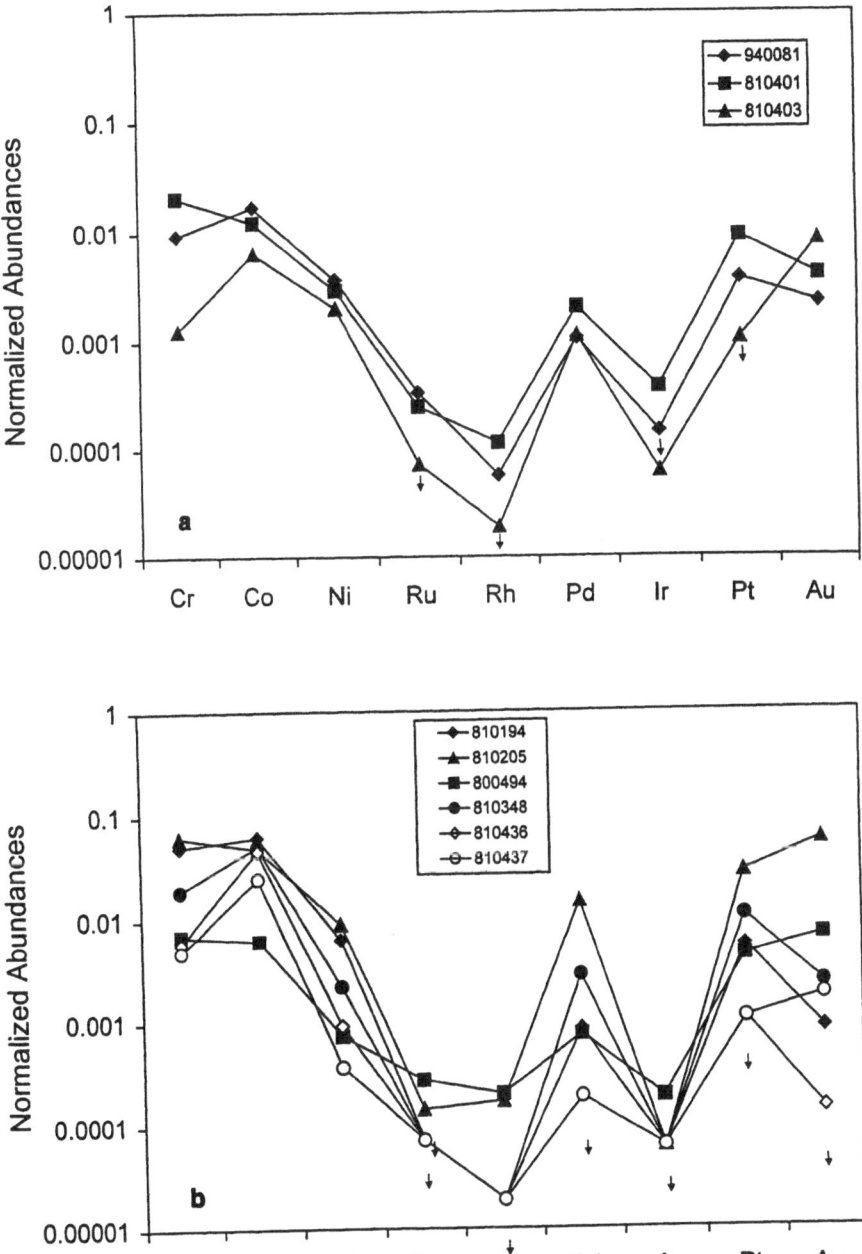

Fig. 2. Normalized abundances of the platinum group elements in samples from Isua. a) BIF samples, C1-normalized; b) samples other than BIF, C1-normalized. Normalization factors from Palme et al. (1981).

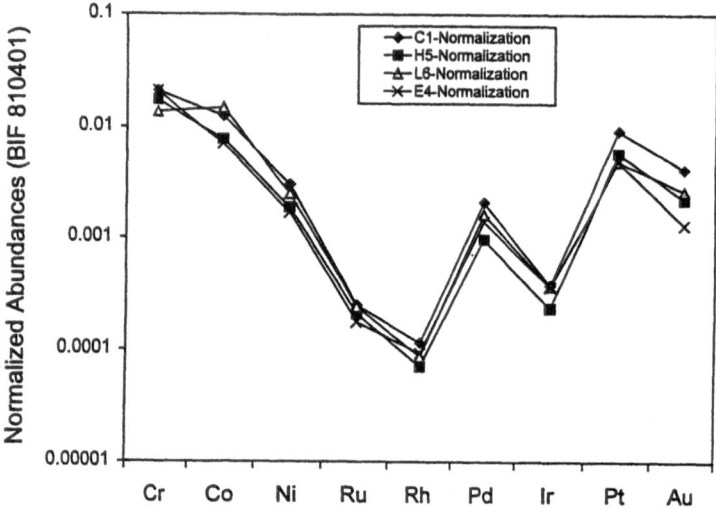

Fig. 3. Variation of normalized abundance patterns, depending on the chondrite type used for normalization, demonstrated on the example of BIF sample 810401. Shown are normalizations to C1, H5, L6, and E4 chondrites. Normalization factors from Palme et al. (1981) and Mason (1979).

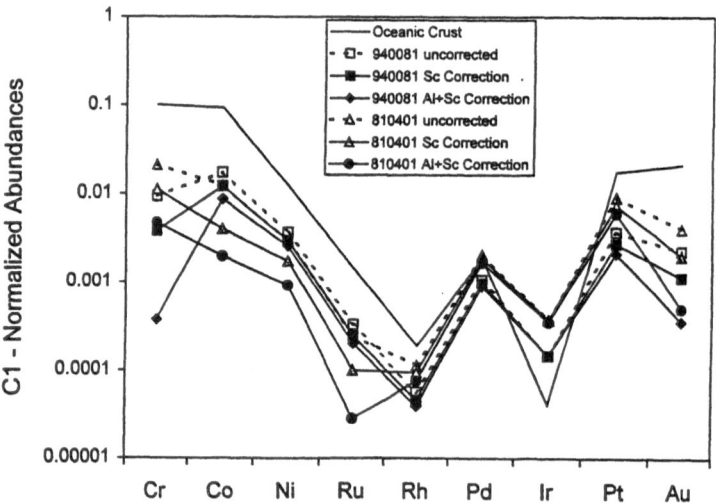

Fig. 4. Chondrite (C1)-normalized abundance patterns for BIF samples 940081 and 810401 (open symbol), compared to the pattern of average oceanic crust (with Pd and Au data for bulk crust); data from Taylor and McLennan (1985). Also displayed are attempts to correct the siderophile element abundances for an indigeneous component (see Koeberl, 1998, for discussion). The abundances of Al and Sc were taken as indicators of a possible addition of material similar in composition to oceanic crust, resulting in two corrected abundance patterns per sample (solid symbol), one based only on the Sc abundance for correction, and one based on both Al and Sc contents; the latter resulted in a more substantial correction.

Summary and Conclusions

As discussed by Ryder et al. (1999), there are at least three possible strategies to search for evidence of pre-3.8 Ga bombardment history on the early Earth: a) search for shocked minerals that have formed as a result of large-scale impact events; b) search for chemical evidence of an increase in the overall flux of extraterrestrial matter onto the early Earth (e.g., Mojzsis et al. 1997; Arnold et al. 1998; Ryder and Mojzsis 1998; it may also be possible to employ the methods of Koeberl and Shirey 1993, or Farley et al. 1998); and c) detection of strongly altered and metamorphosed remnants of impact ejecta that might have been incorporated into early Archean rock formations. In the present study of rocks of presumably sedimentary or metasedimentary origin from Isua, Greenland, which represent the oldest known sedimentary rocks on Earth, we mainly tried out approaches a) and c), both without any clear success. A search for shocked zircons in rocks of probably sedimentary origin yielded only obviously unshocked zircons, and a geochemical search in sedimentary Isua rocks, using sensitive PGE analyses with sub-ppb detection limits, yielded three samples with (compared to the other samples and to average continental crust) elevated siderophile element contents (including Ir). However, non-chondritic siderophile element patterns and strongly variable patterns between different samples do not allow the unambiguous interpretation of these enrichments are being due to the presence of an extraterrestrial component.

There can be several reasons for the failure to obtain evidence for a LHB on the Earth from the present sample suite. In the case of the zircon study, analysis of only two samples (even if yielding several hundred zircons each) is statistically still insufficient. Koeberl and Sharpton (1988) have already attempted to find evidence for shocked quartz in Isua rocks, but did not find any evidence either, which could also be because of annealing of planar deformation features in quartz. It would be desirable, though, to perform a petrographic search for shocked zircons extended to the detrital zircons of known age (3.8 to 4.27 billion years old), and a statistically significant number of samples from different locations should be studied. However, as is indicated by new calculations of Cintala and Grieve (1998), it is also possible that large-scale impact events produce relatively larger amount of melt compared to smaller impact events, which would immediately anneal most shocked minerals (except maybe zircon) in autochthonous rocks, leaving only distal ejecta with traces of shock metamorphism. Such a scenario would result in significantly lower amounts of shocked minerals than expected before. In this case the absence of shocked minerals may not be evidence against a LHB, but we still expect zircon to be more stable against resetting than most other minerals. The search for geochemical evidence was broader (cf. also Ryder and Mojzsis, 1998, and Arnold et al., 1998), but there are several possibilities to explain the negative or ambiguous results as well. The samples chosen for analysis might not have been ideal for such a search (e.g., they may not represent sediments that would show an increase in general extraterrestrial flux, or not include any impact ejecta), or the Isua rocks do not overlap in age with the late heavy bombardment on the Moon. In the latter case it is not even necessary to follow the radical arguments of Moorbath and co-workers; it would be enough if the rocks date to 3.80 Ga, whereas the LHB recorded on the Moon may have had a very short half-life and may have ceased at 3.84 Ga. Clearly, this important aspect of early Earth evolution (cf. Maher and Stevenson 1988; Sleep et al. 1989; Zahnle and Sleep 1997), which must have influenced the development of the earliest life forms (Mojzsis et al. 1996), deserves more attention.

Acknowledgements

This research is supported by the Austrian Fonds zur Förderung der wissenschaftlichen Forschung (FWF), grant Y58-GEO (to C.K.). We thank S. Farrell (Univ. Witwatersrand) and H. Huber (Univ. Vienna) for help with XRF and INAA, respectively. We appreciate helpful and constructive reviews of this manuscript by F.T. Kyte and G. Pearson, as well as comments by editor I. Gilmour. The IMPACT program of the European Science Foundation is thanked for logistical support. This paper is University of the Witwatersrand Impact Cratering Research Group contribution No. 11.

References

Allaart JH (1976) The pre-3760 my old supracrustal rocks of the Isua area, central West Greenland, and the associated occurrence of quartz-banded ironstone. In: Windley BF (ed) The Early History of the Earth. Wiley, London, pp 177-189

Alvarez LW, Alvarez W, Asaro F, Michel HV (1980) Extraterrestrial cause for the Cretaceous-Tertiary extinction. Science 208: 1095-1108

Amelin Y, Lee D-C, Halliday AN, Pidgeon RT (1999) Nature of the Earth's earliest crust from hafnium isotopes in single detrital zircons. Nature 399: 252-255

Appel PWU (1979) Cosmic grains in an iron-formation from the Early Precambrian Isua supracrustal belt, West Greenland. Journal of Geology 87: 573-578

Arndt N, Chauvel C (1991) Crust of the Hadean Earth. Bull. geol. Soc. Denmark 39: 145-151

Arnold G, Anbar A, Mojzsis SJ (1998) Iridium and platinum in early Archean metasediments: Implications for sedimentation rate and extraterrestrial flux. Geol. Soc. Am., Abstracts with Programs 30, No. 7, A82-A83

Baldwin RB (1971) On the history of lunar impact cratering: The absolute time scale and the origin of planetesimals. Icarus 14: 36-52

Baldwin RB (1974) Was there a "Terminal Lunar Cataclysm" 3.9-4.0x10^9 years ago? Icarus 23: 157-166

Baldwin RB (1981) On the origin of the planetesimals that produced the multi-ring basins. In Multi-ring basins. Proceedings, Lunar and Planetary Science Conference 12A: 19-28

Baldwin RB (1987) On the relative and absolute ages of seven lunar front face basins. II. From crater counts. Icarus 71: 19-29

Bogard DD, Garrison DH (1991) $^{40}Ar/^{39}Ar$ ages of achondrites: Evidence for a lunar-like cataclysm? Meteoritics 26: 320

Bohor BF, Betterton WJ, Krogh TE (1993) Impact-shocked zircons: discovery of shock-induced textures reflecting increasing degrees of shock metamorphism. Earth and Planetary Science Letters 119: 419-424

Bowring SA, Housh T (1995) The Earth's early evolution. Science 269: 1535-1540

Cameron AGW, Benz W (1991) The origin of the Moon and the single impact hypothesis IV. Icarus 92: 204-216

Carlson RW, Lugmair GW (1988) The age of ferroan anorthosite 60025: Oldest crust on a young Moon? Earth and Planetary Science Letters 90: 119-130

Cintala MJ, Grieve RAF (1998) Scaling impact melting and crater dimensions: Implications for the lunar cratering record. Meteoritics and Planetary Science 33: 889-912

Compston W, Pidgeon RT (1986) Jack Hills, evidence of more very old zircons in Western Australia. Nature 291: 193-196

Davies GF (1985) Heat deposition and retention in a sold planet growing by impacts. Icarus 63: 45-68

Deutsch A, Schärer U (1994) Dating terrestrial impact events. Meteoritics 29: 301-322

Dimroth E (1982) The oldest rocks on Earth: stratigraphy and sedimentology of the 3.8 billion years old Isua supracrustal sequence. In: Sidorenko AV (ed) Sedimentary Geology of the Highly Metamorphosed Precambrian Complexes, Nauka, Moscow, pp 16-27

Dymek RF, Klein C (1988) Chemistry, petrology and origin of banded iron-formation lithologies from the 3800 Ma Isua supracrustal belt, West Greenland. Precambrian Research 39: 247-302

Farley KA, Montanari A, Shoemaker EM, Shoemaker CS (1998) Geochemical evidence for a comet shower in the late Eocene. Science 280: 1250-1253

Fegley B Jr, Prinn RG, Hartman H, Watkins GH (1986) Chemical effects of large impacts on the Earth's primitive atmosphere. Nature 319: 305-308

Frei R, Bridgwater D, Rosing M, Stecher O (1999) Controversial Pb-Pb and Sm-Nd isotope results in the early Archean Isua (West Greenland) oxide iron formation: Preservation of primary signatures versus secondary disturbances. Geochimica et Cosmochimica Acta 63: 473-488

Frey H (1980) Crustal evolution of the early Earth: The role of major impacts. Precambrian Research 10: 195-216

Goodwin AM (1976) Giant impacting and the development of the continental crust. In: Windley BF (ed) The Early History of the Earth. John Wiley & Sons, London, pp 77-95

Govindaraju K (1994) 1994 compilation of working values and sample descriptions for 383 geostandards. Geostandards Newsletter 18: 1-154

Green DH (1972) Archean greenstone belts may include terrestrial equivalents of lunar maria? Earth and Planetary Science Letters 15: 263-270

Grieve RAF (1980) Impact bombardment and its role in proto-continental growth on the early Earth. Precambrian Research 10: 217-247

Gruau G, Rosing M, Bridgwater D, Gill RCO (1996) Resetting of Sm-Nd systematics during metamorphism of >3.7-Ga rocks: implications for isotopic models of early Earth differentiation. Chemical Geology 133: 225-240

Harper CL, Jacobsen SB (1992) Evidence from coupled ^{147}Sm-^{143}Nd and ^{146}Sm-^{142}Nd systematics for very early (4.5-Gyr) differentiation of the Earth's mantle. Nature 360: 728-732

Hartmann WK (1965) Secular changes in meteoritic flux through the history of the solar system. Icarus 4: 207-213

Hartmann WK (1966) Early lunar cratering. Icarus 5: 406-418

Hartmann WK (1975) Lunar "cataclysm": A misconception? Icarus 24: 181-187

Jackson SE, Fryer BJ, Gosse W, Healey DC, Longerich HP, Strong DF (1990) Determination of the precious metals in geological materials by inductively coupled plasma-mass spectrometry (ICP-MS) with nickel sulphide fire-assay collection and tellurium coprecipitation. Chemical Geology 83: 119-132

Kamber BS, Moorbath S (1998) Initial Pb of the Amitsoq gneiss revisited: implications for the timing of the early Archaean crustal evolution in West Greenland. Chemical Geology 150: 19-41

Koeberl C (1993) Instrumental neutron activation analysis of geochemical and cosmochemical samples: A fast and proven method for small sample analysis. Journal of Radioanalytical and Nuclear Chemistry 168: 47-60

Koeberl C (1998) Identification of meteoritical components in impactites. In: Grady MM, Hutchison R, McCall GJH, Rothery DA (eds) Meteorites: Flux with Time and Impact Effects. Geological Society of London, Special Publication 140, pp 133-152

Koeberl C, Reimold WU (1995) Early Archaean spherule beds in the Barberton Mountain Land, South Africa: no evidence for impact origin. Precambrian Research 74: 1-33

Koeberl C, Sharpton VL (1988) Giant impacts and their influence on the early earth. In: Papers presented to the Conference on "Origin of the Earth", LPI Contribution 681, Lunar and Planetary Institute, Houston, pp 47-48

Koeberl C, Shirey SB (1993) Detection of a meteoritic component in Ivory Coast tektites with rhenium-osmium isotopes. Science 261: 595-598

Koeberl C, Simonson BM, Reimold WU (1999) Geochemistry and petrography of a Late Archean spherule layer in the Griqualand West Basin, South Africa. Lunar and Planetary Science 30: abs. #1755

Kröner A, Layer PW (1998) Crust formation and plate motion in the Early Archean. Science 256: 1405-1411

Lenahan WC, Murray-Smith R de L (1986) Assay and analytical practice in the South African mining industry. The South African Institute of Mining and Metallurgy, Monograph Series M6, 640 pp

Leroux H, Reimold WU, Koeberl C, Hornemann U, Doukhan J-C (1999) Experimental shock deformation in zircon: A transmission electron microscopic study. Earth and Planetary Science Letters 169: 291-301

Lowe DR, Byerly GR, Asaro F, Kyte FT (1989) Geological and geochemical record of 3400-million-year-old terrestrial meteorite impacts. Science 245: 959-962

Maher KA, Stevenson DJ (1988) Impact frustration of the origin of life. Nature 331: 612-614

Mason B (1979) Meteorites. Data of Geochemistry. US Geological Survey Professional Paper 440-B-1, Chapter B, Part 1, 132 pp

Matsui T, Abe Y (1986) Evolution of an impact-induced atmosphere and magma ocean on the accreting Earth. Nature 319: 303-305

McCulloch MT, Bennett VC (1993) Evolution of the early Earth: Constraints from $^{143}Nd/^{142}Nd$ isotopic systematics. Lithos 30: 237-255

McDonald I (1998) The need for a common framework for collection and interpretation of data in Platinum-Group Element geochemistry. Geostandards Newsletter 22: 85-91

McEwen AS (1998) The Phanerozoic impact cratering rate: The lunar record. EOS Transactions, American Geophysical Union 79, No. 45 suppl.: F46-F47

Mojzsis SJ, Arrhenius G, McKeegan KD, Harrison TM, Nutman AP, Friend CRL (1996) Evidence for life on Earth before 3800 million years ago. Nature 384: 55-59

Mojzsis SJ, Ryder G, Righter K (1997) Crustal contamination by meteorites in the early Archean. EOS Transactions, American Geophysical Union 78, No. 46 suppl.: 399-400

Moorbath S, Kamber BS (1998) A reassessment of the timing of early Archaean crustal evolution in West Greenland. Geology of Greenland Survey Bulletin 180: 88-93

Moorbath S, Whitehouse MJ, Kamber BS (1997) Extreme Nd-isotope heterogeneity in the early Archaean - fact or fiction? Case histories from northern Canada and West Greenland. Chemical Geology 135: 213-231

Neukum G, Konig B, Fechtig H, Storzer D (1975) Cratering in the Earth-Moon system - Consequences for age determination by crater counting. Proceedings, Lunar Science Conference 6th, pp 2597-2620

Nutman AP (1986) The Early Archean to Proterozoic history of the Isukasia area, southern West Greenland. Gron Geol Unders Bull 154: 1-80

Nutman AP (1997) The Greenland sector of the North Atlantic Craton. In: de Wit MJ, Ashwal LD (eds) Greenstone Belts. Oxford Monogr Geol Geophys 35, Oxford University Press, Oxford, pp 665-674

Nutman AP, Allaart JH, Bridgwater D, Dimroth E, Rosing M (1984) Stratigraphic and geochemical evidence for the depositional environment of the early Archean Isua supracrustal belt, southern West Greenland. Precambrian Research 25: 365-396

Nutman AP, McGregor VR, Friend CRL, Bennett VC, Kinny PD (1996) The Itsaq Gneiss Complex of southern West Greenland; the world's most extensive record of early crustal evolution (3900-3600 Ma). Precambrian Research 78: 1-39

Nutman AP, Mojzsis SJ, Friend CLR (1997) Recognition of ≥3850 Ma water-lain sediments in West Greenland and their significance for the early Archean Earth. Geochimica et Cosmochimica Acta 61: 2475-2484

Palme H, Suess HE, Zeh HD (1981) Abundances of the elements in the solar system. In: Schaifers K, Voigt HH (eds) Landolt-Boernstein. Springer Verlag, Heidelberg, pp 257-273

Premo WR, Tatsumoto M (1992) U-Th-Pb, Rb-Sr, and Sm-Nd isotopic systematics of lunar troctolitic cumulate 76535: Implications on the age and origin of this early lunar, deep-seated cumulate. Proceedings of Lunar and Planetary Science 22: 381-397

Reimold WU, Koeberl C, Bishop J (1994) Roter Kamm impact crater, Namibia: Geochemistry of basement rocks and breccias. Geochimica et Cosmochimica Acta 58: 2689-2710

Reimold WU, Leroux H, Koeberl C, Hornemann U, Armstrong RA (1998) Optical and transmission electron microscopic analysis of experimentally shock deformed zircon. Meteoritics and Planetary Science 33: A128-A129

Rose NM, Rosing MT, Bridgwater D (1996) The origin of metacarbonate rocks in the Archaean Isua supracrustal belt, West Greenland. Amercian Journal of Science 296: 1004-1044

Rosing MT (1999) ^{13}C-depleted carbon microparticles in >3700-Ma sea-floor sedimentary rocks from western Greenland. Science 283: 674-676

Rosing MT, Rose NM, Bridgwater D, Thomsen HS (1996) Earliest part of Earth's stratigraphic record: a reappraisal of the > 3.7 Ga Isua (Greenland) supracrustal sequence. Geology 24: 43-46

Ryder G (1989) The absence of a heavy early lunar bombardment, the presence of a 3.85 Ga cataclysm, and the geological context of Apollo 14 rock samples. In: Taylor GJ, Warren PH (eds) Moon in Transition: Apollo 14, KREEP, and Evolved Lunar Rocks. LPI Technical Report 89-03, Lunar and Planetary Institute, Houston, pp 107-110

Ryder G (1990) Lunar samples, lunar accretion, and the early bombardment history of the Moon. EOS Transactions, American Geophysical Union 71: 313-323

Ryder G, Mojzsis SJ (1998) Accretion to the Earth and Moon around 3.85 Ga: What is the evidence? EOS Transactions, American Geophysical Union 79, No. 45 suppl.: F48

Ryder G, Koeberl C, Mojzsis SM (1999b) Heavy bombardment of the Earth at ~3.85 Ga: The search for petrographic and geochemical evidence. In: Canup R, Righter K (eds) Origin of the Earth and Moon, University of Arizona Press, Tucson, in press.

Shazali I, Van't Dack L, Gijbels R (1987) Determination of precious metals in ores and rocks by thermal neutron activation/γ-spectrometry after preconcentration by nickel sulphide fire assay and coprecipitaiton with tellurium. Analytica Chimica Acta 196: 49-58

Sleep NH, Zahnle KJ, Kasting JF, Morowitz HJ (1989) Annihilation of ecosystems by large asteroid impacts on the early Earth. Nature 342: 139-142

Smith JV (1981) The first 800 million years of Earth's history. Philosophical Transactions, Royal Society London A301: 401-422

Spudis PD (1993) The Geology of Multi-ring Impact Basins. Cambridge University Press. 263 pp

Taylor SR (1982) Planetary Science: A Lunar Perspective. Lunar and Planetary Institute, Houston, 481 pp

Taylor SR (1989) Growth of planetary crusts. Tectonophysics 161: 147-156

Taylor SR (1992) The origin of the Earth. In: Brown G, Hawkesworth C, Wilson C (eds) Understanding the Earth. Cambridge University Press, Cambridge, pp 25-43

Taylor SR (1993) Early accretion history of the Earth and the Moon-forming event. Lithos 30: 207-221

Taylor SR (1999) Hadean Eon. In: McGraw-Hill Encyclopedia of Science and Technology, McGraw-Hill, in press.

Taylor SR, McLennan SM (1985) The Continental Crust: its Composition and Evolution. Blackwell Scientific Publications, Oxford, 312 pp

Tera F, Papanastassiou DA, Wasserburg GJ (1974) Isotopic evidence for a terminal lunar cataclysm. Earth and Planetary Science Letters 22: 1-21

Tredoux M, McDonald I (1996) Komatiite Wits-1, low concentration noble metal standard for the analysis of non-mineralized samples. Geostandards Newsletter 20, 267-276

Vickery AM (1990) Impacts and atmospheric erosion on the early Earth. Abstracts for the InternationalWorkshop on Meteorite Impact on the Early Earth. Lunar and Planetary Institute, Houston, LPI Contribution No. 746, pp 51-52

Wetherill GW (1975) Late heavy bombardment of the Moon and terrestrial planets. Proceedings, 6th Lunar Science Conference, pp 1539-1561

Wilhelms DE (1984) Moon. In: Carr MH, Saunders RS, Strom RG, Wilhelms DE (eds) The Geology of the Terrestrial Planets. NASA SP-469, pp 107-205

Wilhelms DE (1987) The geologic history of the Moon. U.S. Geological Survey Professional Paper 1348, 302 p

Zahnle KJ, Sleep NH (1997) Impacts and the early evolution of life, In: Thomas PJ, Chyba CF, McKay CP (eds) Comets and the origin and evolution of life. Springer-Verlag, New York., pp 175-208

Zappalà V, Cellino A, Gladman BJ, Manley S, Migliorini F (1998) Asteroid showers on Earth after family breakup events. Icarus 134: 176-179

5 The Oldest Impact Deposits on Earth — First Confirmation of an Extraterrestrial Component

Alexander Shukolyukov[1], Frank T. Kyte[2], Günter W. Lugmair[1,3], Donald R. Lowe[4] and Gary R. Byerly[5]

[1]Scripps Institute of Oceanography, University of California, San Diego, La Jolla CA 92093-0212, USA.
[2]Center for Astrobiology, Institute of Geophysics and Planetary Physics, University of California, Los Angeles, CA 90095-1567, USA.
[3]Max-Planck- Institute for Chemistry, Cosmochemistry, PO 3060, 55020 Mainz, Germany.
[4]Department of Geological and Environmental Sciences, Stanford University, Stanford, CA 94305, USA.
[5]Department of Geology and Geophysics, Louisiana State University, Baton Rouge, LA 70803, USA.

Abstract. The chromium isotopic compositions of samples from an early Archean (3.22 Ga) spherule bed (S4) from the Barberton Greenstone Belt, South Africa, are distinct from that in background rocks and other terrestrial samples. This positively confirms the presence of an extraterrestrial component in this bed and supports hypotheses of an impact origin. The source of the extraterrestrial Cr is most likely to be a carbonaceous chondrite, probably of the CV variety. The estimated extraterrestrial component in this spherule bed is about 15% based on Cr isotopes and platinum group elements. Since this is similar in many regards to the global fallout from the Cretaceous-Tertiary impact, we infer that they have a similar origin. However, the total amount of meteoritic material in the Archean spherule bed is about 30 times greater than in Cretaceous-Tertiary deposits, implying a much larger projectile, possibly 20 km or more in diameter.

Introduction

No direct physical evidence remains on Earth of the Late Heavy Bombardment, which may have ended at ~3.8 Ga (e.g., Taylor 1992). The oldest record of major impact events on Earth may be a number of early Archean (3.5 to 3.2 Ga) spherule beds that have been identified in the Barberton Greenstone Belt, South Africa. Two spherule beds were first recognized as possible ejecta deposits by Lowe and

Byerly (1986a), based largely on textures interpreted as pseudomorphs after quench products, and an absence of evidence indicating a volcanic origin. Based on more detailed field, petrographic and chemical data, Lowe et al. (1989) identified four distinct beds and listed seven criteria that distinguish these beds from typical volcanic and clastic sediments. These criteria include the wide geographic distribution of two beds in a variety of depositional environments, the presence of relict quench textures, absence of juvenile volcaniclasitc debris within the beds, and extreme enrichment of Ir and other platinum group elements (PGE) relative to surrounding sediments. Further detailed analyses of PGE in samples from one of these spherule beds (Kyte et al. 1992) showed that relative abundances of Ir, Pt, and Os are nearly the same as those in chondritic meteorites. Although these samples were somewhat depleted in Pd and highly depleted in Au, relative to the abundances in chondrites, Kyte et al. (1992) pointed out that noble metal abundances in this rock could be reasonably explained only by diagenetic or metasomatic alteration of a preexisting meteoritic component. Despite these reasonably compelling arguments, acceptance of these spherule beds as impact deposits has not been unanimous. Notably, Koeberl et al. (1993) showed that Au and Ir tended to correlate with sulfide content in samples of one spherule bed from the Princeton Mine locality. They argued that the extreme enrichment of siderophile elements is readily explained by gold mineralization. Koeberl et al. (1993) also argued that the concentrations and total amounts of Ir and other siderophiles were much higher than those in any other known impact deposit, making an impact origin for these spherule beds unlikely. Koeberl and Reimold (1995) expand on this based on the study of additional Princeton Mine data as well as analyses of samples from the Sheba Mine and a surface exposure of bed S2. They argue that spherules of any kind are rare in known impact deposits and if present they are not usually associated with a significant PGE anomaly.

In this paper we report on the application of a new method to the analysis of one of these spherule beds. The recent development of methodologies for high precision measurement of the Cr isotopic composition and its application to meteoritic and terrestrial samples has provided a method for unequivocally demonstrating an extraterrestrial component (ETC) in impact ejecta with high concentrations of meteoritic Cr (Lugmair and Shukolyukov 1998; Shukolyukov and Lugmair 1998, 1999). Our results demonstrate that one of the Barberton spherule beds contains a significant amount of Cr that must be derived from extraterrestrial materials.

Samples and Procedures

Samples analyzed in this study were taken from a slab of spherule bed S4 and interbedded clastic sediments that were previously analyzed by Kyte et al. (1992), where details of these samples can be found. Briefly, bed S4 is known from only one locality in the upper part of the Fig Tree Group at 25° 54' 53" S latitude and 31° 01' 08" E longitude. At this locality, bed S4 is found 6.5 m above bed S3. Byerly et al. (1996) review current data on ages of these spherule beds. S3 is likely close to 3243 ±4 Ma and S4 is somewhat younger. The youngest Fig Tree is 3225 ±4 Ma. Bed S4 is 10 to 15 cm thick, containing a mixture of impact-produced components and detrital components. It occurs within a fine-grained black chert layer, 10-20 cm thick and probably representing silicified mud and silt, that lies within a section of coarser sandstone and conglomerate. The spherule bed is composed of up to 50% spherules in some layers, mixed with volcanic and carbonaceous chert detritus, and in a chert matrix (Figure 1). Layers several centimeters in thickness display normal size grading. Spherules range in diameter from about 0.5 to 2.8 mm, though most are between 1-2 mm and most spherules are flattened in the plane of the bedding, many severely. They are somewhat variable in mineralogy, but most are essentially pure cryptocrystalline chlorite (Figure 2A). However, even among those spherules that are nearly pure chlorite, many have accumulations of rutile and apatite at their margins, and a smaller population of spherules contains significant apatite, microcrystalline quartz, and rutile within (Figure 2B). None of the S4 spherules have the quench textures seen in other impact beds (Lowe and Byerly, 1986A). The chlorite, quartz, apatite, and rutile are all likely the products of diagenesis and low-grade metamorphism. The matrix to the spherules is now an alteration assemblage composed largely of chert, sericite, and chlorite. It originally consisted of silt- to sand-sized detrital grains that included rock fragments, angular quartz, possibly ripped-up siltstone and mudstone clasts, spherule fragments, and trace amounts of biotite, apatite, monazite, zircon, and probably carbonaceous matter. However, most of these components are present within interbedded sandstones and conglomerates, and the absence of spherules in these units suggests that any reworking was local.

Two samples from the slab of bed S4 and two background sediment samples were analyzed for their Cr isotopic composition using high-precision methods described by Lugmair and Shukolyukov (1998). Sample C is taken from a powder prepared from a portion of the slab (Lowe's SAF-349-3) that contained 7.5 cm of section. Sample D-4 is a rock chip from a 0.5 cm interval that was selected on the basis of high Ir and Cr concentrations, which were 240 ng/g and 1630 µg/g, respectively, in an adjacent rock chip. Two samples of lithic sandstones from above and below the spherule bed were also analyzed. Sample S21 was collect 0.2 m above the spherule bed and sample S18 was from 2 m below the bed.

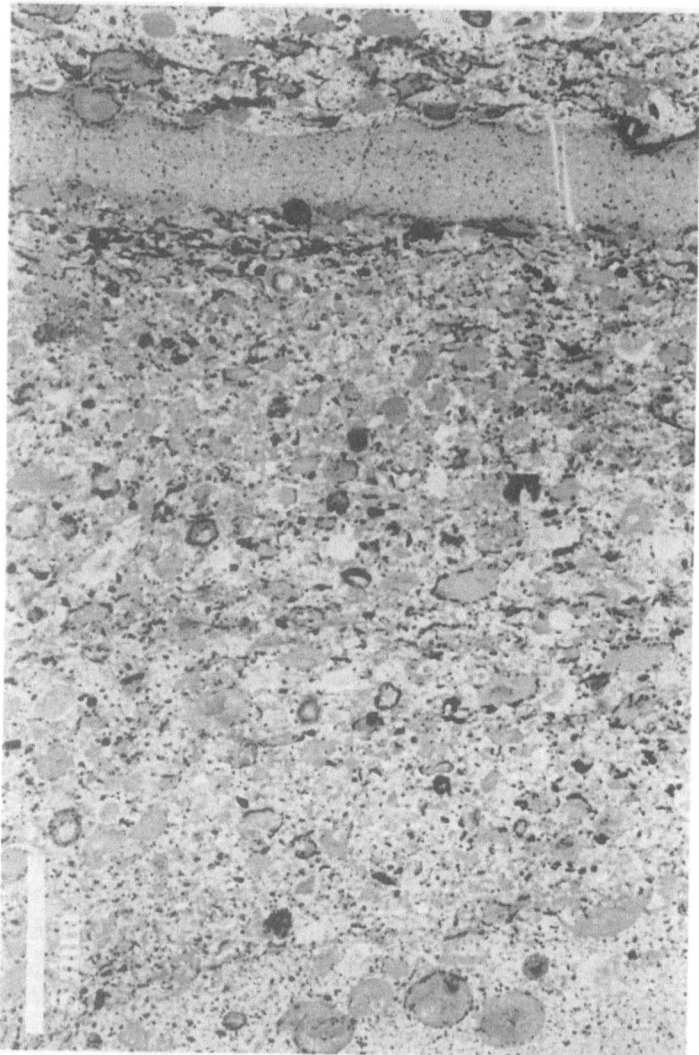

Fig. 1. Photograph of a thin section of a spherule bed S4 illustrating a graded layer of impact debris. Several large impact spherules at base of layer are about 2 mm in diameter. Sizes of spherules range from 0.5 to 2.8 mm. Impact debris is mixed with locally derived volcanogenic sediments and carbonaceous chert.

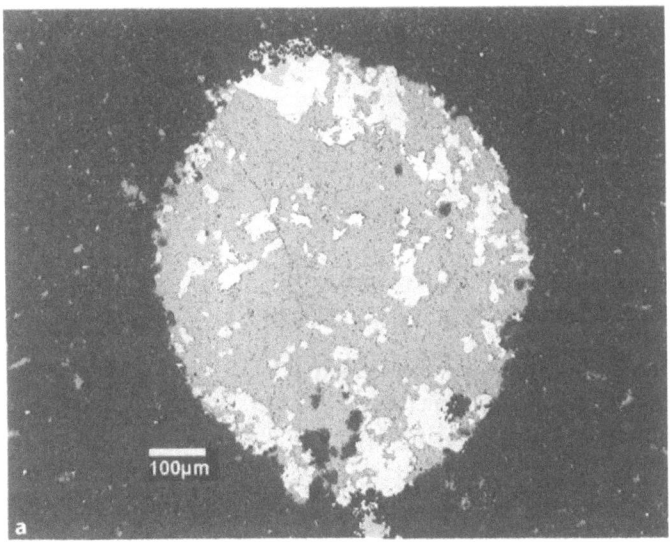

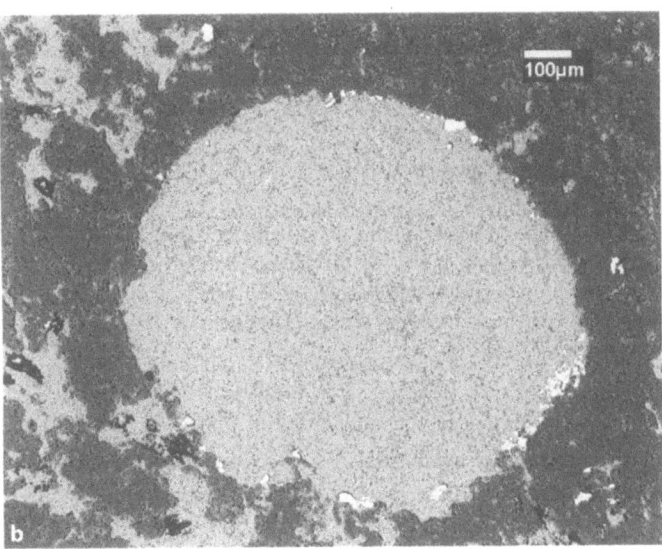

Fig. 2. Backscattered electron images of typical spherules. (A) This 1 mm diameter spherule is internally structureless and composed of very fine-grained chlorite, though it does have concentrations of rutile and apatite along the rim. Surrounding matrix is composed of microcrystalline quartz with patches of chlorite. (B) This 0.75 mm diameter spherule is composed of fine-grained chlorite (dark grey) with lesser amounts of apatite (intermediate grey) and rutile (white). The surrounding matrix is primarily composed of quartz (black). About 10 % of the spherules contain significant concentrations of apatite and rutile.

Table 1. $^{53}Cr/^{52}Cr$ ratios and Cr concentrations in various terrestrial samples, carbonaceous chondrites, and Archean deposit from the Barberton Greenstone Belt, bed S4.

Sample	Cr (ppm)	$^{53}Cr/^{52}Cr$ (ε -units*)
Terrestrial minerals, rock, and sediment		
Laboratory shelf standard	--	≡0
KH-1 Px, Kilbourne Hole, USA (pyroxene)	2500	-0.01±0.08
JAG 89-9, Jagersfontein, South Africa (garnet)	nd	+0.04±0.08
SC Ol, San Carlos Volcanic Field, USA (olivine)	202	-0.03±0.11
MB 81-14, Deccan Traps, India (basalt)	112	-0.04±0.06
ODP 31-302-5-5, Western Pacific (clay)	34	-0.02±0.09
Bulk carbonaceous chondrites		
Allende (CV3)	3540	-0.41±0.09
Orgueil (CI)	2530	-0.43±0.09
Archean spherule bed S4 and background sediments		
SAF-349-3, D-4	1340	-0.32±0.06
SAF-349-3, C	1360	-0.26±0.11
SAF-380-21, 0.2 m above the bed	383	-0.03±0.09
SAF-380-18, 2 m below the bed	261	-0.01±0.10

*1 ε is 1 part in 10^4. †From (Shukolyukov and Lugmair, 1998). An apparent deficit of ^{53}Cr in the carbonaceous chondrites and in the Archean spherule bed S4 samples is due to an excess of ^{54}Cr (see text). Presented uncertainties are based on estimates from the reproducibility of repeat measurements (external precision); 2δ uncertainties would be a factor of two smaller.

Major and minor element analyses of mineral phases in the spherule bed were obtained by electron probe microanalysis. Wavelength dispersive analyses used beam currents of 4 nA and 60 second counting times. A suite of Smithsonian standards was used for calibration (Jarosevich et al. 1980). The ZAF method was used to correct for matrix effects, and a suite of additional well-characterized standards was analyzed as controls. Of particular interest, this method resulted in a lower limit of detection of 0.06 wt% for NiO and 0.07 wt% for Cr_2O_3.

Isotopic Systematics

Recent studies have elucidated the ^{53}Mn-^{53}Cr isotopic systematics in various solar system objects (Lugmair and Shukolyukov 1998; Shukolyukov and Lugmair 1999). The radionuclide ^{53}Mn decays to the stable ^{53}Cr with a half-life of 3.7 Ma. Although ^{53}Mn was present at the time of formation of the first solids in the early solar system and even during formation of early planetesimals it has now fully decayed because of its relatively short half life. The decay products of ^{53}Mn were first discovered in refractory inclusions from the Allende meteorite, a type CV carbonaceous chondrite (Birck and Allègre 1985), and have now been identified in a variety solar system objects (Lugmair and Shukolyukov 1998; Birck and Allègre 1988). The former presence of ^{53}Mn is determined by measuring variations in the relative abundance of the radiogenic daughter, ^{53}Cr. These isotopic variations are measured as the deviations of the ^{53}Cr/^{52}Cr ratios from the standard terrestrial ^{53}Cr/^{52}Cr ratio which are usually expressed in ε-units (1 ε is 1 part in 10^4, or 0.01%). Thus, by definition, the standard terrestrial ^{53}Cr/^{52}Cr is $\equiv 0$ ε. Lugmair and Shukolyukov (1998) developed a technique for high precision mass spectrometric analysis of the Cr isotopic composition in rocks and minerals which allows measurement of ^{53}Cr/^{52}Cr variations of less than 1 ε with an uncertainty of 0.05 to 0.10 ε-units.

Because Earth homogenized long after ^{53}Mn had fully decayed, no variation of ^{53}Cr/^{52}Cr ratios is expected for any terrestrial samples. Indeed, all terrestrial samples examined to date exhibit the same ^{53}Cr/^{52}Cr ratio (~ 0 ε) regardless of their origin (Table 1, Fig. 3). In contrast, all meteorite classes studied so far have ^{53}Cr abundances that are clearly different from terrestrial (Fig. 1). The ordinary chondrites have a characteristic ^{53}Cr excess of ~0.48 ε. The Mn-Cr isotope systematics of the angrites, primitive achondrites, pallasites, and the howardite-eucrite-diogenite association are all consistent with a chondritic Mn/Cr ratio of 0.68 and with ^{53}Cr/^{52}Cr ratios of ~0.5 ε in their bulk parent bodies. The ^{53}Cr/^{52}Cr ratios of *individual* eucrites and diogenites show a range which is due to an early planet-wide Mn/Cr fractionation (Lugmair and Shukolyukov, 1998) in their parent bodies. The ^{53}Cr excesses of the Martian (SNC) meteorites (~0.23 ε) and of the EH-chondrites (~0.16 ε) are intermediate between those of Earth and the chondrites. The carbonaceous chondrites Allende (CV3 type) and Orgueil (CI type) show an apparent deficit of ^{53}Cr (Table 1 and Fig. 3). However, we have found that, in contrast to all other studied meteorites classes, the bulk samples of Allende and Orgueil are characterized by an *elevated* ^{54}Cr/^{52}Cr ratio. This is due to the presence in the carbonaceous chondrites of a presolar component which is enriched in ^{54}Cr (Papanastassiou 1986; Podosek et al. 1997). In our method the

$^{54}Cr/^{52}Cr$ ratio is used for a second order fractionation correction (Lugmair and Shukolyukov 1998) and, therefore, an excess of ^{54}Cr would translate into an apparent deficit of ^{53}Cr. (The application of the second order fractionation correction was discussed by Lugmair and Shukolyukov (1998) and will not be repeated here). The measured "raw" $^{53}Cr/^{52}Cr$ and $^{54}Cr/^{52}Cr$ ratios of the bulk Allende (that is, without application of the second order fractionation correction) are +0.1±0.1 ε and +0.9±0.2 ε (Lugmair and Shukolyukov, in preparation). Preliminary data for bulk Orgueil show that the actual $^{53}Cr/^{52}Cr$ ratio is comparable with that for the other undifferentiated asteroid belt meteorites (i.e., a significant excess of ^{53}Cr) and the $^{54}Cr/^{52}Cr$ ratio is elevated up to ~ +1.5 ε. However, the use of the "raw" $^{53}Cr/^{52}Cr$ ratio would drastically decrease the precision, and thus the resolution of our measurements. For this reason we prefer to apply the second order fractionation correction even for the samples with an elevated $^{54}Cr/^{52}Cr$ ratio (but keeping in mind that in reality the bulk carbonaceous chondrites this ratio varies from ε values close to 0 ε – for Allende – to excess ^{53}Cr- for Orgueil). Regardless of the way in which we express the $^{53}Cr/^{52}Cr$ ratio, the Cr-isotopic composition of the carbonaceous chondrites is unique and allows us to distinguish them from the other classes of meteorites.

Earlier work has shown that the observed distribution of radiogenic ^{53}Cr is unlikely to be due to differences in the bulk Mn/Cr ratios of the parent bodies (Lugmair and Shukolyukov 1998; Shukolyukov and Lugmair 1999). This distribution has been interpreted to reflect an original spatial heterogeneity of ^{53}Mn in the early solar system, which is now revealed as a radial gradient in the radiogenic ^{53}Cr abundance. Regardless of the origin of these variations, what is important in the present context, is that the observed difference in $^{53}Cr/^{52}Cr$ ratios between Earth and meteorites represents a direct experimental fact that does not involve any models or assumptions. This allows us to unequivocally demonstrate an extraterrestrial component in geological samples on Earth that contain a significant proportion of meteoritic Cr, based on measurements of the Cr isotopic composition.

This method was recently used to demonstrate that the Cretaceous/Tertiary (K/T) boundary layer contains an abundant ETC (Shukolyukov and Lugmair 1998). The K/T boundary sediments from Stevns Klint, Denmark, and Caravaca, Spain, were found to have a Cr isotopic signature which is very similar to that of the carbonaceous chondrites: from -0.33 to -0.40 ε, while the background clays were found to have a normal Cr isotopic composition (Fig. 3). The "raw" $^{53}Cr/^{52}Cr$ and $^{54}Cr/^{52}Cr$ ratios of the carbonaceous chondrites and the K/T sediments were also found to be similar. These results indicate that more than 80% of Cr in the K/T sediments originated from a carbonaceous chondrite type impactor.

Results

The Cr isotopic composition of samples from spherule bed S4 are unquestionably non-terrestrial (Table 1, Fig. 3). Sample D-4 was found to have a clearly anomalous normalized $^{53}Cr/^{52}Cr$ ratio of -0.32±0.06 ε. The less precise data on sample C are concordant with this result (-0.26±0.11 ε). The background sediments yield normal terrestrial $^{53}Cr/^{52}Cr$ ratios, indistinguishable from laboratory standards. The "raw" $^{53}Cr/^{52}Cr$ and $^{54}Cr/^{52}Cr$ ratios of the bed S4 samples are –0.1±0.2 ε and +0.4±0.3 ε, respectively. Although the $^{54}Cr/^{52}Cr$ ratio seems to be elevated, due to a smaller number of measurements the uncertainty is larger than that for Allende.

We conducted electron microprobe analyses on three samples of bed S4 and found that Cr was detectable only in the chlorite and rutile. The chlorites are rather typical Mg-Fe chlorites (Table 2). Microprobe totals of ~88 wt% are reasonable for chlorites that typically contain ~12% H_2O. In spite of rather significant differences in color, chlorite grain size, and the number and type of other phases present, the chlorites are quite homogenous in composition. Cr_2O_3 concentrations were observed to vary between 0.30 and 0.61 wt% with an average of 0.43 wt% or 2.9 mg/g Cr. In comparison, the average Cr concentration in 15 samples measured by Kyte et al. (1992) was 1.35 mg/g. This works out to about 45% chlorite in the sample, which is consistent with petrographic observations. Cr_2O_3 concentrations in rutile range from 0.05 to 1.58 wt% with an average of 0.50 wt%. Although this is somewhat greater on average than the Cr content of the chlorite, rutile is modally less than 1% of the total rock and nearly all of the Cr in bed S4 must reside in the chlorite. We also looked for the site of Ni in these samples, since any Ni could be extraterrestrial in origin. No measurable amounts were found in trace pyrite, chalcopyrite, or siderite analyzed. The chlorite contained an average of 0.09 wt% NiO. This is more than enough to account for the average of 290 µg/g Ni measured in this specimen (Kyte et al. 1992) so chlorite may be the primary host for Ni in bed S4.

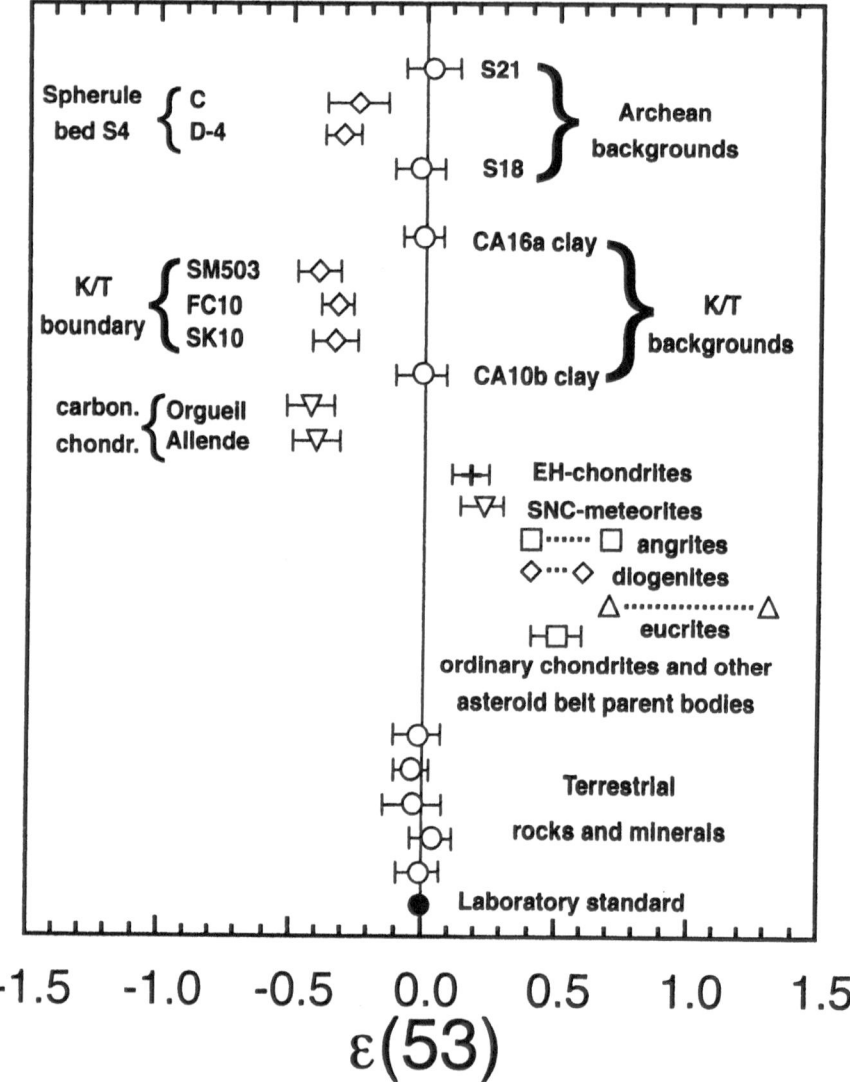

Fig. 3. $^{53}Cr/^{52}Cr$ ratios in all meteorite groups analyzed differ significantly from those in terrestrial samples. Shukolyukov and Lugmair (1998) showed that K/T boundary samples have Cr ratios similar to those in carbonaceous chondrites, while background sediments have isotopic compositions indistinguishable from terrestrial values. In this study we also find that the Cr-isotopic composition of samples from spherule bed S4 are similar to those in carbonaceous chondrites, while background sediments have normal terrestrial compositions. We estimate from these data that ~60% of the Cr in spherule bed S4 is meteoritic. An apparent deficit of ^{53}Cr in the carbonaceous chondrites, in the K/T sediments, and in the Archean spherule bed S4 samples is due to an excess of ^{54}Cr (see text).

Table 2 Chlorite compositions from three samples of the S4 spherule bed. Five points from each of three spherules were analyzed from each sample. The average composition for all analyzed spherules is included at the right.

	179/1	179/2	179/3	306/1	306/2	306/3	349/1	349/2	349/3	avg
SiO_2	24.98	24.94	25.64	26.12	26.12	26.07	25.91	25.68	23.53	25.44
TiO_2	0.05	0.04	0.04	0.06	0.06	0.05	0.04	0.04	0.03	0.05
Al_2O_3	21.86	22.20	21.88	22.50	22.50	21.82	22.37	22.17	21.42	22.08
Cr_2O_3	0.38	0.30	0.34	0.61	0.61	0.30	0.45	0.47	0.53	0.44
FeO	25.77	24.94	23.18	27.47	27.47	27.86	27.29	27.27	28.23	26.61
MnO	0.14	0.07	0.04	0.12	0.12	0.04	0.04	0.04	0.10	0.08
NiO	0.09	0.12	0.14	0.01	0.01	0.08	0.06	0.14	0.12	0.09
MgO	13.87	13.92	14.38	14.31	14.31	14.72	14.61	13.90	13.63	14.18
CaO	0.03	0.03	0.16	0.00	0.00	0.05	0.05	0.06	0.04	0.05
Na_2O	0.00	0.00	0.02	0.00	0.00	0.00	0.00	0.00	0.01	0.00
K_2O	0.03	0.04	0.05	0.03	0.03	0.05	0.03	0.04	0.03	0.04
Total[††]	87.19	86.60	85.85	91.24	91.24	91.04	90.86	89.81	87.68	89.06

[†] All Fe measured as FeO
[††] Total excludes H_2O

Discussion

We consider this to be clear evidence that at least one of the early Archean spherule beds contains a significant ETC and an origin as impact ejecta is its most likely source. The 'second order corrected' isotopic signature of the spherule bed S4 samples is similar to that of the carbonaceous chondrites, and, as such, would imply a carbonaceous chondrite type impactor, as identified for the K/T boundary event. A small excess of ^{54}Cr observed in the "raw" isotopic data is consistent with this interpretation.

The concentration of Cr in chlorites within S4 provides new insights into the host phase(s) of the ETC. Kyte et al. (1992) discovered a few µm-sized nuggets of platinum-group elements (PGEs) within a single grain of pyrite in bed S4. Based on the rarity of these nuggets, it was impossible for them to be the primary host phase for the PGEs, but Kyte et al. (1992) inferred that a significant fraction of the PGEs might be hosted by the trace sulfides. Koeberl et al. (1993) argue that sulfides are an important carrier of the PGEs in their samples and that mineralization has played an important role in their distribution. Because the PGEs cannot be detected by electron microprobe analyses at such low concentrations, these inferences were not directly testable. Our results indicate that in addition to the fact that a large fraction of the Cr is extraterrestrial in origin, it is concentrated in the chlorite in S4, which is the phase that has pseudomorphed the presumed impact-derived spherules. Although we cannot prove it with the current data, it is reasonable to suspect that a significant fraction of the PGEs is also contained in the spherules. This interpretation is supported by the moderate covariance of Ir and Cr concentrations in this sample (Kyte et al. 1992) and the fact that Cr concentrations should vary with spherule content in individual samples.

No Evidence for a Terrestrial Origin

There is no evidence that mineralization or other terrestrial processes played a role in concentrating meteoritic Cr or PGEs in spherule bed S4. S4 shows no sulfide or carbonate mineralization as do the Princeton Mine samples of Koeberl et al. (1993), which commonly contained siderite or pyrite as the most abundant mineral phase. In bed S4, as well as in surrounding sediments, carbonates and sulfides are trace components (<1%). Also lithic sandstones interbedded with S4 contain background Ir concentrations (Kyte et al. 1992) and terrestrial Cr-isotopic compositions, which argues strongly against mineralization as a source of the PGE and Cr anomalies. The pervasive silicification common to virtually all sedimentary rocks, including the Fig Tree spherule beds, apparently took place shortly after deposition (Lowe and Byerly 1986b; Hanor and Duchac 1990), whereas gold mineralization in the Barberton Greenstone Belt occurred between 3126 and 3084 Ma (de Ronde et al. 1991), over 100 Ma after deposition of the Fig Tree spherule beds. The gold mineralization affected a variety of rock types, including the older Onverwacht Group volcanics, Fig Tree and younger Moodies Group sediments. The suggestion that these regionally extensive spherule beds have Ir anomalies related to this gold mineralization (Koeberl et al. 1993) seems implausible.

Throughout the central part of the Barberton Greenstone belt, shearing has been partitioned into less competent strata and volcanic units: silicified layers, such as S4, show few or no shear effects. There is no evidence that S4 has been thickened through shearing, folding or mylonitization as described for Princeton Mine samples (Koeberl and Reimold, 1995).

We also consider it unlikely that sediment reworking is responsible for the high spherule content of bed S4. This layer is interbedded within a sequence of sands and conglomerates deposited by high-energy flood events, probably by streams. The bed itself managed to survive erosion but there was some mixing with finer-grained clastic sediment. However, reworking must have been slight and the distance of transport short because there was so much clastic material available that extensive reworking would have diluted the less abundant spherules and no coherent, recognizable spherule layer would have remained. If surface processes have affected S4, they have acted to dilute rather than concentrate the impact-produced fall debris.

The Extraterrestrial Component

We can only roughly estimate the total fraction of extraterrestrial Cr in bed S4, because of uncertainties in the isotopic analyses, questions with regard to how representative our samples are of the entire bed, and how representative analyses of Allende and Orgueil are of the population of carbonaceous asteroids. It is reasonable to expect that a significant fraction of the total Cr in this sample could be terrestrial, since potential impact target rocks include komatiites and mafic volcanics, which can have Cr concentrations ranging up to 4 mg/g (Lowe et al. 1989), more than twice the average Cr concentration in S4. Even early Archean shales have high Cr concentrations, approaching 1 mg/g (Taylor and McLennan, 1985). The Cr isotopic signature of sample C, which represents a 7 cm section from bed S4, implies ~60% of carbonaceous chondrite Cr in this bed. The "raw" $^{53}Cr/^{52}Cr$ and $^{54}Cr/^{52}Cr$ ratios of the bed S4 samples (-0.1±0.2 ε and +0.4±0.3 ε, respectively) are consistent with those one would expect from ~50% of a terrestrial Cr component mixed with ~50% of Allende type Cr (+0.1±0.1 ε and +0.9±0.2 ε; Lugmair and Shukolyukov, in preparation). On the other hand, a mixture containing ~60% of Orgueil-type Cr with a $^{54}Cr/^{52}Cr$ ratio of ~+1.5 ε would result in a "raw" $^{54}Cr/^{52}Cr$ ratio in bed S4 of ~+0.9 ε, which is not the case. Thus, we tentatively infer that the projectile was more likely composed of a CV-type material rather than of a CI-type material. Using the average Cr concentration of 1.35 mg/g measured in bed S4 by Kyte et al. (1992), the portion of the cosmic Cr in the bed S4 of ~60%, and the Cr concentration in CV chondrites of 3.61 mg/g (Kallemeyn et al. 1981), we estimate an ETC component in bed S4 of ~20%. This value, although rather uncertain, is consistent with more precise ETC calculation based on Ir and the other PGE, as discussed below.

In estimating the ETC in bed S4, we believe the best approach is to take the use the average Ir of 15 samples analyzed by INAA by Kyte et al. (1992). These are the most precise PGE data and their average is the most representative available measure of the meteoritic content of the bed. Extremely high or low concentrations in individual samples may due to heterogeneous distribution of ejecta phases or possible chemical redistribution by post-depositional diagenesis or metasomatism. For this reason, even extremely high concentrations of Ir, Cr, or Au in an individual sample such as those described in bed S2 (Koeberl and Reimold, 1995) cannot be used to argue against a meteoritic origin when the average Ir content of the bed is only a fraction of the concentration in chondrites. By comparison Schmitz (1988) reported a K/T boundary sample with 460 ng/g Ir, Robin et al. (1993) report spherules in K/T boundary sediments with up to 610 ng/g Ir, and Kyte (1998) found a fossil meteorite in K/T boundary sediments with Au concentrations of 213,000 ng/g! These extreme values do not invalidate the impact origin of the K/T boundary layer and they are not useful in estimating the amount of meteoritic material in K/T boundary sediments.

The average Ir concentration in S4 is 116 ng/g (Kyte et al. 1992), about 15% of the value for CV chondrites (Kallemeyn et al. 1981). This is close to the rough ETC estimate of 20% obtained using Cr isotopes. Similar, but slightly lower ETCs would be obtained using Os or Pt concentrations, which correlate strongly with Ir. An ETC of 15% is not extreme, when compared to K/T boundary sediments where ETCs ranging from 5 to 20% have been noted at a number of localities (e.g., Kyte et al. 1985; Strong et al. 1987). We note that the ETC for the K/T boundary at Caravaca Spain is about 30% based on Cr isotopes, if we use the Cr concentration in CV chondrites.

Impact Origin for the Spherule Beds

There are very few well-preserved and widely accepted impact deposits in the geologic record, but virtually all contain spherules. These include microtektites deposits from the Australasian, Ivory Coast and North American strewnfields (e.g., Glass et al. 1979), clinopyroxene-bearing spherules in late Eocene sediments (e.g., Glass et al. 1985), ejecta from the oceanic impact of the Eltanin asteroid (Margolis et al. 1991), and the impact deposits at the K/T boundary. In the case of the K/T boundary, Ir-poor spherules similar to microtektites are found in thick deposits near the Chicxulub crater (Hildebrand et al. 1991) in Haiti and around the Gulf of Mexico (e.g., Smit 1999). Spherules are also ubiquitous in the Ir-rich global fallout layer of the K/T boundary (e.g., Smit and Klaver 1981; Smit and Kyte 1994; Robin et al. 1993; Kyte et al. 1996; Smit 1999).

Overall, the similarity between the global fallout layer at K/T boundary and spherule bed S4 is striking in that both are composed largely of spherules, they contain extraterrestrial Cr derived from a carbonaceous chondrite source, and they have extraterrestrial components of about 5 to 20%. For this reason, we consider it most likely that both are global fallout deposits from major impact events. We are unable to conceive of a reasonable alternative. The greatest difference between S4 and the K/T boundary is in the total amount of extraterrestrial material they contain. Typical thickness of the global fallout layer for the K/T boundary layer is <3 mm with Ir fluences of 50 to 100 ng/cm^2 (e.g., Smit and Romein 1985; Kyte et al. 1996; Smit 1999). Spherule bed S4 is 100 to 150 mm thick. We find no reason to argue that this is not representative of the fallout from this impact event. After decades of analysis of over 100 K/T boundary sites, nothing closely resembling this thickness has been found. An estimate of the Ir fluence in S4 can be made assuming 10 cm thickness, 116 ng/g Ir, and a density of 2.6 g/cm^3. This yields an Ir fluence of 3000 ng/cm^2, or about 30 times the amount typically found in the K/T boundary. There are, of course, large uncertainties in these estimates, but if the K/T projectile was ~10 km in diameter (Alvarez et al. 1980), the S4 projectile must have been in excess of 20 km in diameter and could have been considerably larger. Two other spherule beds, S2 and S3, were deposited less than 30 Ma before bed S4 (Kyte et al. 1992) and each of these appear to contain considerably more ejecta than S4. If further work conclusively determines that these are also distal ejecta deposits, then it may be necessary to consider the possibility that a tail to the period of heavy bombardment may have persisted up to about 3.2 Ga.

Acknowledgements

We thank Chris MacIsaac for his help in the lab. Isotopic analyses were supported by NASA grant 5-4145 to G.W. Lugmair and A. Shukolyukov. Other chemical analyses and field work were supported by NSF grant EAR-9418303 to F.T. Kyte, NASA grant NAGT 9-136 to D.R. Lowe and G.R. Byerly, and NASA grant NCA 2-721 to D.R. Lowe.

References

Alvarez LW, Alvarez W, Asaro F, Michel HV (1980)   Extraterrestrial cause for the Cretaceous–Tertiary extinction. Science, 208: 1095-1108

Birck J-L, Allègre CJ (1985) Evidence for the presence of ^{53}Mn in the early solar system. Geophys Res Lett 12: 745-748

Birck J-L, Allègre CJ (1988) Manganese-chromium isotope systematics and development of the early Solar System. Nature 331: 579-584

Byerly GR, Kroner A, Lowe DR, Todt W, Walsh MM (1996) Prolonged magmatism and time constraints for sediment deposition in the early Archean Barberton greenstone belt: evidence from the Upper Onverwacht and Fig Tree groups. Precambrian Research 78: 125-138

de Ronde CEJ, Kamo S, Davis DW, de Wit MJ, Spooner ETC (1991) Field, geochemical and U-Pb isotopic constraints from hypabyssal felsic intrusions within the Barberton greenstone belt, South Africa: Implications for tectonics and the timing of gold mineralization. Precambrian Research 49: 261-280

Glass BP, Burns CA, Crosbie JR, and DuBois DL (1985) Late Eocene North American microtektites and clinopyroxene bearing spherules. J Geophys Res 90: D175-D196

Glass BP, Swincki MB,, Zwart PA (1979) Australasian, Ivory Coast and North American Strewnfields: Size, mass and correlation with geomagnetic reversals and other earth events. Proc Lunar Planet Sci Conf 10: 2535-2545

Hanor JS, Duchac KC (1990) Isovolumetric silification of early Archean komatiites: geochemical mass balances and constraints on origin. J Geol 98: 863-877

Hildebrand AR, Penfield GT, Kring DA, Pilkington M, Camargo AZ, Jacobsen SB, Boynton WV (1991) Chicxulub crater: A possible Cretaceous/Tertiary boundary impact crater on the Yucatán Peninsula, Mexico. Geology 19: 867-871

Jarosewich E, Nelen JA, Norberg JA (1980) Reference samples for electron microprobe analysis. Geostandards Newsletter 4: 43-47

Kallemeyn GW, Wasson JT (1981) The compositional classification of chondrites--I. The carbonaceous chondrite groups. Geochim Cosmochim Acta 45, 1217-1230

Koeberl C, Reimold WU (1995) Early Archean spherule beds in the Barberton Mountain Land, South Aftica: no evidence for impact origin. Precambrian Research 74: 1-33

Koeberl C, Reimold WU, Boer RH (1993) Geochemistry and mineralogy of early Archean spherule beds, Barberton Mountain Land, South Africa: evidence for origin by impact doubtful. Earth Planet Sci Lett 119: 441-452

Kyte FT (1998) A meteorite from the Cretaceous/Tertiary boundary. Nature 396: 237-239

Kyte FT, Smit J and Wasson JT (1985) Siderophile interelement variations in the Cretaceous- Tertiary boundary sediments from Caravaca, Spain. Earth Planet Sci Lett 73: 183-195

Kyte FT, Zhou L, Lowe DR (1992) Noble metal abundances in an early Archean impact deposit. Geochim Cosmochim Acta 56: 1365-1372

Kyte FT, Bostwick JA, Zhou L (1996) The Cretaceous-Tertiary boundary on the Pacific plate: composition and distribution of impact debris. In: Ryder G, Fastovsky D, Gartner S (eds) The Cretaceous-Tertiary event and other catastrophes in Earth history, Geol Soc Amer Spec Pap 307: 389-401

Lowe DR, Byerly GR (1986a) Early Archean silicate spherules of probable impact origin, South Africa and Western Australia. Geology 14: 83-86

Lowe DR, Byerly GR (1986b) Archean flow-top alteration zones formed initially in a low-temperature sulphate-rich environment. Nature 324: 245-248

Lowe DR, Byerly R, Asaro F, Kyte FT (1989) Geological and geochemical record of 3400-million-year-old terrestrial meteorite impacts. Science 245: 959-962

Lugmair GW, Shukolyukov A (1998) Early solar system timescales according to ^{53}Mn-^{53}Cr systematics. Geochim Cosmochim Acta 62: 2863-2886

Margolis SV, Claeys P, Kyte FT (1991) Microtektites, microkrystites and spinels from a Late Pliocene asteroid impact in the Southern Ocean. Science 251: 1594-1597

Papanastassiou DA (1986) Chromium isotopic anomalies in the Allende meteorite. Astrophys J: L27-L30

Podosek FA, Ott U, Brannon JC, Neal CR, Bernatowicz TJ, Swan P, Mahan SE (1997) Thoroughly anomalous chromium in Orgueil. Meteoritics Planet Sci 32: 617-627

Robin E, Froget L, Jehanno C, Rocchia R (1993) Evidence for a K/T impact event in the Pacific Ocean. Nature 363: 615-617

Schmitz B (1988) Origin of microlayering in worldwide distributed Ir-rich Cretaceous/Tertiary boundary clays. Geology 16: 1068-1072

Shearer CK, Papike JJ, Rietmeijer FJM (1998) The planetary sample suite and environments of origin. In: Papike JJ (ed) Planetary Materials, Reviews in Mineralogy, 36: pp 1-1 to 1-28

Shukolyukov A, Lugmair GW (1998) Isotopic evidence for the Cretaceous-Tertiary impactor and its type. Science 282: 927-929

Shukolyukov A, Lugmair GW (1999) The ^{53}Mn-^{53}Cr isotope systematics of the enstatite chondrites. Lunar Planet Sci 30: #1093

Smit J (1999) The global stratigraphy of the Cretaceous-Tertiary boundary impact ejecta. Ann Rev Earth Planet Sci 27: 75-113

Smit J, Klaver G (1981) Sanidine spherules at the Cretaceous-Tertiary boundary indicate a large impact event. Nature 292: 47-49

Smit J, Kyte FT (1984) Siderophile-rich magnetic spheroids from the Cretaceous-Tertiary boundary in Umbria, Italy. Nature 310, 403-405

Smit J and Romein AJT. (1985) A sequence of events across the Cretaceous-Tertiary boundary. Earth Planet Sci Lett 74: 155-170

Strong CP, Brooks RR, Wilson SM, Reeves RD, Orth CJ, Mao X-Y, Quintana LR, Anders E, (1987) A new Cretaceous-Tertiary boundary site at Flaxbourne River, New Zealand: biostratigraphy and geochemistry. Geochim Cosmochim Acta 51: 2769-2777

Taylor SR (1982) Planetary Science: A Lunar Perspective. Lunar and Planetary Inst, Houston, TX, 481 pp

Taylor SR and McLennan SM (1985) The continental crust: its composition and evolution. Blackwell Scientific Publications, Oxford, 312 pp

6 Early Archean Spherule Beds in the Barberton Mountain Land, South Africa: Impact or Terrestrial Origin?

Wolf Uwe Reimold[1], Christian Koeberl[2], Steven Johnson[1], and Iain McDonald[3]

[1]Impact Cratering Research Group, Department of Geology, University of the Witwatersrand, Private Bag 3, Johannesburg 2050, South Africa.
(065wur@cosmos.wits.ac.za)
[2]Institute of Geochemistry, University of Vienna, Althanstrasse 14, A-1090 Vienna, Austria.
(christian.koeberl@univie.ac.at)
[3]School of Earth & Environmental Sciences, University of Greenwich, Chatham Maritime, Kent ME4 4AW, United Kingdom.

Abstract. The origin of multiple spherule-rich layers of millimeter to meter width, all occurring within the transition from the Fig Tree to the Onverwacht Group of the Barberton Greenstone Belt in South Africa, has been strongly debated during the last decade. One school subscribes to an origin by large meteorite impact, whereas others have preferred terrestrial processes. In particular, strong enrichments in siderophile elements, especially Ir, and chondrite-like PGE patterns for spherule layer samples have been cited as evidence favoring an impact origin. Recently, Cr isotopic signatures obtained for samples from two spherule layers have provided further support for this hypothesis. In contrast, our group has emphasized that secondary hydrothermal processes have pervasively overprinted the whole stratigraphy at this transition. Ir concentrations up to 5 times chondritic are suspect as primary impact-produced signatures.

Here, we report new petrographic and geochemical data for samples from spherulitic horizons marking the S2 layer and from interlayered BIF, chert, and mudstone strata. In contrast to earlier work, the new samples were obtained from outside of the gold-sulfide mineralized ore zone on Agnes Mine. Both spherule and country rock samples are enriched in siderophile elements, with up to >1500 ppb Ir. Some of the highest values are related to clearly secondary fault and shear zone deposits. Chrome-spinel in spherule layers is often zoned. A proton microprobe study identified in one case the mineral gersdorffite, of likely secondary origin, as a carrier phase for Ir, whereas in other samples Ir must be contained in matrix silicates. New PGE analyses for more or less sulfide-mineralized samples yielded uniformly flat, near-chondritic patterns.

In conclusion, whereas Cr isotopic and PGE results seem to favor a meteoritic source for Cr and siderophile elements, the complex petrographic, geochemical and stratigraphic findings require further consideration. Questions such as "why PGE data are near-chondritic, whereas other siderophile interelement relationships are complex and clearly represent secondary signatures", or "if the Ir in these samples is of meteoritic origin, why do concentrations in certain samples exceed 5 times chondritic values" need to be resolved, before it might be possible to consider possible deposition or redistribution processes.

Introduction

Spherule layers in the ~3.4 Ga Barberton Greenstone Belt, South Africa, have been interpreted by Lowe and co-workers (e.g., Lowe and Byerly 1986, 1991, 1992; Lowe et al. 1989; Kyte et al. 1992; Byerly and Lowe 1994) as the result of large asteroid or comet impacts onto the early Earth. The enrichment of siderophile elements, especially the platinum group elements (PGEs), in these layers was taken by these authors as key evidence for the presence of an extraterrestrial component (Lowe et al. 1989), although some fractionations were noted (Kyte et al. 1992). Buick (1987), French (1987), Koeberl et al. (1993), and Koeberl and Reimold (1994, 1995) have challenged the interpretation that these spherule layers are the result of impact processes.

The identification of Precambrian impact deposits (especially distal ejecta) is an important and largely unresolved problem. Unfortunately, so far no definitive criteria for the identification of Archean impact deposits are known. Recently, Cr isotopic analyses of several spherule layer samples from the Barberton area (Shukolyukov et al. 1998, 1999) suggested that carbonaceous chondritic projectiles could have been involved in the formation of these layers. This has been taken, by some, as unequivocal evidence for an origin by impact for these spherule horizons.

While we (Koeberl et al. 1993; Koeberl and Reimold 1994, 1995) have not categorically excluded the possibility of the presence of an extraterrestrial component, we have been of the opinion that the arguments presented by Lowe and co-workers have failed to provide convincing evidence for an impact origin of these spherule beds. The results of Cr isotopic analysis seem to favour meteorite impact processes, but they do not explain, for example, the enormous noble metal concentrations that have been determined for selected spherule layer samples as well as some country rock specimens (in this work, the term *country rock* refers to those strata other than spherule layers, which occur interlayered with spherule beds in the studied stratigraphic intervals).

In order to further investigate the S2 spherule layer and its associated lithologies, but in contrast to earlier work that was carried out on strongly sulfide-mineralized intervals only, several series of drillcore intersections obtained outside of the gold- and sulfide-mineralized ore zone of the Agnes Mine have been studied petrographically and geochemically. This included PGE analysis of more or less mineralized spherule and country rock layers, as well as detailed spinel analysis by proton microprobe.

Barberton Spherule Layers - Background

The Barberton Greenstone Belt is located in the Kaapvaal Craton of southern Africa (Fig. 1). The rocks range in age between about 3.5 and 3.1 Ga (e.g., Kröner et al. 1991; Kamo and Davis 1994) and consist of a lower, predominantly volcanic, sequence, the Onverwacht Group, which is overlain by two mainly sedimentary sequences known as the Fig Tree and Moodies Groups (Fig. 2). Altered quartz porphyries and diabasic lavas that are commonly intercalated with chert layers dominate the strata of the Onverwacht Group. Carbonate and volcaniclastic rocks are also present. More detailed accounts of the geology, stratigraphy, and geochronology of the Barberton Greenstone Belt are given by a variety of authors, including de Wit (1982), Anhaeusser (1986), de Ronde et al. (1991a,b), de Ronde and de Wit (1994), de Wit and Ashwal (1997), and Lowe and Byerly (1999).

Several distinct horizons containing spherules have been found throughout both the Onverwacht and Fig Tree Groups. Some of these extensive spherule beds (with spherules of up to 3 mm diameter), forming layers in the Onverwacht Group were initially interpreted as accretionary lapilli (Lowe and Knauth 1977). Earlier, Ramsay (1963) considered that spherules from within the Onverwacht Group represent silicified marine carbonate ooids. These spherules were re-interpreted by Lowe and Knauth (1978) as accretionary lapilli. In addition, spherule layers were found in association with the Msauli chert, which is an about 20 m thick unit marking the boundary between the Swartkoppie Formation, the uppermost formation of the Onverwacht Group, and the bottom of the Fig Tree Group (e.g., Stanistreet et al. 1981). The Msauli chert contains several spherule-bearing units, mainly occurring in the coarse basal part of the graded unit. These units contain spherules that are usually <3 mm (rarely up to 10 mm) in size, have a concentric appearance with sometimes distinct nuclei, and consist of mostly microquartz (chert) and sericite (e.g., Stanistreet et al. 1981).

As for the other spherule occurrences, several formation hypotheses have been proposed. They have been suggested to represent accretionary lapilli deposited in a shallow water environment (e.g., Lowe and Knauth 1978), or accretionary lapilli that were later redeposited (e.g., Stanistreet et al. 1981). Heinrichs (1984) suggested that they are accretionary lapilli that were not deposited in a subaerial or shallow-water environment - as previously advocated by Lowe and Knauth (1978), whereas Stanistreet et al. (1981) proposed that the spherules formed near subaerial hydrothermal vents.

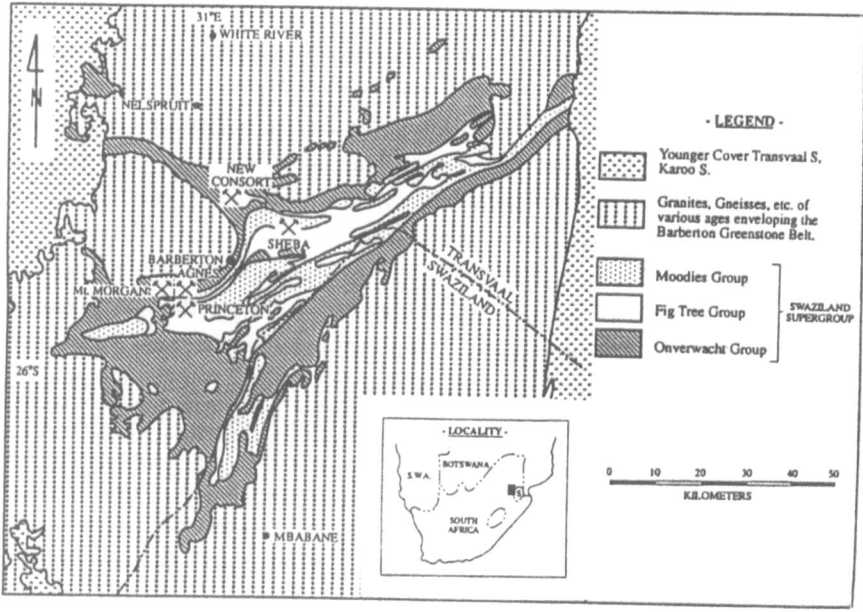

Fig. 1. Geological map of the Barberton Greenstone Belt, Mpumalanga Province, Republic of South Africa. Locations of the Sheba, Mount Morgan and Princeton mines near Barberton are shown (after Anhaeusser, 1986).

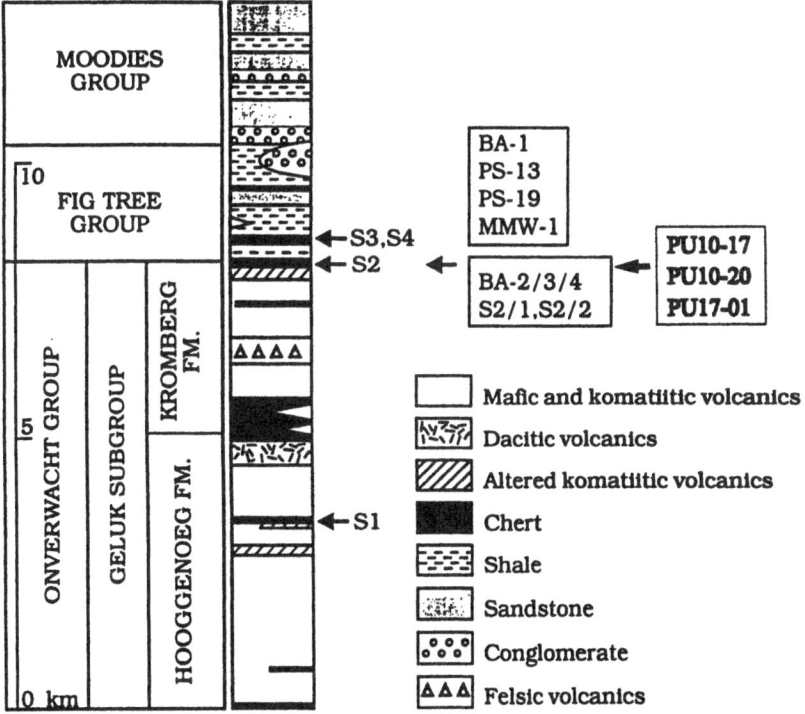

Fig. 2. Stratigraphy of the upper part of the Barberton Greenstone Belt, with locations of the spherule beds S1-S4 after Lowe et al. (1989). Also given are the stratigraphic positions of the samples mentioned in this and previous studies, after Koeberl and Reimold (1995). Samples BA-1, PS-13, PS-19, PU10-17, PU10-20 and PU17-10 originate from the Agnes mine, Princeton section. MMW-1 originates from the Mount Morgan mine and BA-2/3/4 were collected from the Sheba mine. Samples S2/1 and S2/2 are from a surface exposure described by Lowe and Byerly (1991).

Arguments For and Against an Impact Origin

In addition to the spherule layers mentioned above, Lowe and Byerly (1986) described several "unusual" spherule beds from localities near Barberton and in Australia. The spherule layers from the Barberton Greenstone Belt occur near the base of the Fig Tree Group and contain silicate spherules with diameters of 0.1-4 mm, consisting of intergrown microcrystalline quartz and sericite. Thicknesses of the spherule layers described by Lowe et al. (1989) range from a few centimeters to 140 centimeters. Lowe and Byerly (1986) interpreted the texture of some spherules as pseudomorphs of quench-textures, similar to those observed in chondrules. Their petrological observations, and the occurrence of spherules in a massive layer, led to the suggestion that they could be melt droplets that formed during large Archean meteorite impact events (Lowe and Byerly 1986). In contrast, de Wit (1986) concluded that these spherules represent ocelli or variolites derived from weathered volcanics.

Later, Lowe et al. (1989) and Kyte et al. (1992) studied the distribution of the platinum group elements (PGEs) in samples from the Fig Tree spherule layers. They found very high concentrations of iridium and the other PGEs in some of the samples and concluded that these observations supported an impact origin of the spherule beds. In addition to the one spherule layer (termed S2) observed by Lowe and Byerly (1986), Lowe et al. (1989) found three additional spherule beds. The other spherule beds are in the Onverwacht Group (S1) and in the Fig Tree Group (S3 and S4 layers) (Fig. 2). Lowe et al. (1989) and Kyte et al. (1992) did not only find high PGE contents, but also that the PGEs in the spherule bed samples had roughly chondritic interelement ratios. Lowe et al. (1989) observed a good positive correlation between the elements Cr and Ir for samples from several of these spherule beds; they concluded that these elements were either both hosted by the chloritic matrix of their spherule samples or in chrome-rich spinel.

The interpretation of the Barberton spherule layers and related interpretation of petrographic and geochemical evidence as the result of asteroid or comet impact has been questioned by Koeberl et al. (1993) and Koeberl and Reimold (1994, 1995) on the basis of detailed mineralogical and geochemical measurements. These authors studied a large number of spherule bed samples from drill cores and underground exposures. They separated individual layers and units (some only 1 mm thick) from several of the spherule beds and adjacent country rock sequences to determine if there is a correlation between the extremely high PGE abundances and abundances of specific minerals, such as secondary sulfides, as well as a distinct relationship between these effects and spherule layers only. The highest Ir concentrations found by Koeberl et al. (1993) in sub-samples of the spherule layers range up to 2700 ppb. Compared to an average chondritic Ir abundance of about 600 ppb, such a value represents almost five times the chondritic abundance. It cannot be argued that the source could have been an iron meteorite (which would in a few rare cases reduce the meteoritic contribution in the samples to about 100% - still too high to be reasonable!), as in such a case the other siderophile elements would not occur in roughly chondritic proportions. All known impact deposits, especially distal deposits (including those from Acraman [e.g., Gostin et al. 1986; Williams 1986] or from distal K-T boundary deposits [Smit 1999]) usually contain much less than 1%, rarely up to 10%, of a meteoritic contribution - not close to 500%! Distal impact ejecta of late Eocene age contain no more than 0.3 ppb Ir (Pierrard et al. 1998). Thus, the extremely high PGE abundances in the Barberton spherule samples can not represent primary chondritic signatures.

Mineralogical studies by Koeberl et al. (1993) and Koeberl and Reimold (1995) showed that high siderophile element abundances are correlated with the occurrence of secondary pyrite, arsenopyrite, and other sulfides. Therefore, it appeared reasonable to assume that the PGEs were redistributed. It seems likely that, due to different mobilities of the PGEs in hydrothermal processes (e.g., Wallace et al. 1990; Wood 1991; Colodner et al. 1992), the redistribution of the PGEs during formation of secondary sulfides is likely to result in changes of the PGE interelement ratios. Thus, it is unlikely that any interelement ratios (e.g., Pd/Ir) observed in Barberton spherule samples today would be identical to the values of the initial materials before alteration.

Glikson (1994) calculated an average Ir/Au ratio for eight of our samples and compared the result with chondritic values. However, as we noted above, the usage of bulk values is a flawed approach, because the elemental ratios show very wide variations between individual subunits (separate shale or banded iron formation and spherule layers). Siderophile element enrichments are not confined to the spherule layers, but the variation of elemental abundances in shale samples was found to be as large as those of the spherule samples.

Byerly and Lowe (1992, 1994) mentioned the presence of Ni-rich chrome-spinels in the Barberton spherule samples as further evidence for an impact origin. In contrast, Koeberl et al. (1993) and Koeberl and Reimold (1995) pointed out that there are significant differences between the compositions of Ni-rich spinels from known impact deposits (e.g., the K-T boundary – Robin et al. 1992) and those from Barberton. Glikson (1994) suggested that the oxidation state of Cr-spinels, which are found in Barberton samples, and which show an oxidation state that is unlike that of meteoritic or impact-derived spinels, might have been altered during weathering. The oxidation state of the spinels reflects the oxygen fugacity during their formation, and the terrestrial atmosphere in the Archean is commonly thought to have been nearly devoid of oxygen (e.g., Holland, 1998); although Ohmoto (1998) argues for similar oxygen levels in the Archean as in the present-day atmosphere. Accepting the low oxygen fugacity model for the Archean terrestrial atmosphere, it can certainly be expected that the oxidation state of the Barberton spinels, if produced during an impact event in the atmosphere, will be lower than those of K-T boundary spinels, for example.

However, there are several other features of the Barberton spinels that set them apart from any spinels produced during an impact event (or any meteoritic spinels). First, their oxidation state varies over a wide range, while meteoritic and impact spinels have very well defined oxidation states that show much less variation. For example, Ni-rich spinels from the late Eocene impact horizon in Massignano (Italy) have a restricted range of 68 to 90 atom% Fe^{3+}/Fe_{tot} (Pierrard et al. 1998). Second, the chemical composition of the Barberton spinels is different from the known impact-derived spinels: the major cations in the K-T boundary spinels are Mg, Al, Fe, and Ni, with NiO contents between about 0.2 and 10 wt%, and Cr_2O_3 contents between <0.1 and 1.5 wt% (e.g., Smit and Kyte 1984; Bohor et al. 1986; Kyte and Smit 1986); in contrast, the major cations in the Barberton spinels are Fe, Cr, and Ni, with most NiO contents between about 10 and 15 wt%, and Cr_2O_3 between about 34 and 57 wt%. Also, Reimold et al. (1998) demonstrated that chrome-spinel from S2 layer samples is often strongly, multiply zoned. It was also shown that in these samples Ir is not exclusively related to these complex chrome-spinels, but is hosted by a number of different phases.

The mineralogical and geochemical arguments of Koeberl et al. (1993) and Koeberl and Reimold (1995) can, thus, be listed as follows: a) if the high siderophile element abundances in the spherule beds were derived from meteoritic sources, a correlation between the abundances of these elements (e.g., between Ni and Ir) would be expected; this has only been shown for some bulk samples, to a certain degree, but not for thin (and mineralogically distinct) layers of spherule bed samples; b) in contrast, there is a positive correlation between siderophile element abundances and the contents of chalcophile elements, such as S, As, Sb, and Se; c) there is no difference in the content of Ir and other siderophile elements between spherule layers and adjacent country rock layers (on a mm to cm scale); d) no such massive spherule beds with extremely high siderophile element abundances are associated with confirmed impact craters; e) the siderophile element contents are higher by several orders of magnitude compared to any known proximal or distal impact ejecta; and f) Ni-rich chrome-spinels in Barberton spherule samples have compositions that are distinctly different from those found in known impact ejecta and meteorites.

Glikson (1994) compared the Barberton spherule layers with the Bunyeroo ejecta in Australia (Gostin et al. 1986), which were derived from the Precambrian Acraman impact structure (Williams 1986), and noted that the Acraman ejecta contain fragments of shatter cones, shock metamorphosed minerals, tektite-like impact spherules, and siderophile element anomalies. Koeberl and Reimold (1994) pointed out that there are several key differences between the Acraman ejecta and the Barberton spherule beds. Most notable is a total lack of any evidence of shock metamorphism in the Barberton spherule layers. Shock-deformed minerals (which can fairly easily be told apart from tectonically deformed minerals) are the commonly accepted definitive criterion for an origin by impact (e.g., Stöffler 1972; Stöffler and Langenhorst 1994).

The lack of any shocked minerals in samples from the Barberton spherule layers was already noted by French (1986) and Buick (1986), who took this observation as evidence against an impact origin. We examined a large number of samples and did not find even circumstantial evidence of shock effects (Koeberl et al. 1993; Koeberl and Reimold 1995; this work), which is in agreement with Lowe and co-workers, who also did not find any shocked minerals. In contrast, even in 2-Ga-old impact structures, such as Vredefort and Sudbury, abundant evidence of shock metamorphism has been found (see, e.g., Leroux et al. 1994; Kamo et al. 1996; Reimold and Gibson 1996).

Most recently, the impact hypothesis received a major boost when Cr isotopic analysis (Shukolyukov et al. 1998, 1999) indicated that this element in the Barberton spherule samples could be derived from a CV chondritic source.

Geological Setting of the Sample Localities

To compare the data of Koeberl et al. (1993), Koeberl and Reimold (1995), and the present study with data of Lowe and Byerly (1986) and Lowe et al. (1989), it is important to analyze the same type of spherule layers. The sample locality descriptions of Lowe and Byerly (1986), Lowe et al. (1989), and Kyte et al. (1992) are not detailed enough to allow finding their sample locations in the field. Lowe and Byerly (1986) only noted that they "have identified silicate spherules in a regionally extensive, 30-cm- to 2-m-thick bed near the base of the largely sedimentary Fig Tree Group". Lowe et al. (1989) report that "the lowest of the Fig Tree beds, S2, ranges from a few centimeters to over 1 m thick, occurs in the basal 1 to 3 m of Fig Tree strata, and is present across virtually the entire greenstone belt." Kyte et al. (1992) described a single occurrence of another band, the S4 layer. However, in none of the three publications the authors give an exact geographical location, or an exact tectono-stratigraphic position, for their sampling sites. Furthermore, in their stratigraphic columns, Kyte et al. (1992) place the combined S3 and S4 layers in a stratigraphic position that is at least 0.5 km above that of Lowe et al. (1989). However, we have recently been appraised that this placement was caused by a drafting error (F. Kyte, pers. commun. 1999).

Most recently, Shukolyukov et al. (1999), in their presentation of Cr isotopic measurements, reported that their two samples were derived from a spherule bed in Sheba Mine, which previously had been designated as S2, but these authors now give the origin as "...probably S3". Besides the lack of precise sample localities, there seems to be some problem with the exact stratigraphic placement of the S2-4 spherule beds. In the following section we, therefore, provide detailed descriptions of localities and tectono-stratigraphic placement of our samples.

Drill core specimens were collected from Princeton and Mount Morgan mines (Pepper and Du Plessis 1991) for the studies of Koeberl et al. (1993) and Koeberl and Reimold (1995). In addition, hand specimens were obtained from underground exposures in Princeton and Sheba gold mines, all near Barberton (Fig. 1; also Anhaeusser, 1986). In addition, two samples from the type surface exposure described by Lowe and Byerly (1991) were analyzed in order to compare specimens from heavily sulfide- (and gold-) mineralized settings in gold mines with a sample from a less mineralized locality. Figure 2 shows the stratigraphic positions of our samples that represent either the S2 (BA-2,3,4 from Sheba mine, and the surface sample S2/1) or S3 or S4 (BA-1, PS-13, and PS-19 from 1053 level, and depths of 307.20 m and 490.30 m in Princeton, respectively, and drill cores MMW-1 and MMW1-D1 [deflection] from 187.90 and 183.50 m depths in Mt. Morgan mine) spherule beds. According to the regional geological discussion in Koeberl and Reimold (1995), the Sheba, Mt. Morgan and Princeton samples most likely represent the S2 layer after Lowe et al. (1989) at the contact between Fig Tree and Moodies Groups.

Several sulfide mineral separates or massive sulfide bands from reef intersecting drill cores PS-13, PS-19 and MMW1-D1, from 282.50, 462.25 and 175.71 m depths, respectively, were analyzed in order to allow study of the chemical characteristics of the regional mineralization in comparison with the chemical signatures of spherule beds sampled meters to tens of meters from the reef intersections. Furthermore, sulfide- and chromite-enriched separates from a number of spherule bed samples were analyzed, as it was of further interest to identify the host phase of PGEs and other siderophile elements enriched in the spherule layers.

Recently, one of us (SJ) collected 45 additional samples for analysis. Three spherule layer-intersecting drill cores (PU10-17, PU10-20 and PU17-01) from the Princeton section of Agnes Mine were sampled. These specimens are collectively referred to as the PU-series. A further sub-set of 17 samples from the BA-2 series, as described earlier (Koeberl et al. 1993; Koeberl and Reimold 1994, 1995), was also analyzed. The detailed logs for these drillcore sections are shown in Figure 3.

All new samples are derived from the so-called S2 bed (Stanistreet et al. 1981; Lowe and Byerly 1986; Lowe et al. 1989; Kyte et al. 1992; Koeberl et al. 1993; Koeberl and Reimold 1994, 1995) at the immediate contact between the Fig Tree and Onverwacht Groups. Previous work by our group was carried out on samples from within the gold-sulfide mineralised zone of the Agnes Gold Mine, as well as several surface samples from the Powerline Locality many kilometers south of the mining district near Barberton town. The new drillcores containing spherule layers were taken at localities removed by at least 25 and up to 40 meters from the major gold mineralisation of Princeton section of the gold mine (25°49'50"S/30°57'E). The three PU boreholes were drilled from north to south, from the surface, through Moodies sediments to intersect the Fig Tree-Onverwacht contact. The boreholes form roughly a straight line, ca. 150 m in length, along the NE-SW strike of the Princeton mining zone (PS19N reef). Samples were chosen to demonstrate that at the expected stratigraphic level, spherule beds are not uniform (e.g., with regard to thickness), and that the stratigraphy in their direct environs is also not uniform, whereas the K/T boundary ejecta layer can be pinpointed over distances of thousands of kilometers (e.g., Smit 1999). The BA-2 borehole core originates from Sheba Gold Mine (25°52'10"S/31°09'25"E).

A number of sulfide-rich hand specimens were specifically taken for PGE analyses. Of these 13 samples, 7 were collected in the gold-sulphide mineralized mining areas of the Princeton section, Agnes mine, and are denoted as the Barb-5X series. The other five samples came from various locations: the PU-10 sample is from the PU10-17 borehole core, the SJ03 and 04 samples are from a less-mineralized area of the Princeton section, and the BA-1A,B and BA-2 samples are from a non-mineralized area of Sheba Mine. These samples comprise both spherule layer (BA/1A and BA-1B) and country rock samples (Barb-51 to 57, PU-10), as well as specimens which could not be separated into distinct lithologies due to the analytical requirement for large sample sizes (i.e., samples SJ03 and 04, as well as BA-2 comprise combinations of spherule and country rock material). As for the INAA samples, the PGE sample suite was subdivided into strongly (i.e., >5 vol% sulfide: Barb-51 to 57) and less mineralized (< 5 vol%) groups.

Analytical Methods

All samples were studied by transmitted and reflected light microscopy. The 45 new samples, 28 samples from the PU series and 17 samples of the BA-2 series, were thin sectioned for petrographic description, scanning electron microscopy (SEM) at the Museum of Natural History, Vienna, electron microprobe analysis (EMPA) at the Rand Afrikaans University, Johannesburg, and proton-induced X-ray emission (PIXE) analysis at the National Accelerator Centre, Faure, South Africa. X-ray fluorescence spectrometry was done at the Department of Geology, University of the Witwatersrand, and indicated which samples warranted further

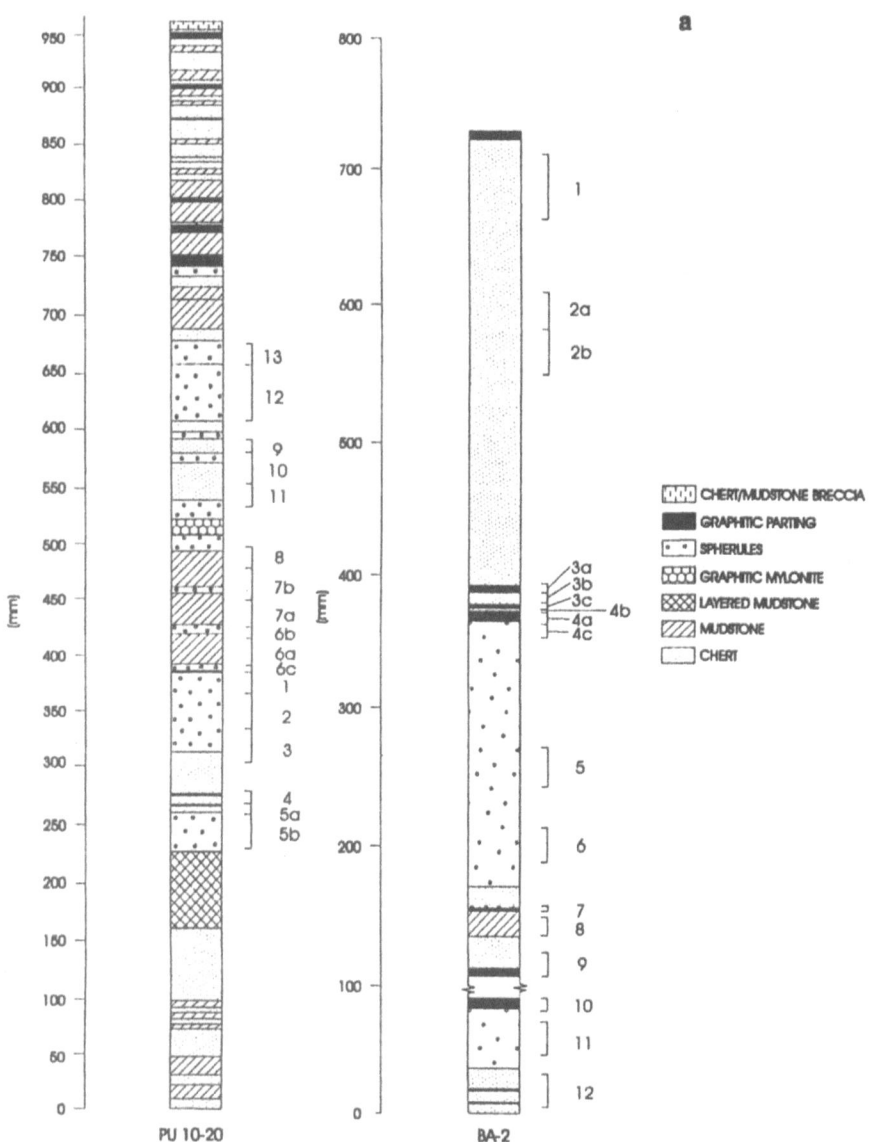

Fig. 3. (above and opposite) Schematic stratigraphic columns for the drillcore sections PU10-20/BA-2 (a) and PU10-17/PU-17-01 (b) sampled for this study. Sample positions are indicated. Vertical scale in millimeters.

b

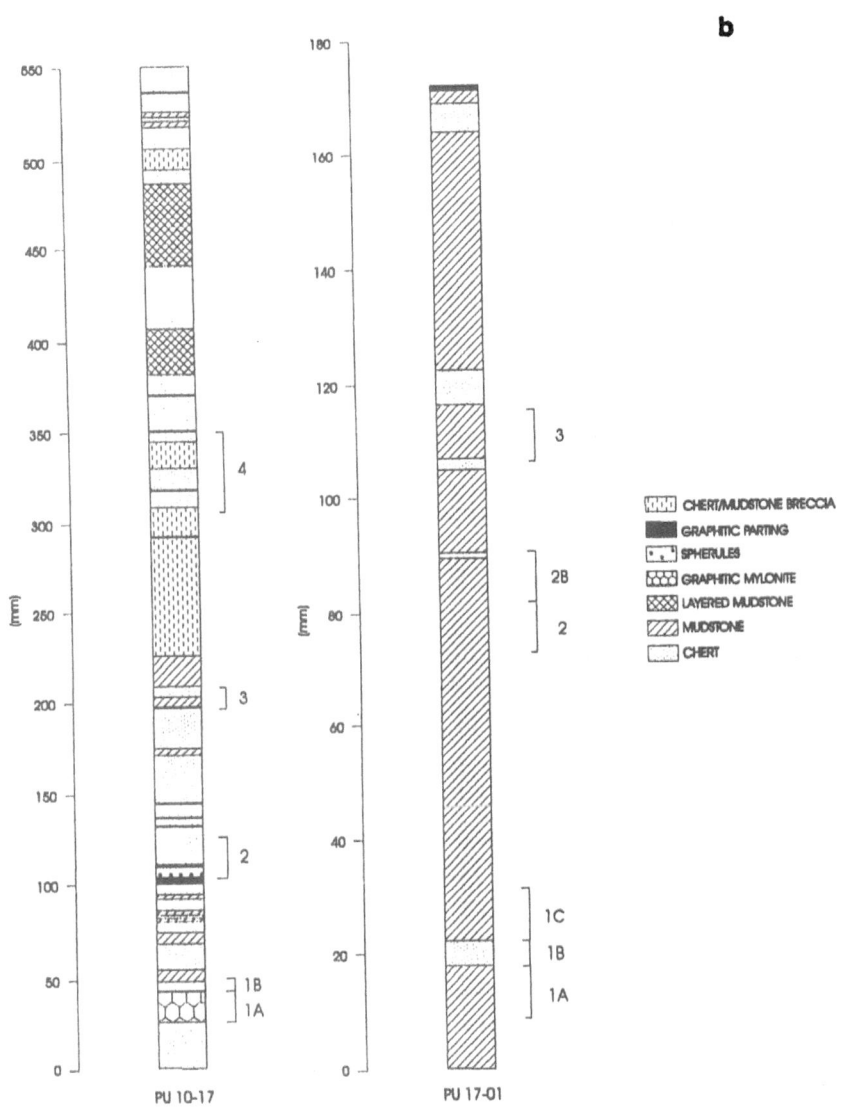

study by, for example, ICP-MS at the Bundesversuchs- und Forschungsanstalt Arsenal-Geotechnik, Vienna. To analyse for elements of extremely low concentrations (ppm-ppb), instrumental neutron activation analysis at the Institute for Geochemistry, University of Vienna, Austria was employed, following procedures described by Koeberl (1993).

Scanning electron microscopy (SEM) was carried out at the Natural History Museum in Vienna, Austria, using a JEOL-6400 SEM and a Kevex energy dispersive X-ray analysis system for mineral identification. By using the PIXE technique at the National Accelerator Centre in the Western Cape, South Africa, a closer examination of interesting areas indicated by SEM was undertaken, especially focusing on the identification of the carrier phase(s) of siderophile elements and iridium.

PIXE elemental maps and point analyses were made using the NAC nuclear microprobe (Churms et al. 1993; Prozesky et al. 1995). A focused 3 MeV proton beam was used - rastered over the desired areas or used for point analyses. The beam current ranged from 0.5 to 2 nA to keep the count rate at about 2000 c/s, with the beam diameter varying between 3 and 5 µm. X-ray energy spectra were registered using a Link Pentafet Si(Li) detector positioned at 135° to the incident beam. A 150 µm Al filter was used to shield the detector from backscattered protons and to reduce peak pile-up effects. The average total accumulated charge for point analyses was between 3 and 5 µC, with the total accumulated charge for the mapped areas ranging from 10 to 30 µC. The PIXE spectra produced were analysed using the GeoPIXE suite of programs (Ryan et al. 1990), using sulphide mineral analyses obtained from EMP measurements for matrix corrections. The rapid matrix transform method of Dynamic Analysis (DA) was used for on-line production of true elemental image maps (Ryan and Jamieson 1993; Ryan et al. 1995; Ryan et al. 1996; IDL User's Manual 1993). The scanned areas were divided into 64 by 64 pixels, with the map data being presented as contours linking areas of equal concentrations. The usual approach when using the DA technique is to build the matrix transform from a fit to a reference spectrum. A brief preliminary scan of a region of interest, or a similar one, is sufficient for this purpose. The distribution of all identified elements can later be studied. This approach was impossible for Ir, as its presence was not found prior to the mapping experiment. Therefore, a reference spectrum had to be constructed by adding a spectrum of pure Ir to a typical spectrum from the polished thin section. Selected spectra were, in addition, processed with the GUPIX software (Maxwell et al. 1989).

Noble metal concentrations were determined using a Ni sulfide fire-assay technique with Te co-precipitation and ICP-MS procedure described by, for example, Koeberl et al. (2000). Method detection and quantitation limits (3 and 10 standard deviations of the blank, respectively) are given in Table 3. Analytical accuracy (Table 3) was determined by analysis of the certified reference materials Peridotite-WPR1 (Govindaraju, 1994) and Komatiite-Wits1 (Tredoux and McDonald 1996; Tredoux et al. in prep.). The high Au concentration in the second aliquot of WPR-1 is believed to reflect contamination of that aliquot by traces of one of the very Au-rich samples, but – as can be seen from Table 3 – PGE concentrations are not significantly affected. Unknown samples were run in duplicate, and the concentrations reported represent the means of two corresponding aliquot values. The reported error is the standard deviation (McDonald 1998). Noble metal data for spherule layer and other samples are given in Table 4.

New Results

Stratigraphy

In contrast to the observations of Koeberl et al. (1993) and Koeberl and Reimold (1994, 1995), the latest drill core examination revealed unambiguous evidence that, at least, some spherule layers have been tectonically duplicated on a centimeter to decimeter scale (Fig. 4a). The country rocks, particularly the more competent chert layers, exhibit strong fracturing and, in part, are brecciated (Fig. 4b). Many less competent mudstone layers have experienced plastic deformation around more competent chert fragments (Fig. 4c). Extensive veining is seen in all parts of the drill cores. Both bedding parallel and cross-cutting veins were observed and represent precipitation from hydrothermal fluids, which extensively altered and/or replaced the original mineral assemblage(s) (Figs. 4d and e). These veins may be filled with quartz and/or magnesio-siderite. Based on the cross-cutting relationships observed, at least three generations of hydrothermal phases can be inferred. The spherules themselves are generally less than 2 millimeters in diameter, but, rarely, they may reach up to 3 millimeters in size. Several of the larger spherules exhibit compositional zonation (Fig. 4f). The spherule beds range in character from single spherule bands (Fig. 4f), or single spherules located at the contact between two contrasting country rock types, to beds of almost 100% spherules, which – in our sample suites - can obtain up to 10 centimeter thickness. The single spherule layers are sometimes associated with a graphitic parting along minor shears (Fig. 4f). The single spherules may be slightly deformed (extended) or may not show any sign of deformation. We have not been able to detect movement indicators along such single-spherule horizons, which could be assumed to indicate lateral tectonic transport of a single spherule. When a single spherule is oval, its long axis is usually parallel to the layering.

Microscopic Observations

Pervasive hydrothermal alteration of all observed lithologies (Fig. 5a) makes determination of the original mineralogical compositions impossible. Not only is the matrix to the spherule layers completely converted to secondary mineral assemblages; but, in addition, multiple generations of quartz and magnesio-siderite veins exist - as determined by their cross-cutting relationships (Figs. 5b-d). As mentioned, spherule beds may also be plastically deformed around brecciated chert clasts (Fig. 4e). In this case, the spherules are often matrix-supported. Some massive spherule beds, composed almost entirely of spherules, exhibit more or less pronounced grading from larger to smaller spherule sizes (Fig. 4a) – indicating a sorting process resulting in differential settling of specific and continuous grain size fractions.

Tectonic deformation is extreme in some cases. Graphite-filled micro-shears or graphite-coated fracture planes, mylonitization, and flattening of spherules due to compression are common (Figs. 6a-b). In other, apparently less deformed samples, more competent mineralogical compositions or spherules located in pressure shadows allowed some spherules to escape deformation (Fig. 6c).

As previously reported from sampling of the gold mineralized zones in Agnes and Sheba Mines, as well as from the Powerline surface exposure, the pervasive hydrothermal alteration of the new samples also includes sulfide mineralization. This occurs mainly, but not exclusively, in the matrix material of spherule layers (Figs. 7a-b). Sulfides are often precipitated around the rims or in the center of spherules; in the latter case it is sometimes obvious that sulfide totally replaces the original nucleus phase (Figs. 7c-d). These nuclei are always destroyed by the hydrothermal alteration, but it is often still possible to see quartz or siderite pseudomorphs in the centers of spherules (Figs. 8a-b). Spherules may also be zoned, either due to original formation processes, such as accretion of material onto smaller spherules, or by secondary mineral growth around their outer rims. Chromite often shows dendritic texture; however, in some cases, zoning of chrome-spinel is seen under the microscope (Figs. 8c-d). The zoning of spinels is also clearly observed at the SEM scale (Fig. 9d).

SEM observations also showed the deposition of very fine-grained sulfides in the matrix, with larger sulfide grains surrounding or growing inside the spherules

Table 1. Electron microprobe spot analyses on a 100 μm chrome-spinel grain from the PU 10-17-2 sample. Points 1 to 8 represent a radial traverse, from rim to core, of approximately equidistant spots.

Weight (%)	1	2	3	4	5	6	7	8
Al_2O_3	16.80	7.80	3.44	0.04	0.06	0.09	0.11	0.81
TiO_2	0.60	0.90	1.00	0.10	0.20	0.10	b.d.l.	0.10
Cr_2O_3	35.70	43.70	46.70	50.30	52.00	51.70	46.60	43.10
MnO	b.d.l.	0.20	0.20	0.10	0.10	0.20	0.10	0.20
MgO	0.10	0.04	0.14	0.22	0.31	0.41	0.31	0.48
Fe_2O_3	25.80	30.70	31.60	35.90	36.50	35.80	37.60	41.90
CoO	0.10	0.10	b.d.l.	0.60	0.70	0.70	0.70	0.60
NiO	0.30	0.70	0.80	8.90	9.40	10.20	11.10	12.10
CuO	b.d.l.	0.10	b.d.l.	b.d.l.	b.d.l.	b.d.l.	0.30	b.d.l.
ZnO	13.10	10.50	7.30	0.60	0.30	0.10	0.30	0.10
V_2O_5	2.30	3.00	2.30	1.40	1.60	1.00	2.40	2.10
Total	94.80	97.74	93.48	98.16	101.17	100.30	99.52	101.49

Table 2 a-c. (on the following 6 pages) Combined analytical results from INAA, XRF, and ICP-MS analysis. Samples PU10-20-5B and BA2-11 did not undergo INAA analysis; the BA2 series of samples did not undergo XRF trace element analysis. Elements which were not determined are marked n.d. and data below detection limits are marked b.d. Abbreviations as follows: Sph-Spherules, Ms-Mudstone, S-Sulphides, Gr-Graphite, shrd - sheared, Mylon Gr-Mylonitic Graphite. Samples marked Sph are >80 % spherules with mudstone matrix (if any). Major element data in wt%, trace element data in ppm, Au and Ir data in ppb.

Table 2a. Combined analytical results from INAA, XRF, and ICP-MS analysis.

Sample	BA2-1	BA2-2a	BA2-2b	BA2-3a	BA2-3b	BA2-3c	BA2-4a	BA2-4b
Type	Chert	Chert	Qtzite	Mylon Gr and Sph	Shrd BIF and Mylon Gr	Shrd BIF	Gr and S rich Sph	Chert
SiO_2	86.4	82.2	50.9	57.1	24.3	86.7	41.8	80.9
TiO_2	0.05	0.06	0.52	0.40	1.16	0.14	0.74	0.02
Al_2O_3	0.93	1.89	17.5	7.13	15	2.54	19.4	0.41
Fe_2O_3	6.73	8.05	12.7	11.7	31.2	5.41	22.4	1.64
MnO	0.14	0.09	0.08	0.08	0.10	0.06	0.1	0.03
MgO	0.97	0.96	2.03	1.59	2.89	0.71	3.26	0.26
CaO	0.54	0.19	0.33	0.28	0.58	0.24	0.51	0.40
Na_2O	0.04	0.08	0.51	0.29	0.31	0.09	0.40	0.56
K_2O	0.25	0.43	4.38	1.75	3.86	0.62	4.49	0.09
P_2O_5	0.03	0.02	0.13	0.07	0.34	0.12	0.26	0.004
LOI	4.49	4.92	9.87	18.9	18.9	3.66	7.06	15.3
Total	100.569	98.89	98.95	99.3	98.64	100.29	100.43	99.61
Li	2	4	23	10	16	4	17	21
Be	0.40	0.60	4.90	2.60	4.00	0.80	5.90	5.50
Sc	2.12	1.56	4.75	15.9	31.3	3.67	54.7	28.7
V	15	19	49	210	1302	73	1846	169
Cr	60.3	55.8	35.9	854	5196	125	8531	185
Co	102	82.0	15.0	48.0	495	54.0	76.0	29.0
Ni	43.7	13.0	68.9	312	2968	79.5	2807	100
Cu	n.d.	n.d.	n.d.	n.d.	n.d.	n.d.	n.d.	n.d.
Zn	28.9	25.8	42.4	36.8	b.d.l.	18.9	b.d.l.	119
Ga	2	3	21	27	63	9	85	16
As	72.5	46.2	50.1	932	6000	334	5056	323
Se	0.17	0.26	3.00	4.10	63.1	2.26	14.4	1.55
Br	n.d.	n.d.	n.d.	n.d.	n.d.	n.d.	n.d.	n.d.
Rb	7.00	16.0	160	159	159	21.00	222	144
Sr	n.d.	n.d.	n.d.	n.d.	n.d.	n.d.	n.d.	n.d.
Y	n.d.	n.d.	n.d.	n.d.	n.d.	n.d.	n.d.	n.d.
Zr	n.d.	n.d.	n.d.	n.d.	n.d.	n.d.	n.d.	n.d.
Nb	n.d.	n.d.	n.d.	n.d.	n.d.	n.d.	n.d.	n.d.
Mo	0.60	<0.5	<0.5	2.20	10.30	0.80	2.50	<0.5
Cd	0.10	<0.1	0.20	<0.1	0.10	<0.1	0.10	<0.1
Sb	6.90	8.12	7.36	53.9	461	27.10	348	10.6
Cs	0.81	1.69	12.2	6.36	11.8	2.05	18.0	9.59
Ba	24	48	484	212	456	66	616	404
La	5.93	10.3	49.3	2.76	11.8	0.85	37.6	8.95
Ce	9.83	14.5	103	5.88	31.6	0.63	76.7	22.5
Nd	5.75	4.67	33.2	4.13	12.6	4.55	44.1	9.79
Sm	1.07	0.69	5.43	0.94	4.81	0.24	11.2	2.02
Eu	0.37	0.19	1.86	0.30	1.23	0.13	4.01	0.70
Gd	0.83	0.53	4.16	1.04	3.90	0.97	8.99	1.89
Tb	0.08	0.11	0.61	0.16	0.98	<0.32	1.47	0.28
Tm	0.06	<0.12	0.19	0.06	0.30	<0.078	0.29	0.10
Yb	0.23	0.17	0.69	0.39	2.46	b.d.l.	1.76	0.74
Lu	0.03	0.02	0.09	0.05	0.12	0.01	0.22	0.16
Hf	0.30	0.42	4.97	2.18	1.74	0.54	2.33	2.00
Ta	<0.42	0.20	0.87	0.23	<0.4	0.13	0.49	0.46
W	n.d.	n.d.	n.d.	n.d.	n.d.	n.d.	n.d.	n.d.
Os	b.d.l.	b.d.l.	b.d.l.	b.d.l.	205	b.d.l.	b.d.l.	b.d.l.
Ir	1.06	7.08	4.09	42.1	1051	12.3	1518	3.16
Au	11.8	8.70	<9.13	25.5	355	13.40	76.3	2.01
Tl	0.10	0.20	2.90	1.30	5.30	0.40	4.30	2.80
Pb	7	16	11	22	95	27	18	6
Bi	<0.5	<0.5	<0.5	<0.5	0.70	<0.5	<0.5	<0.5
Th	0.54	1.62	14.1	2.2	0.67	0.16	0.69	1.41
U	<1.75	<1.83	1.56	1.10	2.22	<0.95	2.07	<1.43

Table 2a (cont.).

BA2-4c S poor Sph	BA2-5 Sph	BA2-6 Sph	BA2-7 Sph on Gr shear	BA2-8 Shear & Sph layer	BA2-9 Bif on Sph contact	BA2-10 Gr fault plane	BA2-11 Sph Layer	BA2-12 Chert with Gr mylon
43.3	26.3	55.1	51.7	38.3	n.d.	23.6	n.d.	91
0.54	1.19	1.36	1.79	0.67	n.d.	0.28	n.d.	0.05
18.6	16.4	22.1	25.8	17.8	n.d.	10.43	n.d.	1.23
25.6	28.4	6.71	5.27	20.3	n.d.	32.6	n.d.	3.94
0.14	0.79	0.03	0.02	0.10	n.d.	0.74	n.d.	0.07
3.80	2.95	1.29	1.08	2.94	n.d.	5.01	n.d.	0.51
0.46	0.59	0.16	0.10	0.47	n.d.	1.10	n.d.	0.26
0.02	0.30	0.35	0.42	0.48	0.04	0.21	0.36	0.03
4.32	4.17	6.39	6.88	4.04	n.d.	2.66	n.d.	0.31
0.22	0.37	0.05	0.002	0.24	n.d.	0.06	n.d.	0.03
1.64	17.6	6.34	6.12	14.2	n.d.	22.6	n.d.	2.7
98.64	99.07	99.88	99.19	99.54		99.29		100.13
2	16	19	23	22	3	13	13	2
0.30	4.90	6.10	7.10	5.80	0.60	3.50	4.60	0.30
0.83	36.0	46.6	46.4	17.4	2.31	19.0	n.d.	1.25
17	1297	369	436	140	34	187	340	20
48.2	4364	410	347	136	110	2052	n.d.	33.1
47.0	391	12.0	9.0	21.0	93.0	69.0	15.0	81.0
15.7	1903	38.3	74.9	73.6	23.1	386	n.d.	70.0
n.d.	n.d.	n.d.	n.d.	n.d.	n.d.	n.d.	n.d.	n.d.
27.6	528	64.1	19.0	63.8	50.1	84.0	n.d.	58.0
2	70	22	24	17	3	17	15	2
39.1	4132	181	122	250	156.0	1317	n.d.	64.1
0.45	48.5	1.14	1.58	1.74	1.24	4.42	n.d.	0.27
n.d.	n.d.	n.d.	n.d.	n.d.	n.d.	n.d.	n.d.	n.d.
5.00	175	238	268	153	13	109	196	11.0
n.d.	n.d.	n.d.	n.d.	n.d.	n.d.	n.d.	n.d.	n.d.
n.d.	n.d.	n.d.	n.d.	n.d.	n.d	n.d.	n.d.	n.d.
n.d.	n.d.	n.d.	n.d.	n.d.	n.d.	n.d.	n.d.	n.d.
n.d.	n.d.	n.d.	n.d.	n.d.	n.d.	n.d.	n.d.	n.d.
<0.5	8.40	<0.5	<0.5	1.00	0.50	6.90	<0.5	<0.5
0.10	0.10	0.10	<0.1	0.20	<0.1	0.10	0.20	0.20
6.08	332	9.15	6.22	7.56	11.9	49.8	n.d.	4.67
0.25	11.8	10.8	11.3	7.40	0.99	6.09	n.d.	0.63
18	488	644	801	423	39	313	540	32
0.53	28.6	5.08	4.47	16.8	1.68	2.57	n.d.	1.22
0.21	50.2	12.2	9.54	39.8	1.42	5.21	n.d.	2.70
b.d.l.	22.4	6.34	4.13	15.7	0.91	3.39	n.d.	0.62
0.22	6.03	1.24	0.69	3.33	0.25	1.31	n.d.	0.21
0.11	2.32	0.33	0.18	1.09	0.13	0.39	n.d.	0.08
0.35	5.81	0.84	0.27	1.95	1.10	<4.81	n.d.	0.21
0.04	1.08	0.13	0.13	0.65	0.43	0.23	n.d.	0.03
<0.067	0.35	<0.17	0.13	0.19	0.05	<0.39	n.d.	0.02
0.14	1.70	0.36	0.38	0.93	0.46	1.05	n.d.	0.10
0.01	b.d.l.	0.05	0.06	0.13	0.00	0.16	n.d.	0.02
0.08	1.11	2.43	2.60	2.99	0.15	1.09	n.d.	0.08
0.12	0.18	0.31	0.19	0.34	0.07	0.67	n.d.	0.05
n.d.	n.d.	n.d.	n.d.	n.d.	n.d.	n.d.	n.d.	n.d.
b.d.l.	1815	b.d.l.	b.d.l.	b.d.l.	b.d.l.	b.d.l.	n.d.	b.d.l.
3.57	849	2.88	3.42	<4.17	6.93	<8.27	n.d.	0.52
2.29	246	2.90	30.4	6.85	6.44	<20.5	n.d.	5.22
0.10	5.40	5.10	5.40	3.00	0.20	2.20	3.60	0.20
13	71	3	3	4	19	27	2	9
<0.5	0.70	<0.5	<0.5	<0.5	<0.5	<0.5	<0.5	<0.5
0.19	b.d.l.	0.63	0.70	2.26	0.24	0.95	n.d.	0.35
<0.82	<2.36	0.57	0.28	1.22	0.37	1.15	n.d.	0.70

Table 2b. Combined analytical results from INAA, XRF, and ICP-MS analysis.

Sample	PU10-17-1A	PU10-17-1B	PU10-17-2	PU10-17-4	PU17-01-1A	PU17-01-1B	PU17-01-1C	PU17-01-02
Type	Gr	Chert and Ms Mylon shear	Ms + Sph	Ms + chert	Ms	Chert	Ms	Ms
SiO_2	34.7	88.0	36.4	35.1	34.0	88.5	36.6	n.d.
TiO_2	0.65	0.19	0.55	0.43	0.60	0.13	1.37	n.d.
Al_2O_3	16.4	2.24	16.1	17.9	18.5	2.25	21.7	n.d.
Fe_2O_3	23.3	4.58	23.2	22.6	22.2	4.36	17.0	n.d.
MnO	0.16	0.07	0.16	0.14	0.20	0.08	0.08	n.d.
MgO	3.29	0.73	3.22	3.10	3.44	0.67	2.60	n.d.
CaO	0.52	0.50	0.47	0.44	0.60	0.51	0.71	n.d.
Na_2O	0.15	0.37	0.28	0.08	0.16	0.02	0.15	0.15
K_2O	5.04	0.65	4.91	5.23	5.30	0.53	5.85	n.d.
P_2O_5	0.34	0.06	0.32	0.27	0.36	0.06	0.05	n.d.
LOI	15.4	3.19	15.3	15.0	15.2	3.30	14.1	n.d.
Total	100.02	100.56	100.83	100.33	100.5	100.4	100.14	n.d.
Li	12	10	12	4	11	2	8	7
Be	2.30	3.50	4.50	1.30	5.90	0.90	4.80	4.90
Sc	18.0	23.3	47.0	47.0	3.40	0.52	3.27	3.22
V	277	195	544	62	34	7	41	47
Cr	1501	332	1103	250	27.3	12.4	20.6	28.8
Co	121	17.0	95.0	18.0	4.00	2.00	4.00	3.00
Ni	2191	336	1002	244	<61.5	14.9	32.3	56.8
Cu	6.00	2656	7.00	<0.20	5.00	<2.00	8.00	13.0
Zn	51.0	427	43.0	44.0	46.0	23.0	53.0	42.2
Ga	22	16	36	9	21	3	21	20
As	2625	388	1129	268	35.3	23.2	17.5	10.3
Se	15.8	3.78	10.0	0.60	2.79	1.03	0.19	0.62
Br	n.d.	n.d.	n.d.	n.d.	0.35	0.31	b.d.l.	0.54
Rb	83.0	112	219	50.0	151	18.0	154	134
Sr	50	285	50	49	26	14	29	n.d.
Y	13.0	7.00	11.0	12.0	15.0	3.00	14.0	n.d.
Zr	349	44	374	252	313	23	311	n.d.
Nb	20.0	15.0	21.0	19.0	19.0	6.00	17.0	n.d.
Mo	3.90	0.50	1.10	<0.5	<0.50	<0.50	<0.50	<0.50
Cd	0.10	0.10	<0.10	<0.10	0.40	<0.10	0.20	0.40
Sb	222	38.6	101	21.4	7.67	4.44	7.86	7.83
Cs	7.16	7.04	11.6	2.86	4.98	0.65	7.31	4.13
Ba	162	265	396	100	1019	113	1009	868
La	2.53	10.2	7.33	9.37	31.7	5.87	61.2	52.0
Ce	7.58	22.7	17.4	20.2	64.8	11.3	119	101
Nd	5.84	13.5	10.9	9.64	30.4	4.74	48.0	47.7
Sm	1.57	3.21	2.23	2.32	5.90	0.70	7.29	6.72
Eu	0.74	1.21	0.74	0.87	0.72	0.12	1.64	0.90
Gd	2.02	3.34	2.05	2.09	11.7	1.07	6.04	7.57
Tb	0.40	0.64	0.27	0.37	0.52	0.08	0.59	0.38
Tm	0.26	0.3	0.13	0.17	0.19	0.03	0.21	0.21
Yb	2.02	1.71	0.69	0.93	0.88	0.14	0.87	0.98
Lu	0.31	0.24	0.1	0.14	0.16	0.02	0.08	0.11
Hf	1.53	1.72	2.19	2.33	8.14	0.91	7.62	8.58
Ta	0.29	0.18	0.17	0.20	1.33	0.50	1.12	1.23
W	21.4	9.26	12.5	6.23	18.5	2.35	17.9	15.1
Os	161	107	124	<237	b.d.l.	b.d.l.	b.d.l.	b.d.l.
Ir	297	12.8	116	0.16	5.87	3.03	<0.58	2.52
Au	112	34.5	34.5	20.5	39.8	9.82	49.9	14.7
Tl	1.70	1.90	5.10	1.10	1.70	0.20	1.70	1.50
Pb	67	32	40	5	19	10	8	7
Bi	<0.50	<0.50	<0.50	<0.50	<0.50	<0.50	<0.50	<0.50
Th	1.52	1.77	0.83	1.07	20.8	2.52	21.6	23.8
U	0.81	0.57	0.32	0.77	7.41	0.48	2.99	3.31

Table 2b (cont.).

PU17-01-2B Ms + Chert	PU17-01-03 Ms	PU17-01-04 Ms	PU10-20-1 Sph & chert clast	PU10-20-2 Gr Sph and chert	PU10-20-3 Sph + S and Gr	PU10-20-4 Chert	PU10-20-5A Chert
36.1	35.5	34.7	33.0	45.7	37.4	94.7	96.0
0.52	0.46	0.38	0.86	0.65	1.09	0.12	0.15
16.4	16.7	18.0	14.1	10.9	13.8	0.66	0.78
22.7	23.0	22.3	27.4	21.8	24.7	2.08	3.08
0.13	0.13	0.12	0.08	0.06	0.07	0.04	0.04
3.30	3.24	3.14	3.00	2.75	3.12	0.19	0.23
0.51	0.56	0.47	0.24	0.24	0.17	0.12	0.03
0.14	0.14	0.15	0.41	0.42	0.42	0.02	0.01
4.62	4.76	0.22	3.35	2.33	3.37	0.08	0.19
0.32	0.36	0.27	0.14	0.14	0.08	0.02	0.01
15.3	15.4	15.0	17.2	14.1	16.0	n.d.	0.59
100.06	100.27	94.69	99.84	99.03	100.23	98.02	101.15
6	6	10	11	9	10	<1	<1
4.50	4.40	5.40	5.20	3.70	4.50	<0.30	<0.30
2.86	2.93	3.05	24.2	19.9	26.7	1.72	0.92
35	29	30	553	355	439	21	17
23.1	21.1	19.6	2099	1284	1868	107	72.8
4.00	4.00	3.00	331	103	115	5.00	8.00
53.5	113	51.0	2942	1019	1478	82.4	84.1
3.00	<2.00	2.00	377	160	104	5.00	83.0
45.0	41.0	43.0	143	84.0	122	13.0	15.0
20	20	23	55	41	50	4	3
13.8	12.4	6.77	3941	1228	1640	108	76.7
0.12	1.21	0.49	54.4	21.1	16.3	1.59	1.56
0.41	0.41	0.43	<0.47	<0.69	0.62	0.19	0.19
133	137	149	117	87.0	121	6.00	4.00
23	25	26	52	49	50	7	35
11.0	11.0	13.0	16.0	8.00	9.00	<3.00	<3.00
288	260	248	68	53	79	9	10
16.0	16.0	16.0	5.00	7.00	8.00	5.00	5.00
<0.50	<0.50	<0.50	4.90	2.20	2.60	0.50	0.70
0.30	0.30	0.30	0.20	<0.10	0.20	<0.10	<0.10
5.50	4.78	5.56	234	81.3	100	12.2	9.26
3.60	4.23	4.28	4.70	3.40	4.25	0.58	0.25
857	896	994	267	197	252	17	11
48.9	50.7	46.2	6.10	1.91	3.65	0.26	1.14
97.2	100	86.3	16.6	5.25	9.07	0.53	2.17
39.9	36.7	32.1	8.45	3.45	6.00	0.54	0.84
5.89	6.76	5.94	2.60	1.19	1.41	0.09	0.22
0.86	0.95	0.77	0.75	0.39	0.33	0.04	0.05
6.54	8.80	9.09	1.33	1.25	1.04	<0.07	0.15
0.37	0.47	0.41	0.33	0.24	0.21	0.019	0.02
0.18	0.21	0.21	0.15	0.11	0.12	0.02	0.01
0.72	0.84	0.83	1.06	0.59	0.77	0.03	0.10
0.09	0.11	0.13	0.14	0.08	0.13	2.50	<0.01
7.13	7.44	7.30	1.41	1.13	1.58	0.08	<0.09
0.93	1.08	1.06	0.04	<0.16	0.27	0.02	0.02
10.5	9.37	7.28	7.63	4.92	10.9	0.07	0.80
b.d.l.	b.d.l.	b.d.l.	37.1	210	260	14.3	170
1.99	1.72	1.83	425	196	291	8.12	6.01
13.7	14.2	15.8	346	96.2	81.3	4.36	8.99
1.50	1.60	1.70	4.00	2.30	2.40	0.10	<0.10
7	5	6	235	74	70	19	13
<0.50	<0.50	<0.50	<0.50	<0.50	<0.50	<0.50	0.80
18.2	20.8	25.0	1.68	0.89	1.75	0.07	0.16
1.84	3.22	4.85	0.77	0.72	1.61	0.07	0.07

Table 2c. Combined analytical results from INAA, XRF, and ICP-MS analysis.

Sample Type	PU10-20-5B Sph + S Gr	PU10-20-6A Ms + S	PU10-20-6B Ms + S minor Sph	PU10-20-6C Sph	PU10-20-7A Sph + S	PU10-20-7B Ms + Sph
SiO_2	34.4	19.8	26.1	39.3	30.8	22.7
TiO_2	1.16	0.22	0.62	1.36	1.01	0.15
Al_2O_3	16.3	5.08	9.00	19.9	14.6	5.33
Fe_2O_3	23.4	43.7	36.1	17.8	28.7	41.1
MnO	0.07	0.08	0.06	0.05	0.06	0.09
MgO	2.65	2.67	1.84	2.44	2.25	2.90
CaO	0.25	0.42	0.23	0.10	0.27	0.31
Na_2O	n.d.	0.13	0.27	0.49	0.38	1.10
K_2O	4.18	1.16	2.27	5.20	3.76	1.16
P_2O_5	0.15	0.24	0.15	0.05	0.17	0.20
LOI	15.7	25.3	22.3	12.7	17.6	25.0
Total	98.17	98.74	98.84	99.31	99.54	100
Li	11	4	7	14	10	5
Be	5.30	2.00	3.00	6.20	4.60	1.70
Sc	n.d.	11.3	21.4	49.2	34.0	12.9
V	625	175	336	469	536	272
Cr	n.d.	267	1024	606	1714	292
Co	149	915	566	14.0	469	896
Ni	n.d.	6001	5935	489	4189	6239
Cu	409	908	779	28.0	652	730
Zn	222	93	104	29.0	132	76.0
Ga	60	19	37	67	51	20
As	n.d.	6948	4047	701	5630	7702
Se	n.d.	121	208	3.17	119	0.14
Br	n.d.	2.62	1.23	<0.19	<0.85	b.d.l.
Rb	153	42.0	91.0	187	136	45.0
Sr	52	40	46	56	50	16
Y	21.0	31.0	21.0	12.0	20.0	19.0
Zr	94	25	50	97	70	21
Nb	8.00	<3.00	4.00	9.00	5.00	<3.00
Mo	6.50	12.7	8.30	<0.50	6.30	9.30
Cd	0.30	0.20	0.30	0.10	0.20	0.10
Sb	n.d.	664	391	40.9	407	816
Cs	n.d.	2.19	3.52	5.53	5.05	4.56
Ba	332	109	214	419	315	112
La	n.d.	7.75	12.5	1.91	10.4	5.17
Ce	n.d.	17.4	27.3	5.48	25.3	12.2
Nd	n.d.	9.28	13.2	4.29	14.2	6.47
Sm	n.d.	2.73	3.49	0.98	3.85	2.26
Eu	n.d.	0.95	1.14	0.23	0.99	1.18
Gd	n.d.	2.23	0.63	<1.66	1.53	2.54
Tb	n.d.	0.51	0.44	0.17	0.48	0.43
Tm	n.d.	0.28	0.18	<0.07	0.23	0.22
Yb	n.d.	1.43	0.9	0.53	1.05	1.26
Lu	n.d.	0.23	0.11	0.09	0.15	0.14
Hf	n.d.	0.25	0.89	2.35	1.54	0.33
Ta	n.d.	<0.27	<0.29	0.64	0.31	0.91
W	n.d.	0.61	0.09	7.63	7.14	6.34
Os	n.d.	328	b.d.l.	b.d.l.	210	185
Ir	n.d.	94.5	265	5.64	339	73.2
Au	n.d.	1595	999	6.78	6.52	2123
Tl	3.60	3.70	10.9	3.20	6.60	2.70
Pb	181	777	845	6	476	832
Bi	<0.50	1.20	<0.50	<0.50	<0.50	1.80
Th	n.d.	0.93	0.61	0.53	0.68	1.02
U	n.d.	1.29	<1.27	0.53	1.61	0.12

Table 2c (cont.).

PU10-20-8 Sph	PU10-20-10 Sph + S	PU10-20-11 Sph + S	PU10-20-12A Sph	PU10-20-12B Sph	PU10-20-13 Sph
24.4	73.1	33.8	36.7	45.6	49.4
0.90	0.39	1.03	1.04	1.63	0.78
12.5	5.80	16.1	17.2	24.7	15.8
33.7	10.5	25.2	22.5	9.65	15.8
0.07	0.04	0.08	0.08	0.05	0.14
2.38	1.30	2.33	3.33	1.46	2.26
0.50	0.18	0.32	0.10	0.72	0.27
0.27	0.19	0.25	0.55	0.63	0.41
3.14	1.27	3.72	3.99	6.39	3.89
0.34	0.09	0.20	0.05	0.02	0.12
20.3	6.87	17.0	14.9	8.7	11.4
98.53	99.66	99.99	100.4	99.49	100.2
7	5	11	15	17	13
3.90	1.50	4.30	6.40	6.90	4.70
3.4	10.4	26.9	38.3	60.1	25.8
750	194	700	277	695	227
3964	696	3538	321	999	354
759	82.0	506	26.0	21.0	14.0
6361	910	6716	441	460	246
1190	109	1092	22.0	16.0	129
493	40.0	440	34.0	27.0	40.0
50	21	44	26	57	18
7956	1193	7878	561	730	b.d.l.
138	22.4	79.0	3.97	2.19	1.69
<1.73	b.d.l.	<1.23	0.41	b.d.l.	b.d.l.
118	46.0	127	138	233	136
49	14	52	53	40	33
35.0	7.00	29.0	8.00	11.0	24.0
55	31	65	73	114	100
3.00	6.00	4.00	7.00	8.00	9.00
11.7	1.30	7.80	<0.50	0.70	0.70
0.30	0.10	0.20	0.10	0.10	0.10
519	75.3	485	29.6	44.6	b.d.l.
4.46	3.39	<0.59	4.06	12.4	8.30
274	104	319	321	561	335
17.6	2.11	6.34	3.43	1.22	11.1
44.4	5.45	21.1	8.02	3.28	25.3
27.5	3.74	16.5	4.57	1.98	12.3
6.22	0.95	3.87	1.28	0.68	3.23
1.79	0.39	1.29	0.29	0.18	1.17
5.52	0.62	3.37	0.7	0.72	2.62
0.83	0.12	0.60	0.15	0.14	0.52
0.35	0.09	0.24	0.09	0.14	0.19
1.46	0.42	1.76	0.47	0.43	1.08
0.10	0.05	0.14	0.05	0.06	0.19
0.55	0.52	0.99	1.40	2.62	2.07
0.15	0.08	0.41	0.18	0.16	0.30
6.25	2.54	8.89	5.70	15.8	n.d.
1519	64.9	204	154	b.d.l.	b.d.l.
837	134	798	6.35	10.5	8.85
39.8	72.4	339	5.05	12.7	16.4
7.40	1.40	3.60	2.30	3.80	2.30
578	75	304	6	11	16
0.6	<0.50	<0.50	<0.50	<0.50	<0.50
<0.74	0.55	1.04	0.38	0.51	2.17
2.01	b.d.l.	0.83	<0.30	0.58	0.88

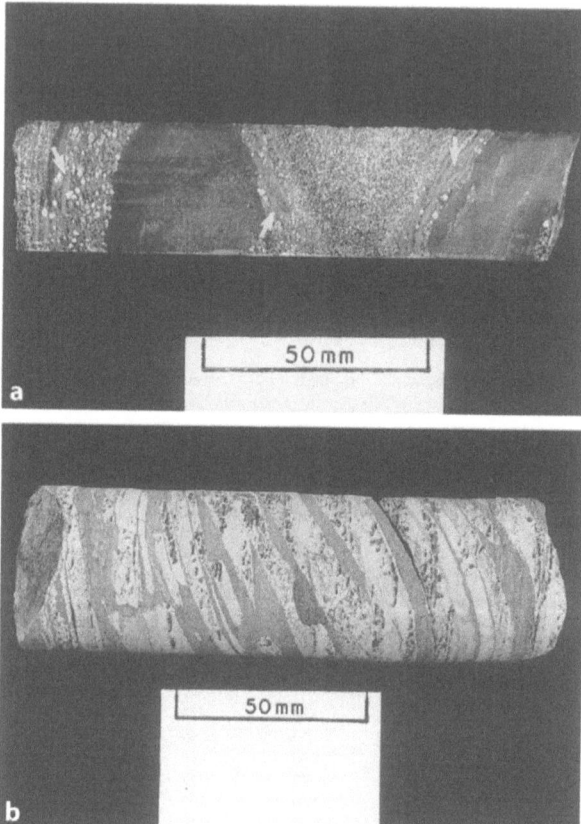

Fig. 4. (above and opposite) Macroscopic features of spherule layers in typical chert-mudstone (metapelite) host rock. a) Spherule layer in the PU 10-20 drill-core, clearly folded around fine-grained metapelite (dark grey) and coarser grained metapelite (light grey). The spherule layer grades from large (3 mm) to small (0.5 mm) spherules; this is particularly apparent in the part of the layer at the left margin of the photograph. Fine-grained sulfide "boudins" (arrows) are often associated with the spherule layer. b) Note the brecciation of the more competent chert bands, with the mudstones plastically filling the voids. c) Relatively undeformed spherules in a highly brecciated (brittle deformation, for a change) mudstone-chert matrix. Cores from the PU series. d) Drill-core through a typical mudstone-chert assemblage. Quartz veins cutting the layer perpendicularly, with graphite-fill in veins along layer-parallel shears (At right end of core). Cores form PU series. Shear planes in the samples from the PU series are often characterized by graphite-coated partings (black). e) These shear zones are often sited along the contact between different rock types, but may also cross them (as in this case of a spherule layer). The spherules are often associated with the shear zones, possibly due to their mudstone matrix making them less competent than the chert bands. f) Bands of individual large (up to 3 mm) spherules are often associated with graphitic partings; smaller spherules (less than 1 mm) occur in mudstone matrix (arrows).

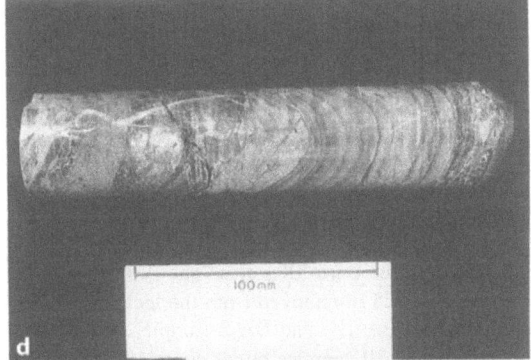

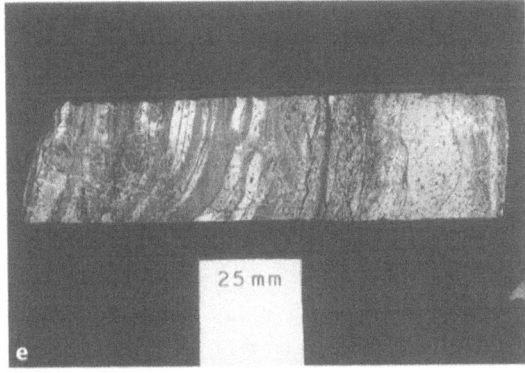

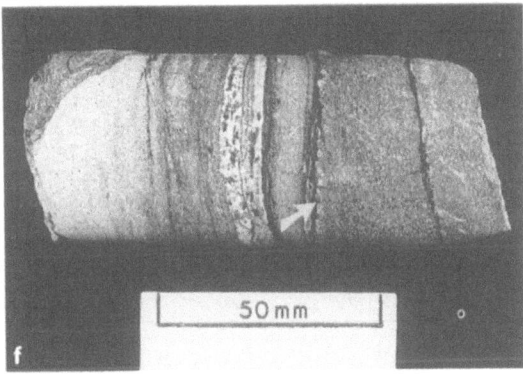

Table 3. Summary of detection/quantitative limits and noble metal concentrations in blanks and standard reference materials used for PGE analysis; all data in ppb.

(ppb)	Ir	Ru	Rh	Pt	Pd	Au
Detection Limit	0.03	0.06	0.03	0.15	0.13	0.02
Quantitative Limit	0.09	0.18	0.09	0.5	0.42	0.07
Blank	0.12	<0.15	n.d. <0.02	<0.29	<0.22	0.39
WPR1-1	13.9	23.1	14.3	265	232	38.0
WPR1-2	13.4	21.8	13.5	256	237	185
Recorded Values	13.5	21.3	13.4	285	235	42.0
WITS1-1	1.40	4.20	0.98	6.10	4.60	4.40
WITS1-2	1.20	3.50	0.99	4.90	4.70	7.90
WITS1-3	1.20	3.30	1.10	5.20	4.20	4.90
Recorded Values	1.4±0.2	3.5±0.4	1.1±0.2	5.2±0.5	4.5±0.8	4.9±1.5

Table 4. Noble metal data for sulphides in spherules layer and country rock samples. Samples taken from the strongly gold mineralized zone in the Princeton section of the Agnes mine are represented by the Barb-5X series. The PU-10 sulphide sample was taken from the PU-10 borehole, some 25 m removed from the location where the Barb-5X series was collected. The SJ03/SJ04, BA-1X, and BA-2 samples were collected in areas of no significant gold mineralization. All samples either contain or are closely associated with spherules. All data in ppb.

Sample	Ir	Ru	Pt	Rh	Pd	Au
Barb-51	0.44 ± 0.04	0.46 ± 0.08	< 0.50	0.12 ± 0.02	1.1 ± 0.1	1850 ± 80
Barb-52	0.46 ± 0.06	0.60 ± 0.10	0.91 ± 0.08	0.15 ± 0.02	2.0 ± 0.2	1670 ± 90
Barb-53	0.18 ± 0.02	0.43 ± 0.07	0.52 ± 0.14	< 0.09	1.4 ± 0.1	1450 ± 50
Barb-54	1.2 ± 0.1	1.3 ± 0.2	1.2 ± 0.2	0.24 ± 0.03	1.5 ± 0.2	3330 ± 270
Barb-55	0.21 ± 0.03	0.41 ± 0.03	< 0.50	0.09 ± 0.01	0.89 ± 0.16	570 ± 40
Barb-56	0.40 ± 0.05	0.49 ± 0.06	0.64 ± 0.06	0.14 ± 0.01	1.8 ± 0.2	4720 ± 350
Barb-57	0.40 ± 0.01	0.44 ± 0.09	< 0.50	0.11 ± 0.01	1.3 ± 0.1	2250 ± 180
PU-10	0.41 ± 0.03	0.51 ± 0.02	1.2 ± 0.2	0.17 ± 0.01	1.9 ± 0.3	11 ± 2
SJ03	34 ± 2	46 ± 4	69 ± 4	9.1 ± 0.5	23 ± 2	16 ± 4
SJ04	8.5 ± 0.7	12 ± 1	18 ± 2	2.4 ± 0.3	7.1 ± 0.8	22 ± 6
BA-1A	0.96 ± 0.10	4.1 ± 0.5	6.1 ± 0.5	1.1 ± 0.1	3.5 ± 0.4	2.3 ± 0.7
BA-1B	4.9 ± 0.2	7.4 ± 0.5	9.6 ± 0.8	1.5 ± 0.1	9.1 ± 1.5	6.0 ± 2.0
BA-2	178 ± 10	232 ± 13	280 ± 15	44 ± 3	77 ± 6	2.6 ± 0.4
Quant Limit	0.08	0.18	0.5	0.09	0.42	0.08
(in ppb)	Ir	Ru	Pt	Rh	Pd	Au
Orgueil	480	690	1050	130	530	140
C1	480	683	982	140	560	148

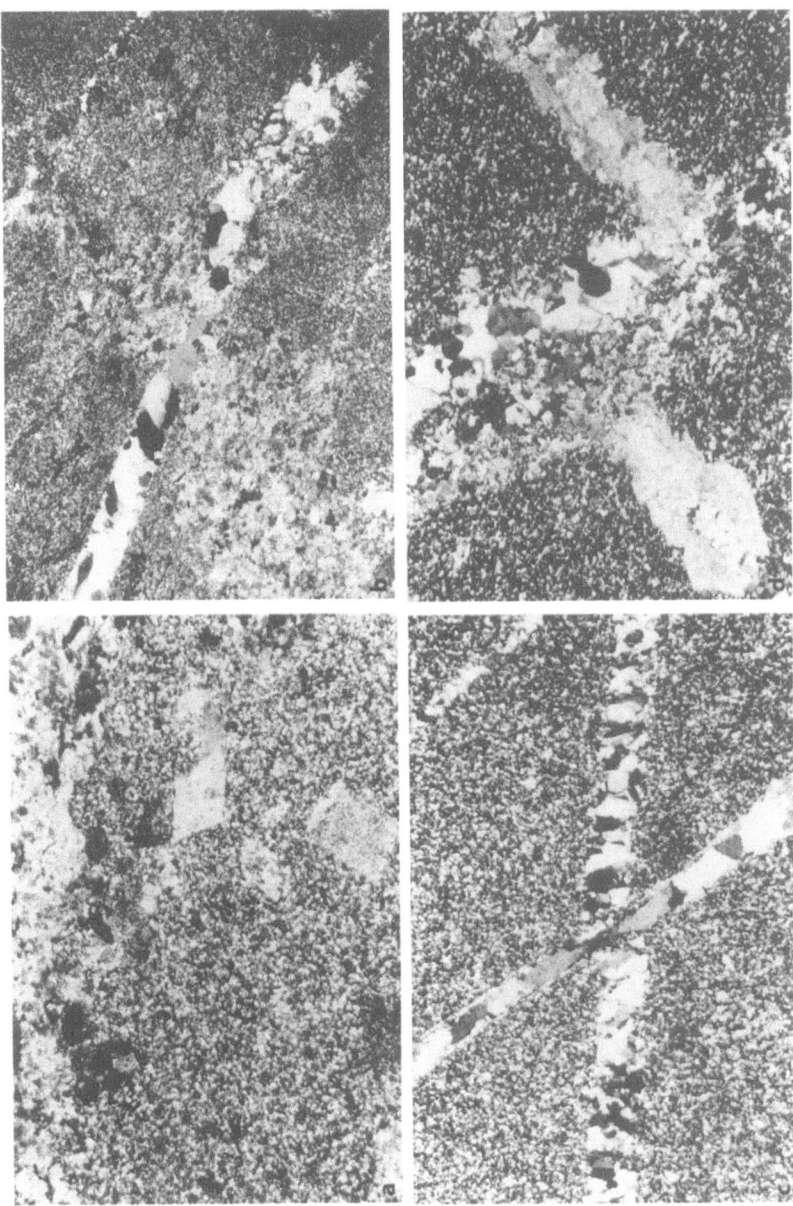

Fig. 5. Cross-cutting vein relationships provide evidence for multiple hydrothermal events. a) Euhedral siderite rhombohedrons in a chert matrix. Sample PU17-01-1. Photograph 3 mm wide; crossed nichols. b) Later quartz vein cutting an earlier siderite vein in a chert matrix. Sample PU10-20-3. Photograph 2 mm wide; crossed nichols. c) Multiple cross-cutting quartz veins in a chert matrix. Sample PU 10-17-4. Photograph 4 mm wide; crossed nichols. d) Later quartz vein cutting an earlier siderite vein (NE-SW trending) along a microfault. Sample PU 10-17-4. Photograph 2 mm wide; crossed nichols.

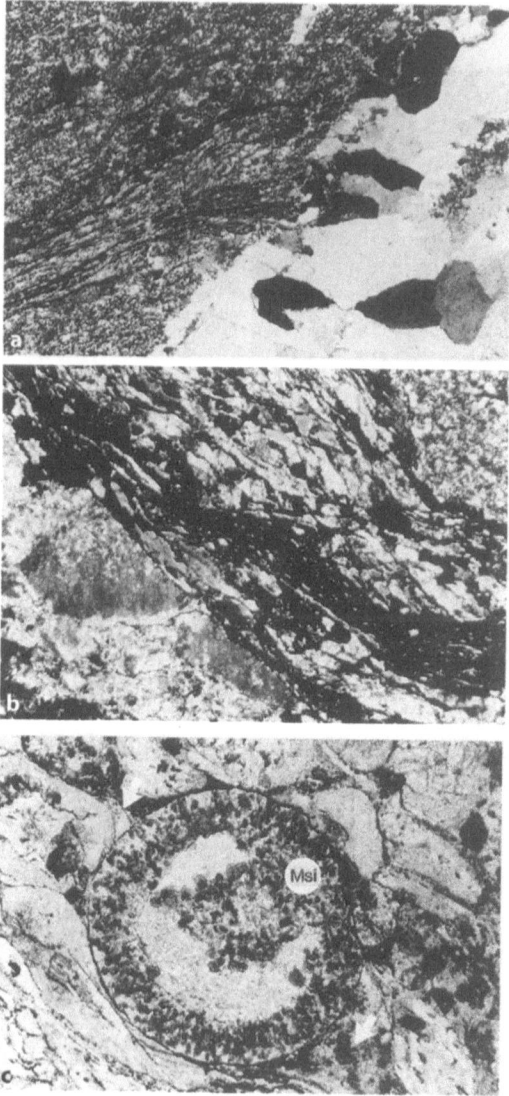

Fig. 6. a) Evidence for multiple vein generations. First generation of quartz (outer vein margin) cut by shearing (shear trending NE-SW); second generation of quartz utilizing original vein path. Sample PU 10-17-4. Photograph 1.5 mm wide; crossed nichols. b) Sample PU 10-17-4: Two cross-cutting mylonitic and more (black) or less graphitic shear zones (note: fragment of older mylonite in the younger breccia, at lower right. Photograph 1.5 mm wide, crossed nichols.c) An undeformed spherule with pressure shadows (arrows) in a mylonitic spherule bed. Core and rim material have been replaced by magnesio-siderite (Msi). The spherule center is replaced by fine-grained quartz. The original euhedral nucleus, possibly pyrite, has been destroyed by hydrothermal alteration. Photograph 4 mm wide; parallel nichols.

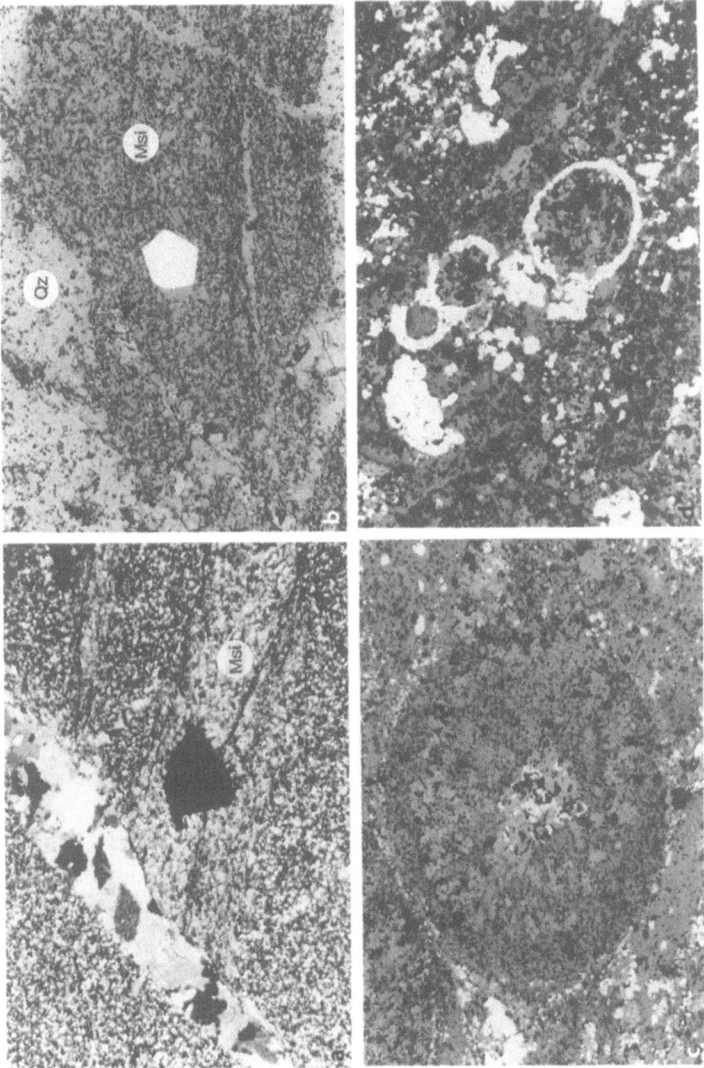

Fig. 7. a) Euhedral, late stage pyrite grain in a magnesio-siderite (Msi) vein (trending NE-SW), which, in turn, is cut by a quartz vein. Sample PU 10-17-4. Photograph 1 mm wide; reflected light. b) Euhedral late stage pyrite grain in matrix of magnesio-siderite and fine-grained quartz (Qz). Matrix crosscut by several generations of quartz veins. Sample PU 10-09-04. Photograph 1 mm wide; reflected light. c) Relatively undeformed spherule in a quartz-magnesio-siderite matrix. Reflected light, 2 mm wide. Pyrite mineralization restricted to the zone just outside of the outer margin of the spherule. Partially resorbed pyrite and chalcopyrite replace the original nucleus. This spherule too, displays a radial growth pattern of magnesio-siderite. Sample PU 10-20-8. d) Undeformed spherule boundaries clearly defined by pyrite mineralization. Scattered grains of pyrite in the matrix. Sample PU 10-20-8. Photograph 5 mm wide; reflected light.

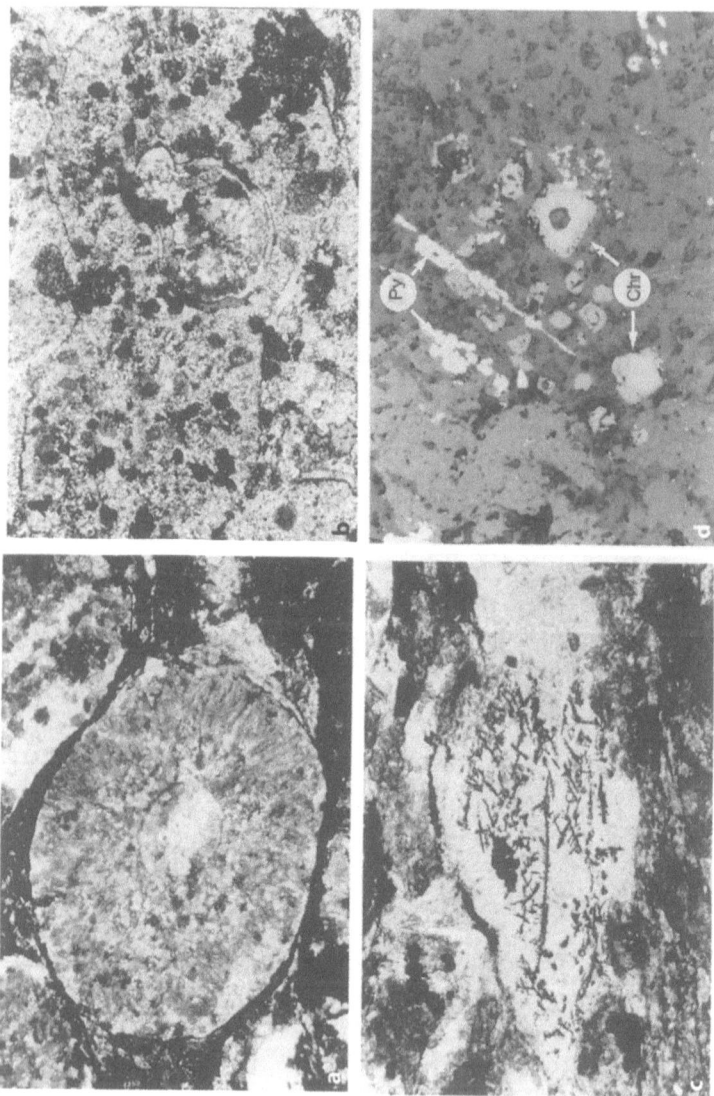

Fig. 8. a) An isolated ovoid spherule in a mylonitic spherule bed in sample PU 10-20-9. Note radial growth pattern of magnesio-siderite within the spherule. The spherule contains a nucleus, possibly a smaller spherule. Photograph 3mm wide; crossed nichols. b) Dumbbell shaped spherule displaying radial or barred-chondrule growth pattern within a larger dumbbell shaped spherule. Spherule interiors composed completely of quartz and magnesio-siderite. Sample PU 10-20-13. Photograph 4 mm wide; crossed nichols. c) Dendritic growth of Cr-spinel, in a magnesio-siderite rich vein of sample PU 10-20-1. Parallel nicols, photograph 2 mm wide. d) Obvious zonation in two large Cr-spinel (Chr) grains. Smaller grains clearly show secondary spinel overgrowth on anhedral spinel cores. Pyrite (Py) intergrown with the spinel crystals. Sample PU 10-20-2. Reflected light, ~1.5 mm wide.

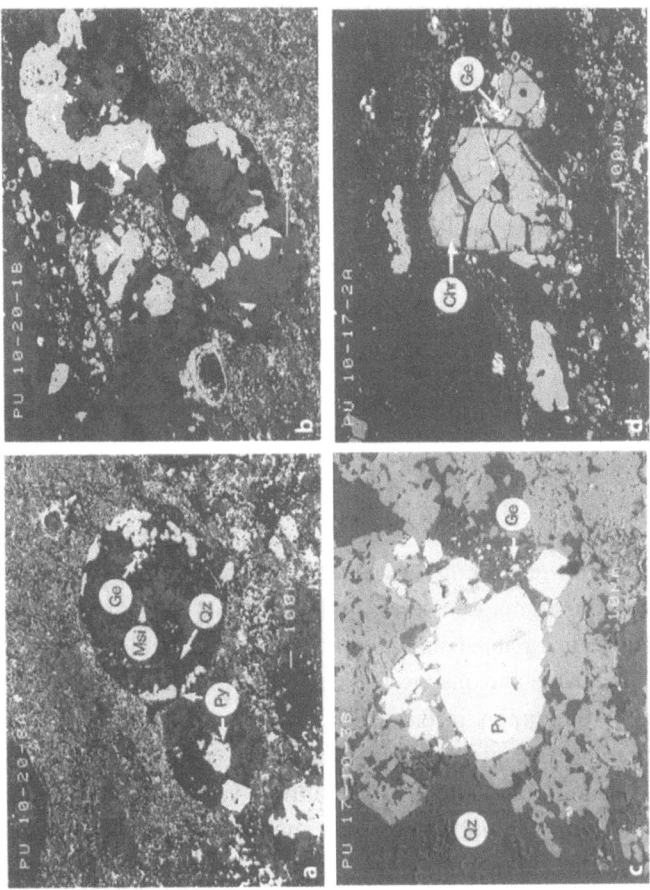

Fig. 9. a) Backscattered electron (BE) image of pyrite (Py) and gersdorffite (Ge) mineralization along inside of spherule margin and sub to euhedral pyrite grains inside spherules, in sample PU 10-20-8A. Pyrite contains minor inclusions of chalcopyrite and gersdorffite. Fine-grained sulfides in the matrix are predominantly pyrite and gersdorffite. Spherule matrix is quartz (Qz) and magnesio-siderite (Msi). Scale bar = 100 m. b) has a fractured spinel associated with gersdorffite grains. These spinels are typically located outside the spherule margins, in association with secondary sulfides. Sample number PU 10-20-1B; Scale bar = 100 m. c) Backscattered electron (BE) image of multiple sulfide generations showing obvious overgrowth patterns. Matrix of quartz (Qz) and magnesio-siderite intimately associated with the large pyrite (Py) grain. The pyrite grain has overgrown a mineral that has since been replaced by lead. Bright specks are gersdorffite (Ge). Sample PU 17-10-3B. Scale bar ⁻ 10 m. d) Backscattered electron (BE) image of a large Cr-spinel (Chr) with intergrown gersdorffite (Ge) in a quartz matrix in sample PU 10-17-2A. The fracturing, related to late stage brittle deformation, post-dates the mineralization event, as the rim zonation of the spinel does not continue along the fracture edges into the core of the grain. Scale bar = 100 m.

(Figs. 9a-b). The sulfide grains are often overgrown by successive generations of hydrothermal deposits (Fig. 9c). Sulfide mineralogy is dominated by pyrite, but gersdorffite and arsenopyrite do occur prominently as well. De Ronde et al. (1992, 1994) have also discussed evidence of hydrothermal processes in the Barberton Greenstone Belt.

As in earlier studies (Koeberl et al. 1993; Koeberl and Reimold 1994, 1995), no evidence of shock deformation could be identified in any of the samples of this study. The spherule layers contain rare small angular quartz clasts, in none of which any form of characteristic shock deformation, such as planar deformation features (PDFs), could be observed – confirming the results of previous studies. Co- and Ni-rich Cr-spinels from the Bon Accord nickel deposit in the Barberton Greenstone Belt have been described by de Waal (1978).

Mineral Chemistry

Scanning electron microscopy, together with energy-dispersive X-ray analysis (EDXA), was carried out to investigate the nature of the matrix to spherule layers and to identify fine-grained sulfides in the matrix. The matrix is predominantly composed of quartz and magnesio-siderite. Pyrite and chalcopyrite form the majority of sulfide grains in the matrix, with gersdorffite (NiAsS) often being present as well. In general, the sulfide mineralogy is very similar to that described by Koeberl and Reimold (1995). A sulfide mineral paragenesis frequently observed comprises gersdorffite-chalcopyrite-tetrahedrite-pyrite±sphalerite. However, in other samples, these minerals may belong to several generations of sulfides, consistent with the overall finding that several hydrothermal events have comprehensively overprinted the whole region throughout the stratigraphic section investigated.

Various workers (Byerly and Lowe 1992, 1994; Koeberl et al. 1993; Koeberl and Reimold, 1994, 1995) have described nickel-rich chrome-spinels (chromites) in Barberton spherule layers. The previous work was, however, restricted to spot analysis on spinel crystals. Recent work by S. Johnson showed, through combined EPMA spot and PIXE Dynamic Mapping analysis, that the spinels in PU and BA spherule samples are zoned.

EPMA spot analyses from a large spinel crystal (Fig. 8d) show that the zonation observed microscopically along the rim of the spinel is not merely derived from oxidation of the spinel, but is due to a change in its chemical composition. Several spot analyses on the rim, the core, and in transitional areas were carried out (Table 1). The spinel exhibits a threefold zonation, with central, middle, and rim zones. The center is relatively enriched in Ni; the middle zone is Ni-rich (12.1-10.2 wt%

NiO), whereas the rim zone is low in Ni (0.8-0.3 wt% NiO). Other enriched elements in the core, compared to the middle and rim zones, include MgO, FeO, and CoO. Elements with abundances that increase from the core to the rim include Al_2O_3, ZnO, and TiO_2. The Cr_2O_3, TiO_2, and V_2O_5 contents are slightly variable: the PIXE trace elemental maps show slight decreases towards the core and rim for the abundances of these elements. The Zn enrichment of the rims is a strong indicator of secondary alteration (Bjerg et al. 1993).

Dynamic PIXE mapping and spot analyses were employed in an attempt to locate the carrier phase(s) of iridium in several spherule layer samples. In most cases investigated, it was not possible to identify the Ir carrier phase, due to Ir being most likely distributed in matrix mineral phases, at concentrations below the sensitivity of the proton microprobe. However, in one case of Dynamic Mapping it was found that gersdorffite (NiAsS) hosts iridium. PIXE point analysis confirmed that there are indeed small inclusions of Ir in gersdorffite, which occurs closely associated with Cr-spinel (Fig. 9d) but is considered a late, secondary sulfide phase. On the basis of these results it must be assumed that much, if not the overwhelming amount, of Ir in the spherule layers occurs in matrix minerals and secondary phases in spherules, at concentrations below the detection limit of PIXE analysis (in our experiments at circa 155 ppm).

Trace Element Chemistry

In this chapter, data for the siderophile, chalcophile, and – especially - highly mobile elements analyzed in the new sample suites (Tables 2a-c) will be discussed. Numerous interelement relationships were evaluated for these data. Figures 10-15 are representative for our findings and illustrate the most important results.

On first glance, the abundances of Cr and Ir show a reasonable and positive correlation (Fig. 10a). Apparent exceptions are a number of country rock samples of very low Ir content. However, it must be emphasized that several of the samples with highest Cr and Ir contents are also country rock samples (e.g., chert), *not spherule layer samples!* Also indicated are the trends after Lowe et al. (1989) for spherule and komatiite samples, respectively. These authors analyzed samples from three spherule layers, but in their publication did not indicate which data refer to S2 spherule layer samples – which otherwise might have been compared to our data. Quite obviously these trends do not coincide with the arrays determined for our sample suites.

A weak correlation between Cr and Zn, a highly mobile element that has been strongly redistributed by the multiple alteration events that affected the region, is evident from Figure 10b. In contrast, Cr and Au (Fig. 10c), another element which most certainly has been hydrothermally redeposited, is highly mobile, and is related to secondary sulfide mineralization, do not have a distinct correlation, and neither do iridium and gold (Fig. 10d).

Nickel contents, which could be either of meteoritic origin or derived from regional mafic or ultramafic rocks, are clearly not related to Ir abundances (Fig. 11a). The Ni/Ir ratios for this sample suite vary between about 22 000 and 84 000 and, thus, are clearly different from meteoritic values. Our highest Cr and Ir values were obtained for a chert sample, which is not distinguished by a remarkably high Ni content. Co and Ni, both of which could be of meteoritic origin (Fig. 11b), show a weak correlation for most of the sample suite, but notably for both spherule layer *and* country rock samples. However, the average Ni/Co ratio for the relatively well correlated samples is of the order of 8, more than a factor of 2 different from chondritic values. Cobalt and iridium contents (Fig. 11c) are not correlated, nor are those of Ni and Cr (Fig. 11d; Ni/Cr ratios vary from 0.4 to 2.7, with meteoritic values falling between about 2.5 and 4) or Co and Cr (Fig. 11e; Cr/Co ratios vary between about 0.5 and 4000, whereas meteoritic values are between about 4 and 8).

The iridium abundances were also plotted against those of As, resulting in a great scatter (Fig. 12a). The Ir versus Zn plot (Fig. 12b) gives a similar result. However, when comparing Ir and Sb abundances, a significant correlation becomes evident (Fig. 12c). Gold and As, as well as Au and Sb are not well correlated (Figs. 12d-e), which is highly surprising with regard to the regional gold-sulfide association. One explanation for this lack of correspondence could be seen in the fact that repeated hydrothermal overprint is a phenomenon that affected the whole region on every scale (compare, for example, the section on microscopic observations), and, thus, obscured perhaps originally existant correlations.

As has been shown in earlier studies by our group, rather good correlations are seen in the comparisons of Ni with As and Sb data, respectively (Figs. 13a-b). Obviously a major part of the nickel is located in arsenic-rich phases (gersdorffite), but the PIXE results (cf. above) have demonstrated that late-stage deposition of nickel (for example, at rims of chrome-spinel) is not negligible. Figures 13c and d illustrate the cobalt versus arsenic and antimony systematics, which are very similar and demonstrate the secondary nature of the cobalt distribution, too, with both country rock and spherule layer samples showing partial correlation. As the PIXE analyses showed, only some of the iridium can so far be linked with gersdorffite mineralization.

In Figures 14a to f some chemical characteristics of strongly (> 5 vol% sulfides) and less sulfide-mineralized samples (i.e., both country rock and spherule layer samples) are compared, in an attempt to further investigate whether the apparent correlations between elements observed represent a primary signature or are the result of secondary chemical overprint. For the examples shown here, as well as many other correlations evaluated, scatter as well as apparent correlations, as described above, are not reserved for either the well mineralized or the barely mineralized sample groups. This represents confirmation of the petrographic result that the whole sample suite has been thoroughly hydrothermally overprinted and confirms the view that the chemical signatures observed are not primary ones.

Diagrams of chondrite-normalized rare earth element patterns (Figs. 15a-f) for spherule layer and country rock samples, respectively, display generally similar patterns characterized by slight enrichments of the light REE, but representing a rather wide range of REE abundances, from near-chondritic to highly enriched (La > 100) total REE values. Obviously this variation must be related to the sample mineralogy, with those samples with a large quartz content (chert and some BIF samples) having lower total REE than other samples. Only a few banded iron formation and several spherule layer samples display near-horizontal patterns, whereas many others have highly varied anomalies of different magnitudes. These effects are not correlated with major mineral variations and are regarded as the result of either the presence of small but significant trace mineral abundances or of the pervasive secondary alteration.

Platinum Group Element Analysis

Relatively less sulfide-mineralized samples BA2, SJ3 and 4, and BA-1B contain concentrations of Ir and Ru that are in excess of bulk mantle or any non-mineralized mafic mantle melt, such as kimberlites or komatiites (e.g., McDonald et al. 1995; Pattou et al. 1996). When the PGE data for the non-mineralized samples are normalized against CI chondrites (Pd from Anders and Grevesse 1989, other metals from Jochum 1996), the resulting patterns (Fig. 16) show a number of consistent features. Whereas the absolute concentrations of PGE in the BA2, SJ3 and 4, and BA-1B samples vary by a factor >35, the normalized patterns from Ir to Pt are essentially flat (i.e., near-CI chondritic). Pd and Au are moderately fractionated, either slightly enriched or slightly depleted relative to CI, which indicates that these elements have probably been more mobile than Ir, Ru, Rh, and Pt.

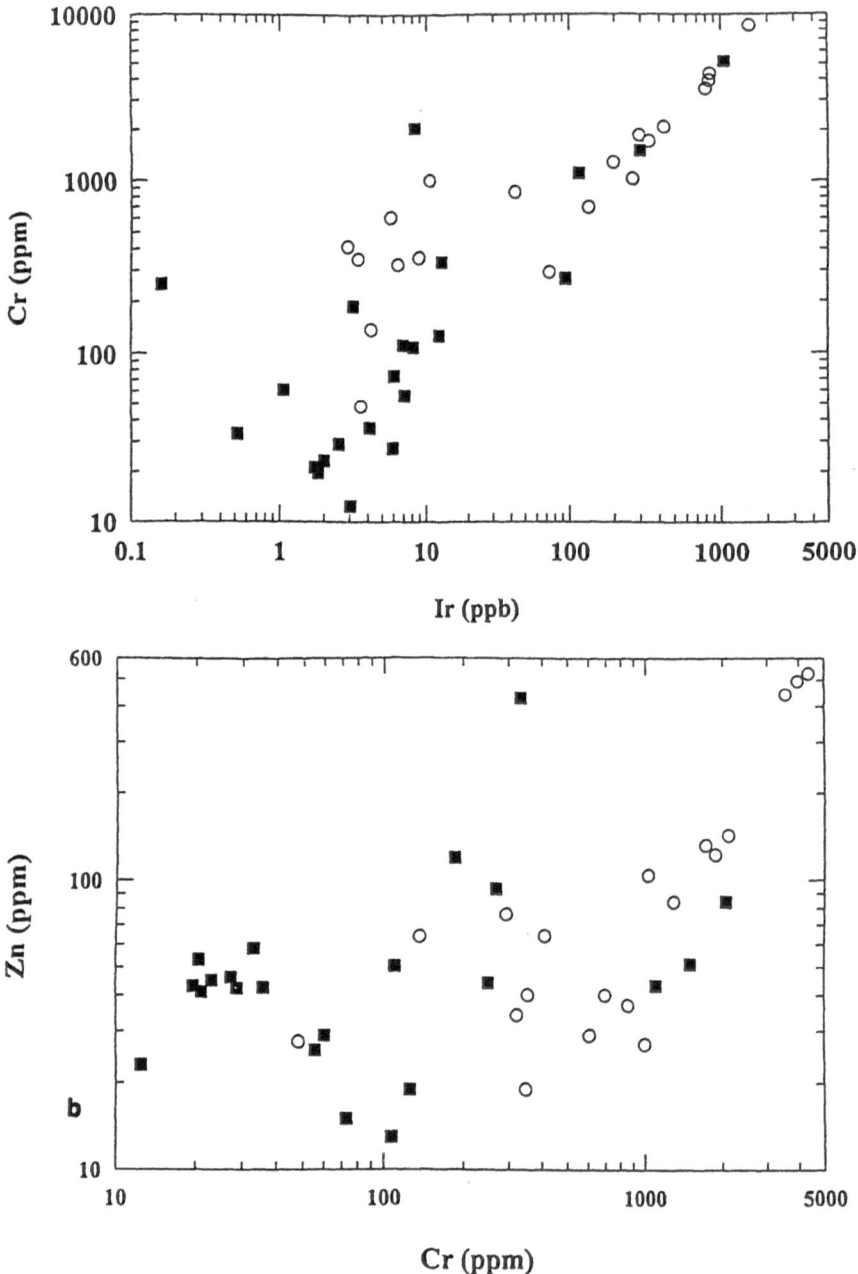

Fig. 10. a-d (above and opposite) Diagrams illustrating the respective relationships (or lack thereof) between Cr and Ir, Zn, and Au in all BA-2 and PU samples (compare Tables 2a-c). The samples have been divided into those which contain spherules (open circles) and do not contain spherules (country rocks, solid squares). Note that scales vary from diagram to diagram.

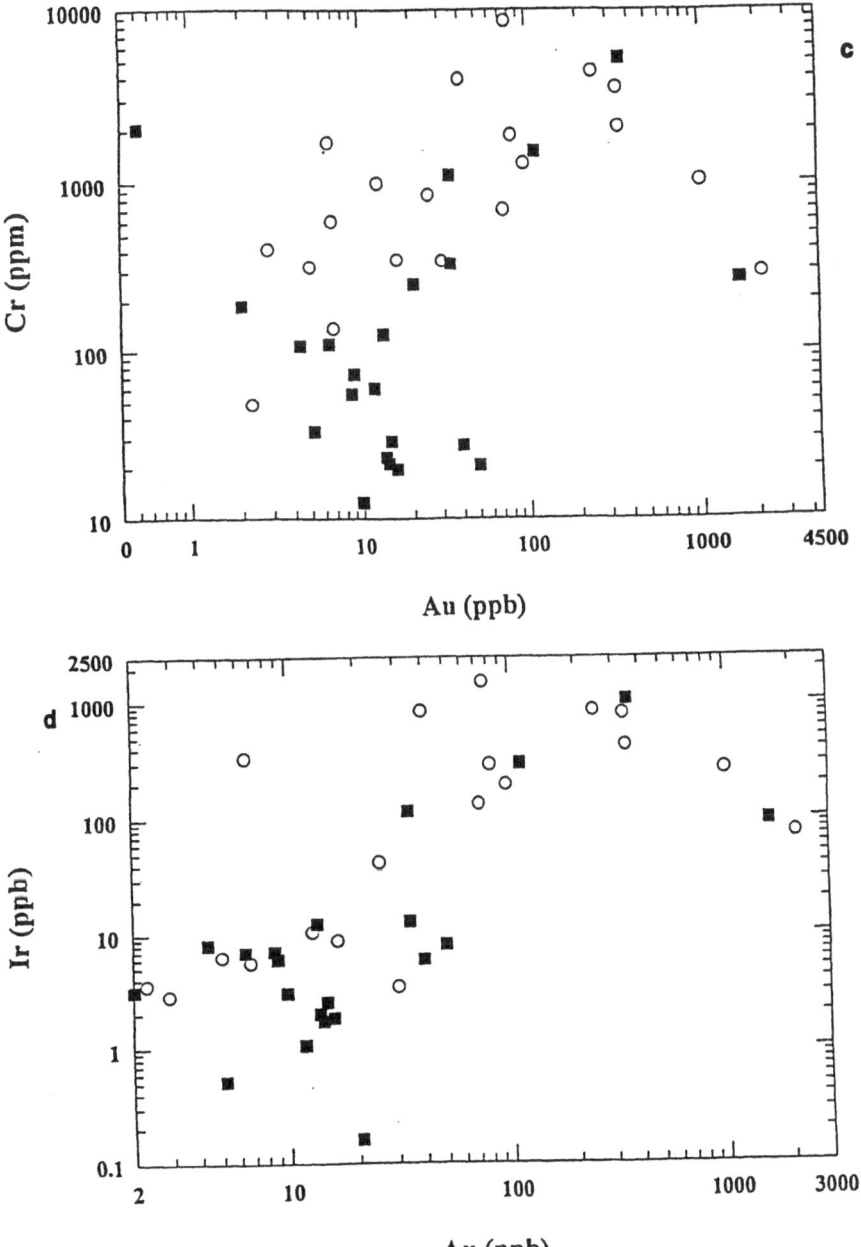

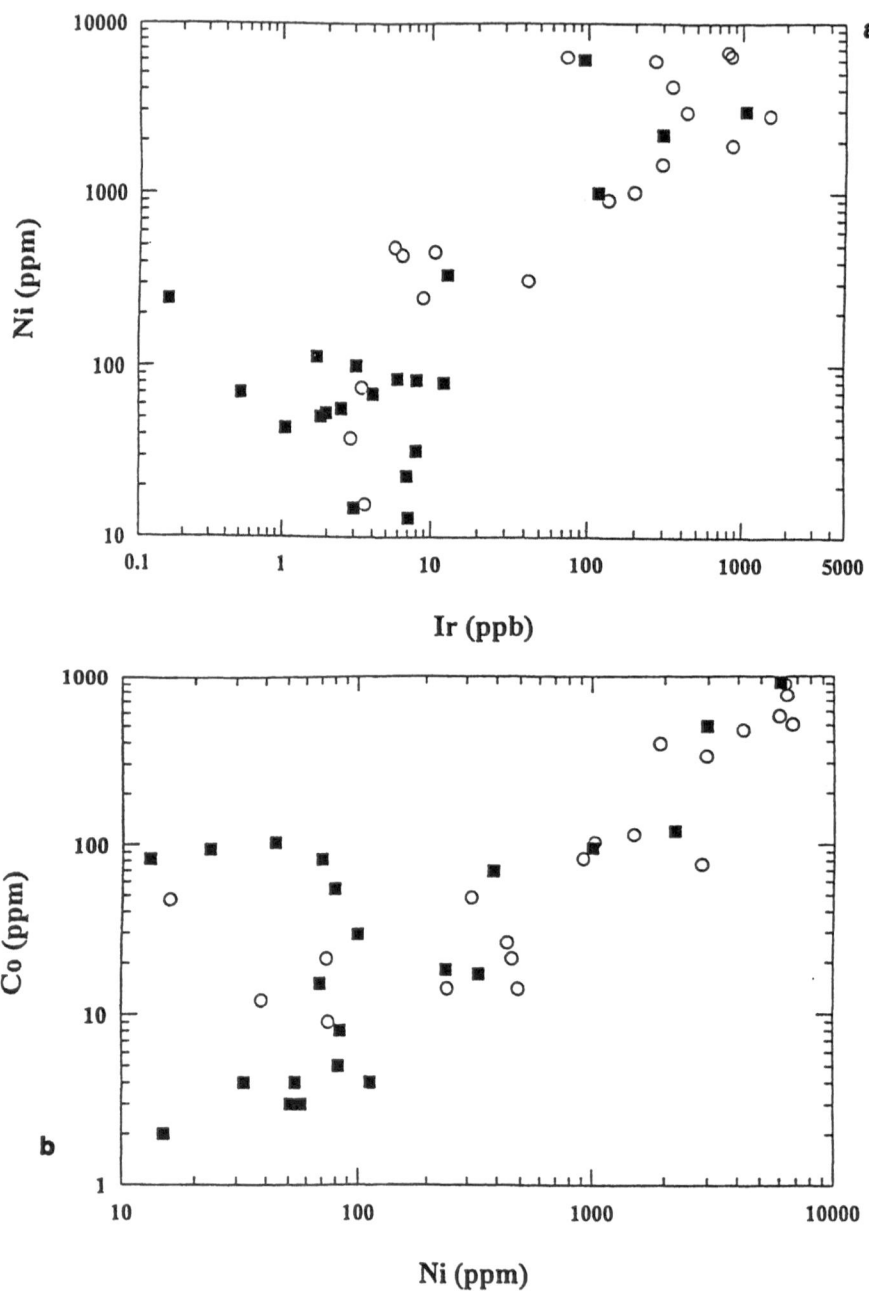

Fig. 11. (above and opposite) Systematics of Ni, Co, and Ir abundances, as well as Co and Cr in all BA-2 and PU samples given in Tables 2a-c. (a) nickel versus iridium contents, (b) cobalt versus nickel contents, (c) cobalt versus iridium, (d) nickel versus iridium, and (e) cobalt versus chromium. Spherule layer samples – open circles; country rock samples – solid squares.

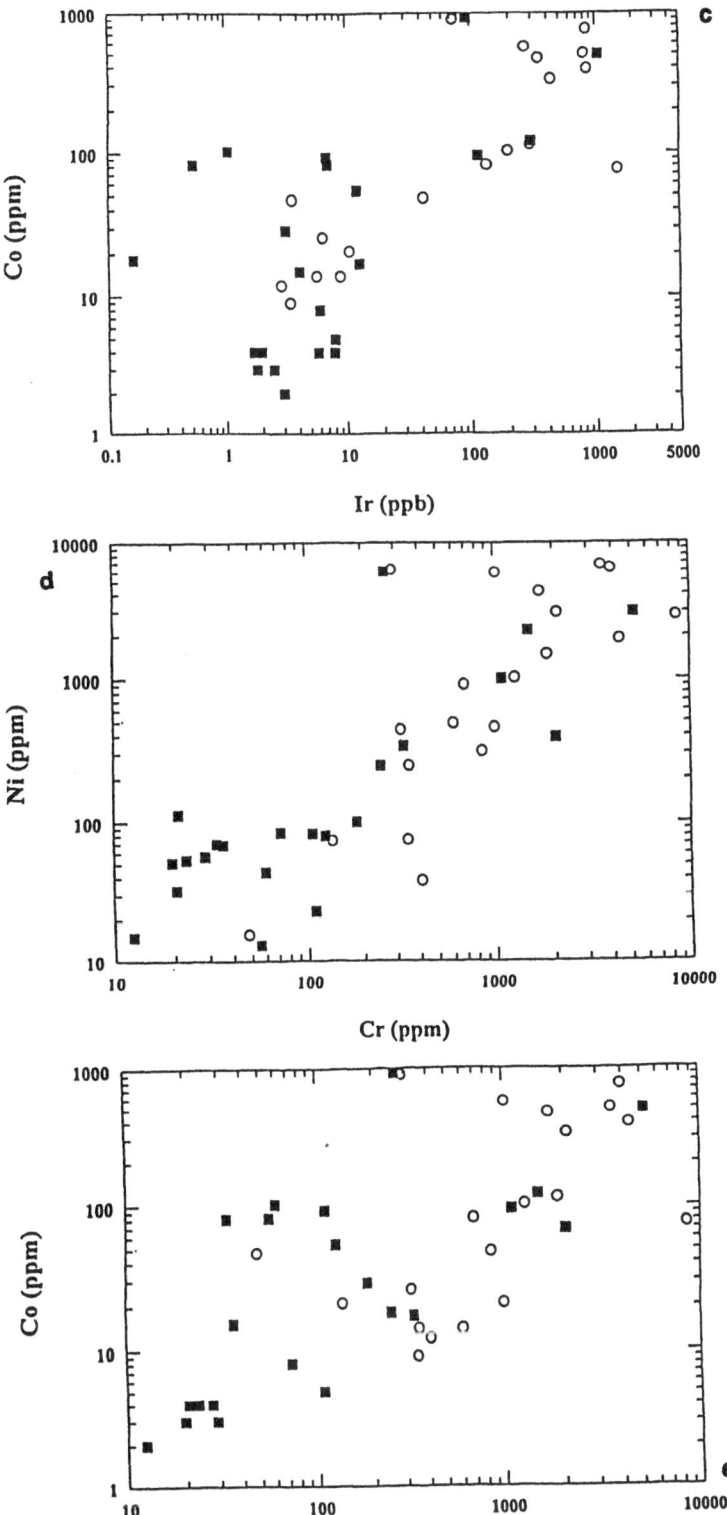

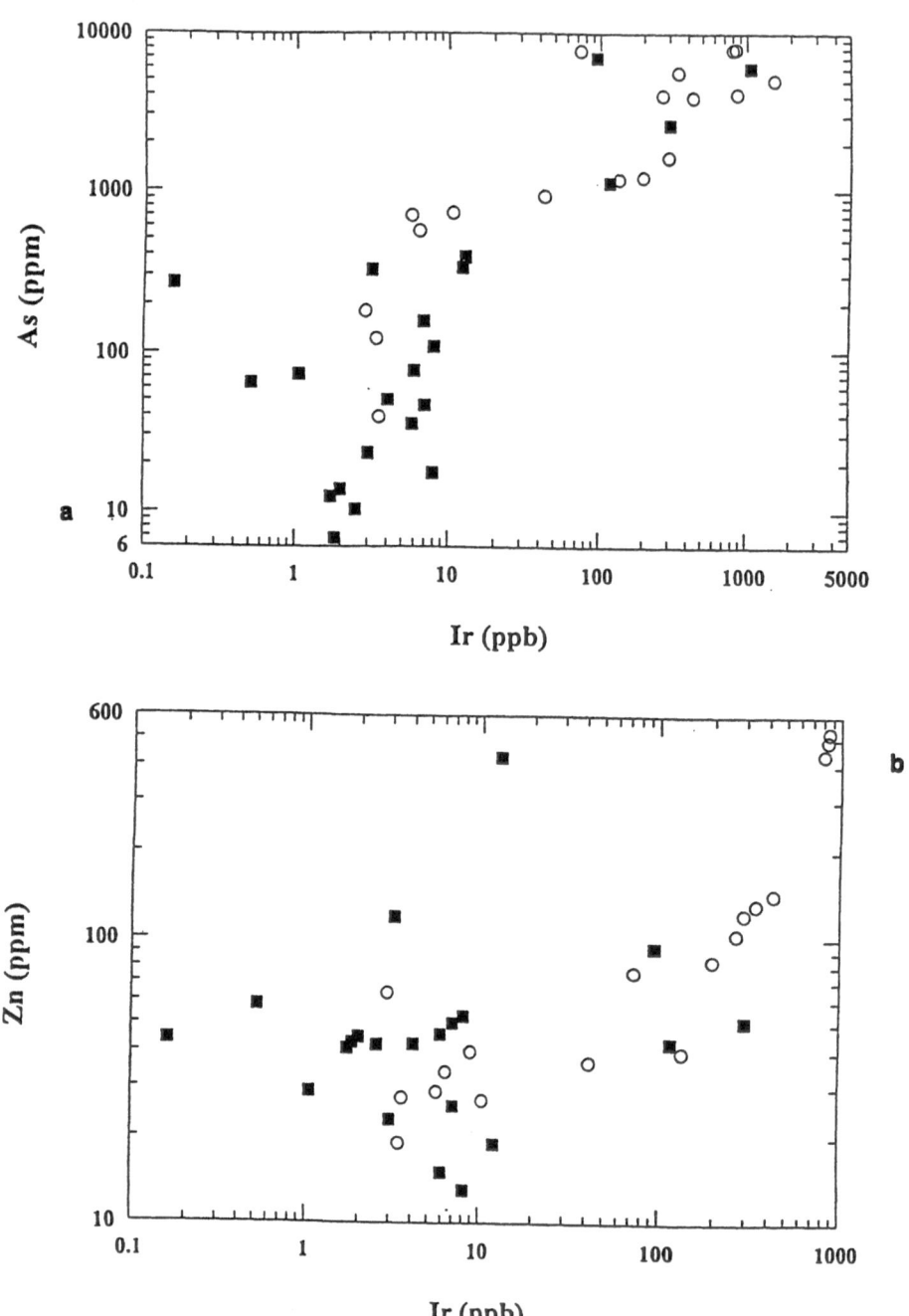

Fig. 12. (above and opposite) Abundances of Ir and Au plotted against those for the mobile elements As, Sb, and Zn in all BA-2 and PU samples given in Table 2a-c. (a) As versus Ir, (b) Sb versus Ir, (c) Zn versus Ir, (d) As versus Au, and (e) Sb versus Au. Spherule samples – open circles; country rocks – solid squares.

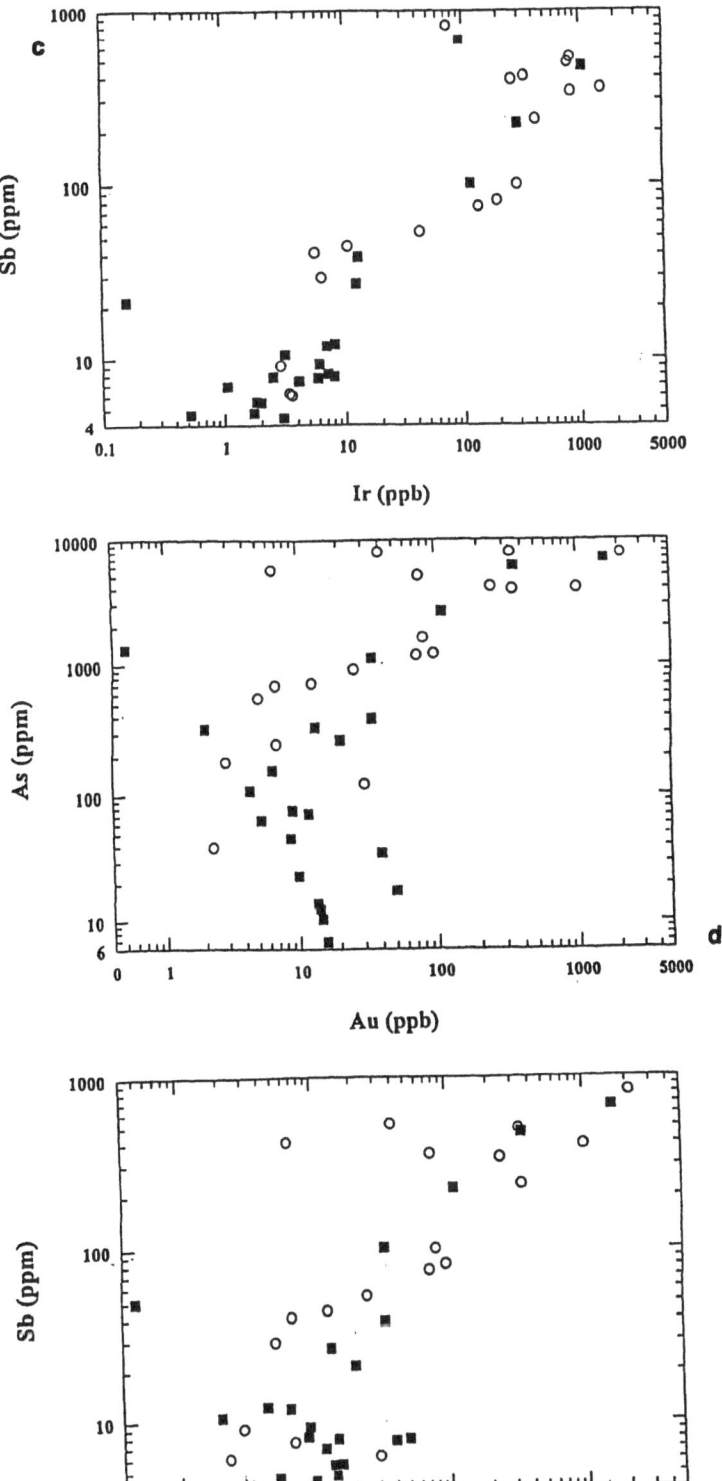

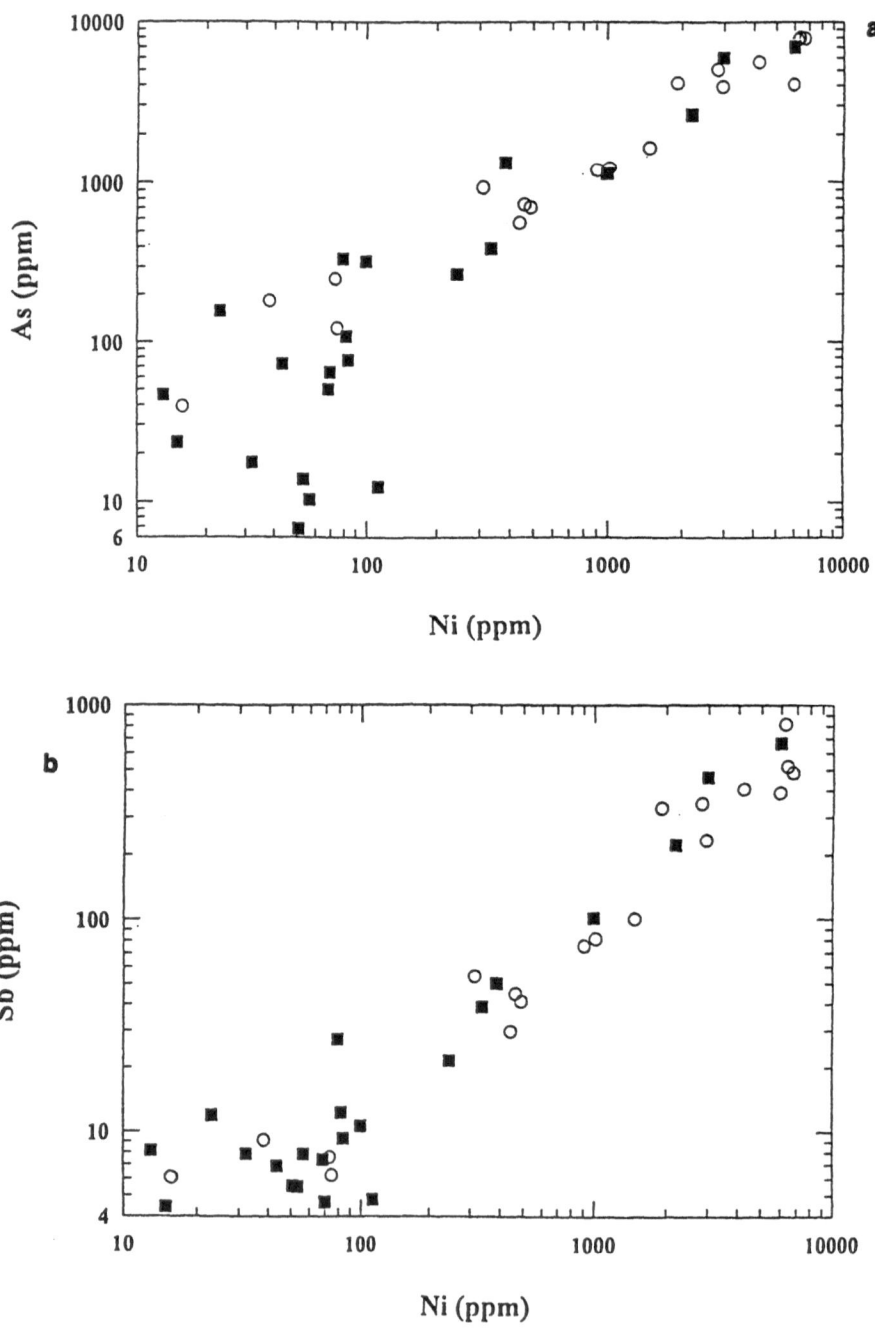

Fig. 13. a-e (above and opposite) Abundances of nickel and cobalt versus those of arsenic and antimony, respectively, in all BA-2 and PU samples given in Tables 2a-c. Spherule samples – open circles; country rocks – solid squares.

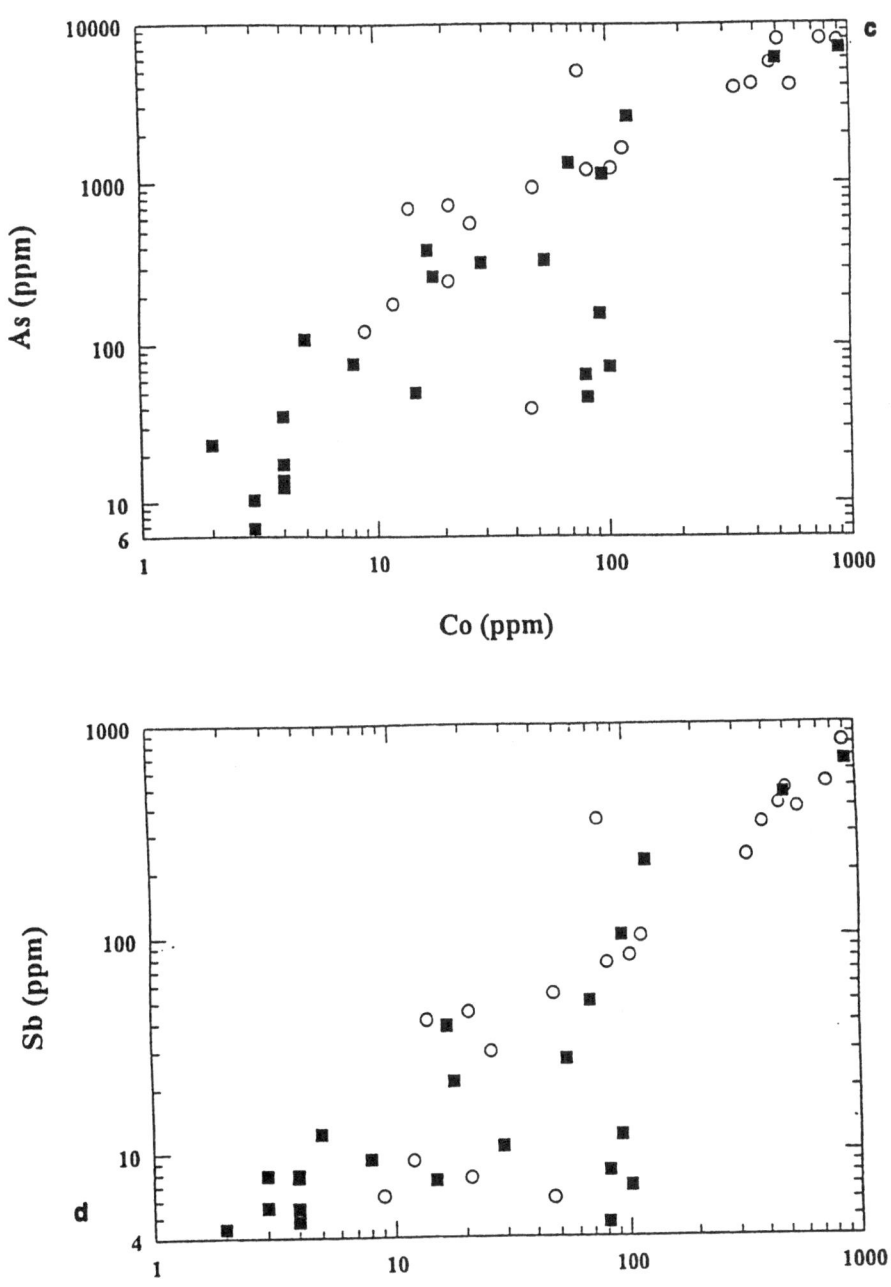

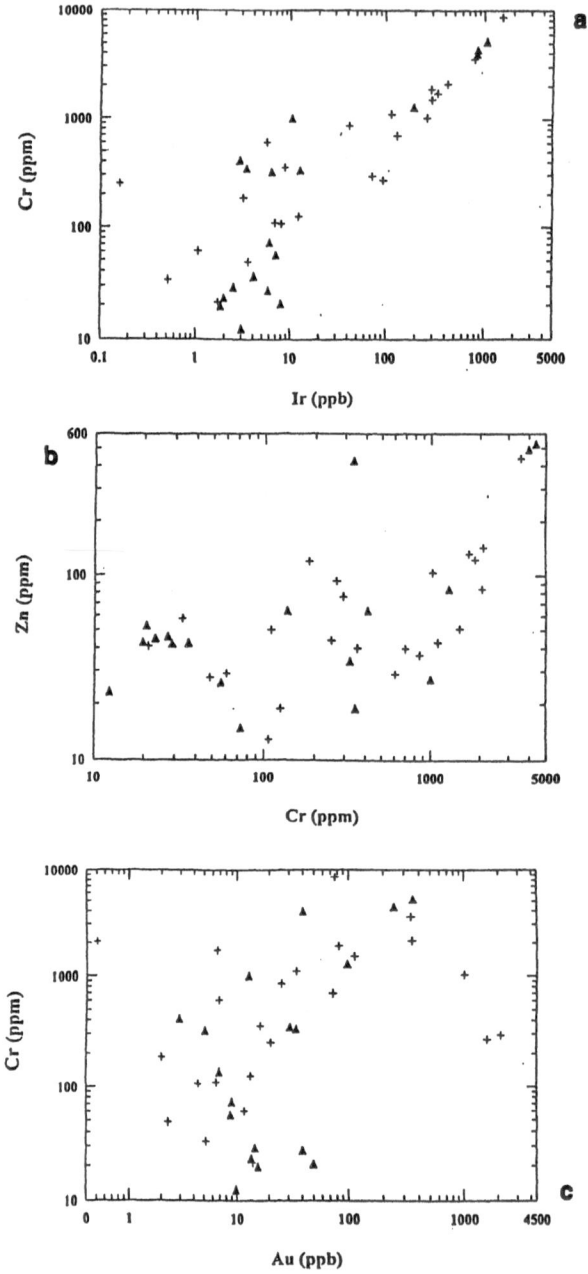

Fig. 14. a-f (above and opposite) Diagrams illustrating the respective relationships (or lack thereof) between Cr and Ir, Zn, Ni, Co, and Au in all BA-2 and PU samples given in Tables 2a-c. Note that scales vary for these diagrams. The sample suite has been divided into those which contain appreciable amounts of sulphide mineralization (>5 %; marked by cross) and those samples which do not (marked by solid triangles).

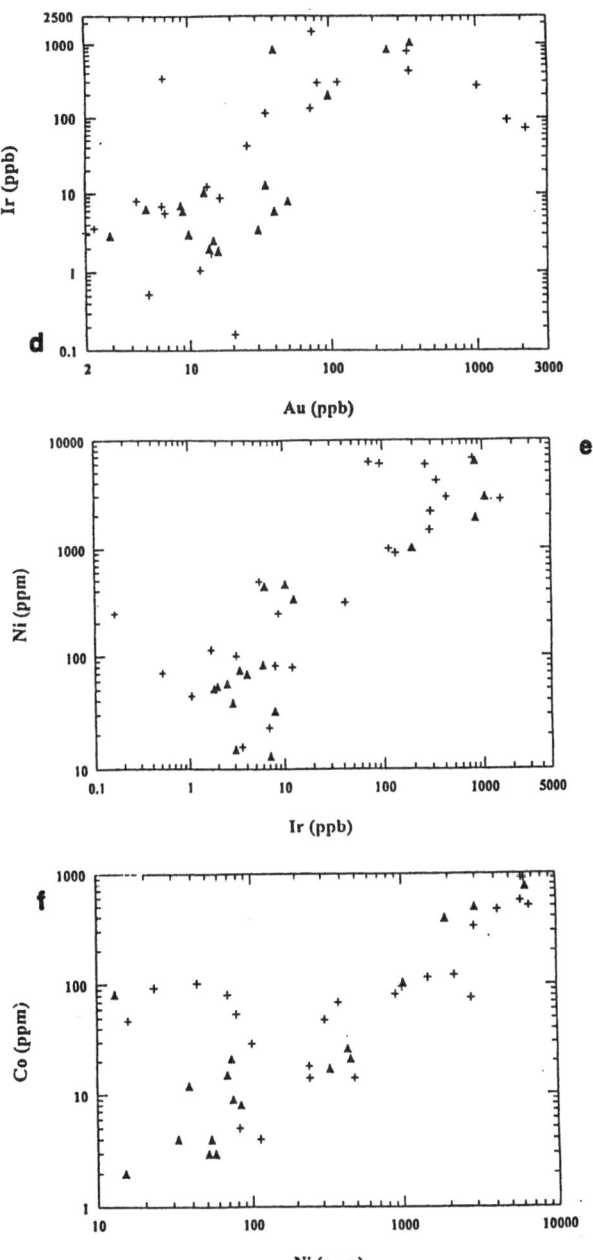

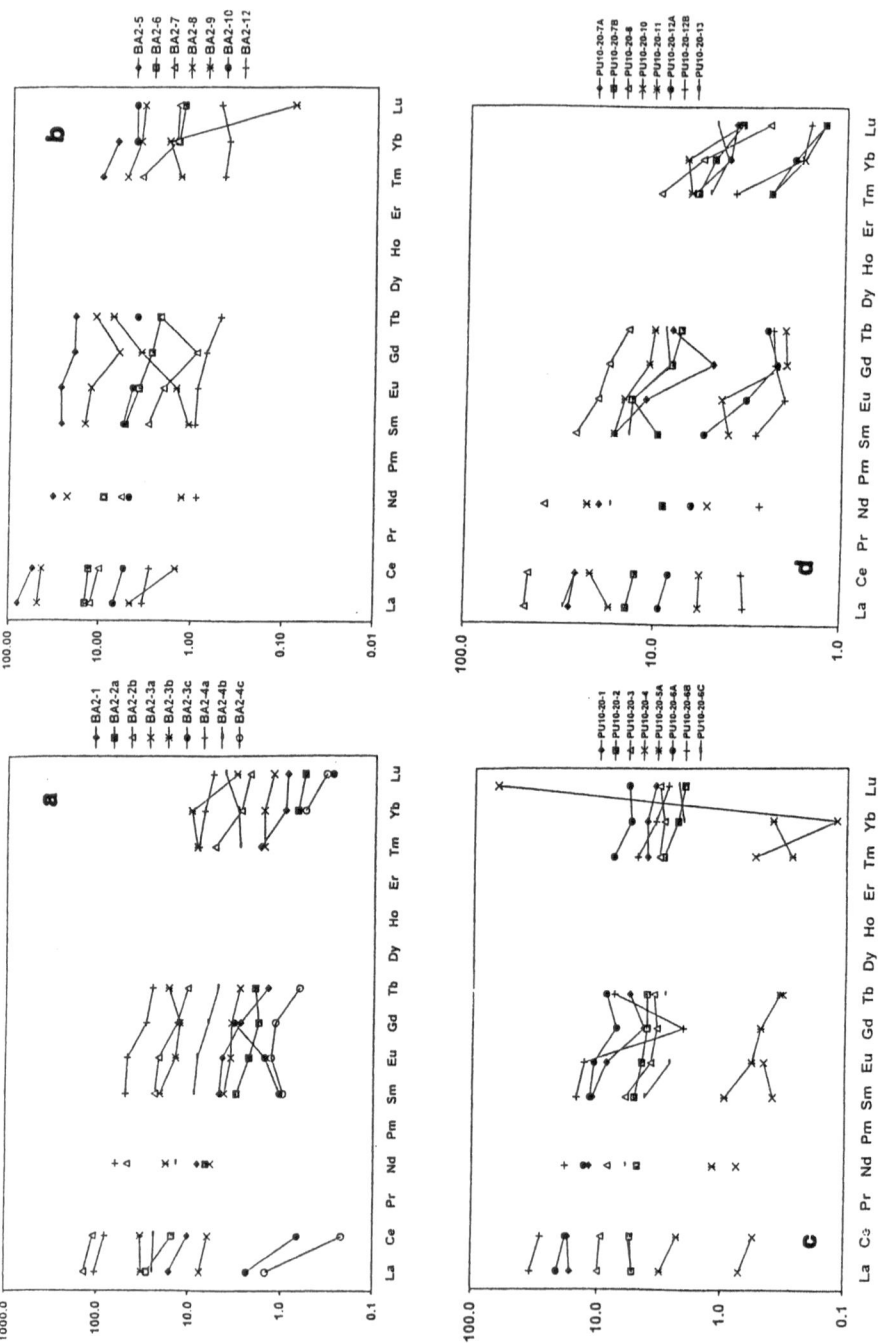

Fig. 15. a-f (above and opposite) Normalized rare earth element abundances for all samples given in Tables 2a-c. Samples are normalized relative to C1 chondrite abundances after Evensen et al. (1978).

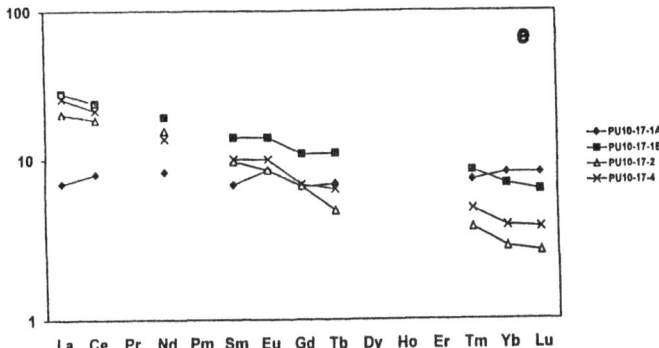

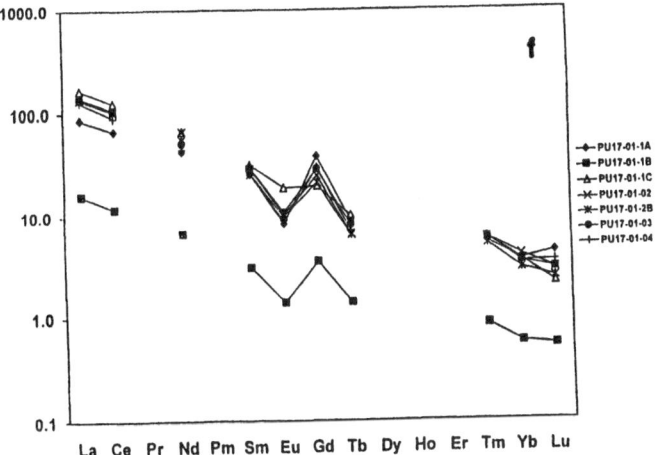

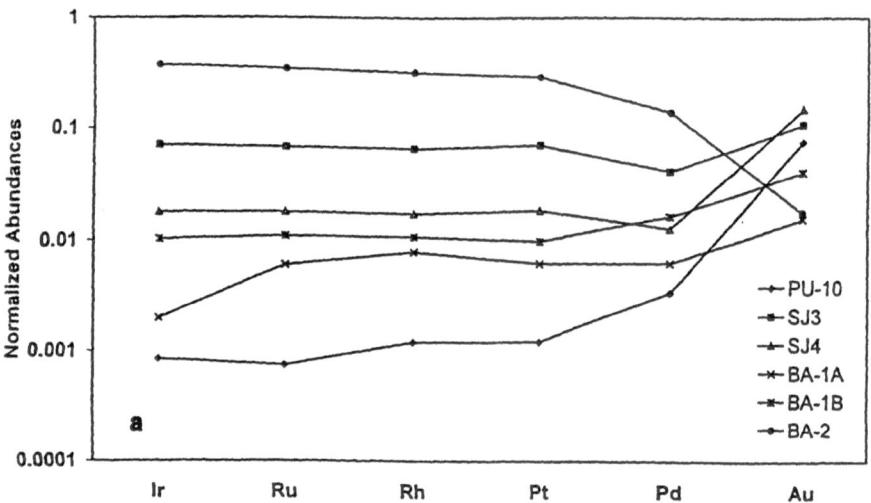

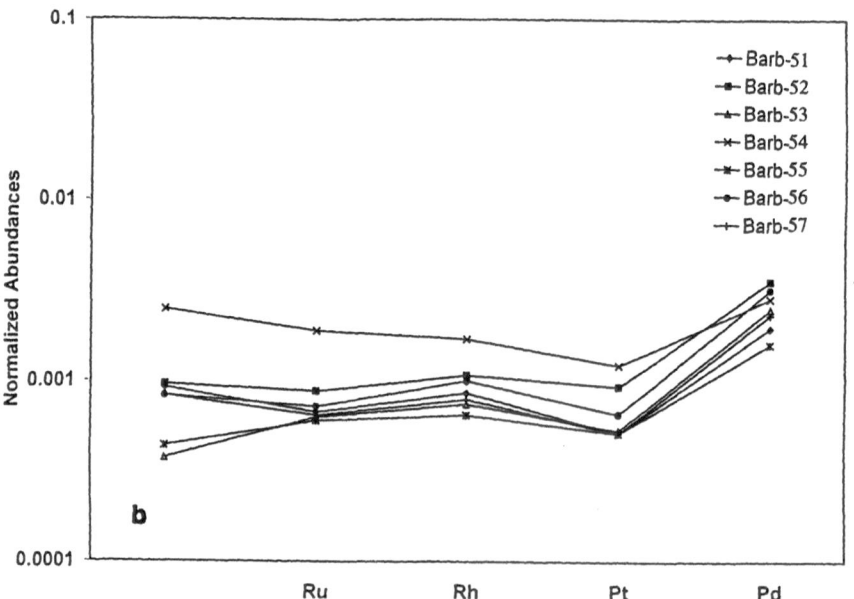

Fig. 16. a-d (above and opposite) Platinum group element abundances normalized to Orgueil CI or average CI compositions (after Jochum, 1996 or Palme et al. 1981, respectively). Compare data in Table 3. Mineralized and non-mineralized samples classified according to gold-sulphide mineralization content.

Orgueil normalized - Non-mineralized samples

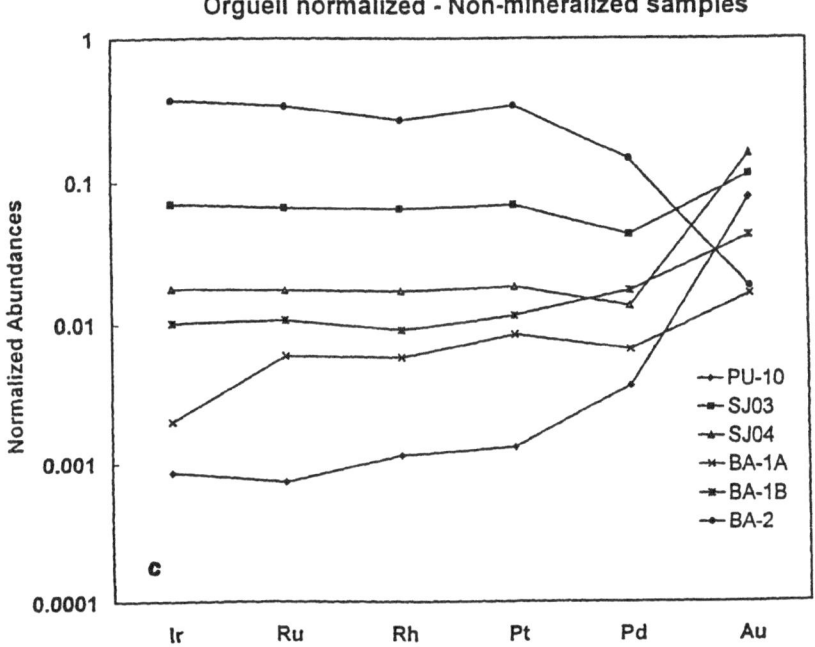

Orgueil normalised - Mineralized samples (no Au)

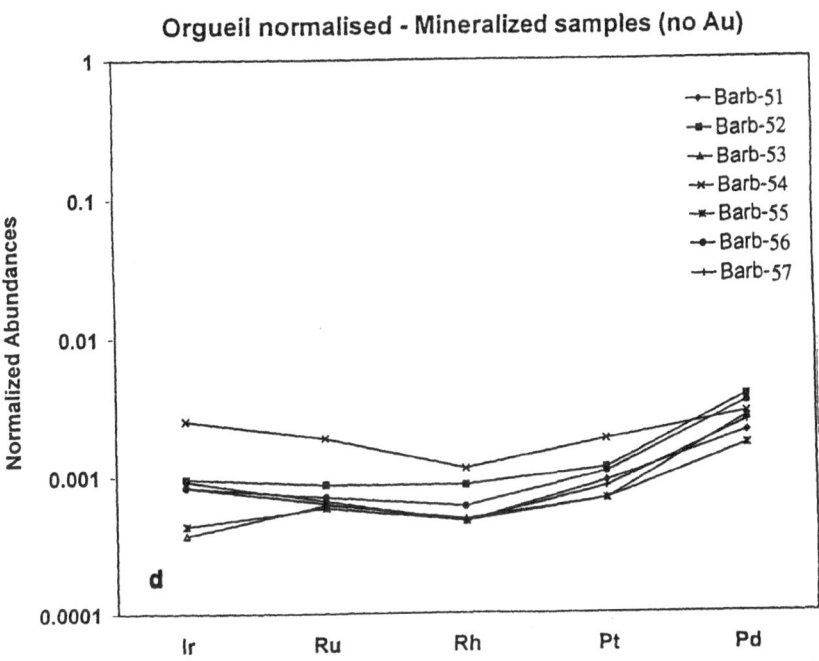

Samples BA-1A and PU-10 have significantly lower Ir concentrations than the previous four samples. Their Ir contents (0.41 and 0.96 ppb) fall within or just below the typical range of komatiites (0.5-3.5 ppb – Dowling and Hill 1992; Zhou 1994), and this – together with a gentle positive slope from Ir to Pd in the normalized patterns – could be regarded as consistent with derivation of some or all of the PGE in these samples from eroded komatiites. These two samples also show variable enrichment in Au, similar to the four samples discussed above.

The B51-57 samples, taken from closer to the ore zone than the samples discussed earlier, contain very high concentrations of gold (up to 4720 ppb) but do not show any evidence of matching PGE enrichment. With the exception of sample B54, these samples contain PGE amounts that are slightly higher than those found in many sediments, which have not been impact-derived (e.g., Huber et al. 1999; Koeberl et al. this volume). When the data are normalized against CI chondrites (Fig. 16), the patterns have a slight positive slope (with enrichment of Pd) and a consistent negative anomaly for Pt. These features are known from komatiites (Dowling and Hill 1992; Tredoux et al. 1995), and it is possible that the moderately high levels of PGE in these rocks could reflect input from a komatiitic source. The pattern for sample B54 is flatter than the rest, and this – together with the generally higher concentrations of Ir, Ru, Rh, and Pt, and the low Pd concentration – could be taken for a hint at the presence of a non-terrestrial component in this sample. Gold in the samples of this group is significantly enriched.

Discussion and Conclusions

New sample suites consisting of both spherule layers and various country rocks from the S2 horizon in Princeton section of Agnes Gold Mine near Barberton have been studied petrographically and chemically. The stratigraphic position of the analyzed intervals is defined as good as the complex regional geology and mining constraints permit. The studied section seems to coincide with the S2 horizon as defined by Lowe et al. (1989). In contrast to the earlier investigations by Koeberl et al. (1993) and Koeberl and Reimold (1995) which concentrated mainly on strongly sulfide-mineralized samples, this new study focused on samples from localities removed by significant distances from the main ore zone of this mine. However, it was found that these new samples, too, have all been pervasively hydrothermally overprinted. This alteration is of regional extent and any interpretation of chemical data for samples from this region *must* remain conscious of this fact!

Not only is evidence for pervasive alteration present, but these new sample suites also indicate tectonic activity. Duplication of some strata, including spherule layers, has been observed. Hydrothermal activity must be linked to both syn-tectonic and post-tectonic phases. Primary mineral parageneses can no longer be studied. Whether sulfide (and associated gold) mineralization can be directly linked to a specific, pervasive alteration event is not clear, but it appears from morphological and textural evidence that more than one event contributed to the pervasive alteration.

In the light of recent geochemical evidence for the presence of extraterrestrial chromium in spherule samples from the S3(?) and S4 horizons (Shukolyukov et al. 1998, 1999), the petrographic observation of multiply zoned chrome-spinel in our samples is surprising. EPMA and PIXE analysis demonstrated that multiple chemical zonation is present in at least some chrome-spinel, and the chemical systematics observed strongly favor hydrothermal overgrowth as the process responsible. This leaves the question whether possibly existing cosmic chromium remains in a regional - generally open - chemical system, without significant fractionation (for example, secondary crystallization processes). Another question to be answered is whether chromium in the S2 layer will show a cosmic isotopic signature as well – or whether each spherule layer in the Barberton stratigraphy needs to be considered at its own merit.

Extensive multi-element geochemical analysis of both spherule layer and country rock samples shows complex relationships between critical elements (critical with regard to the discussion about the origin of the spherule layers and the siderophile element signatures of these stratigraphic intervals). Ir, Ni, Cr, and Co abundances show, at best, weak interelement correlations and - in comparison to data for other elements – inconsistent degrees of better or worse correlation. More confusing is that relative enrichments of samples do not conform to specific sample types, such as spherule layer or country rock types, or groups of well or less mineralized samples, respectively. Overall, an image of complex secondary hydrothermal overprint emerges, which does not allow the identification of primary chemical identities of spherule horizons.

Whereas some of the high iridium-bearing country rock samples occur in close proximity to spherule beds, others are separated from the next spherule layer (for example, borehole PU-17/01) by at least 1.5 meter. This does not corroborate a suggestion by F. Kyte (pers. commun. 1999) that such enriched country rock strata could be considered part of complex, inter-stratified impact ejecta layers. The recently made observation that Ir enrichment (against regional background) is also found in mylonitic or graphite-coated shear and fault plane settings further emphasizes that this element must have been mobilized during post-depositional events. By implication, it must be expected that other PGEs have also been mobilized.

It must be noted that it has not been possible to successfully carry out stratigraphic correlation between the four intervals studied by us in drillcore. In other words, the complex vertical and lateral stratigraphy of this part of the Barberton Greenstone Belt, obviously a consequence of poly-tectonic deformation, does not allow the use of individual siderophile element-enriched layers as stratigraphic markers. It is very difficult to ascertain how many spherule layers actually occur in this interval, and in exactly which stratigraphic position(s). As shown by Koeberl et al. (1993) and Koeberl and Reimold (1994, 1995), many narrow spherule layers may occur within 20-30 cm intervals alone. To assume that they originally represented a single depositional layer and all have been split or duplicated due to later tectonic or sedimentary overprint is speculative. Another handicap for chemo-stratigraphic work is the fact that spherule and country rocks are enriched in Ir and other siderophile elements, at highly variable proportions.

In contrast to these last thoughts, the observation that four of six little sulfide-mineralized samples contain Ir, Ru, Rh, and Pt in chondritic relative abundances and at concentrations in excess of any terrestrial melt, which could provide a primary source rock, is difficult to explain by invoking any simple terrestrial mechanism. Even if the source material for these layers was PGE-rich sulfide from a mineralized komatiite (e.g., Kambalda-type sulfides – Lesher 1989) or even the exotic Bon Accord Ni-Fe-PGE material described from a section of the Onverwacht Group by Tredoux et al. (1989a), it must be noted that both these materials are characterized by distinctive fractionation of PGE relative to CI chondrites. It would be highly coincidental if originally fractionated sulfide or metal would be altered back to almost CI chondritic signature. This sort of explanation for the high PGE abundances and overall signature must be considered unlikely. Whereas there is good evidence that Au and, to a lesser extent, Pd have been mobile (enriching or depleting individual samples), the simplest explanation for the relative enrichment of the other PGEs would be that they are impact-derived and, on a scale of 10-20 g samples, have remained relatively immobile compared to Au, Pd, Cr, and Ni, during post-depositional metamorphism and alteration.

The effect of gold mineralization on the B51-57 samples has been demonstrated. Gold is significantly enriched but, if one assumes that PGE in these rocks are derived from weathered mafic/ultramafic rocks, the fact that the PGE concentrations in the mineralized samples are almost exactly the same as those of non- (or less) mineralized samples containing similar terrestrial PGE would argue that the PGE – especially Ir, Ru, and Rh – have not been strongly enriched or depleted by the mineralizing fluids. Silicification/carbonatisation may have lowered the absolute PGE concentrations in some layers, but there is no evidence for any large-scale fractionation of the PGEs, as indicated by the chondrite-normalized patterns. Whilst there is extensive evidence for metal mobility at the millimeter to centimeter scale (Koeberl et al. 1993; this study), on the scale of larger samples, as those taken for the noble metal analyses discussed here, the PGE (excluding Pd!) may have largely remained immobile.

When comparing samples containing spherules (either as a major component or as an associated component) and those assumed to be made up primarily of country rock, a contrast in PGE signatures is revealed. Spherule-bearing samples, with a possible exception being the BA-1A portion of sample BA-1 (Table 4), contain high concentrations of Ir, Ru and Rh, and normalized PGE patterns are consistent with a non-terrestrial PGE input. Spherule-poor or -lacking samples contain PGE in proportions more consistent with a local, possibly komatiitic source.

However, whilst the presence of an extraterrestrial aspect to the Barberton spherule layers is seemingly supported by the PGE signatures, as well as by the recent Cr isotopic measurements, the remarkably high PGE concentrations (with respect to Ir also in some country rocks and clearly secondary fault/shear deposits) and the thicknesses of some of the spherule layers and what this could indicate about the size of a projectile remain controversial. As noted earlier, most well preserved distal impact ejecta contain < 1% meteoritic component. Only some K-T boundary sites, such as Stevns Klint, do contain high PGE concentrations (Ganapathy 1980; Tredoux et al. 1989b; Evans et al. 1993) indicative of up to, or - in some cases - in excess of, 10% meteoritic component. Shukolyukov et al. (1998, and this volume) have argued that the high PGE concentrations (up to 420 ppb Ir and the 100-150 mm thickness of the S4 spherule layer (after Kyte et al. 1992; note that we have measured still higher concentrations in some samples!) must indicate derivation from a massive projectile. Whilst it is true to say that average PGE concentration in the S4 layer is high, closer analysis of the data by Kyte et al. (1992) reveals more complex variations. There are at least two high-PGE zones (their subsamples D3-D4 and D11-D13) with two low-PGE zones (sub-samples D5-D10 and D14-D15) immediately above them. Sub-samples D1 and D2 may also define a thin zone with elevated PGEs and a thin low-PGE layer immediately below these. A cruder separation into high- and low-siderophile concentration layers, coincident with the variation according to PGE concentrations, is also evident in the Cr and Ni data. It could be argued that the PGE might have been mobilized to distinct bands from a more homogeneous layer, but this would require mobilization of PGEs over several centimeters. This is considered unlikely, given the known differences in mobility of the various PGEs during alteration. One could also imagine dilution due to silicification/carbonatisation in certain layers only – which sounds reasonable considering the mineralogical compositions of the spherule layers, but is not well constrained.

Alternatively, if the PGEs have been as immobile as suggested above and were originally associated with phases such as spinel or alloys, the high and low PGE zones could reflect a primary feature such as grading of dense PGE-rich material towards the base of depositional units. In this interpretation, the S4 layer, where it has been studied by Kyte et al. (1992), may have been produced by multiple cycles of deposition. The high PGE zones could be considered the product of sorting and redistribution of specific minerals during redistribution of primary material (i.e., S4 could be termed a 'paleoplacer'). This might explain the remarkably high PGE concentrations found in certain 5 mm thick subsamples of S4, which – if taken at face value – could imply meteoritic contributions between 50 and 70%; this is, of course, almost an order of magnitude higher than the most enriched K-T boundary sample ever analyzed. Besides, such high contributions are, in no way, feasible, as they should be reflected, at least in the case of some immobile elements, in the bulk rock chemical composition of these samples. To be cautious – this discussion is fictitious and neither supported by sedimentological nor any other evidence. What is more, how can the much higher – up to > 2500 ppb - Ir concentrations determined by our group in individual samples from the BA sample suite, or the up to >1500 ppb values in country rock samples from this study be integrated into such an impact/redistribution process? Would it be sufficient – or satisfactory - to draw on convenient, but unconstrained secondary alteration?

If the S4 layer contains multiple cycles of redistributed ejecta, its current thickness may have nothing to do with how material was spread and deposited by an initial impact event, and its thickness can, therefore, not be used to model the volume of ejecta involved and to deduce the size of an impactor. In addition, if the PGEs were simply mechanically distributed/concentrated, and the initial distribution of PGEs in relatively dense ejecta was not dramatically altered by redeposition, then the relative proportions of more immobile PGEs (Ir, Ru, Rh and Pt) could still reflect original meteoritic composition. However, the absolute concentrations of the metals could not be used to reconstruct the original mass of the projectile. Studies that attempt to reconstruct the projectile must obviously establish that the whole S4 layer (or any other layer, for that matter!) actually reflects a single depositional event. It is also not clear why the PGE and other siderophile element (e.g., Cr, Ni, Co) data do not correlate well for our S2 sample suite. This immediately raises the question whether the PGE signatures described here are indeed of primary value or are the result of – probably complex – secondary processes?

In Figure 10a, apparent trends correlating Ir and Cr are shown for both our own spherule layer and country rock sample data and the results for spherule layer and komatiite samples by Lowe et al. (1989). It is obvious that the slopes of the four trends are distinctly different – a first order observation that argues against a mutual and uniform origin of these two elements in the two spherule layer sample suites. Alternatively, pervasive open-system behavior of at least one of these elements is needed to achieve this complex systematic. However, due to the fact that the Lowe et al. (1989) trend also incorporates some S2 samples, this alternative does not explain why samples from the same layer should plot into both spherule layer trends. All other chemical evidence (except the PGE data) presented here indicates that the siderophile element signatures observed are of secondary origin.

Finally, we again failed to detect any direct evidence of impact in the form of shock metamorphism. Whilst a significant number of clastic, including quartz, particles could be detected in our spherule layer samples, no shock metamorphic effects could be observed.

The reason for the complete absence of such evidence from the Barberton spherule samples is unknown (if they really are of impact origin), although recently Simonson et al. (1998) proposed that impacts into deep ocean basins may not produce readily recognizable shocked ejecta. The study of Simonson et al. (1998) dealt with Late Archean spherule beds from the Hamersley Basin in Western Australia. Simonson (1992), Hassler and Simonson (1995), Simonson and Davies (1996), and Woodhead et al. (1998) described at least two (possibly four) distinct spherule layers from different locations in the Hamersley Basin and elsewhere with ages of about 2.54 Ga. The spherules are supposedly similar to those from the K-T boundary impact layer or the late Eocene clinopyroxene spherules (e.g., Glass et al. 1985; Jéhanno et al. 1992), but they have characteristics that are distinctly different from those of the Barberton spherules. The Hamersley Basin spherules occur within turbidite or debris flow deposits. Geochemical analyses (Simonson et al. 1998) have shown that they are associated with minor Ir anomalies that average 0.5 to 0.6 ppb Ir (with maximum values of 1.7 ppb) and elevated contents of Pd and Pt, but with non-chondritic interlelement ratios or patterns of the siderophile elements. Simonson et al. (1998) interpreted the non-chondritic PGE patterns as evidence of re-distribution after an impact event, not unlike what has been found for the PGEs in Acraman ejecta. These authors also suggested that the total absence of shocked quartz indicates an oceanic impact. While the low quartz content of oceanic rocks make it less likely that shocked quartz is formed, the total absence of any shocked minerals (including shocked feldspar) is somewhat unexpected.

Smit (1999) has given a detailed account of the inter-stratigraphy and lateral (from proximal – near the Chicxulub impact structure - to distal facies) variations of the K-T boundary layer(s). It is clear with regard to both the distribution of siderophile elements laterally away from the source of meteoritic contamination, the relative positions of such chemical meteoritic signatures and spherule concentrations within the K-T horizon, and the relative and regional enrichment of siderophile elements in this layer, that the K-T boundary layer is highly variable in its internal characteristics. Furthermore, nothing definite is known about the carrier phases of the PGEs in the K/T boundary material. In the absence of a complete lateral record for even this horizon and of any detailed depositional information for other relatively recent impact ejecta layers, it must be considered speculative, at best, to utilize the spurious information from currently known Archean layers of possible impact affinity to conclude on aspects such as distance from impact site or possible size of impactors (e.g., Shukolyukov et al. 2000). As long as there is still strong debate about fractionation of PGE and other siderophile elements in spherule layers and about the completely unexplained observation that certain country rock samples from the Barberton area are as enriched as spherule bed samples in apparently unfractionated (chondritic) PGEs, any discussion of primary, impact related features is premature.

The interpretation of the origin of the Barberton spherule layers poses a number of problems. Whereas the recent Cr isotopic results and PGE studies seem to be in favor of an origin by impact (Lowe et al. 1989; Kyte et al. 1992; Shukolyukov et al. 1998), several objections to that interpretation remain. The Barberton spherule beds are characterized by the absence of primary minerals and the total lack of any evidence of shock metamorphism. The textures of the spherules can be explained as growth patterns and do not necessarily represent quench textures in impact-generated melts. Very high PGE abundances were found (e.g., up to 2700 ppb Ir). This is in stark contrast with abundances in chondritic meteorites (the meteorite type favored by the Cr isotopic results of Shukolyukov et al. 1998, 1999 for the presumed 'Barberton impactor(s)'), with Ir abundances of about 400 to 600 ppb. Almost all known impact deposits contain less than 1 % of such a meteoritic component (with somewhat elevated values in some K-T boundary clay samples, which nevertheless never even come close to chondritic values and could well be the result of secondary processes). Thus, the extremely high values in the Barberton samples cannot be primary, but must be due to alteration and redeposition, probably, formation of secondary sulfide mineralization. It is difficult to comprehend (but, of course, not totally impossible) that PGE interelement ratios could remain at primary chondritic values during such pervasive reworking, as the individual PGEs have different mobilities.

The composition of Ni-rich Cr-spinels found in samples from the Barberton spherule layers is distinctly different from that of spinels found in known impact deposits or meteorites. Their chemical zonation also favors a multi-stage formation and chemical evolution and is not consistent with a primary, impact-related origin. In many cases they are intergrown with secondary sulfides and may, thus, themselves be of secondary origin. The Barberton spinels have generally well-crystallized, euhedral shapes and range in size from <10 to >150 μm – in contrast to generally much smaller and often dendritic spinels from other confirmed impact ejecta horizons, such as the K-T boundary or in the late Eocene.

In only one sample has it been possible so far to confirm the nature of the host phase of Ir. We have shown that this element is hosted in presumably secondary gersdorffite, whereas in other cases (such as the sulfide separates analyzed by Koeberl and Reimold 1995) iridium seems to be disseminated in matrix silicates of spherule layers (as also assumed by Lowe et al. 1989). However, these authors proposed that Cr and Ir were correlated and hosted in the same phase (chrome-spinel), which is not supported by the findings of the present study.

Glikson (1994) explained the absence of shock features in the spherule beds by deposition far away from the impact location(s). However, in such a case it is difficult to understand why such massive spherule deposits would be present. Any known large-scale impact events, including the K-T boundary or the late Eocene spherule layers, are associated with only very thin distal spherule layers, and no massive proximal spherule layers either. However, all these deposits contain shocked minerals (e.g., quartz or feldspar with PDFs), notably in the K/T boundary clay with global distribution. Of course, and as discussed above, it can not be excluded that an Archean impact failed to 'hit' into felsic crust containing those minerals, such as quartz or zircon, which could preserve relics of shock effects for billions of years.

It has been suggested to us by F. Kyte (pers. commun. 1999) that the observation that country rock samples may be enriched in siderophile elements to the same degree as spherule layer samples could be explained by incorporating such strata into an internal stratification of an impact ejecta layer. Taking our PU10-17 section (Figure 3) as an example, it is noted that the sampled section was removed laterally from the nearest spherule layer by at least several tens of meters and that – vertically - no spherule layer was detected in this drillcore over a distance of at least 40 meters on either side of the sampled section. Clearly, the argument that it is likely that our subsamples are all part of a complexly stratified impact ejecta layer is untenable. The small-scale layering in the BIF-mudstone-chert sequences characteristic of the regional transition from the Fig Tree to underlying Onverwacht Group is so intricate (on a millimeter to centimeter scale, however, sometimes extending to the meter scale) and multiple that no comparison to the only reasonably well studied impact ejecta layer, namely that of the K-T boundary is possible. The K-T horizon is characterized over large distance by a double-layer which locally contains a zone of a few centimeters thickness of spherule-rich material and another part that is enriched in siderophile elements (Smit 1999). Thickness of the spherule-rich zone increases towards (relative) proximity of the Chicxulub structure, but has not been described to attain a thickness of >1 meter. Naturally, the pervasive hydrothermal overprint on Barberton strata and frequently observed strong deformation (shearing, mylonitisation, cataclasis, tectonic duplication) complicates small-scale stratigraphic studies, especially when only drillcore material is available. In conclusion, much more detailed and laterally extensive work is required with regard to stratigraphic correlation within the interval straddling the Fig Tree-Onverwacht transition.

Archean impact deposits must exist, as impact events must have occurred over the whole geological history of the Earth, and may have been more frequent in pre-3.5 Ga times than now (e.g., Frey 1980; Glikson 1993; Grieve 1980; Koeberl and Sharpton 1988; Koeberl et al. 2000). Several horizons containing spherules (other than the Barberton spherules) have been found within Archean rocks and have been interpreted as having formed through impact processes (e.g., Simonson 1992; Simonson et al. 1997, 1998; Koeberl et al. 1999). However, due to possible problems associated in generating impact-characteristic shock deformation in oceanic impact events (considering that in early Archean times the ratio of oceanic crust to continental crust was much higher than today), and the possibility of severe alteration during the last several billion years, the identification of such early impact deposits remains a difficult problem. In the future, much more detailed studies are needed to define specific criteria for the identification of such early Archean impact deposits.

Acknowledgments

We thank D. Scott, M. van den Berg and C. Robus, Eastern Transvaal Consolidated Mines Limited (Barberton), and their staff for providing samples, assistance in the field, unpublished information on local geology, and geological interpretations of the stratigraphic control of the samples. S.J. would like to thank

Mrs. N. Day for assistance with operating the EMP at the Rand Afrikaans University, and G. Kurat for the use of and F. Brandstätter for the operation of the SEM instrument at the Natural History Museum, Vienna, Austria. Woijtek Przybylowicz and the staff at the National Accelerator Center in Faure (South Africa) made the proton microprobe study possible. Rudy Boer's assistance in obtaining the new samples is gratefully acknowledged. We also acknowledge the European Science Foundation (ESF) program on "Response of the Earth System to Impact Processes". This research was supported by the South African FRD (now NRF) (to W.U.R.) and by the Austrian FWF, project Y58-GEO (to C.K.). A very thorough and candid review by Frank Kyte is appreciated. This paper represents University of the Witwatersrand Impact Cratering Research Group Contribution No. 12.

References

Anders E, Grevesse N (1989) Abundances of the elements: Meteoritic and solar. Geochim Cosmochim Acta 53: 197-214

Anhaeusser C.R (1986) Archean gold mineralization in the Barberton Mountain Land. In: Anhaeusser CR, Maske S (eds) Mineral Deposits of Southern Africa. Geol Soc S Afr, Johannesburg, pp 113-154

Bjerg EA, de Brodtkorb MK, Stumpfl EF (1993) Compositional zoning in Zn-chromites from the Cordillera Frontal Range, Argentina. Min Mag 57: 131-139

Bohor BF, Foord EE, Ganapathy R (1986) Magnesioferrite from the Cretaceous-Tertiary boundary, Caravaca, Spain. Earth Planet Sci Lett 81: 57-66

Buick R (1987) Comment on "Early Archean silicate spherules of probable impact origin, South Africa and Western Australia". Geology 15: 180-181

Byerly GR, Lowe DR (1992) Exotic nickel-chromites in impact spherules from the Archean Barberton Greenstone Belt. Lunar Planet Sci 23: 193-194

Byerly GR, Lowe DR (1994) Spinel from Archean impact spherules. Geochim Cosmochim Acta 58: 3469-3486

Churms CL, Pilcher JV, Springhorn KA, Tapper UAS (1993) A VAX and PC-based data acquisition system for MCA, scanning and LIST-mode analysis. Nucl Instr Meth B 77: 56-61

Colodner DC, Boyle EA, Edmond JM, Thomson J (1992) Post-depositional mobility of platinum, iridium and rhenium in marine sediments. Nature 358: 402-404

de Ronde CEJ, de Wit MJ (1994) Tectonic history of the Barberton greenstone belt, South Africa: 490 million years of Archean crustal evolution. Tectonics 13: 983-1005

de Ronde CEJ, Hall CM, York D, Spooner, ETC (1991a) Laser step-heating $^{40}Ar/^{39}Ar$ age spectra from early Archean (~3.5 Ga) Barberton greenstone belt sediments: A technique for detecting cryptic tectono-thermal events. Geochim Cosmochim Acta 55: 1933-1951

de Ronde CEJ, Kamo S, Davis DW, de Wit MJ, Spooner ETC (1991b) Field, geochemical and U-Pb isotopic constraints from hypabyssal felsic intrusions within the Barberton greenstone belt, South Africa: Implications for tectonics and the timing of gold mineralization. Precambr Res 49: 261-280 de Ronde CEJ, Spooner, ECT, de Wit, MJ, Bray CJ (1992) Shear zone-related, Au quartz vein deposits in the Barberton Greenstone Belt, South Africa: Field and petrographic characteristics, fluid properties and light stable isotope geochemistry. Econ Geol 87: 366-402

de Ronde CEJ, de Wit MJ, Spooner ETC (1994) Early Archean (>3.2 Ga) Fe-oxide-rich, hydrothermal discharge vents in the Barberton greenstone belt, South Africa. Geol Soc Amer Bull 106: 86-104

de Waal SA (1978) Nickel minerals from Barberton, South Africa: VII. The spinels Co-chromite and Ni-chromite and their significance for the origin of the Bon Accord nickel deposit. Bull BRGM II (2): 223-230

de Wit MJ (1982) Gliding and overthrust nappe tectonics in the Barberton Greenstone Belt. J Struct Geol 4: 117-136

de Wit MJ (1986) A possible origin for chondrule-like particles in the 3.6-3.3 Ga Barberton greenstone belt, South Africa. Lunar Planet Sci 17: 182-183

de Wit MJ, Ashwal LD (eds) (1997) Greenstone Belts. Oxford Mon Geol Geophys 35, Oxford Science Publications, 809 pp

Dowling SE, Hill RET (1992) The distribution of PGE in fractionated Archaean komatiites, Western and Central Ultramafic Units, Mt. Keith region, Western Australia. Austral J Earth Sci 39: 349-363

Evans NJ, Gregoire DC, Grieve RAF, Goodfellow WD, Veizer J (1993) Use of platinum-group elements for impactor identification: Terrestrial impact craters and Cretaceous-Tertiary boundary. Geochim Cosmochim Acta 57: 3737-3748

Evensen MN, Hamilton PJ, O'Nions RK (1978) Rare-earth abundances in chondritic meteorites. Geochim Cosmochim Acta 42: 1199-1212

French BM (1987) Comment on "Early Archean silicate spherules of probable impact origin, South Africa and Western Australia". Geology 15: 178-179

Frey H (1980) Crustal evolution of the early earth: The role of major impacts. Precambr Res 10: 195-216

Ganapathy R (1984) A major meteorite impact on the earth 65 million years ago: Evidence from the Cretaceous-Tertiary boundary clay. Science 209: 921-923

Glass BP, Burns CA, Crosbie JR, DuBois DL (1985) Late Eocene North American microtektites and clinopyroxene bearing spherules. Proc Lunar Planet Sci Conf 16, J Geophys Res 90: D175-196

Glikson AY (1993) Asteroids and early Precambrian crustal evolution. Earth-Sci Rev 35: 285-319

Glikson AY (1994) Archaean spherule beds: Impact or terrestrial origin? A discussion of the paper 'Geochemistry and mineralogy of Early Archean spherule beds, Barberton Mountain Land, South Africa: Evidence for origin by impact doubtful' by C Koeberl, WU Reimold, R Boer. Earth Planet Sci Lett 126: 493-496

Gostin VA, Haines PW, Jenkins RJE, Compston W, Williams IS (1986) Impact ejecta horizon within late Precambrian shales, Adelaide Geosyncline, south Australia. Science 233: 198-200

Govindaraju K (1994) 1994 compilation of working values and sample descriptions for 383 geostandards. Geostand Newslett 18: 1-154

Grieve RAF (1980) Impact bombardment and its role in protocontinental growth of the early earth. Precambr Res 10: 217-248

Hassler SW, Simonson BM (1995) Depositional history of a resedimented impact layer in the Early Precambrian Hamersley Group, Western Australia. In: Montanari A, Coccioni R (eds) Abstracts, The Effects of Impacts on the Atmosphere and Biosphere with

Regard to Short- and Long-Term Changes, 4th Int Worksh, ESF Sci Network Impact Cratering and Evolution of Planet Earth (Ancona, May 1995), pp 88-89

Heinrichs T (1984) The Umsoli chert: Turbidite testament for a major phreatoplinian event at the Onverwacht/Fig Tree transition (Swaziland Supergroup, Archaea, South Africa). Precambr Res 24: 237-283

Holland H (1998) Atmospheric oxygen between 3.5 and 1.5 Ga. Min Mag 62A: p 644

Huber H, Koeberl C, McDonald I, Reimold WU (2000) Use of γ-γ coincidence spectrometry in the geochemical study of diamictites from South Africa. J Radioanal Nucl Chem: in press

IDL User's Manual (1993) Interactive Data Language. Res. Systems Inc., 777 29th Str, Boulder, USA, 2 volumes

Jéhanno C, Boclet D, Froget L, Lambert B, Robin E, Rocchia R, Turpin L (1992) The Cretaceous-Tertiary boundary at Beloc, Haiti: No evidence for an impact in the Caribbean area. Earth Planet Sci Lett 109: 229-241

Jochum KP (1996) Rhodium and other platinum-group elements in carbonaceous chondrites. Geochim Cosmochim Acta 60: 3353-3357

Kamo SL, Davis DW (1994) Reassessment of Archean crustal development in the Barberton Mountain Land, South Africa, based on U-Pb dating. Tectonics 13: 167-192

Kamo SL, Reimold WU, Krogh TE, Colliston WP (1996) A 2.023 Ga age for the Vredefort impact event and a first report of shock metamorphosed zircons in pseudotachylitic breccias and Granophyre. Earth Planet Sci Lett 144: 369-387

Koeberl C (1993) Instrumental neutron activation analysis of geochemical and cosmochemical samples: A fast and proven method for small sample analysis. J Radioanal Nucl Chem 168: 47-60

Koeberl C, Reimold WU (1994) Archean spherule beds: Impact or terrestrial origin? Reply to the comment by A Glikson. Earth Planet Sci Lett 126: 497-499

Koeberl C, Reimold WU (1995) Early Archaean spherule beds in the Barberton Mountain Land, South Africa: no evidence for impact origin. Precambr Res 74: 1-33

Koeberl C, Sharpton VL (1988) Giant impacts and their influence on the early earth. In: Papers presented to the Conference on "Origin of the Earth", Lunar Planet Inst, Houston, pp 47-48.

Koeberl C, Reimold WU, Boer RH (1993) Geochemistry and mineralogy of Early Archean spherule beds, Barberton Mountain Land, South Africa: Evidence for origin by impact doubtful. Earth Planet Sci Lett 119: 441-452

Koeberl C, Simonson BM, Reimold WU (1999) Geochemistry and petrography of a Late Archean spherule layer in the Griqualad West Basin, South Africa. Lunar Planet Sci 30: abs. #1755

Koeberl C, Reimold WU, McDonald I, Rosing M (2000) Search for petrographic and geochemical evidence for the late heavy bombardment on Earth in Early Archean rocks from Isua, Greenland. In: Gilmour I, Koeberl C (eds) Impacts and the Early Earth. Lecture Notes in Earth Sciences, 91, Springer, Heidelberg-Berlin, pp 73-96

Kröner A, Byerly GR, Lowe DR (1991) Chronology of early Archean granite-greenstone evolution in the Barberton Mountain Land, South Africa, based on precise dating by single zircon evaporation. Earth Planet Sci Lett 103: 41-54

Kyte FT, Smit J (1986) Regional variations in spinel compositions: An important key to the Cretaceous/Tertiary event. Geology 14: 485-487

Kyte FT, Zhou L, Lowe DR (1992) Noble metal abundances in an Early Archean impact deposit. Geochim Cosmochim Acta 56: 1365-1372

Leroux H, Reimold WU, Doukhan JC (1994) A T.E.M. investigation of shock metamorphism in quartz from the Vredefort dome, South Africa. Tectonophys 230: 223-239

Lesher CM (1989) Komatiite-associated nickel-sulphide deposits. In: Whitney JA, Naldrett AJ (eds) Ore Deposition Associated with Magmas (Rev Econ Geol 4). Soc Econ Geol, El Paso, Texas, pp 45-101

Lowe DR (1991) Geology of the Barberton Greenstone Belt: An overview. In: Ashwal LD (ed) Two Cratons and an Orogen - Excursion Guidebook and Review Articles for a Field Workshop through selected Archean Terranes of Swaziland, South Africa and Zimbabwe. Univ. Witwatersrand, Johannesburg, pp 47-58 and 156-159

Lowe DR, Byerly GR (1986) Early Archean silicate spherules of probable impact origin, South Africa and Western Australia. Geology 14: 83-86

Lowe DR, Byerly GR (1991) Day 7: Powerline Road section across the central Barberton greenstone belt. In: Ashwal LD (ed) Two Cratons and an Orogen - Excursion Guidebook and Review Articles for a Field Workshop through selected Archean terranes of Swaziland, South Africa and Zimbabwe, IGCP Project 280. Univ. Witwatersrand, Johannesburg, pp 154-163

Lowe DR, Byerly GR (1992) Depositional mechanics of impact-produced debris in the Archean Barberton Greenstone Belt, South Africa. Lunar Planet Sci 23: 811-812

Lowe DR, Byerly GR (eds) (1999) Geologic evolution of the Barberton Greenstone Belt, South Africa. Geol Soc Amer Spec Pap 329, 319 pp

Lowe DR, Knauth LP (1977) Sedimentology of the Onverwacht Group (3.4 billion years), Transvaal, South Africa, and its bearing on the characteristics and evolution of the early earth. J Geol 85: 699-723

Lowe DR, Knauth LP (1978) The oldest marine carbonate ooids reinterpreted as volcanic accretionary lapilli, Onverwacht Group, South Africa. J Sed Petrol 48: 709-722

Lowe DR, Byerly GR, Asaro F, Kyte FT (1989) Geological and geochemical record of 3400-million-year-old terrestrial meteorite impacts. Science 245: 959-962

McDonald I (1998) The need for a common framework for collection and interpretation of data in Platinum-Group Element geochemistry. Geostand Newslett 22: 85-91

McDonald I, de Wit MJ, Smith CB, Bizzi LA, Viljoen KS (1995) The geochemistry of the platinum-group elements in Brazilian and southern African kimberlites. Geochim Cosmochim Acta, 59: 2883-2903

Ohmoto H (1998) Evidence in trace elements and Fe^{3+}/Fe^{2+} ratios of Archean and early Proterozoic shales for the early development of oxic atmosphere. Min. Mag. 62A: 1106

Palme H, Grieve RAF, Wolf R (1981) Identification of the projectile at the Brent crater, and further considerations of projectile types at terrestrial craters. Geochim Cosmochim Acta 45: 2417-2424

Pattou L, Lorand JP, Gros M (1996) Non-chondritic platinum-group element ratios in the Earth's mantle. Nature 379: 712-715

Pepper S, du Plessis E (1991) Princeton Section, Agnes gold mine. In: Ashwal LD (ed), Two Cratons and an Orogen - Excursion Guidebook and Review Articles for a Field Workshop through Selected Archean Terranes of Swaziland, South Africa and Zimbabwe, IGCP Project 280. Univ Witwatersrand, Johannesburg, pp 174-177

Pierrard O, Robin E, Rocchia R, Montanari A (1998) Extraterrestrial Ni-rich spinel in upper Eocene sediments from Massignano, Italy. Geology 26: 307-310

Prozesky VM, Przybylowicz WJ, van Achterbergh E, Churms CL, Pineda CA, Springhorn, KA, Pilcher JV, Ryan CG, Kritzinger J, Schmitt H, Swart T (1995) The NAC nuclear microprobe facility. Nucl Instr Meth B104: 36-42

Reimold WU, Gibson RL (1996) Geology and evolution of the Vredefort impact structure, South Africa. J Afr Earth Sci 23: 125-167

Reimold WU, Koeberl C, Johnson S, McDonald I (1998) Archean and Proterozoic spherule layers – remnants of distal ejecta from ancient impact events? In: Abstracts, ESF Sci Network Workshop on Impacts and the Early Earth, Cambridge, Dec 1998, p 29

Robin E, Bonté P, Froget L, Jéhanno C, Rocchia R (1992) Formation of spinels in cosmic objects during atmospheric entry: A clue to the Cretaceous-Tertiary boundary event. Earth Planet Sci Lett 108: 181-190

Ryan CG, Jamieson DN (1993) Dynamic Analysis: On-line Quantitative PIXE Microanalysis and its Use in Overlap-Resolved Elemental Mapping. Nucl Instr Meth B 77: 203-214

Ryan CG, Cousens DR, Sie SH, Griffin WL (1990) Quantitative Analysis of PIXE Spectra in Geoscience Applications. Nucl Instr Meth B49: 271-276

Ryan CG, Jamieson DN, Churms CL, Pilcher JV (1995) A New Method for On-Line True Elemental Imaging using PIXE and the Proton Microprobe. Nucl Instr Meth B104: 157-165

Ryan CG, van Achterbergh E, Jamieson DN, Churms CL (1996) Overlap corrected on-line PIXE imaging using the Proton Microprobe. Nucl Instr Meth B109/110: 154-160

Sharpton VL, Dalrymple GB, Marin LE, Ryder G, Schuraytz BC, Urrutia-Fucugauchi J (1992) New links between the Chicxulub impact structure and the Cretaceous/Tertiary boundary. Nature 359: 819-821

Sharpton VL, Burke K, Camargo-Zanoguera A, Hall SA, Lee S, Marín, LE, Suárez-Reynoso G, Quezada-Muñeton JM, Spudis PD, Urrutia-Fucugauchi J (1993) Chicxulub multiring impact basin: Size and other characteristics derived from gravity analysis. Science 261: 1564-1567

Shazali I, Van't Dack L, Gijbels R (1987) Determination of precious metals in ores and rocks by thermal neutron activation spectrometry after preconcentration by nickel sulphide fire assay and coprecipitation with tellurium. Anal Chim Acta 196: 49-58

Shukolyukov A, Kyte FT, Lugmair GW, Lowe DR (1998) The oldest impact eposits on earth – First confirmation of an extraterrestrial component. In: Abstracts, ESF Sci Network Workshop on Impacts and the Early Earth, Cambridge Dec 1998, p 35

Shukolyukov A, Kyte FT, Lugmair GW, Lowe DR, Byerly GR (1999) Extraterrestrial chromium in Early Archean spherule beds – Further evidence for an impact origin. In: Buffetaut E, Le Loeuff J (eds) Abstracts, Workshop on Geological and Biological Evidence for Global Catastrophes, ESF Impact Programme, Quillan, France, pp 66-67

Shukolyukov A, Kyte FT, Lugmair GW, Lowe DR, Byerly GR (2000) The oldest impact deposits on earth – first confirmation of an extraterrestrial component. In:: Gilmour I, Koeberl C (eds) Impacts and the Early Earth. Lecture Notes in Earth Sciences. 91, Springer, Heidelberg-Berlin, pp 97-114

Simonson BM (1992) Geological evidence for a strewn field of impact spherules in the early Precambrian Hamersley Basin of Western Australia. Geol Soc Amer Bull 104: 829-839

Simonson BM, Davies D (1996) PGEs and quartz grains in a resedimented late Archean impact horizon in the Hamersley Group of Western Australia. Lunar Planet Sci 27: 1203-1204

Simonson BM, Beukes NJ, Hassler SW (1997) Discovery of a Neoarchean impact spherule horizon in the Transvaal Supergroup of South Africa and possible correlations to the Hamersley Basin of Western Australia. Lunar Planet Sci 27: 1323-1324

Simonson BM, Davies D, Wallace M, Reeves S, Hassler SW (1998) Iridium anomaly but no shocked quartz from Late Archean microkrystite layer: Oceanic impact ejecta? Geology 26: 195-198.

Smit J (1999) The global stratigraphy of the Cretaceous-Tertiary boundary impact ejecta. Ann Rev Earth Planet Sci 27: 75-113

Smit J, Kyte FT (1984) Siderophile-rich magnetic spheroids from the Cretaceous-Tertiary boundary in Umbria, Italy. Nature 310: 403-405

Stanistreet IG, de Wit MJ, Fripp REP (1981) Do graded units of accretionary spheroids in the Barberton Greenstone Belt indicate Archaean deep water environment? Nature 293: 280-284

Stöffler D (1972) Deformation and transformation of rock-forming minerals by natural and experimental shock processes: 1. Behaviour of minerals under shock compression. Fortschr Mineral 49: 50-113

Stöffler D, Langenhorst F (1994) Shock metamorphism of quartz in nature and experiment: I. Basic observations and theory. Meteoritics 29: 155-181

Tredoux M, McDonald I (1996) Komatiite Wits-1, low concentration noble metal standard for the analysis of non-mineralized samples. Geostand Newslett 20: 267-276

Tredoux M, de Wit MJ, Hart RJ, Armstrong RA, Lindsay NM, Sellschop JPF (1989a) Platinum group elements in a 3.5 Ga nickel-iron occurrence: Possible evidence of a deep mantle origin. J Geophys Res 94: 795-813

Tredoux M, de Wit MJ, Hart RJ, Lindsay NM, Verhagen B, Sellschopp JPF (1989b) Chemostratigraphy across the Cretaceous-Tertiary boundary and a critical assessment of the iridium anomaly. J Geol 97: 585-605

Tredoux M, Lindsay NM, Davies G, McDonald I (1995) The fractionation of platinum-group elements in magmatic systems, with the suggestion of a novel causal mechanism. S Afr J Geol 98: 157-167

Wallace MW, Gostin VA, Keays RR (1990) Acraman impact ejecta and host shales: Evidence for low-temperature mobilization of iridium and other platinoids. Geology 18: 132-135

Williams GE (1986) The Acraman impact structure: Source of ejecta in late Precambrian shales, South Australia. Science 233: 200-203

Wood SA (1991) Experimental determination of the hydrolysis constants of Pt^{2+} and Pd^{2+} at 25°C from the solubility of Pt and Pd in aqueous solutions. Geochim Cosmochim Acta 55: 1759-1767

Woodhead JD, Hergt JM, Simonson BM (1998) Isotopic dating of an Archean bolide impact horizon, Hamersley Basin, Western Australia. Geology 26: 47-50

Zhou M-F (1994) PGE distribution in 2.7 Ga layered komatiite flows from the Belingwe greenstone belt, Zimbabwe. Chem Geol 118: 155-172

7 Particles in Late Archean Carawine Dolomite (Western Australia) Resemble Muong Nong-type Tektites

Bruce M. Simonson[1], Miriam Hornstein[1]. Scott Hassler[2]

[1]Department of Geology, Oberlin College, Oberlin, OH 44074-1044 U.S.A. (bruce.simonson@oberlin.edu)
[2]Department of Geological Sciences, California State University, Hayward, CA 94542 U.S.A. (shassler@ccnet.com)

Abstract. In addition to craters and related structures, large impacts may create strewn fields of ejecta that have the potential to be incorporated into the stratigraphic record. Nine layers rich in sand-size spherules of former silicate melt interpreted as impact ejecta have been reported from early Archean to early Paleoproterozoic successions. The layer in the late Archean Carawine Dolomite (Western Australia) is the only one of the nine which contains larger, more irregular particles of former silicate melt. The layer which contains the irregular particles is a ~25 meter-thick composite of three coarse-grained, poorly sorted carbonate debris flows. The irregular particles are concentrated at the base of the lowest flow, whereas the spherules are most abundant in the finer, sandy upper parts of the flows, indicating they were segregated hydrodynamically. The spherules and irregular particles are both largely replaced by fibrous K-feldspar crystals. These crystals generally radiate inwards from the edges of the spherules, whereas they are commonly randomly oriented to parallel-fibrous in the irregular particles. The irregular particles also differ from the spherules in being larger and more angular on average, and in showing banding, abundant vesicles, and greater textural heterogeneity internally. A few of the larger irregular particles appear to have spherules embedded inside them, indicating they formed during the same event. Fine quartz crystals with rare candidates for relict planar deformation features are also included in a few of the irregular particles. While the irregular particles are unlike any of the clasts in the volcanoclastic layers in associated units, they are texturally similar to Muong Nong-type tektites, which generally have irregular blocky shapes, are layered, and contain relict minerals as inclusions. However, the irregular particles differ from the Muong Nong-type tektites in that they are smaller, more rounded, and may have been partially crystalline. If the irregular particles were originally tektites, they would be the oldest yet found, and the coarsest distal ejecta found to date in an early Precambrian succession. It appears, however, that they are not a perfect match for Muong Nong or any other type of tektite, which suggests some strewn fields may contain additional types of glassy ejecta that do not fit neatly into existing categories. Their existence also suggests that textures in tektite-like bodies can be stabilized by replacement during diagenesis and survive essentially indefinitely, and the irregular particles are probably products of an impact in continental rather than oceanic crust.

Introduction

When large extraterrestrial bodies strike the earth, they create two things with the potential to be incorporated into the geologic record: 1) craters and related structures formed at the site of impact, and 2) strewn fields of ejecta, which is material derived largely from the target rocks, accelerated to high velocities, then deposited far from the impact site. No more then 10% of the 150 known impact structures listed in Grieve et al. (1995) are Precambrian in age, and none of these formed during the Archean, which represents the first 43% of earth history. Archean rocks only constitute 7% of modern continental crust (Stanley 1999); it is thus unlikely that large numbers of Precambrian impact structures will ever be available for study. If we wish to understand the history and effect of impacts on early earth, it therefore seems prudent to search for impact ejecta as well as impact structures in Precambrian successions. Precambrian impact ejecta can and do get preserved, as evidenced by the strewn field from the Acraman structure of South Australia (Gostin et al. 1986; Wallace et al. 1989, 1990a, 1996). Layers rich in distinctive sand-size spherules of former silicate melt interpreted as replaced microtektites and/or microkrystites have also been identified in early Precambrian successions of South Africa and Western Australia. In this paper, we describe the only reported occurrence of particles that resemble Muong Nong-type tektites in any of these older layers.

A minimum of nine spherule layers have been reported from stratigraphic units in the early Precambrian of Western Australia (WA) and South Africa (SA). These are located in the early Archean Barberton greenstone belt (SA) and Pilbara craton (WA) (Lowe and Byerly 1986; Lowe et al. 1989) and the late Archean to Paleoproterozoic Hamersley (WA) and Griqualand West (SA) basins (Simonson 1992; Simonson and Hassler 1997; Simonson et al. 1998, 1999). In addition to spherules, one of the nine layers also contains larger, more irregular particles of former silicate melt. This layer is in the Carawine Dolomite of the Hamersley basin (es. 1,2), and it appears to be the only one that contains such particles. The shapes and internal textures of these particles are similar to Muong Nong-type tektites, suggesting they are the coarsest impact ejecta discovered to date in an Archean succession. Simonson (1992) gave a brief account of these larger particles. This paper presents the results of more detailed petrographic studies which demonstrate that the particles are texturally distinct from, yet genetically related to, the spherules found in this and other early Precambrian layers.

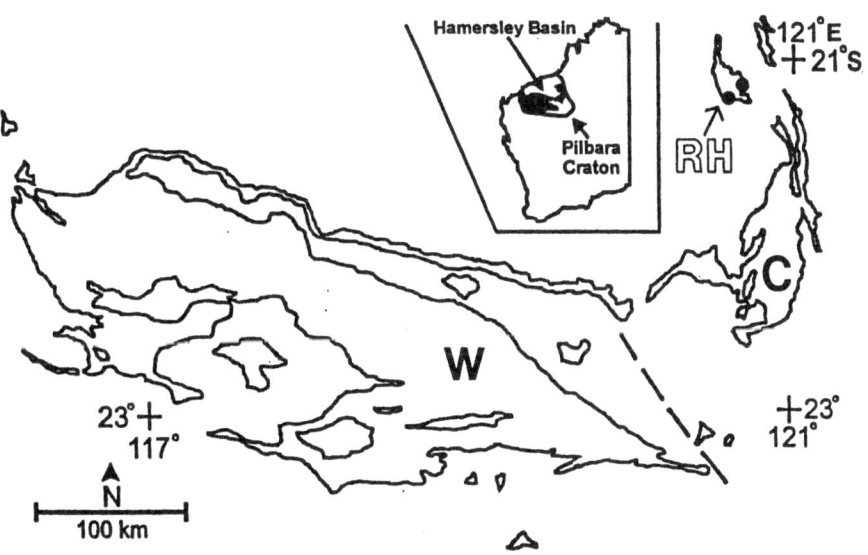

Fig. 1. Outline of areas of occurrence of strata of the Hamersley Group. Carawine Dolomite is restricted to area northeast of dashed line (indicated by C), whereas Wittenoom and Brockman Iron Formations are both restricted to area south and west of it (indicated by W). Solid dots in Ripon Hills (RH) indicate where layer in Carawine Dolomite, which contains both spherules and irregular particles has been observed (Simonson and Hassler 1997). Inset shows location of Hamersley basin within outline of Western Australia.

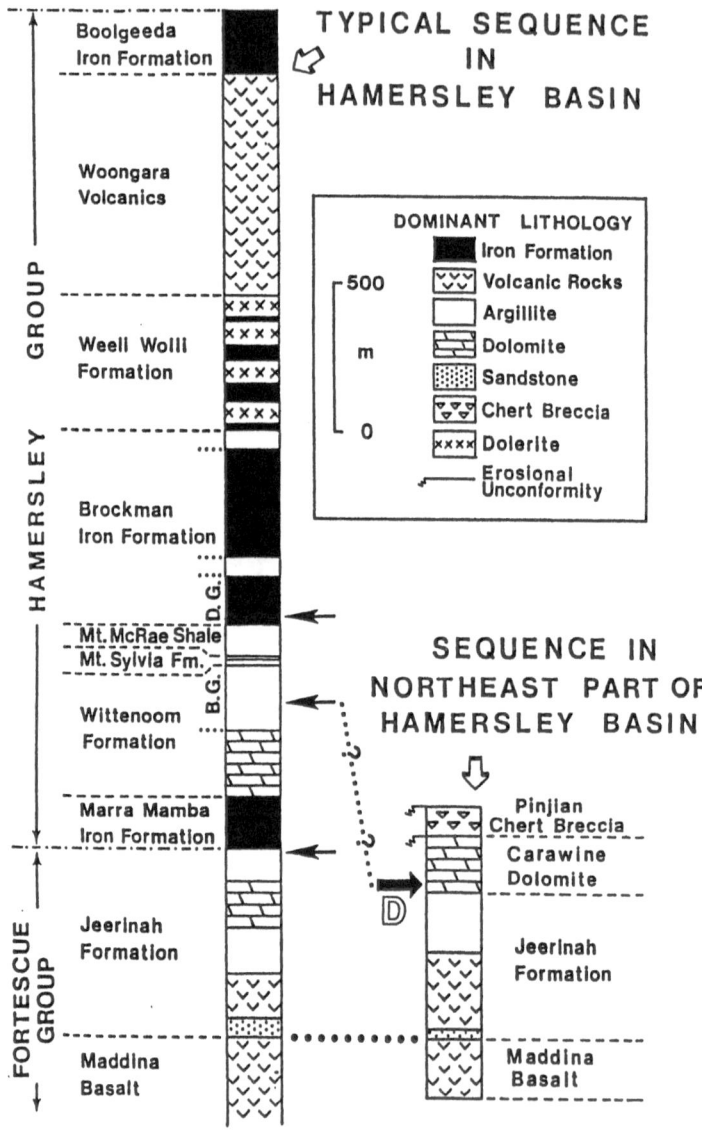

Fig. 2. Generalized stratigraphic columns showing contrast between strata of Hamersley Group and uppermost Fortescue Group in northeastern part vs. rest of Hamersley basin (after Simonson et al. 1998). Thick horizontal arrow (labeled D for dolomixtite) shows stratigraphic position of layer with spherules and irregular particles within Carawine Dolomite. Thin horizontal arrows indicate stratigraphic positions of three spherule-rich layers units in main part of Hamersley basin; from bottom to top, they are located in Jeerinah Formation, Bee Gorge Member (B.G.) of Wittenoom Formation, and Dales Gorge Member (D.G.) of Brockman Iron Formation. Dotted line indicates Simonson and Hassler's (1997) proposed correlation of layers in Carawine Dolomite and Wittenoom Formation.

Geological Background

The Carawine Dolomite occurs in the northeastern part of the Hamersley basin of Western Australia (Fig. 1), which is one of the best preserved and most extensive successions of late Archean to Paleoproterozoic sedimentary strata on earth (Trendall 1983, 1990). The Carawine Dolomite is at least 500 m in thick and consists mostly of platformal dolomite, i.e., carbonate strata with depositional features such as wave ripples, oolites, and pseudomorphs of evaporite crystals indicative of deposition in shallow-water environments (Simonson et al. 1993a). However, such structures are lacking in a minority of the strata in the Carawine Dolomite. What these strata show instead are structures typical of basinal or deeper-water deposition, such as roll-up structures (Simonson and Carney 1999), interbedding with thinly laminated black shale, and the presence of thin, normally graded tuff beds (Simonson et al. 1993a,b).

Within such deeper-water strata near the stratigraphic base of the Carawine Dolomite (Fig. 2) is an unusual layer ~25 m thick that consists of a poorly sorted mixture of dolomite clasts ranging from boulders as much as 2.2 m long down to sand-size and possibly finer detritus. The vast majority of the pebble- to boulder-size clasts are tabular bodies of local provenance that clearly formed as rip-up clasts. The layer also contains rare clasts of platformal carbonates, such as a block of cross-bedded oolite ca. 20 cm across, that we estimate were transported for distances on the order of tens of kilometers. The finer detritus in this layer consists largely of intraclastic dolomite sand with lesser amount of chert, and the particles described in this paper are one constituent of this finer fraction. Simonson (1992) and Simonson and Hassler (1997) named this layer the dolomixtite and interpreted it as the product of three large carbonate debris flows that were deposited in rapid succession, probably in a matter or hours to several days at the most. The dolomixtite crops out intermittently in low scarps for ca. 10 km along the southern edge of the Ripon Hills (Fig. 1). It was also intersected by a core drilled nearby to the north (Simonson and Hassler 1997). It has not, however, been detected in any other exposures of the lower part of the Carawine Dolomite, which Simonson (1992) suggested may mean the debris flows which formed it were channelized.

All of the available isotopic dates strongly suggest the Carawine layer was deposited near the end of the Archean. Woodhead et al. (1998) reported a Pb-Pb isotopic age of 2548 +26/-29 Ma from carbonates ot the Carawine Dolomite, including some samples from the layer itself. Jahn and Simonson (1995) reported a Pb-Pb isochron age of 2541 ± 32 Ma for Carawine Dolomite samples, but argued that this date was more likely to represent diagenetic events than depositional ones. Trendall et al. (1998) obtained an age of 2597 ± 5 Ma via SHRIMP U-Pb dating of zircons in a tuff near the top of the Marra Mamba Iron Formation, which directly underlies the Carawine Dolomite in some areas. Ages of 2479 ± 3 and 2463 ± 5 Ma were also obtained via SHRIMP U-Pb dating of zircons in tuffs from the Brockman Iron Formation (Nelson et al. 1999). Although believed to be a bit younger than the Carawine Dolomite, this cannot be tested directly because the Brockman Iron Formation's area of occurrence does not overlap that of the Carawine Dolomite (Figs. 1, 2).

Carbonate debris flow layers occur elsewhere in the Hamersley Group, although few are as thick or as coarse-grained as the dolomixtite. The one thing that appears to set the dolomixtite apart from any other layer in the Carawine Dolomite, and also other carbonate debris flow deposits in the Hamersley basin, is the presence of particles of former silicate melt. Many of these particles take the form of millimeter-size spherules that consist largely of K-feldspar and display cooling and devitrification textures. Simonson (1992) interpreted these spherules as impact ejecta because of their similarity to well-documented impact spherules retrieved from the moon (Lofgren 1971a), the Cretaceous-Tertiary boundary layer (Smit and Klaver 1982; Montanari 1991; Bohor and Glass 1995; Martínez Ruíz et al. 1997) and Cenozoic strewn fields of microtektites or microkrystites (John and Glass 1974; Glass 1984, 1990; Glass et al. 1985; Glass and Burns 1987).

Spherules like those in the dolomixtite also occur in three other formations in the Hamersley basin, namely the Jeerinah Formation of the Fortescue Group and the Wittenoom Formation and Brockman Iron Formation of the overlying Hamersley Group (Fig. 2). Simonson and Hassler (1997) argued that the dolomixtite and spherule layer in the Wittenoom Formation were deposited simultaneously during an anomalously high energy event, presumably in the aftermath of an impact. They based their interpretation on the following lines of evidence: 1) the two formations in which these spherule layers occur occupy similar stratigraphic positions in different parts of the Hamersley basin, but do not overlap geographically (Fig. 1, 2) the dolomixtite and the layer in the Wittenoom Formation each contain the only spherules in their respective formations; and 3) together, the dolomixtite and the spherule layer in the Wittenoom Formation display a continuum of sedimentary structures consistent with deposition from several large sediment gravity flows moving south and west across the Hamersley basin in rapid succession. This interpretation was bolstered by Woodhead et al. (1998), who obtained a Pb-Pb isotopic age of 2541 +18/-15 Ma on carbonate samples collected from the Wittenoom spherule layer and immediately adjacent strata. This compares quite favorably with their date for the Carawine Dolomite, but the correlation of the dolomixtite with the Wittenoom spherule layer has yet to be proven. One reason we still question this correlation is the apparent absence from the Wittenoom spherule layer of larger, more irregular particles like those found in the dolomixtite and described here.

The interpretation of the spherules and other particles of former silicate melt in both the dolomixtite and the Wittenoom spherule layer as impact ejecta is supported by significant iridium enrichment. The Ir content of five samples from the dolomixtite averages 0.596 ppb and reaches a maximum of 1.54 ppb, whereas the Ir concentration in eleven samples from the surrounding shales and dolomites averages only 0.033 ppb and reaches a maximum of 0.13 ppb (Simonson et al. 1998). With the exception of Ru, other siderophile elements (Pd, Pt, Au, Cr, Co, and Ni) display no significant anomalies relative to the background sediments. Ir, Ru, and Ni are correlated and have near-chondritic interelement ratio, whereas Co displays a correlation with Ir, but the interelement relationship is not chondritic. The geochemical patterns associated with the ejecta layer from the ca. 590 Ma Acraman impact structure of South Australia are very similar. The highest Ir concentrations in bulk samples of the Acraman ejecta are ~2 ppb vs. an average value of ~ 0.02 ppb for the surrounding red shales (Gostin et al. 1989), and the siderophile elements of the ejecta depart significantly from chondritic ratios. This has been attributed to differential mobilization of Ir and other platinoids during low-temperature diagenesis under reducing conditions (Wallace et al. 1990b), and the same could be true in the Hamersley Basin as the shales associated with the dolomixtite are pyritic and carbonaceous (Simonson et al. 1998).

The larger, more irregular particles that resemble Muong Nong-type tektites are concentrated at the base of the lowest of the three debris flows that make up the dolomixtite. In contrast, the spherules are most abundant in the finer-grained sediment at the tops of each of the three debris flows, particularly the uppermost one. We attribute this to hydrodynamic segregation during transport and deposition, indicating both the spherules and irregular particles are reworked, rather than being part of a primary impact deposit.

Description of Particles Interpreted as Ejecta

General Characteristics and Methods

The external shapes and internal textures of all particles of possible impact ejecta (a total of 917) were measured in twenty thin sections from thirteen samples of the dolomixtite. Their shapes were quantified in two ways. First, the lengths of the longest dimension (= major axis) and the dimension perpendicular to it (= minor axis) were measured in each cross-section. Second, each cross-section was assigned a roundness value using the scale in Table 14 of Pettijohn (1949). The endpoints of this roundness scale are 0 (for particles with many small, angular projections) and 1.0 (for particles with nothing but broadly curved, smoothly rounded edges). We estimated the roundness of each particle on this scale to the nearest tenth by visual comparison with a template of silhouettes like those in Figure 24 of Pettijohn (1949). The internal textures of all of the particles were also studied in transmitted light with a polarizing microscope. We examined some of the thin sections more cursorily using a Nuclide ELM-2E Luminoscope operating at approximately 7 kV. Finally, an average of 1,470 points were counted on each of ten thin sections from eight of the particle-bearing samples, and an average of 190 points were counted on each of six individual particles that contained vesicles.

11% of the particles we interpret as former impact melt were too altered by weathering and/or diagenetic replacement to measure accurately and are not considered further in this paper. Based on related differences in their external shapes and internal textures, the remaining 813 particles were subdivided into two main categories informally referred to as spherules (40%) and irregular particles (60%). The latter are the particles that resemble Muong Nong-type tektites. Although the spherules are not the main focus of this paper, they are described here to permit comparison. About two-thirds of the spherules and the irregular particles combined consist largely of finely crystalline K-feldspar. The rest mostly consist of either coarsely crystalline carbonate or a chlorite-like sheet silicate. The chlorite(?) is yellow to greenish brown in plane polarized light with first order gray birefringence, whereas particles replaced by K-feldspar are generally tan or darker brown in plane polarized light.

Based on the point counts, dolomite makes up around 93% by volume of the samples on average. This dolomite is all coarsely crystalline, obviously replacive in origin, and represents framework-supporting clasts with lesser amounts of intergranular cement. The remaining 7% by volume of these samples consists of non-carbonate material, primarily intraclastic chert pebbles, with lesser amounts of diagenetic pyrite crystals, medium to coarse quartzose sand grains, and the particles of former silicate melt. The spherules and the irregular particles each make up about 1% on average of the samples examined, although locally they can occur in much greater abundance. One unusual sample had around 29% irregular particles by volume, which is an order of magnitude greater than the norm. Given that the most spherule-rich material was selected in the field, the actual abundances of the spherules and irregular particles in the dolomixtite as a whole are probably lower. Simonson et al. (1998) presented arguments that the quartzose sand in the dolomixtite originated as epiclastic detritus rather than as impact ejecta, so it is not considered further here.

Irregular Particles

The irregular particles have major axes that range up to 11 mm, and they tend to be more elongated than the spherules. Their axial ratios (= minor axis ÷ major axis) range from 1.0 to 0.1, with a mean of 0.65. The irregular particles also tend to be more angular than the spherules. Roundness values measured on irregular particles range from a minimum of 0.1 (angular) to a maximum of 0.6 (rounded to well rounded). The mean roundness value is 0.35, which falls near the mid-point of Pettijohn's (1949) subrounded category. Despite the fact that they are more angular on average, a few of the smoother and more rounded of the so-called "irregular" particles appear to be classic splash forms such as teardrop shapes (Fig. 3). None of the irregular particles, however, show the perfectly rounded spheroid to ovoid shapes that typify the spherules. In addition, the perimeter of virtually every one of the larger particles has been invaded by dolomite rhombs, which resulted in margins that are pocked-marked rather than smooth (Fig. 3A).

The internal textures displayed by the irregular particles clearly indicate they were originally molten. First, many of the irregular particles contain planar to swirling bands elongated parallel to their long axes that resemble flow banding or schlieren (Figs. 3B, 4; Plate 1). In plane polarized light, variations in the generally tan color of the particles give definition to the banding. In some particles, the banding is also evident between crossed polarizers because the K-feldspar crystals form parallel-fibrous textures with their fast optic axes aligned parallel to banding (Plate 1). Still other irregular particles show hybrid textures with the K-feldspar crystals in optical alignment in some areas, but not others.

About 95% of the irregular particles also contain one or more vesicles (Fig. 5). These vesicles are generally evenly distributed and circular in cross-section, although larger vesicles tend to be elongated parallel to the long axes of particles, particularly in areas of pronounced banding. The abundance of vesicles varies considerably between individual particles. Vesicles make up a maximum of 23.5% and an average of 13.8% by volume of six of the most prominently vesicular irregular particles. The vesicles are generally filled with coarsely crystalline quartz or carbonate and/or finely crystalline muscovite. The quartz crystals do not show appreciable cathodoluminescence, so they probably have a low-temperature, diagenetic origin (Marshall 1988).

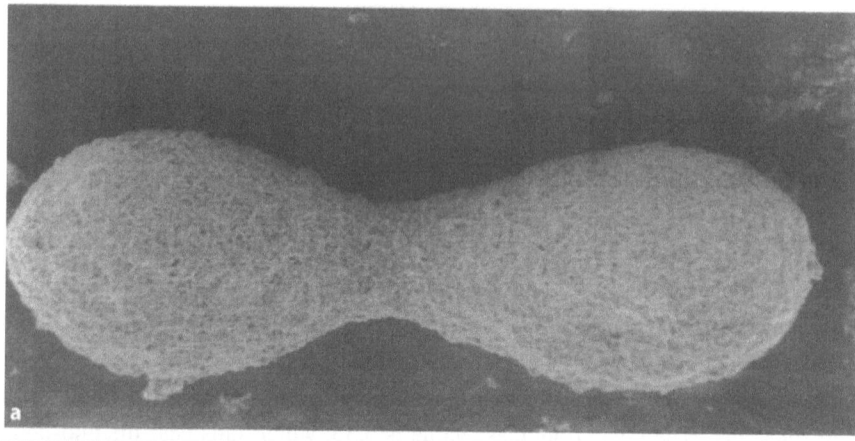

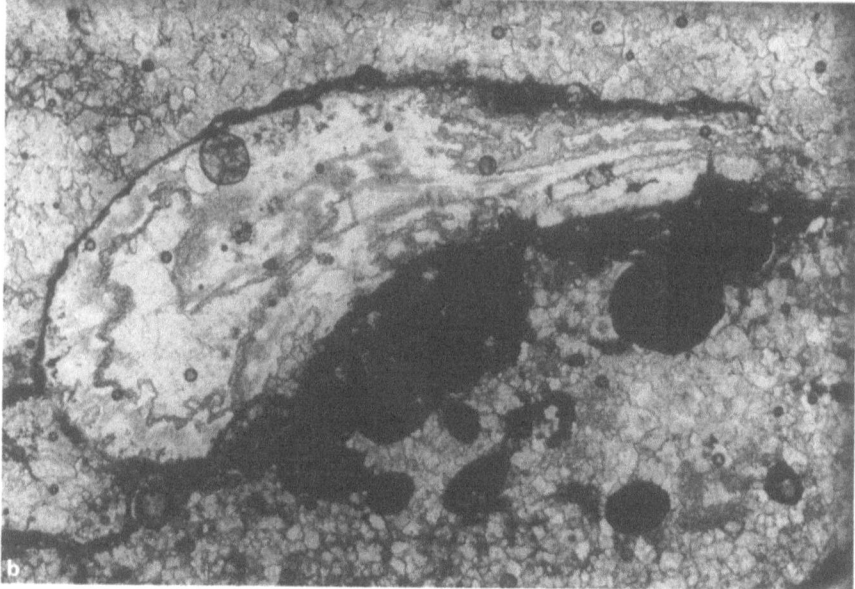

Fig. 3. (above and opposite) Particles with non-spheroidal splash-form shapes. A) Scanning electron photomicrograph of peanut-shaped particle made of K-feldspar and liberated from dolomite matrix by dissolution in HCl. Internal texture of particle is not known. Small pits on surface of particle are molds of diagenetic crystals of dolomite that invaded outer rim of particle. Long axis of particle is ~2.3 mm. B) Photomicrograph in plane polarized light of teardrop-shaped particle that consists largely of finely crystalline K-feldspar with lesser amounts of coarsely replacive carbonate. Matrix is mostly dolomite, and opaque bodies are probably diagenetic pyrite microconcretions oxidized by surface weathering. Long axis of field of view is 2.76 mm. C) Photomicrograph in plane polarized light of peanut-shaped particle classified as a spherule. Dark rim consists of fibrous K-feldspar and clear center consists of coarsely crystalline quartz. Matrix is again dolomite. Long axis of field of view is 1.38 mm.

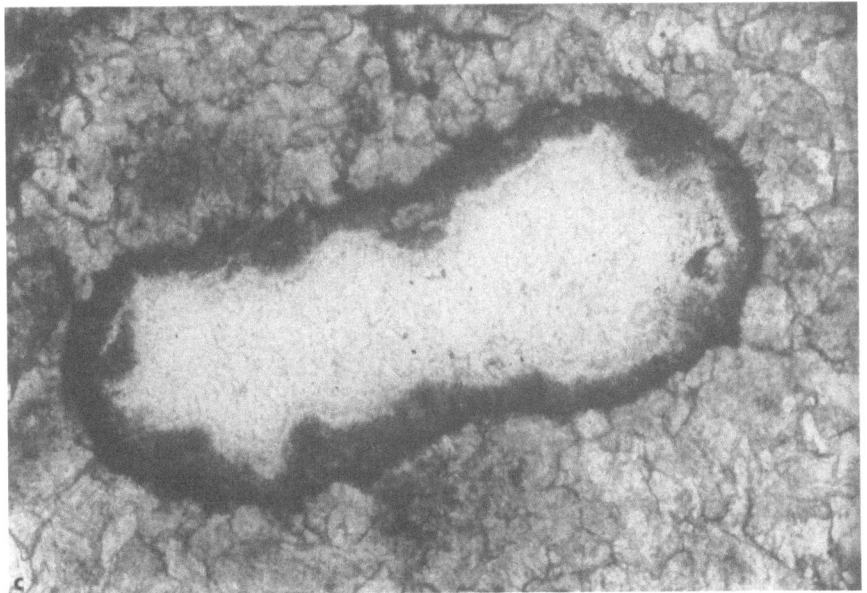

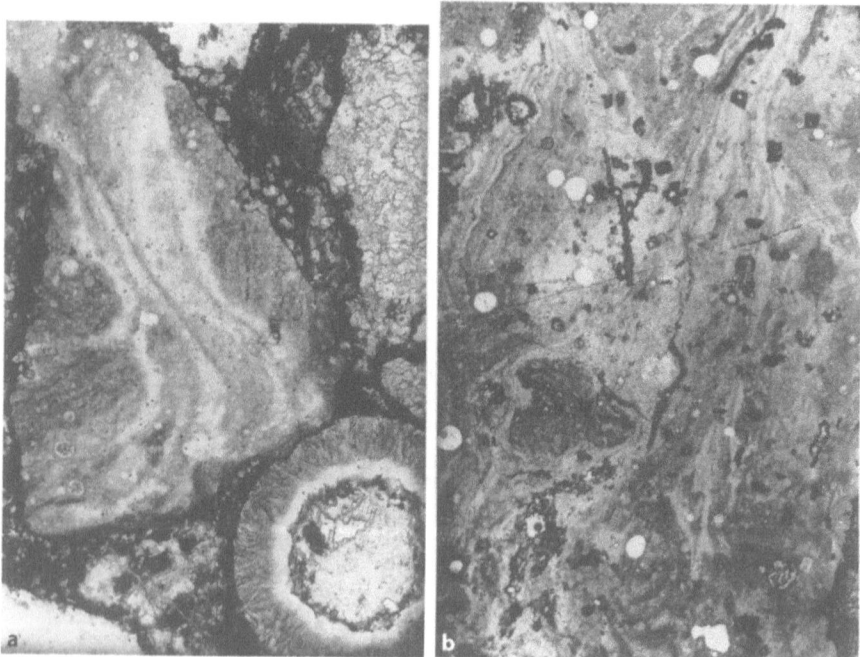

Fig. 4. (above and opposite) Photomicrographs in plane polarized light of irregular particles with banding. A) Heterogeneous irregular particle with contorted banding (upper left) next to near-circular spherule with prominent central spot (lower right). Contact between the two is slightly pressure-solved. B) Part of an 11 mm-long irregular particle (the largest one observed) showing prominent banding and heterogeneities. K-feldspar inside particle shows optical alignment locally, but not in general. C) Irregular particle with smooth, gently curved banding and strong optical alignment of the K-feldspar crystals length-fast parallel to same. High-relief inclusions are coarsely crystalline carbonates replacing parts of selected bands and filling vesicles. Same field of view shown in Plate 1C. D) Two-thirds of an irregular particle with highly contorted banding. Same field of view shown in Plate 1D. Long axis of field of view is 5.53 mm in A and B and 2.76 mm in C and D.

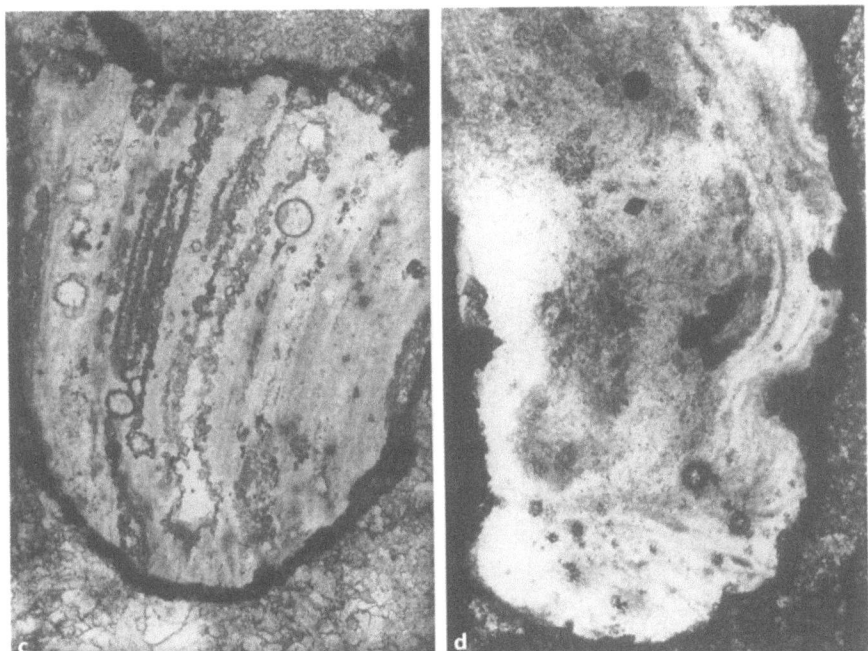

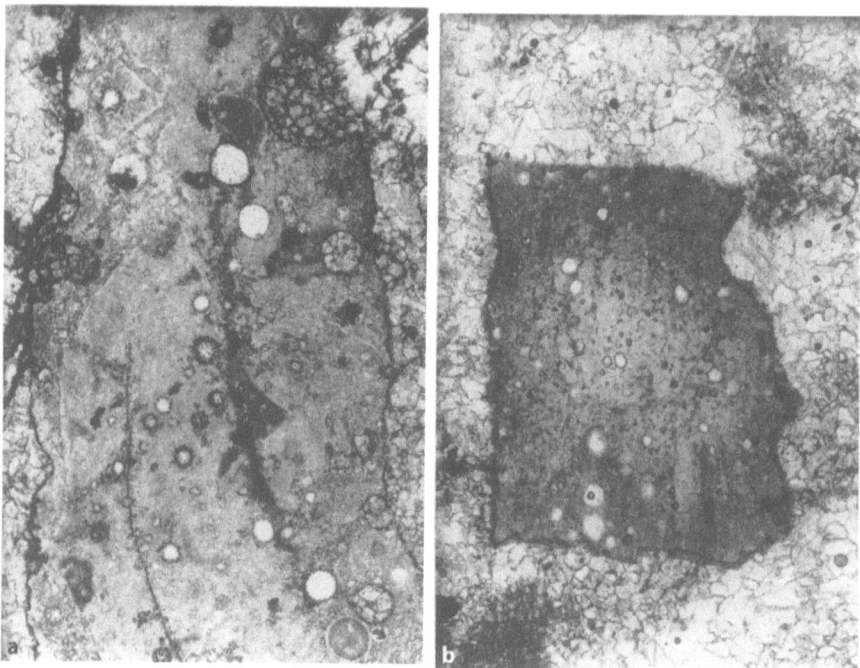

Fig. 5. (above and opposite) Photomicrographs in plane polarized light of vesicular irregular particles. A) Middle half of highly elongated irregular particle with numerous circular vesicles, mostly filled with quartz. B) Angular irregular particle with numerous circular vesicles and mineral inclusions; most if not all of latter are probably diagenetic in origin. C) More irregular particle between crossed polarizers with gypsum plate inserted. Particle consists largely of chlorite(?) with quartz filling most of the vesicles, one of which is highly elongated. D) Highly vesicular irregular particle next to spherule replaced almost entirely by a single crystal of carbonate. Long axis is 2.76 mm in all fields of view.

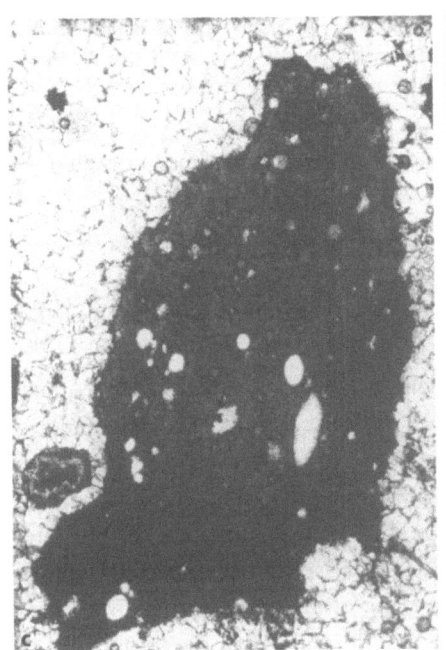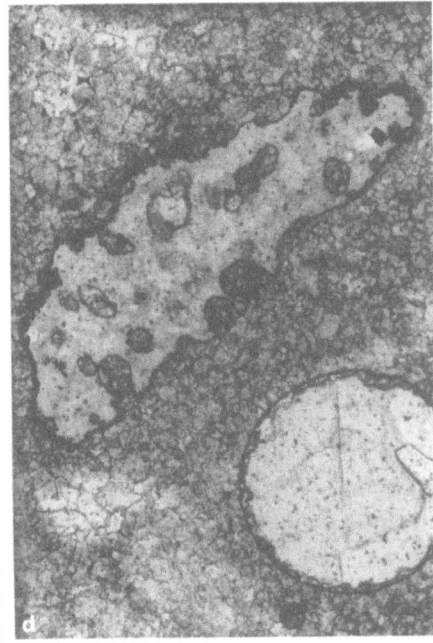

Many of the largest particles show considerable heterogeneity, or even consist of composites with internal boundaries separating areas that clearly differ from one another in the type or abundance of vesicularity, banding, the presence of crystal sprays, and/or other textural characteristics (Fig. 6). At least some of these particles appear to be agglutinates formed by collisions of smaller particles that were still partially molten. Additionally, a minority of the irregular particles contain local pockets or layers rich in sprays of minute lath-shaped crystals that are visible in plane polarized light (Fig. 6B). These sprays closely resemble some of the fabrics observed in the spherules. In a very few instances, the irregular particles have sharply bounded inclusions that show the same internal textures as the spherules (Fig. 6C, D). Such inclusions possess teardrop or oval shapes and range from 0.36 mm to 0.68 mm across.

A few of the irregular particles (especially the more strongly banded and vesiculated ones) also contain cloudy patches or lenses up to at least 0.6 mm long that resemble augen in deformed rocks (Fig. 7A). Such lenses consist mainly of clusters of quartz crystals that vary in size but are ≤ 0.06 mm. Linear trails of minute inclusions that are candidates for relict planar deformation features are present in a few of these quartz crystals (Fig. 7B). Isolated quartz crystals about 0.04 mm across also occur as inclusions in some of the particles (Fig. 7A). The more strongly banded and vesiculated a piece is, the more inclusions of this sort it is likely to have.

Spherules

In cross-section, the spherules range in size from 2.25 to 0.25 mm in diameter. One of the key characteristics of the spherules, their spheroidal shape (Figs. 4A, 5D, 8), is clearly reflected in the fact that the vast majority have roundness value of 0.9 or 1.0, as well as a mean axial ratio of 0.84. The reason this ratio is not higher is that an estimated 13% of the particles included in the spherule category have shapes such as oval, peanuts, teardrops, or crescents (Fig. 3C). Stray spherules with such shapes have axial ratios as low as 0.3. Another of the departures from a simple spheroidal shape consists of a pair of fused bodies in which one part has a completely circular cross-section whereas the second is crescentic (Plate 1A, B). We interpret such particles as agglutinated particles formed when one hot, soft spherule struck another that was already cool enough to be rigid.

Internally, the spherules show a variety of textures, but most are dominated by two elements: 1) fibrous to lath-shaped crystals that are generally tan in plane polarized light, and 2) areas or spots that appear clear in plane polarized light (Fig.

7). All of the fibers and laths consist of K-feldspar (Figs. 4A, 8A, C, Plate 1A, B) or more rarely chlorite(?) (Fig. 8B). The clear spots are more variable, consisting of K-feldspar, carbonate, quartz, and/or finely felted muscovite. The elongated crystals of K-feldspar or chlorite(?) are generally organized into radial sprays that diverge inward from the outer edges of the spherules in botryoidal masses. These sprays vary from discrete laths that diverge individually to acicular crystals that are tightly bunched into hemispherical fans with uniformly sweeping extinction. In a minority of the spherules, the elongated crystals are randomly oriented throughout instead of radially oriented from the edge in. The fact that the largest of the spherules with random textures are the same size as the largest of the spherules with radial textures indicates the former are not simply sections through the edges of the latter. Like all particles of former silicate melt found in the dolomixtite, the spherules have an opaque rim on their outer edge ≤ 0.08 mm thick. In addition, the rims of a minority of the spherules contain fine vesicles and/or more equant crystal inclusions.

The clear spots inside the spherules vary in location and in the sharpness of their boundaries. Some spots are located in the centers of spherules (Fig. 4A), whereas others are eccentric, i. e., located off center (Fig. 8D). Some spots are sharply bounded whereas others have vaguer limits. The margins of the clear spots also vary from nearly perfect circles to scalloped lines consisting of convex-inward arcs. In addition, a minority of the clear spots have non-circular cross-sections including teardrop shapes, in which case the narrow tip of the teardrop is always closest to the exterior margin of the spherule.

The presence of crescentic arc fragments with textures identical to those of intact rims suggests some of these spherules were formerly hollow. Like the intact spherules, the crushed pieces and arc fragments show dominantly radial-fibrous textures consisting of K-feldspar. The fact that multiple pieces are typically still side by side within the rock indicates some spherules were crushed by compaction.

Origins of the Irregular Particles and Spherules

Various lines of evidence indicate the irregular particles and spherules both originated as silicate melts, including splash form shapes (Fig. 3) flow banding (Fig. 4, Plate 1), vesicles (Fig. 5), and crystalline textures formed during cooling and devitrification (Figs. 6, 8). The only ways such large, widespread masses of silicate melt could have formed are in volcanic eruptions or as impact ejecta. First we assess the volcanic hypothesis.

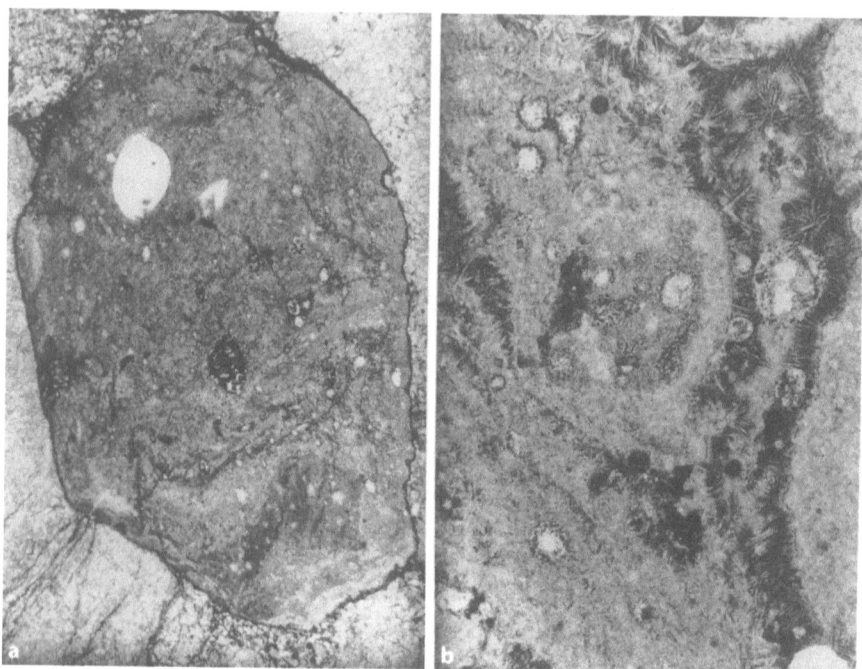

Fig. 6. (above and opposite) Photomicrographs in plane polarized light of miscellaneous textures in irregular particles. A) Irregular particle with one large quartz-filled vesicle and numerous smaller ones.   Internal heterogeneity and banding are present, but not as pronounced as in some irregular particles. B) Pocket within an irregular particle with abundant sprays of small fibrous crystals of K-feldspar. Note small vesicles. C) Large ovoid irregular particle. Truncated oval on edge of particle in lower left is candidate for spherule embedded inside particle. D) Closer view of candidate for embedded spherule. Long axis of field of view is 5.53 mm in A and C, 2.76 mm in B, and 1.38 mm in D.

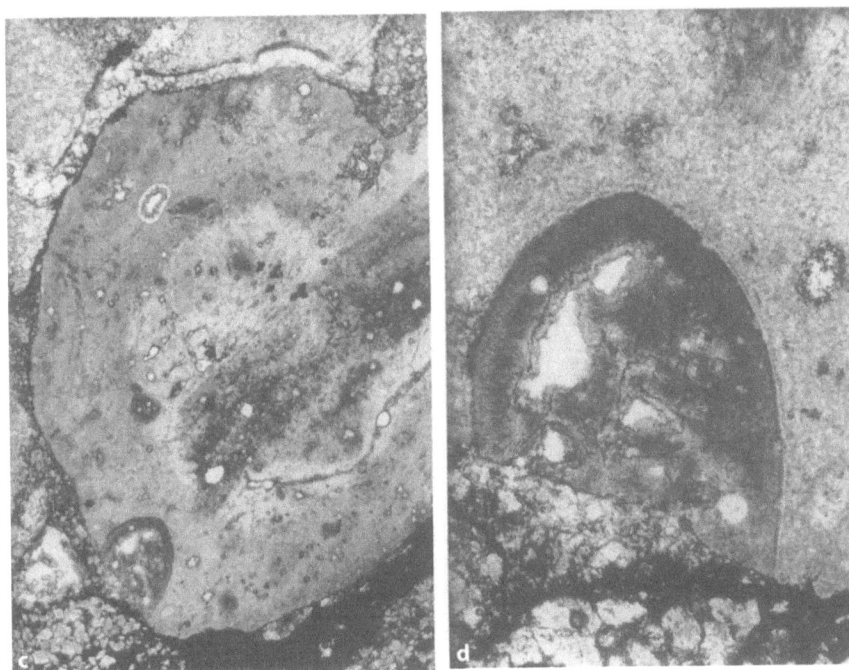

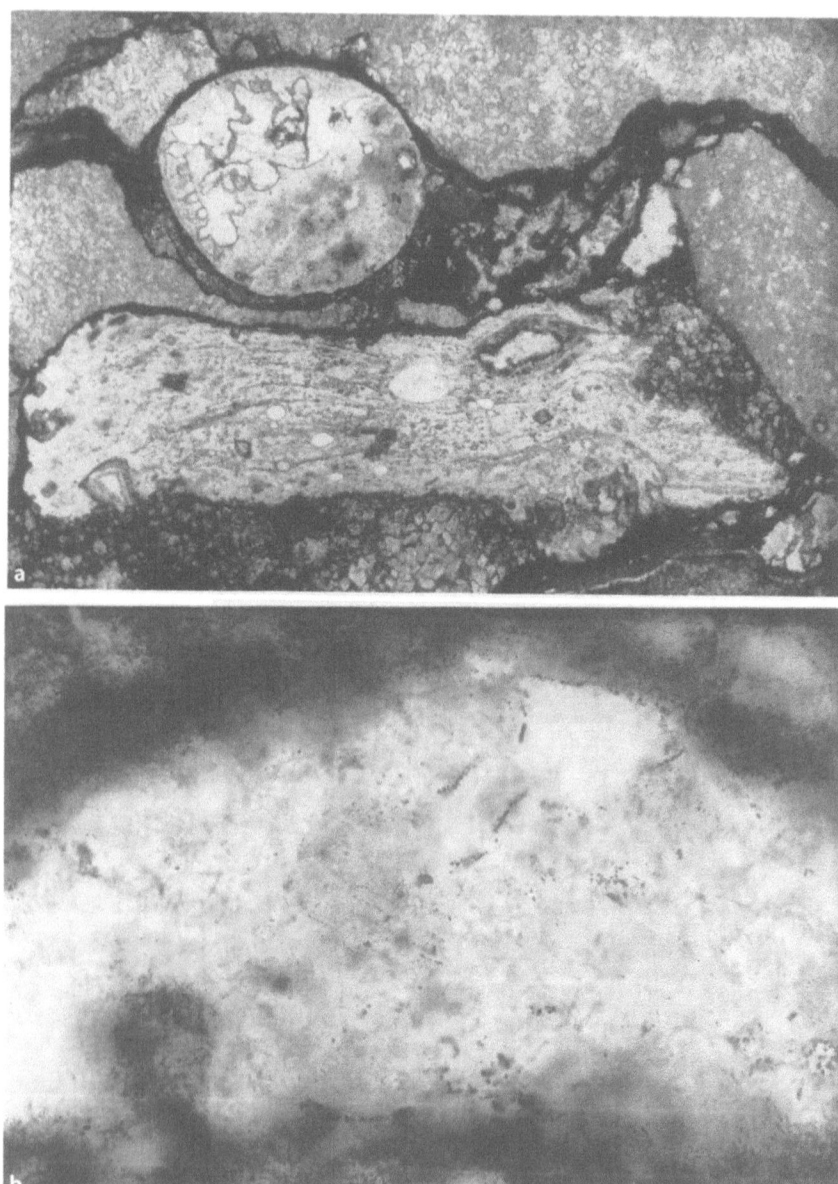

Fig. 7. Photomicrographs in plane polarized light of miscellaneous textures in irregular particles. A) Highly elongated irregular particle showing strong longitudinal banding and elongated, quartz-filled vesicles. More irregular lens with dark rim in upper right part of particle is a heterogeneous pocket of quartz crystals. Both irregular particle and spherule above consist mainly of K-feldspar. B) Closer view of part of irregular lens in A. Quartz crystal in center has parallel lines of minute inclusions, which are candidates for relict PDFs. Long axis of field of view is 5.53 mm in A and 0.35 mm in B.

volcanoclastic rocks are present at various levels in the Hamersley basin, but petrographically, they are consistently different from both the spherules and irregular particles described here. Tuffaceous interbeds occur in both the Jeerinah and Wittenoom Formations (Fig. 2), but not in the Carawine Dolomite. These tuffs are thin and consist mainly of angular vitric and lithic pyroclasts typical of tuffs formed via phreatomagmatic eruptions of basaltic to intermediate magmas (Hassler 1993; Simonson et al. 1993b; Trendall et al. 1998). Such pyroclasts lack the textures described above for the irregular particles. The coarsest tuffs contain some accretionary and armored lapilli, which is in keeping with a hydrovolcanic origin. These lapilli are spheroidal, but internally they consist of fine-grained ash with crude concentric banding (e.g. fig. 3C in Hassler 1993) rather than crystallization textures and flow-banding. We found no sign of accretionary lapilli, pyroclasts, or any other bona fide volcanoclastic materials in either the thin sections or HCl-insoluble residues of several large samples from the Carawine layer.

Neither the spherules nor the irregular particles compare well with volcanic materials, but both are a good petrographic match for certain types of impact ejecta. Numerous similarities have already been noted between spherules from the dolomixtite and microtektites and microkrystites in Phanerozoic strewn fields (Simonson 1992; Simonson et al. 1998, 1999). However, the larger, more irregular particles have much more in common with Muong Nong-type tektites. Irregular, blocky shapes are typical of the Muong Nong-type tektites, whereas splash-form tektites, microtektites, and microkrystites generally have smoother, more rounded shapes (Glass 1984, 1990; Schnetzler 1992; Koeberl 1992). Likewise, Muong Nong-type tektites typically have a pronounced compositional layering or banding which is more prominent than the schlieren of splash-form tektites (Glass 1984, 1990; Koeberl 1992; Fiske 1996). Vesicles are also common in Muong Nong-type tektites (e.g., Koeberl 1992, fig. 2), and some of these vesicles are elongated parallel to the banding via deformation while flowing as either a melt or hot glass (Fiske 1996). Lastly, unlike splash-form tektites, Muong Nong-type tektites commonly contain relict minerals as inclusions (Glass 1984; Koeberl 1992).

Despite the numerous similarities, there also appear to be important differences between the irregular particles and Muong Nong-type tektites. First, the irregular particles are smaller and more rounded on average than typical Muong Nong-type tektites. This could simply reflect abrasion and hydrodynamic sorting of the irregular particles during transport. Other differences are not so easily explained. These include the presence within some of the irregular particles of crystals that may have formed during cooling of the melt or devitrification (Fig. 6B) possible planar deformation features in quartz crystals (Fig. 7B). Neither these features nor splash forms have been reported from Muong Nong-type tektites. On the other hand, the lechatelierite particles that are ubiquitous in Cenozoic tektites (Glass 1984, 1990) have not been observed in the irregular particles in the dolomixtite. Further discussion of the significance of the similarities and differences between Muong Nong-type tektites and the irregular particles in the dolomixtite is deferred to the final section.

Several lines of evidence strongly suggest the spherules and irregular particles formed simultaneously during a single event. First, they are both restricted to the same layer within the Carawine Dolomite, the dolomixtite. Second, they consist of the same minerals and show similar parageneses. Third and most compelling, individual spherules appear to be embedded in some of the larger irregular particles (Fig. 6C, D). This would require that the irregular particles were molten while the spherules were cool enough to be rigid, and that one or both were moving at high speeds during the same event. Even though the spherules and irregular particles are both replaced by essentially the same minerals, the textures they show are consistently different. Radial-fibrous textures dominate the spherules, whereas the fibrous crystals in the larger, more irregular particles are generally smaller and have either a parallel-fibrous or random orientation. We are not sure what causes this discrepancy, but we can suggest two possibilities. One is that the spherules and irregular particles represent target rocks melted to different degrees during an impact. Lofgren (1977) demonstrated experimentally that submicroscopic crystallites which persist in partially melted rocks can strongly influence the textures that subsequently form during cooling. Perhaps crystallites aligned by flow are responsible for the preferred orientation of the K-feldspar crystals in the banded irregular particles (Plate 1). The other possibility stems from the observation that Muong Nong-type tektites are enriched in volatiles relative to splash-form tektites (Koeberl 1990, 1992). Either, or both, of these factors could be responsible for helping crystals to grow more readily in the irregular particles. In fact, the textures in some of Lofgren's (1977, figs. 5 D, E) partially melted experimental charges strongly resemble those of the irregular particles. In contrast, the scarcity of volatiles and/or the lack of unmelted crystalites may have limited the nucleation of crystals to the exterior surfaces of the spherules, which in turn would lead to a dominance of radial-fibrous textures. If submicroscopic crystallites were present in the irregular particles at the time of deposition, this too would mark a departure from Muong Nong-type tektites.

In Cenozoic strewn fields, Muong Nong-type tektites are associated with microtektites rather than microkrystites (Glass 1984, 1990; Koeberl 1986; Glass et al. 1995). This suggests the spherules in the dolomixtite are probably altered microtektites, despite Simonson et al.'s (1998) referring to them as microkrystites. It is also consistent with the fact that many of the radial-fibrous textures in the spherules are comparable to textures formed during devitrification (Lofgren 1971a,b), which would mean they were glass at the time of deposition. However, some of the spherules with randomly oriented crystals closely resemble crystallization textures produced experimentally in basaltic melts by Lofgren (1977). If the crystals formed during cooling and prior to quench, these spherules are best termed microkrystites (Glass and Burns 1988). We therefore believe that both microtektites and microkrystites are present in the dolomixtite. This situation may be analogous to ejecta from both the Eltanin impact structure (Gersonde et al. 1997) and the North American strewn field, which include both microtektites and microkrystites (Margolis et al. 1991; Glass 1989).

One last feature that deserves attention are the clear spots in the spherules. We believe these arose in two different ways. The clear spots with smooth circular to oval cross-sections and sharply defined margins appear to represent former vesicles or gas bubbles that were subsequently filled with diagenetic crystals. Such vesicles occur in both tektites (e.g., Koeberl 1992, Fig. 2) and microtektites (e.g., Smit et al. 1992, fig. 6). In contrast, the clear spots with scalloped rims bear a striking morphological resemblance to the glass cores bounded by spherical concavities typical of partially replaced tektites from the Cretaceous-Tertiary boundary layer (Sigurdsson et al. 1991; Izett 1991; Bohor and Glass 1995). Therefore, we do not believe the clear spots with scalloped rims represent gas bubbles or vesicles; instead, we interpret them as glass cores that were replaced separately from the rims later in diagenesis via either direct devitrification or dissolution and precipitation of new phases.

Implications

One important aspect of the irregular particles in the dolomixtite is simply their presence in rocks so old. Based on the isotopic dates reviewed above, the Carawine Dolomite is most likely late Archean in age. If our suggestion that the irregular particles in the dolomixtite were originally similar to Muong Nong-type tektites, they would represent the oldest tektites discovered to date. They are also likely to be closer to their source than any other early Precambrian ejecta because they are the coarsest. This is consistent with their being Muong Nong-type tektites, whose textural and geochemical characteristics indicate they have neither been heated to temperatures as great as other varieties of tektites nor traveled as far from their point of origin in most cases (Koeberl 1992). We also emphasize

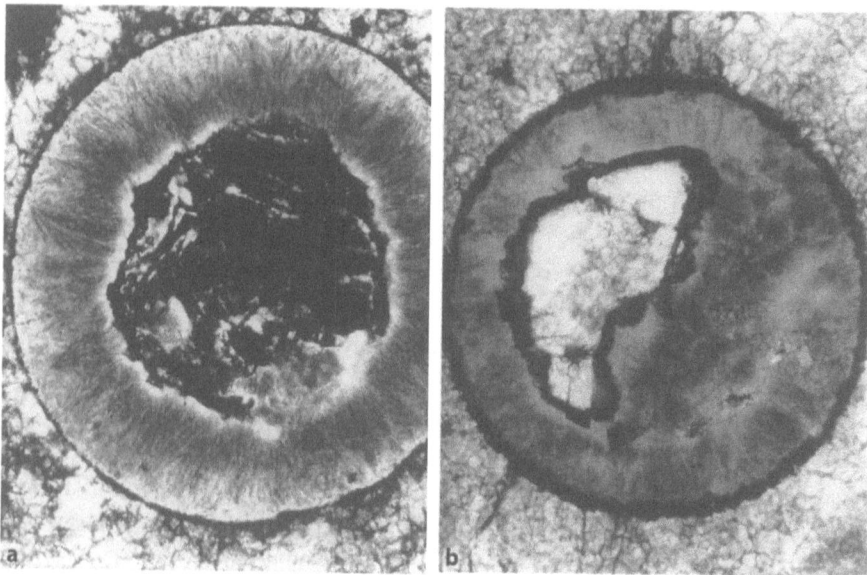

Fig. 8. (above and opposite) Photomicrographs of spherules in plane polarized light. A) Spherule whose cross-section is nearly a perfect circle. Outer rim consists of inward-radiating sprays of needle-like K-feldspar crystals; dark central region is largely oxide-stained carbonate. B) Spherule consisting of inward-radiating rind of chlorite(?) with eccentric clear spot consisting of coarsely crystalline carbonate. C) Spherule with inward-radiating sprays of thicker K-feldspar crystals. Clear band just inside sprays is also K-feldspar. Central spot has scalloped margins and is filled with iron oxide-stained carbonate (dark material) and finely felted muscovite (lighter material). D) Spherule with tan, inward-radiating sprays of K-feldspar crystals and an eccentric clear spot (probably a former vesicle) filled with quartz and K-feldspar. Long axis is 1.38 mm in all fields of view.

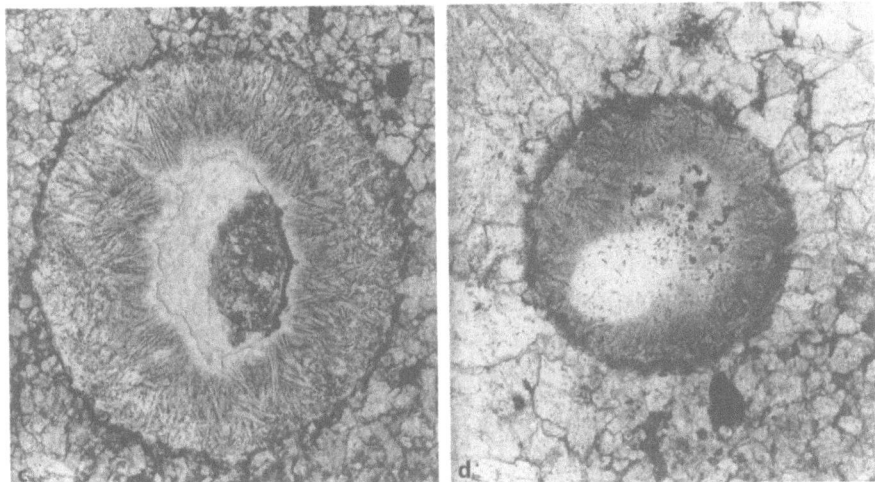

that the irregular particles are large enough to be easily recognizable in core or outcrop and identifiable with a hand lens. This bodes well for the identification of additional occurrences of impact ejecta in early Precambrian successions. In fact, we believe ejecta layers may actually be easier to identify in Precambrian successions than they are in Phanerozoic ones. Given their thinness, ejecta strewn fields are rarely preserved as discrete, recognizable layers except in low-energy environments where they can escape physical reworking, e.g., below wave base on continental margins. However, such thin layers are usually dispersed in ambient sediment by burrowing metazoans (Glass 1969) and therefore hard to recognize. Biogenic disturbance of ejecta layers was not an issue prior to the late Precambrian, based on the trace fossil record (Crimes and Droser 1992).

Another important aspect of the irregular particles in the dolomixtite is that, even though Muong Nong-type appear to be the closest match, there are still important differences between the two. If the irregular particles originally were glassy impact ejecta, the fact that they do not fit neatly into the Muong Nong-type (or any other) tektite category suggests ancient strewn fields may contain some additional types of glassy ejecta which do not fit neatly into existing categories. We believe this is a definite possibility, given the fact that, until recently, most of what we know about tektites, microtektites, and microkrystites originated from in-depth studies of only four Cenozoic strewn fields (Glass 1984, 1990; Koeberl 1990). This "data base" has expanded via studies of ejecta within the Cretaceous-Tertiary boundary layer (Izett 1991; Sigurdsson et al. 1991; Bohor and Glass 1995) and associated with the Eltanin structure (Margolis et al. 1991; Gersonde et al. 1997). If our interpretation of the irregular particles in the dolomixtite is correct, the range of possibilities for glassy impact ejecta in strewn fields may need to be expanded even further. None of the categories currently in use encompass particles > 1 mm across that contain crystals formed during cooling.

If the irregular particles in the dolomixtite were originally like Muong Nong-type tektites, another important implication is that some textures in tektite glass can sometimes be stabilized and preserved essentially indefinitely in diagenetic environments via replacement by other phases. As Glass et al. (1997) noted, it was widely believed until recently that, unlike obsidian, tektite glass did not undergo hydration. Without this critical step, it was therefore unlikely that internal textures in tektite glass would be preserved via replacement. Such ideas were challenged by the discovery of tektites in the Cretaceous-Tertiary boundary layer that were replaced by smectites and other minerals, yet still preserved original textures such as flow banding and vesicles (Izett 1991; Smit et al. 1992; Bohor and Glass 1995). By analogy with volcanic rocks, Bohor and Glass (1995) suggested the margins of the Cretaceous-Tertiary tektites were palagonitized when hot glass hit seawater, then preferentially replaced later in diagenesis. Glass et al. (1997) subsequently demonstrated hydration and palagonitization are widespread in tektites and microtektites deposited in marine environments. We believe the existence of well-defined flow banding and vesicles in the irregular particles from the dolomixtite provide additional evidence that it is possible to replace glassy impact ejecta volume-for-volume, and thereby stabilize and preserve internal textures for posterity. Like the examples described by Bohor and Glass (1995) and Glass et al. (1997), the Carawine Dolomite was deposited in a marine environment, so this may be a key element. However, the replacement rims on the Cretaceous-Tertiary tektites attributed to palagonitization by Bohor and Glass (1995) do not appear to show any of the radial or fibrous textures which are so common in the early Precambrian spherules.

Since the dolomixtite is only one of at least three temporally distinct layers that contain purported impact ejecta in the Hamersley basin (Fig. 2), these occurrences can also be used to estimate the rate at which impacts were taking place in the late Archean. Interpolating between zircon-bearing tuffs that have been dated isotopically, we estimate the lowest and highest spherule layers (in the Jeerinah Formation and Dales Gorge Member) were deposited ~2.63 Ga and ~2.49 Ga, respectively. This means three strewn field-forming impacts took place in roughly 140 million years. Less certain is how representative they are of impacts on a global scale. The preponderance of spherules with radial textures which we interpret as altered microtektites and the similarities between the irregular particles and Muong Nong-type tektites both suggest the particles in the dolomixtite are part of a regional strewn field rather than a global one. Using the largest known strewn field in which both occur (the Australasian) is a guide (Glass 1984), the ejecta in the dolomixtite would belong to a strewn field covering roughly 10% of the earth's surface. Interpreting the other layers in the Hamersley basin similarly and extrapolating to the whole surface of the earth would imply the formation of 30 strewn fields within 140 million years, or roughly 1 every 5 million years. This compares favorably with the Cenozoic since six strewn fields of microtektites and/or microkrystites are known to have formed in the last ~36 million years of earth history, i.e., about 1 every 7 million years. However, spherules are much more abundant per unit area in the dolomixtite and the other layers in the Hamersley basin than they are in any of the Cenozoic strewn fields. The only accumulations of essentially pure impact spherules that are comparable volumetrically to those in the Hamersley basin are found in the Cretaceous-Tertiary boundary layer (Smit et al. 1992; Norris et al. 1999). This suggests the spherule layers in the Hamersley basin are more likely to represent global affairs rather than regional strewn fields. If so, the sizes of the impacts responsible increase dramatically, while the inferred rate at which the strewn fields formed decreases to 3 in 140 million years, or roughly 1 every 70 million years. Given that there may be other late Archean spherule layers that have yet to be discovered, we feel this value is a maximum and that a lower number is likely.

Lastly, Simonson et al. (1998, 1999) argued that all of the spherule layers in the Hamersley basin were products of oceanic impacts. This is in keeping with Glikson's (1999) estimate that 80% of all projectiles impacted on time-integrated oceanic crust since the late heavy bombardment. However, almost all of the tektite strewn fields known from the Phanerozoic are the products of impacts into continental crust, including all from which Muong Nong-type tektites have been reported. If the irregular particles in the dolomixtite were like Muong Nong-type tektites, this suggests they were generated by an impact in continental rather than oceanic crust. Since Muong Nong-type tektites contain relict mineral inclusions, a more in-depth study of the irregular particles in the dolomixtite might yield intact relicts of the target material that could confirm or refute this interpretation. However, irregular particles like those in the dolomixtite have not been observed in other early Precambrian spherule layers, including those formed in the early Archean (G. Byerly personal communication 1999). Moreover, the dolomixtite does not contain the symmetrical bedforms Hassler et al. (in press) and Simonson et al. (in press) interpret as the products of impact-induced tsunami waves. Therefore we still believe most or all of the early Precambrian spherule layers are the products of oceanic impacts, which is markedly different from the Phanerozoic. Why this discrepancy? It may simply be a bias in the data available, given the small number of both Precambrian spherule layers and Phanerozoic strewn fields discovered to date, in which case lots of evidence for Phanerozoic marine impacts has yet to discovered. On the other hand, it could also reflect in part a higher areal ratio of oceanic to continental crust in the Archean relative to the Phanerozoic (Nisbet 1987).

Acknowledgements

Field work was supported by the National Geographic Society and Oberlin College. Many Oberlin students helped us in both the field and the lab, most notably Melissa Berke and Paul Harnik. Bill Glass and Philippe Claeys read the manuscript and offered many valuable suggestions.

References

Bohor BF, Glass BP (1995) Origin and diagenesis of K/T impact spherules - From Haiti to Wyoming and beyond. Meteoritics 30: 182-198

Crimes TP, Droser ML (1992) Trace fossils and bioturbation: The other fossil record. Ann Rev Ecol Systematics 23: 339-360

Fiske PS (1996) Constraints on the formation of layered tektites from the excavation and analysis of layered tektites from northeast Thailand. Meteoritics Planet Sci 31: 42-45

Gersonde R, Kyte FT, Bleil, U, Diekmann, B, Flores JA, Gohl K, Grahl G, Hagen R, Kuhn G, Sierro FJ, Völker D, Abelmann A, Bostwick JA (1997) Geological record and reconstruction of the late Pliocene impact of the Eltanin asteroid in the Southern Ocean:. Nature 390: 357-363.

Glass BP (1969) Reworking of deep-sea sediments as indicated by the vertical dispersion of the Australasian and Ivory Coast microtektite horizons. Earth Planet Sci Letters 6: 409-415

Glass BP (1984) Tektites. J Non-Crystalline Solids 67: 333-344

Glass BP (1989) North American tektite debris and impact ejecta from DSDP Site 612. Meteoritics 24: 209-218

Glass BP (1990) Tektites and microtektites: key facts and inferences. Tectonophys 171: 393-404

Glass BP, Burns CA, Crosbie JR, DuBois DL (1985) Late Eocene North American microtektites and crystal-bearing spherules. Proc 16th Lunar Plan Sci Conf Part 1, J Geophys Res 90, Supplement: D175-D196

Glass BP, Burns CA (1987) Late Eocene crystal-bearing spherules: two layers or one? Meteoritics 22: 265-279

Glass BP, Burns CA (1988) Microkrystites: A new term for impact-produced glassy spherules containing primary crystallite. Proc 18th Lunar Plan Sci Conf, pp 455-458

Glass BP, Koeberl C, Blum JD, Senftle F, Izett GA, Evans BJ, Thorpe AN, Povenmire H, Strange RL (1995) A Mong Nong-type Georgia tektite. Geochim Cosmochim Acta 59: 4071-4082

Glass BP, Muenow DW, Bohor BF, Meeker GP (1997) Fragmentation and hydration of tektites and microtektites. Meteoritics Planet Sci 32: 333-341

Glikson AY (1999) Oceanic mega-impacts and crustal evolution. Geology 27: 387-390

Gostin VA, Haines PW, Jenkins RJF, Compston W, Williams IS (1986) Impact ejecta horizon within Late Precambrian shales, Adelaide Geosyncline, South Australia. Science 233: 198-200

Gostin VA, Keays RR, Wallace MW (1989) Iridium anomaly from the Acraman impact ejecta horizon: Impacts can produce sedimentary iridium peaks. Nature 340: 542-544

Grieve R, Rupert J, Smith J Therriault A (1995) The record of terrestrial impact cratering. GSA Today 5: 189-196

Hassler SW (1993) Depositional history of the Main Tuff Interval of the Wittenoom Formation, Late Archean-Early Proterozoic Hamersley Group, Western Australia. Precamb Res 60: 337-359

Hassler SW, Robey HF, Simonson BM (in press) Bedforms produced by impact-generated tsunami, ~2.6 Ga Hamersley basin, Western Australia. Sed Geol

Izett GA (1991) Tektites in Cretaceous-Tertiary boundary rocks on Haiti and their bearing on the Alvarez impact extinction hypothesis. J Geophys Res 96: 20,879-20905

Jahn B-m, Simonson BM (1995) Carbonate Pb-Pb ages of the Wittenoom Formation and Carawine Dolomite, Hamersley Basin, Western Australia (with implications for their correlation with the Transvaal Dolomite of South Africa). Precamb Res 72: 247-261

John C, Glass BP (1974) Clinopyroxene-bearing glass spherules associated with North American microtektites. Geology 2: 599-602

Koeberl C (1986) Muong Nong type tektites from the Moldavite and North American strewn fields? Proc 17th Lunar Plan Sci Conf Part 1, J Geophys Res 91, B13: E253-E258.

Koeberl C (1990) The geochemistry of tektites: an overview. Tectonophys 171: 405-422

Koeberl C (1992) Geochemistry and origin of Mong Nong-type tektites. Geochim Cosmochim Acta 56: 1033-1064

Lofgren GE (1971a) Devitrified glass fragments from Apollo 11 and Apollo 12 lunar samples. Proc 2nd Lunar Sci Conf 1: 949-955

Lofgren GE (1971b) Spherulitic textures in glassy and crystalline rocks. J Geophys Res 76: 5635-5648

Lofgren GE (1977) Dynamic crystallization experiments bearing on the origin of textures in impact-generated liquids. Proc 8th Lunar Sci Conf: 2079-2095

Lowe DR, Byerly GR (1986) Early Archean spherules of probable impact origin, South Africa and Western Australia. Geology 14: 83-86

Lowe DR, Byerly GR, Asaro F, Kyte FT (1989) Geological and geochemical record of 3400-million-year-old terrestrial meteorite impacts. Science 245: 959-962

Margolis SV, Claeys P, Kyte FT (1991) Microtektites, microkrystites, and spinels from a Late Pliocene asteroid impact in the southern ocean. Science 251: 1594-1597

Marshall DJ (1988) Cathodoluminescence of Geological Materials. Unwin Hyman, Boston

Martínez Ruíz F, Ortega Huertas M, Palomo I, Acquafredda P (1997) Quench textures in altered spherules from the Cretaceous-Teritary boundary layer at Agost and Caravaca, SE Spain. Sed Geol 113: 137-147

Montanari A (1991) Authigenesis of impact spheroids in the K/T boundary clay from Italy: New constraints for high-resolution stratigraphy of terminal Cretaceous events. J Sed Petrol 61: 315-339

Nelson DR, Trendall AF, Altermann W (in press) Chronological correlations between the Pilbara and Kaapvaal cratons. Precamb Res

Nisbet EG (1987) The Young Earth - An introduction to Archaean geology. Allen and Unwin, Boston

Norris RD, Huber BT, Self-Trail J (1999) Synchroneity of the K-T oceanic mass extinction and meteorite impact: Blake Nose, western North Atlantic. Geology 27: 419-422

Pettijohn FJ (1949) Sedimentary Rocks (1st ed.). Harper & Brothers, New York

Schnetzler CC (1992) Mechanism of Muong Nong-type tektite formation and speculation on the source of Australiasian tektites. Meteoritics 27: 154-165

Sigurdsson H, D'Hondt S, Arthur MA, Bralower TJ, Zachos JC, van Fossen M, Channell JET (1991) Glass from the Cretaceous/Tertiary boundary in Haiti. Nature 349: 482-487

Simonson BM (1992) Geological evidence for a strewn field of impact spherules in the early Precambrian Hamersley Basin of Western Australia. Geol Soc Amer Bull 104: 829-839

Simonson BM, Carney KE (1999) Roll-up structures: evidence of *in situ* microbial mats in late Archean deep shelf environments. PALAIOS 14: 13-24

Simonson BM, Hassler SW (1997) Revised correlations in the Early Precambrian Hamersley Basin based on a horizon of resedimented impact spherules. Austral J Earth Sci 44: 37-48

Simonson BM, Schubel KA, Hassler SW (1993a) Carbonate sedimentology of the early Precambrian Hamersley Group of Western Australia. Precamb Res 60: 287-335

Simonson B M, Hassler S W, Schubel K A (1993b) Lithology and proposed revisions in stratigraphic nomenclature of the Wittenoom Formation (Dolomite) and overlying formations, Hamersley Group, Western Australia. Geol Surv Western Austral Rept 34, Prof Papers: 65-79

Simonson BM, Davies D, Wallace M, Reeves S, Hassler SW (1998) Iridium anomaly but no shocked quartz from Late Archean microkrystite layer: oceanic impact ejecta? Geology 26: 195-198

Simonson BM, Hassler SW, Beukes NJ (1999) Late Archean impact spherule layer in South Africa that may correlate with a layer in Western Australia. In: Dressler BO, Sharpton VL (eds) Impact Cratering and Planetary Evolution II. Geol Soc Amer Spec Paper 339

Smit J, Alvarez W, Montanari A, Swinburne N, Van Kempen TM, Klaver GT, Lustenhouwer WJ (1992) "Tektites" and microkrystites at the Cretaceous Tertiary boundary: two strewn fields, one crater? Proc Lunar Plan Sci 22: 87-100

Smit J, Klaver G (1982) Sanidine spherules at the Cretaceous-Tertiary boundary indicate a large impact event. Nature 292: 47-49

Stanley SM (1999) Earth System History. W H Freeman and Co, New York

Trendall AF (1983) The Hamersley Basin. In: Trendall AF, Morris RC (eds) Iron-formations: Facts and Problems. Elsevier, Amsterdam, pp 69-129

Trendall AF (1990) Hamersley Basin. In: Geology and Mineral Resources of Western Australia, Geol Surv Western Austral Mem 3, pp 163-190

Trendall AF, Nelson DR, de Laeter JR, Hassler SW (1998) Precise zircon U-Pb ages from the Marra Mamba Iron Formation and Wittenoom Formation, Hamersley Group, Western Australia. Austral J Earth Sci 45: 137-142

Wallace MW, Gostin VA, Keays RR (1989) Discovery of the Acraman impact ejecta blanket in the Officer Basin and its stratigraphic significance. Austral J Earth Sci 36: 585-587

Wallace MW, Gostin VA, Keays RR (1990a) Spherules and shard-like clasts from the Late Proterozoic Acraman impact ejecta horizon, South Australia. Meteoritics 25: 161-165

Wallace MW, Gostin VA, Keays RR, (1990b) Acraman impact ejecta and host shales: Evidence for low-temperature mobilization of iridium and other platinoids. Geology 18: 132-135.

Wallace MW, Gostin VA, Keays RR (1996) Sedimentology of the Neoproterozoic Acraman impact-ejecta horizon, South Australia. AGSO J Austral Geol & Geophys 16: 443-451

Woodhead JD, Hergt JM, Simonson BM (1998) Isotopic dating of an Archean bolide impact horizon, Hamersley Basin, Western Australia. Geology 26: 47-50

Plate 1. (opposite) Photomicrographs of irregular particles and spherules. A) Angular irregular particle with gently curved banding and a few vesicles in plane polarized light next to spherule with a "figure 8" shape and internal textures indicating it is a composite of two agglutinated particles. B) Same field of view as A between crossed polarizers with gypsum plate inserted. Colors reflect contrasting radial alignment of K-feldspar crystals in spherule vs. parallel alignment in irregular particle. C) Same field of view as Figure 4C between crossed polarizers with gypsum plate inserted. Consistent blue color indicates strong optical alignment of the K-feldspar crystals length-fast parallel to the banding. High birefringence inclusions are coarse crystals of carbonates replacing parts of selected bands and filling vesicles. D) Same field of view as Figure 4D between crossed polarizers with gypsum plate inserted. Consistent patchs of yellow and blue color indicates strong optical alignment of the K-feldspar crystals parallel to the banding even where it is contorted. Long axis is 2.76 mm in all four fields of view.

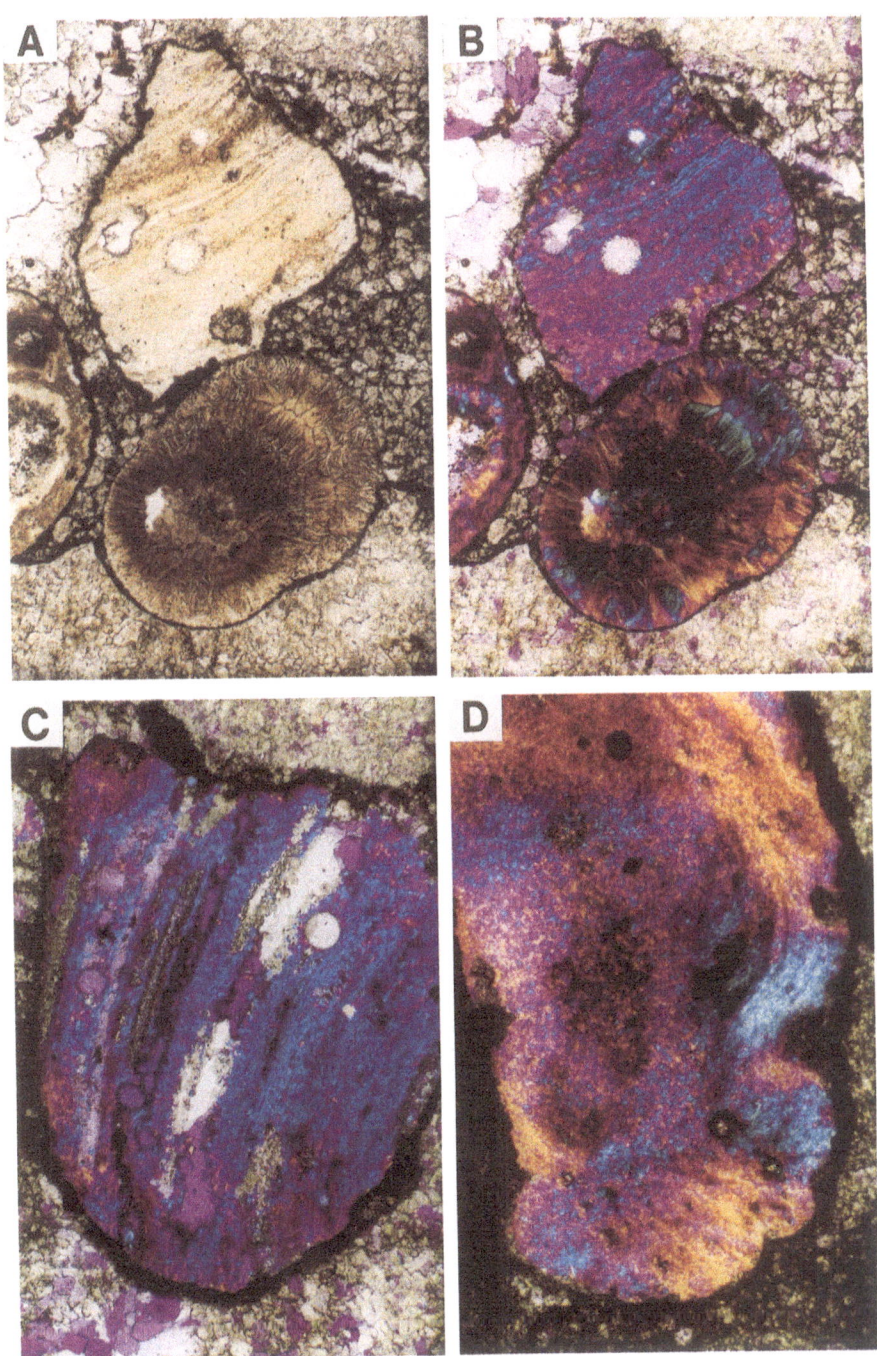

8 Fossil Micrometeorites from Finland - Basic Features, Scientific Potential, and Characteristics of the Mesoproterozoic Host Rocks

Dirk Kettrup[1,2], Alexander Deutsch[1], Pekka Pihlaja[2], and Lauri J. Pesonen[2]

[1]Institute for Planetologie, D-48149 Münster, Germany.
[2]Geological Survey of Finland, FIN-02150 Espoo, Finland.

Abstract. The oldest known micrometeorites occur in the up to 1800 m thick Mesoproterozoic Satakunta sandstone in SW-Finland. This typical red bed formation covers a graben with the dimensions of about 15 x 100 km². The Satakunta formation correlates with the Jotnian sandstone, overlaying at several locations in Fennoscandia basement rocks, which are part of the about 1.8 to 1.9 Ga old Svecofennian orogenic belt. The age of the Satakunta formation s.s. is not well constrained: Sedimentation may have already begun 1.65 Ga ago, and ended prior to the intrusion of the post-Jotnian diabases (1.26 Ga). The depositional environment of the Satakunta sediments was primarily mostly fluvial. In arkose sandstones of this formation, over 60 cosmic spherules (melted micrometeorites) have been identified. They belong to different mineralogical types, and display unaltered mineralogical and chemical features, including the presence of a still glassy matrix. Moreover, this cosmic dust lacks clear signs of mechanical transport. So far, the reasons for the excellent preservation of the micrometeorites are enigmatic. Conceivable factors that generally may have influenced the relatively high abundance of the micrometeorites in the Satakunta formation are (i) distinct concentrations mechanisms acting prior to the embedding into the host sediment, (ii) settling of the spherules at low energy, and lack of further transport in the sedimentological environment, (iii) only minor diagenetic compaction of the host sediments at rather reducing conditions, and (iv) a quite specific time-integrated temperature history for over a billion years. In this contribution, we outline sedimentological characteristics of both, barren, and spherule-carrying Satakunta lithologies. In addition, we discuss possible scenarios for deposition, and survival of this ancient cosmic dust. Understanding of these processes is of prime importance as red beds are quite common lithologies in the Earth's history, and hence, may represent a major and easily accessible reservoir for fossil micrometeorites. Finding cosmic spherules in red beds of other locations and age may allow reconstruction of the flux rate of micrometeorites over the geological time scale. In addition, the characterization of micrometeorite populations can provide model-independent information on the evolution of the atmospheric fO_2.

Introduction

Terrestrial sediments or sedimentary rocks may contain natural spherules of non-volcanic and non-authigenic origin. Such objects originate either during cratering processes (ejected "impact melt glass") or by melting of extraterrestrial matter during hypervelocity entry in the Earth's atmosphere, followed by quenching during deceleration. These so-called "cosmic spherules" represent the most abundant group of micrometeorites (Brownlee et al. 1997). Micrometeorites are collected almost exclusively in specific environments. Their major sources are polar ice (Maurette et al. 1987; Taylor et al. 1998), deep-sea sediments (Brownlee et al. 1984), and the stratosphere (Brownlee 1985). While impact glass droplets occur in several geological horizons (for a review, see Grieve 1998), e.g., in the Late Eocene or at the Cretaceous/Tertiary (K/T) boundary, cosmic spherules usually have a short terrestrial residence time and, therefore, are mostly Quaternary in age. Seemingly unique "old" exceptions have been discovered in Eocene sediments (Taylor and Brownlee 1991), in Jurassic hardgrounds (Jéhanno et al. 1988), and in the ~1.4 Ga old (=Mesoproterozoic) Jotnian Satakunta sediments, Finland (Marttila 1969; Deutsch et al. 1998). For what reasons these cosmic spherules have been preserved is unknown so far. In this contribution, we make a first attempt to solve this problem

Geological Setting of the Satakunta Formation

The Satakunta graben is located in SW Finland (Fig. 1). This part of the Fennoscandian shield is characterized by the ~1.88 Ga old Svecofennian folding (Kohonen et al. 1993), a following long period of erosion, and intrusion of post-orogenic Rapakivi granites, and associated porphyries. Deposition of the red coloured Jotnian sediments, 1.6 - 1.3 Ga ago, was seemingly restricted to graben-zones in the deeply eroded Svekokarelian basement. The mineralogy, especially the heavy mineral spectrum (Marttila 1969) of these clastic sediment series reflects the respective Svecofennian source rocks, i.e., gneisses and granites. The

Jotnian Satakunta formation continues into the Bothnian Sea, where, according to geophysical data, its thickness is estimated to reach 1800 m. Using the evolution of well-studied graben zones as analogues, the 1800 m of red bed sediments may correspond to a time interval of 20 to 100 Ma. The general interpretation of the depositional history of the Satakunta formation is hampered by the fact that due to the omnipresent glacial cover in SW Finland, only a few outcrops are available for investigations, and most of those are less than some tens of meters in size. Notwithstanding, some lithologies of this red bed are remarkable - they carry seemingly uniquely ancient cosmic dust, i.e. cosmic spherules (Marttila 1969; Deutsch et al. 1998; Kettrup et al. 1998; Pesonen et al. 1998; Robin et al. 1998).

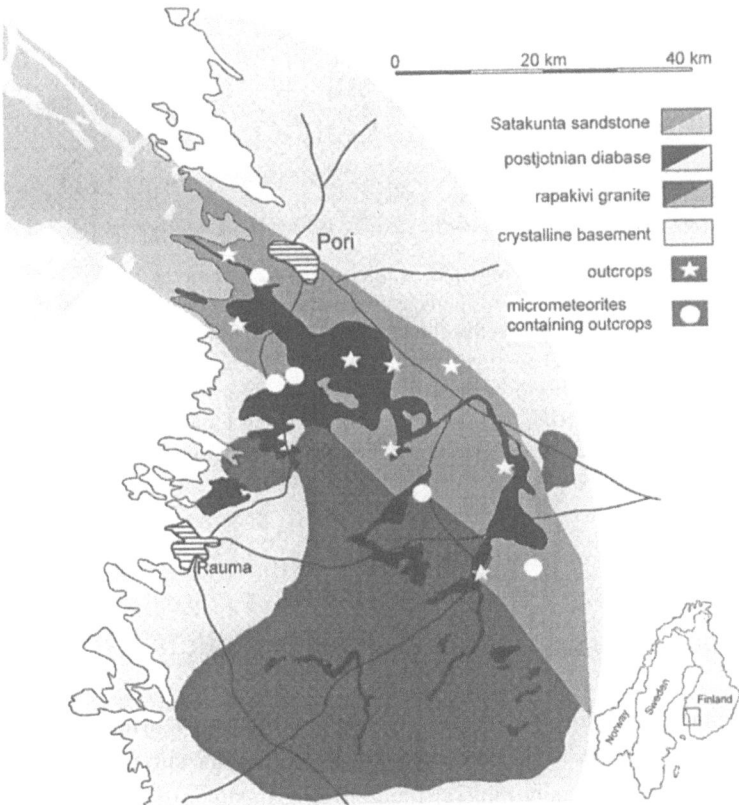

Fig. 1. Geological sketch map of the Satakunta area. Outcrops are given as stars; white dots mark outcrops where lithologies with cosmic spherules occur. Intrusion of the 1.26 Ga old post-Jotnian diabase set a time limit for cessation of the sedimentation in the Satakunta graben. Modified after Kohonen et al. (1993).

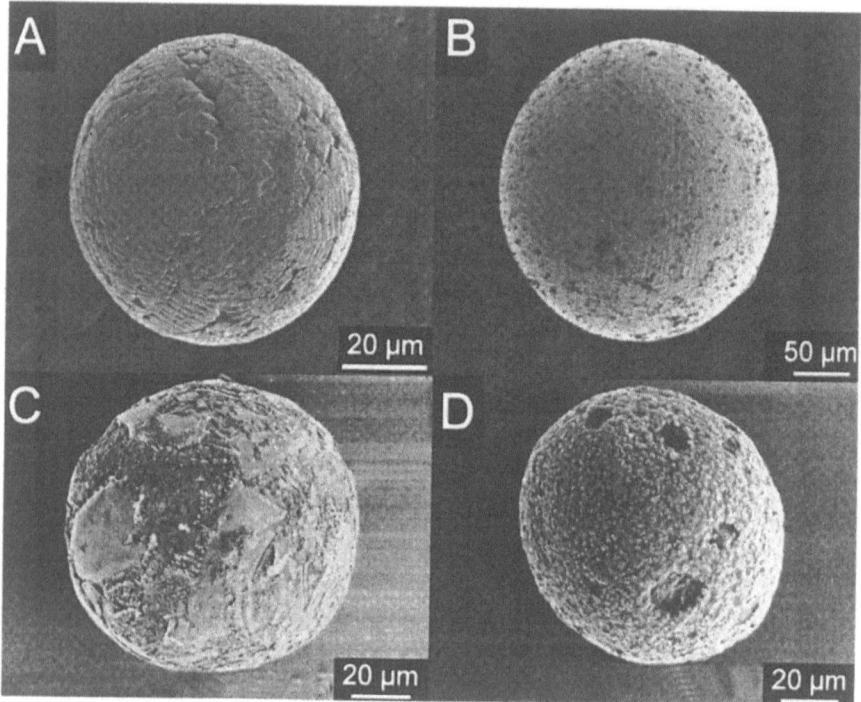

Fig. 2. Secondary electron micrographs of micrometeorites from the Satakunta formation. The chemical and textural type of the cosmic spherules shown with (**A**) from the locality Sahankoski, and (**B**) from Naskalinkallio is yet unknown. The objects in (**C, D**) are from Murronmäki and have a (**C**) porphyritic, and (**D**), and a barred-olivine texture. Note skeletal magnetite crystals, which are connected in a typical chain-like fashion in (**B, C**), and octahedral magnetite in (**D**), and the total lack of signs for transport and mechanical abrasion. Uncoated samples; field emission SEM JEOL 6300F; operating conditions: 1 kV acceleration voltage, 60 pA beam current; Inst. f. Planetologie

Micrometeorites in the Satakunta Formation

Approximately 60 cosmic spherules (melted micrometeorites in terms of Brownlee 1985; Robin 1998) have been recovered from the strongly magnetic part of the 60 to 125 µm heavy mineral fractions of Satakunta rocks. The extremely time-consuming separation of the extraterrestrial objects which is at the moment semi-quantitative at best, includes the following steps: crushing of up to 10 kg rock in a jaw crasher, sieving, cleaning with H_2O^{dest} and ethanol p.A., density separation with bromoform ($\rho = 2.87$ gcm^{-3}), rinsing with acetone p.A., magnetic separation, and hand-picking. We identify the strongly magnetic opaque objects under reflecting light by their spherical morphology, their metallic shine, and/or characteristic shaped surfaces (Fig. 2). The abundance of the micrometeorites in the 60 to 125 µm size fraction is on the order of 1/500 000 000 of the starting mass. According to Marttila (1969), much larger "metallic" spheres occur in Satakunta rocks, yet we have been unable to detect such spherules.

Nearly all recovered micrometeorites display a sculptured surface with accentuation of octahedral or skeletal magnetite crystals (Fig. 2). These crystals are connected in a typical chain-like fashion. It is unknown if this sculptured surface reflects slight weathering of the spherules in the host sandstone with preferential removal of the mesostasis as most cosmic spherules from the atmosphere, and Antarctica show similarly impressive surfaces. The Satakunta micrometeorites generally lack clear signs for mechanical abrasion (Fig. 2a, b) although some objects show deep pits on their surfaces.

The analyzed cosmic spherules of the Satakunta formation (Deutsch et al. 1998; Robin et al. 1998) belong to various groups that have been defined in collections of recent cosmic dust (e.g., Brownlee et al. 1997; Maurette et al. 1987; Robin 1988; Taylor et al. 1998). The detailed investigated Satakunta micrometeorites are "melted particles" of either (i) the fine-grained barred-olivine (Fig. 2c), or (ii) the porphyritic textured (Fig. 2d) type (Deutsch et al. 1998). Varying Cr/Fe, Co/Fe, Ni/Fe and Ir/Fe ratios (Robin et al. 1998) indicate the presence of at least four other types of cosmic spherules (SA-SB, VA-VD, I, and G in the terminology of Robin 1988). The porphyritic spherules contain ≤ 20 µm sized olivine, sometimes with a Mg-rich core surrounded by a Fe-rich rim, and ≤ 5 µm, skeletal, cruciform, or equant spinels in a Fe-rich matrix that is glassy at the resolution of the level of the scanning electron microscope (SEM). In addition, relic phases are present: These are MgO-rich olivine cores, high in Ca, and with FeNi metal inclusions, and Al, Cr-rich spinels. Barred-olivine type spherules are composed of anhedral to subhedral, ≤ 5 µm olivines and dendritic spinels, embedded in a poorly characterized, glassy matrix. The spinels vary widely in grain sizes (maximum, 5 µm) within different regions of one specimen, suggesting a high thermal gradient during cooling.

The textural features, mineral composition, and bulk composition of the Satakunta spherules (Deutsch et al. 1998) fall within the range shown by micrometeorites collected in Greenland, the Antarctica, and from deep-sea sediments (for a review, see Brownlee et al. 1997). The data indicate that the Satakunta spherules are exceptionally fresh despite the billion years long terrestrial residence.

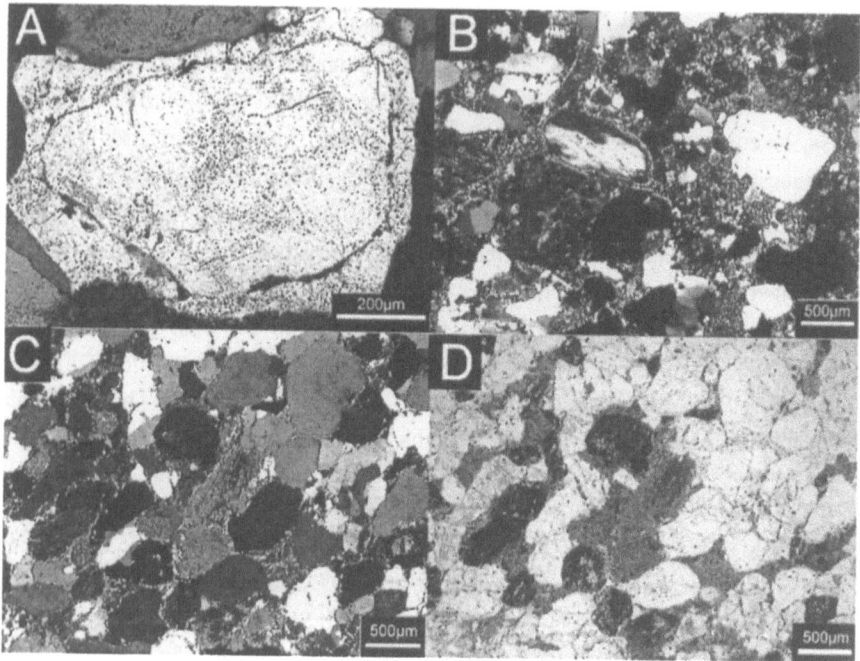

Fig. 3. Photomicrographs of arkose rocks from the Satakunta formation - (**A**) clastic quartz grain with an authigenic quartz rim; Knapernummi; (**B**) high matrix content, and low grade of sorting, the clasts are sub-angular to rounded; Murronmäki; (**C, D**) better sorting of the rounded to well rounded clasts, alteration of feldspars to kaolinite is considered as post-depositional process. The distinction of these pseudomorphs from the fine aggregates forming the matrix is not always clear-cut; Knapernummi; (**A, B, D**) - plane, (**C**) - crossed polarizers.

Sedimentology of the Satakunta Formation

The Satakunta formation is a typical clastic red bed deposit; their source rocks were crystalline lithologies of the Svecofennian belt rather than Rapakivi granites. The apparent polar wander path curve compiled by Pesonen (1992) for the cratonic NE Europe indicates that during sedimentation of the Jotnian sandstone Fennoscandia lay at a latitude of about 0-20° south. At this paleogeographic position, we expect deep chemical, and physical weathering as well as intense staining of the deposited material by iron oxides and hydroxides.

According to Kohonen et al. (1993), extended parts of the Satakunta formation have been deposited in the sedimentological environment of poorly channeled braided streams. For a general introduction to the geology of fluviatile deposits, we refer to Miall (1996). Braided rivers typically show an interlaced network of several channels of low sinuosity. Sandy and pebbly components dominate the sediments. Fine grained silts, however, are deposited only during flood periods on top of the bars and between the active channels; this material is usually poorly preserved, and occurs only subordinate. In the Satakunta case, several types of lithofacies which characterize channelled braided streams, are present, ranging from structure-less fine gravel conglomerates (Gm and Gp; in the nomenclature of Maill 1996); trough cross-bedded (St), planar cross-bedded (Sp), horizontally bedded or laminated (Sh), and massive structureless sandstones (Sm), to horizontally laminated mud- and siltstones (Fl). From the lithological point of view, arkose, and subarkose dominate over quartz arenite (Tab. 1; Fig. 3); whereas conglomerates, mud- and siltstones occur only subordinately. Chemically, the Satakunta sandstones display only restricted variations. The lithologies show relatively high alkali contents, and match in their overall chemical composition thick arkose accumulations in other geological formations (Tab. 2).

Table. 1. Chemical composition of Satakunta and other arkose sandstones

[wt%]	Satakunta range		Köylio[*] Finland	Torridonian[*] Skye	Old Red[*] Scotland	Auvergne[*] France
SiO_2	78.2	93.3	79.3	75.8	73.3	76.6
TiO_2	0.04	0.26	0.22	0.15	-	0.6
Al_2O_3	3.21	11.40	9.94	11.74	11.31	12.4
Fe_2O_3TOT	0.41	2.53	1.72	1.9	4.26	0.9
MgO	0.11	1.43	0.56	0.54	0.24	0.3
CaO	0.02	0.48	0.38	0.38	1.53	0.4
Na_2O	0.10	1.75	2.21	2.40	2.34	0.3
K_2O	1.72	5.24	4.35	4.51	6.16	3.8
Ni [ppm]	3	20	-	-	-	-

* Data from Pettijohn et al. (1975).

The main clastic components of the sediments are quartz (45-60 vol%; Fig. 3a to d), microcline and minor multiple twinned plagioclase (20-40%; Fig. 3b to d), rock fragments, and detrital muscovite and biotite (Tab. 1). The heavy mineral fractions include in order of decreasing abundance of magnetite, hematite, zircon, garnet, epidote, tourmaline, apatite and monazite (Marttila 1969); in addition, exotic phases such as uraninite have been detected. The fine grained intergranular matrix (Fig. 3b), consisting minute flakes of various sheet silicates (clays, mixed layers, chlorite) and authigenic quartz (Fig. 3a), amounts up to 20 vol%. Secondary quartz grows quite frequently over the fabric. The quartz forms rims on the clastic quartz (Fig. 3a), and fills the pore space. According to Füchtbauer (1988), silicic cementation takes place quite early in the diagenesis due to the circulation of fluids oversaturated with SiO_2. This SiO_2 derives from weathering as well as from pressure solution.

For separation of micrometeorites, we have investigated, and sampled twelve outcrops in the Satakunta region (Fig. 1). In the rocks from eight outcrops, we had no success in the search for ancient cosmic dust, whereas samples from the localities Murronmäki, Naskalinkallio, Kallionpää, and Sahankoski (Fig. 1; for coordinates, see Kohonen et al. 1993) carry fossil spherules. At these four outcrops, the extraterrestrial objects are not restricted to a specific bed yet occur in several samples collected in different stratigraphic levels. The spherule containing differ in several macroscopic and microscopic aspects such as sorting, grain size, and modal composition. Yet arkose is the main lithology in all studied outcrops. Murronmäki, in Köylio, is even listed by Pettijohn et al. (1975) as reference locality for arkose. As shown in Table 1, the Satakunta arkoses contain quite varying amounts of matrix. They are either "arkosic arenites" (less than 15 vol% matrix), or arkosic wackes (Pettijohn et al. 1975). The latter, also known as "residual arkose", originate by in situ disintegration of granitic pre-cursor material. Transport is very limited, hence, this material is characterized by a low grade of sorting. In contrast, rivers have reworked the arkosic arenites. This process yields a better sorting, partial to complete loss of the matrix, and a good rounding of the individual grains. In addition, the modal abundance of quartz increases as feldspars are preferentially decomposed.

At Naskalinkallio (Fig. 4, upper), the beds consists of very coarse grained, partly pebbly, subarkose sandstones with lensoid, more gravely parts. Internal structures are rarely visible. In contrast, at Murronmäki (Fig. 4, lower) a moderately sorted arkose dominates which is occasionally intercalated by thin beds of greenish siltstones. However, layers with a low degree of sorting are also present in this outcrop (Fig. 3b). Altogether, the textural features of the spherule-containing arkose samples from the Satakunta formation point towards transport over only short distances, quick deposition, followed by immediate silicification.

Fig. 4. Typical spherule-containing arkose samples from (upper) Naskalinkallio and (lower) Murronmäki. The upper specimen is more coarsely grained, indicating a higher flow velocity in the river. The Murronmäki sample displays silt layers and intercalation of fine and medium grained layers.

On the Preservation of Micrometeorites in the Satakunta Formation

At present, it is unconstrained (i) why the Satakunta formation carries micrometeorites, and (ii) what are the reason(s) for their excellent preservation. During geologic history, a variety of factors may have contributed to the presence of quite "pristine" cosmic dust in these red beds. In historical order, these are (i) specific concentrations mechanisms for the in-falling micrometeorites, (ii) specific low energy mechanisms for embedding of this material into the host sediment, and lack of further transport in the sedimentological environment, (iii) only minor diagenetic compaction of the host sediments at rather reducing conditions, and (iv) a quite specific time-integrated temperature history for over a billion years to account for the still glassy nature of the matrix of the melted spheres.

(i) Although we approximately know the abundance of cosmic spherules per kg of crushed sediment, to date we have been unable specify the locations of the micrometeorites in the sedimentary mineral fabric. Especially, it is unknown if the spherules are concentrated in particular horizons, layers or clay galls. We have to take further into account, that the time-span needed for sedimentation of one sandstone bed is unconstrained. Therefore, it is highly speculative to assess if the abundance of micrometeorites in Satakunta sediments compares within errors with the present flux rate for particles sized between 63 and 125 µm (1 spherule a^{-1} m^{-2}; Taylor et al. 1998). In consequence, a mechanism to concentrate cosmic spherules prior to embedding in the host sediment is not required at the moment.

Fig. 5. Harjavalta is the only large scale outcrop of the Satakunta sandstone; the beds dip with about 35° towards the center of the graben.

(ii) As outlined above, rounding of the individual clasts, and sorting of the arkose lithologies are indicative for only minor transport of the material prior to sedimentation. The observation is supported by the perfectly preserved surfaces of the cosmic spherules. It seems that better sorted sediments (i.e., such whose components have been transported over longer distances) lack the spherules. These objects obviously are not very resistant against mechanical abrasion. A better understanding of the sedimentological features of the Satakunta formation is needed to constrain possible locations of the requested types of arkose. Sedimentological research, however, is hampered as only one large outcrop is present in the whole formation (Fig. 5).

(iii) The diagenetic history of the sediments hosting the fossil micrometeorites is clearly dominated by a quick silicification induced by pore-water, which was oversaturated with SiO_2. This process is documented by the omnipresent newly grown quartz grains: they either form thin seams around detrital components or fill the pore space. Obviously, this cementation prevented further circulation of fluids; moreover, the silicified rocks are very resistant against weathering.

(iv) Because all sampled outcrops are roughly in a similar paleogeographic level (Kohonen et al. 1993) and present topographic level, they underwent most probably a similar post-depositional history. The temperature evolution of the area that hosts the Satakunta formation was not at all in the scientific focus. New preliminary fission track data for apatite of heavy mineral fractions from Murronmäki which carry cosmic spherules, indicate that the temperature of the sandstone at the sampling locations never have exceeded about 100° C since Pre-Cambrian times (pers. communication, U. Glasmacher 1999). This result points to an exceptional stable geologic situation in this part of Fennoscandia since Mesoproterozoic times. To the contrary, unpublished paleomagnetic data (L.J. Pesonen) give evidence that parts of the Satakunta formation have been annealed perhaps up to 550° C by the huge post-Jotnian diabase sheets. We assume that micrometeorites occur only in those parts of the sandstone that escaped this low-grade thermal metamorphism.

Outlook

The Satakunta spherules are the oldest known cosmic spherules in the world. Our present state of knowledge indicates that silicified arkose deposits and comparably "unsorted" sedimentary rocks bear the best chance for discovering additional ancient micrometeorites. Finding extraterrestrial materials with only minor alteration effects in such old sediments, opens exciting research perspectives: A more systematic analysis of abundance and type of the spherules in the Satakunta, and similar sandstones yet of distinctly different age, could yield a better knowledge on whether variations exist in the flux of extraterrestrial material with time, and as discussed by Gayraud et al. (1996) provide model-independent information on the evolution of the atmospheric fO_2. According to the published data for Satakunta spherules (Deutsch et al. 1998), the atmosphere 1.4 Ga ago was more oxygen-rich than assumed in most models (see Kasting 1993, for discussion).

Acknowledgements

This work is supported by a DAAD - The Academy of Finland exchange program to A.D. and L.J.P., grant De 401/16-1 of the German Science Foundation and by the European Science Foundation, Strasbourg. F. Bartschat, M. Flucks, Th. Grund, and U. Heitmann (IfP) are to be thanked for skilful technical assistance. We appreciate discussions with E. Robin, R. Roccia (Gif/s/Yvette), U. Glasmacher (Heidelberg), and H. Bahlburg (Münster), as well as critical comments by M. Genge, and I. Gilmour (Milton Keynes). This work is part of D. Kettrup's doctoral thesis.

References

Brownlee DE, Bates B, Wheelockm MM (1984) Extraterrestrial platinum group nuggets in deep sea sediments. Nature 309: 693-695

Brownlee DE (1985) Cosmic Dust: Collection and research. Ann Rev Earth Planet Sci 13: 147-173

Brownlee DE, Bates B, Schramm L (1997) The elemental composition of stony cosmic spherules. Meteorit Planet Sci 32: 157-175

Deutsch A, Greshake A, Pesonen LJ, Pihlaja P (1998) Unaltered cosmic spheruels in a 1.4-Gyr-old sandstone from Finland. Nature 395: 146-148

Füchtbauer H (1988) Sandsteine. In: Füchtbauer H (ed) Sedimente und Sedimentgesteine. Schweizerbart, Stuttgart, pp 97-184

Gayraud J, Robin E, Rocchia R, Froget L (1996) Formation conditions of oxidized Ni-rich spinel and their relevance to the K/T boundary event. In: Ryder G, Fastovsky D, Gartner S (eds) Geol Soc Am Spec Paper 307: 425-443

Grieve RAF (1998) Extraterrestrial impacts on Earth: the evidence and the consquences. In: Grady MM, Hutchison R, McCall GJH, Rothery DA (eds) Meteorites: flux with time and impact effects, The Geological Society, London, pp 105-132

Jéhanno C, Boclet D, Bonté P, Castellarin A, Rocchia R (1988) Identification of two populations of extraterrestrial particles in a Jurassic hardground of the Southern Alps. Proc Lunar Planet Sci Conf 18: 623-630

Kasting JF (1993) Earth`s Early Atmosphere. Science 259: 920-926

Kettrup D, Marttila E, Pihlaja P, Pesonen LJ (1998) Mesoproterozoic Micrometeorites in Finland II - mode of occurence. "Response of the Earth System to Impact Processes", Workshop Impacts and the Early Earth Cambridge, UK Dec. 1998

Kohonen P, Pihlaja, P, Kujala, H, Marmo, J (1993) Sedimentation of the Jotnian Satakunta sandstone, western Finland. Geol Surv Finl Bull 369: 1-35

Miall AD (1996) The geology of fluvial deposits. Springer, Berlin

Marttila E (1969) Satakunnan hiekkakiven sedimentaatio-olosuhteista. (On the sedimentary environment of the Satakunta sandstone). Dissertation, University of Turku

Maurette M, Jéhanno C, Robin E, Hammer C (1987) Characteristics and mass distribution of extraterrestrial dust from the Greenland ice cap. Nature 328: 699-702

Pesonen LJ (1992) Continental drift of the Fennoscandian shield. In: Koljonen T (ed) The Geochemical Atlas of Finland, part 2: Till. Geological Survey of Finland, Espoo, pp 60-65

Pesonen LJ, Deutsch A, Pihlaja P, Kettrup D (1998) Mesoproterozoic Micrometeorites in Finland I - basic characteristics and scientific potential. "Response of the Earth System to Impact Processes", Workshop Impacts and the Early Earth Cambridge, UK Dec. 1998

Pettijohn EJ, Potter PE, Siever R (1975) Sand and Sandstone. Springer, New York

Robin E, Jéhanno C, Maurette M, Hammer C (1988) A micrometeorite "spectrum" for the mass distribution of well preserved Greenland cosmic dust grains. Lunar Planet Sci, 18: 844-845

Robin E, Bonté P, Froget L, Jéhanno C, Rocchia R (1998) Formation of spinels in cosmic objects during atmospheric entry: A clue to the Cretaceous-Tertiary boundary event. Earth Planet Sci Lett 108: 181-190

Robin E, Rocchia R, Lefevre I, Pierrard O (1998) Abundance and deth profile of Cr, Fe, Co, Ni, and Ir in cosmic spherules from the Satakunta sandstone, Finland. "Response of the Earth System to Impact Processes", workshop Impacts and the Early Earth Cambridge, UK Dec. 1998

Taylor S, Brownlee, DE (1991) Cosmic spherules in the geologic record. Meteoritics 26: 203-211

Taylor S, Lever JH, Harvey RP (1998) Accretion rate of cosmic spherules measured at the South Pole. Nature 392: 899-903

9 Impact Diamonds as Indicators of Shock Metamorphism in Strongly-Reworked Pre-Cambrian Impactites

S.Vishnevsky[1] and J.Raitala[2]

[1]Institute of Mineralogy and Petrology, Russian Academy of Sciences, Novosibirsk, Russia. (svish@math.nsc.ru)
[2]University of Oulu, P.O Box 3000 Fin-90401 Oulu, Finland. (jraitala@sun3.oulu.fi)

Abstract. Pre-Cambrian shields and platforms occupy a large part of the Earth's continental surface. The crystalline rocks of these megastructures are the result of intense metamorphic, igneous and tectonic transformations. It is often difficult to recognize evidence of the past sedimentary, volcanic and igneous activities here. Even more problematic is the identification of old impact structures within Pre-Cambrian shields and platforms. Such pre-metamorphic structures are still unknown in strongly reworked crystalline rocks. However, as the investigations of the Moon and of other celestial bodies show, the impact events at that time played an important role in the evolution of the Earth's crust, including the earliest geologically recorded stages of its development. In searching for evidence of old impact events, which were potentially able to survive the intensive geological activity of the Earth, we would like to highlight the presence of impact diamonds, and to consider, in particular, the possibility of their preservation after exposure to high-grade regional metamorphism.

Impact diamonds are one of the widespread typological minerals of shock origin. Firstly, they originate during impact into various carbon-containing targets and are known from a number of natural objects of both terrestrial (impactites and impactoclastic layers) and cosmic (meteoritic) origin. Secondly, this high-pressure mineral exhibits many characteristic features and is one of the most important among the shock metamorphic criteria.

Impact Diamond Occurrences

Two groups of impact diamonds (IDs) may be divided at this time. 1) Group I, paramorphs of impact diamonds (PIDs) derived from graphite, coals, and other poorly- or non-crystallized forms of carbon; 2) group II, which conventionally covers several different types of IDs, which are condensation/nucleation products, originated from plasma-like fireballs in the explosion cloud of the impact events – a principally new recently found type of condensation/nucleation IDs (CNIDs).

Group I, or paramorphs of impact diamonds (PIDs) are known in various terrestrial and extraterrestrial objects affected by shock metamorphism. Meteoritic diamonds were firstly found in Novo Urei carbonaceous chondrite (Yerofeev and Lachinov, 1888) and in Canyon Diablo iron meteorite (Foote, 1891). At present, impact diamonds of this type are known in many meteorites including ureilites (Urey et al. 1957; Lipschutz 1962; Vdovykin 1967; 1970; 1991; and others) and iron meteorites (Ksanda and Genderson 1939; Clarke et al. 1981). In particular, PIDs from ureilites were derived from poorly or non-crystallized carbon during collisions in outer space. The diamonds of the Antarctic ALHA-77283 iron meteorite originated also in outer space (Clarke et al. 1981), but the PIDs in the Canyon Diablo iron meteorite from Meteor crater, Arizona, formed during the impact (Nininger 1956; Lipschutz and Anders 1961).

PIDs, derived from graphite, are widespread in the impactites of several terrestrial impact sites, such as Popigai (Masaitis et al. 1972; Vishnevsky and Pal'chik 1975) (Fig. 1), Ries (Rost et al. 1978), Puchezh-Katunki (Marakushev et al. 1993), Ukrainian impact structures: Obolon, Ilyinets, Zapadnaya and Terny (Gurov et al. 1995; Val'ter and Er'omenko 1996), and at the Sudbury impact structure, Canada (Masaitis et al. 1997). PIDs, derived from coal, are found at the Kara crater (Ezersky 1982). The latest discovery is at the Lappajarvi impact crater, Finland (Masaitis et al. 1998b). New data (Visnevsky et al. 1999) reveals the Lappajarvi impact diamonds (Fig. 2) as polycrystalline paramorphs of cubic diamond with a preferred orientation of crystallites derived from target rock graphite. They possibly also contain impurities of chaoite and graphite; the hexagonal phase, lonsdaleite, however, still not found here. PIDs were also detected in the "catastrophic layer" of the 1908 Tunguska impact event (Kvasnitsa et al. 1979).

Except for occurrences in impact craters themselves, terrestrial PIDs are also known in a re-deposited state. Impact processes formed one type of re-deposited diamonds; this is a result of proximal and distal dissemination of the high-speed material ejected from impact craters. For example, the strewn field of the IDs around the Popigai crater illustrates this point (Fig. 3). These diamonds can be traced up to a distance of 500 km from the structure, but are characterized by the same features known for the PIDs found in impactites inside the Popigai crater. Another example of this type is represented by microdiamonds from K/T boundary layer at Arroyo El Mimbral, Mexico, related to the Chicxulub impact event (Hough et al. 1997). The energy of large-scale impact events (catastrophic removal of the atmosphere, and a large mass of high-speed ejecta) shows that such strewn fields may be of regional and even of global occurrence. Firstly, it is supported by globally-traced "impactoclastic layers" with iridium anomalies, grains of shocked quartz, microtektites and other impact evidence, found in the Cretaceous-Tertiary and Eocene-Oligocene boundary sedimentary sequences (Alvarez et al. 1995; Koeberl 1993; Montanari et al. 1993; Hildebrand and Boynton 1990). Secondly, such conclusions are supported by the results of numerical simulations of impact events. For example, it was shown for the Popigai crater (Vishnevsky et al. 1996) that a large mass of strongly-shocked target material was excavated from the crater at the speed of 2 to 5.4 km/s (~ 1.5×10^{12} t). The highest speeds of ejected projectile material were estimated to be up to 14-15 km/s due to cumulative effects.

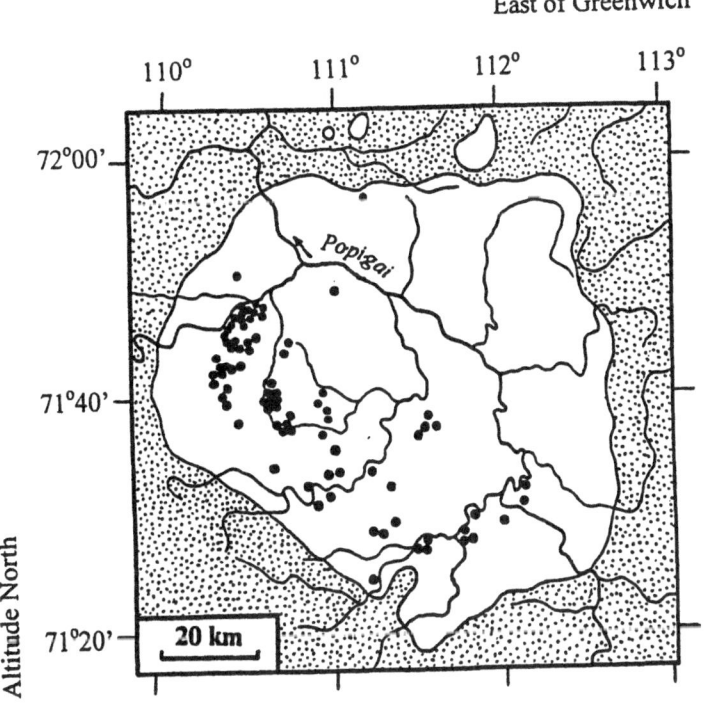

Fig. 1. Diamond occurrences (filled circles) in impactites of the Popigai crater (white areas). After Vishnevsky et al. (1997).

Other types of re-deposited PIDs are related to fluvial and other surface processes; such re-deposited diamonds are known from unconsolidated Cenozoic sedimentary rocks of the Ukraine, Kazakhstan and other regions (Kvasnitsa 1985; Polkanov et al. 1973; Yurk et al. 1973; and others).

Group II, or condensation/nucleation impact diamonds (CNIDs) are found in suevites from the Ries crater (Hough et al. 1995) and in the Cretaceous-Tertiary boundary layer of Central and Northern America (Carlisle and Braman 1991; Gilmour et al. 1992) related to the Chicxulub impact event. Both occurrences result from different formation mechanisms and environments. In case of the Ries, there are microdiamonds and their intergrowths with silicon carbide; in case of the Cretaceous-Tertiary boundary there are isolated nanodiamond crystals.

Nanodiamonds, similar to diamonds from the Chicxulub distal ejecta in a number of ways, are widespread in non-metamorphosed meteorites (Lewis et al. 1987; Newton et al. 1995; and others). The origin of nanodiamonds from these meteorites is under discussion, but some of them may have an impact origin.

Thus, both groups of the IDs, paramorphs and condensation/nucleation products, originated in outer space and on the Earth, are widespread in terrestrial rocks due to constant meteorite influx. But, except for occurrences in impact craters and in some fluvial placers, these diamonds occur in very low concentrations. In a similar way, the pre-Cambrian non-metamorphosed impactites and sedimentary rocks should be characterized by the same distribution.

Fig. 2. Paramorphs of impact diamond derived from parental graphite, from the Lappajarvi impact crater: mostly flattened grains of various colors, with traces of some surface corrosion. Reflected light. Photo by S.Vishnevsky.

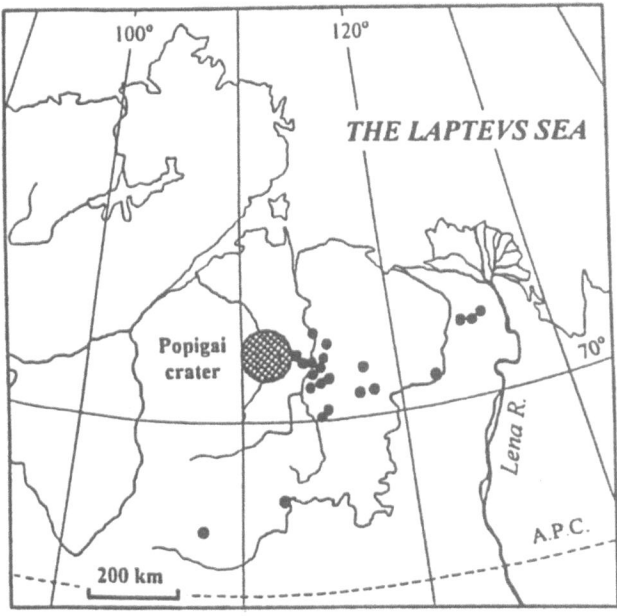

Fig. 3. The traced part of the strewn field of the impact diamonds around the Popigai crater (the occurrences are shown by filled circles). After Vishnevsky et al. (1997).

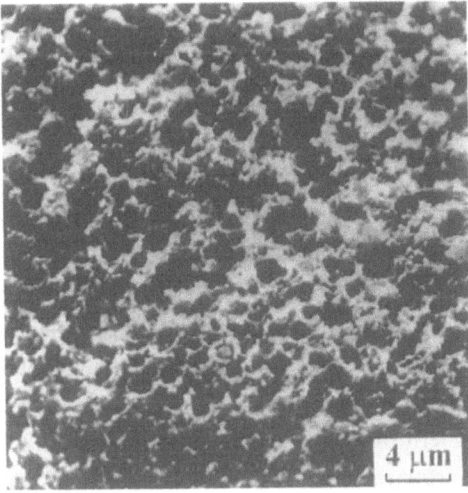

Fig. 4. Etching of diamond grains in Popigai impact melt rocks: superfine surface corrosion. SEM image (Vishnevsky et al. 1997).

The Features of Impact Diamonds and the Conditions of their Origin

Diamonds of group I or PIDs, found in meteorites and impactites, usually have a size from 0.1 to 10 mm, are colored white, gray, yellow-gray and black, and often inherit the morphologic features of the precursor carbon matter from which they were derived. Volume-xenomorphic and angular broken grains are also present; these are the result of mosaic fragmentation of the parental carbon due to polymorphic transformation. The PIDs grains have a fine-grained polycrystalline texture, with crystallite sizes ranging from several nm to 1 μm, and consist either of the cubic phase alone, or of mixtures with the hexagonal phase - lonsdaleite. The PID grains derived from precursor graphite, often have the preferred orientation of crystallites, the crystallographic features of the parental minerals control this. PID grains from the K/T boundary layer of Mexico are polycrystalline aggregates up to 30 μm in size, with individual crystallites from 100 nm to 1 μm; these also display a pseudohexagonal form and may be paramorphs derived from euhedral graphite (Hough et al. 1997).

Explosive synthesis experiments and petrologic estimations show that PIDs can originate both by martensitic and diffusion processes within a shock pressure range of ~30 to 140 GPa and residual temperatures of ~700 to 4500 K. Recent reviews on PIDs can be found in papers by Val'ter et al. (1992), Vishnevsky et al. (1997) and Masaitis et al. (1998a).

PIDs differ from small diamonds of metamorphic (Sobolev and Shatsky 1990) and other endogenous origin common in placers of the Ukraine, Northern Kazakhstan, and other regions in a range of features (the presence of lonsdaleite, preferred orientation of crystallites, fine-grained textures and other features) (Kashkarov and Polkanov 1964; 1972; Kvasnitsa 1985).

Group II of the diamonds, or CNIDs, unites several types of IDs, derived by various formation mechanisms and environments. In general, the origin of CNIDs group of diamonds is still not studied in detail.

The Ries skeletal microdiamonds are polycrystalline aggregates of up to a few micrometers across their largest dimensions, with individual crystals from 10 nm to 2 μm. There is a definite association of the diamonds with silicon carbide, including the presence of intergrowth. Carbon isotopic composition of the diamonds, $\delta^{13}C$, is of terrestrial origin, ranging from –16 to –22   (Hough et al. 1995). Hough et al. (1995) suggested that these diamonds could have originated by chemical vapor deposition (CVD) of carbon atoms from the plasma-like fireball in the Ries explosion cloud. Masaitis et al. (1998a) suggested these diamonds could be the result of PID dissolution in a high-temperature impact melt.

K/T boundary cubic nanodiamonds from Northern and Central America occur as polycrystalline aggregates, with individual crystals up to 6 nm in size (Carlisle and Braman 1991; Gilmour et al. 1992). The carbon isotopic composition of these diamonds varies from "terrestrial" (δ^{13}C ranges from –11 to –19 , Gilmour et al. 1992) to "interstellar" (δ^{13}C ~ –48 , Carlisle and Braman 1991) magnitude. However, as Gilmour et al. (1992) show, the isotopic composition of nitrogen in these diamonds is "terrestrial" (δ^{15}N ranges from –17 to +8.5). Formation in the Chicxulub plasma-like fireball by means of CVD-like mechanism has been suggested for these diamonds (Gilmour, et al 1992).

The Ries microdiamonds and K/T boundary nanodiamonds may also originate by a mechanism of homogenous nucleation described by Burki et al. (1996). The experiments of these authors showed that micro- and nanodiamonds can originate in a broad range of temperatures and pressures, down to ~ 1300 K and 1 bar. R.M.Hough (pers. comm. 1999) has suggested that the K/T nanodiamonds may also be shock-produced similar to detonation experiments. In fact, cubic diamonds of 2 to 200 nm in size can be produced at high pressure in detonation experiments with a negative oxygen balance in the system (Mal'kov 1994; Mal'kov and Titov 1995).

Together with micro- and nanodiamonds, impact-induced processes of condensation/nucleation type may also form carbonados – large polycrystalline cubic diamond aggregates, whose origin is still an open question. An impact origin for carbonados was supposed earlier (Smith and Dawson 1985; Ezersky 1986). CVD may be one of the impact-related mechanisms for carbonados (Hough et al. (1995).

Nanodiamonds, 3-5 nm in size, usually found in meteorites, are present as individual crystals of cubic phase, or rarely as twin intergrowths (Buseck and Barry 1988; Lewis et al. 1987; Huss and Lewis 1995; and others). These diamonds have superlight isotopic composition of carbon, with δ^{13}C of –30 to –40 (Ash et al. 1988; Lewis et al. 1987), which is accompanied by isotopic anomalies of nitrogen and hydrogen. In particular, variations of nitrogen isotopic composition: δ^{15}N can range from to –574 to –1000 (Lewis et al. 1983; 1989). The data of Fedoseev et al. (1971), suggests that the large scale of carbon and nitrogen isotopic fractionation might indicate a condensation origin of diamond. Predominantly, meteoritic nanodiamonds may be of circumstellar condensation origin (Daulton et al. 1997). However, one can not exclude an impact origin for some meteorite nanodiamonds as suggested for CNIDs described above. The hexagonal polymorph of diamond, lonsdaleite, which has been recently identified by Daulton et al. (1996), may be possible evidence of an impact origin for some meteoritic nanodiamonds.

Resistance of Impact Diamonds in Natural Objects and in Experiments

Many PIDs from impact structures bear the traces of dissolution and graphitization, which took place in the high-temperature impact melt. Data by Karklina and Maslakovets 1968; Evans and Santer 1961; Harris and Vance 1974; and others suggests that the destruction was provided by the action of OH-, Na, K and free oxygen. The dissolution was a major destructive process. A superfine etching represents the traces of dissolution of the PIDs, this is in the form of a cell comb-like microrelief on the grain surfaces (Fig. 4). However, sometimes the process was more active, and local parts of the grains, in addition to the superfine etching, were affected by intensive corrosion, with the formation of large cavities and deep cross-cutting penetrations often as complex shapes. The speed of diamond dissolution in the impact melts is estimated to be of ~ several μm/hour at 1800°C (Val'ter et al. 1992), so that the complete dissolution of a 1 mm ID grain could occur in the time period of up to 20 days; the same processes took place in thick buried masses of impact melts. However, the speed of diamond dissolution decreases quickly with the lack of the high temperatures (~2 orders of magnitude, at the cooling of the melt from 1800°C to near subsolidus temperatures, ~ 1000 – 1200°C). The quenched impactites have to provide the best preservation conditions for the IDs to be included. The destructive influence of free oxygen on the preservation of IDs in impact melts was probably very low, because the impactites are usually characterized by a lack of free oxygen (Dolgov et al. 1975; Feldman 1990).

Water is very aggressive in respect to diamond. In experiments on diamond dissolution in basaltic melts with the presence of water, the destruction of the mineral is detected 2 hours after treating at 1100°C and with 2.5 GPa of confining pressure (Pal'yanov et al. 1995). In the experiments where diamond was heated in the presence of free oxygen, the clear graphitization of the mineral still detected at ~ 630°C, and at temperatures of > 850°C the process begins along all of the crystallographic surfaces of the diamond (Evans 1979). The etching of diamond by a combined action of OH-, Na and K (treatment in melts of NaOH and KOH composition without any free oxygen access) becomes evident at >880°C (Karklina and Maslakovets 1968). However, with free oxygen, as these authors pointed out, the detectable etching of diamond begins at lower temperatures, 600-

620°C, in the melts of NaOH, KOH, NaNO₃, KNO₃, Na₂O₂ and their various combinations. It was also found (Karklina and Maslakovets 1968) that the destructive activity of K is several times higher than of Na, with all other conditions being equal. In multi-component systems graphite acts as a buffer with respect to the aggressive components and is important for the preservation of the diamond (Pal'yanov, personal communication).

Preservation of Impact Diamond during Progressive Regional Metamorphism

Experimental data have shown that the preservation of diamond in impactites or in sedimentary placers affected by progressive regional metamorphism (PRM) depends initially on the water and free oxygen activity. Water is a constant participant in any transformations by regional metamorphism. As for free oxygen, its content may vary over a broad range, depending on the presence of various buffers (Dobretsov et al. 1972), although the presence of free oxygen is not required for typical metamorphic reactions. However, considering the problem of the survival of impact diamonds during PRM, we follow the regime of the least favorable conditions and assume that free oxygen, even in small quantities, was present during the metamorphic transformations. Taking into account the geological time scale of the affect, we have to consider that even a small amount of free oxygen is a determinant for diamond survival.

In the last case, the pressure-temperature survival conditions for un-protected diamonds becomes low. These IDs, exposed to the action of aggressive agents, cannot survive PRM of amphibolite and granulite facies (650-1000 °C, 0.3 – 1.5 GPa), although their preservation under conditions of greenschist and andalusite-muscovite facies of PRM (350-650 °C, 0.2 – 1 GPa) is possible (Fig. 5). The discovery of IDs in the Sudbury impactites affected to the greenschist stage of PRM (Masaitis et al. 1997) is in agreement with this conclusion. Taking into account the possible presence of various buffers in metamorphosing strata, the accepted boundary for survival of un-protected IDs is a minimum one.

To further discuss the problem of IDs survival under PRM conditions, the discovery of small metamorphic diamonds in the rocks of the Kokchetav massif (Northern Kazakhstan) is interesting. Following recent investigations (Shatsky and Sobolev 1993, Shatsky et al. 1995), these diamonds (Fig. 6) formed ~530 Ma ago in their stability field at pressures >4 GPa. Then, due to the elevation of the Kokchetav massif, the diamond-bearing rocks were uplifted to a near-surface part of the crust and were affected by regressive regional metamorphism down to the greenschist facies, experiencing, correspondingly, various P-T regimes (Fig. 7). The preservation of the diamonds in these rocks at the high temperatures (600 to 900-1000°C) outside the stability field over geological time scale of ~ 13 Ma becomes possible because they were trapped in garnet or zircon, pyroxene and kyanite, and were protected from action by OH-, free oxygen and other aggressive agents.

The survival of diamond in the rocks of the Kokchetav massif is of great importance for the preservation problem. It shows that IDs, trapped in various rock-forming water-poor minerals like garnet, pyroxene, andalusite/sillimanite/kyanite, and others, may be protected from the action of aggressive agents at high stages of PRM, including amphibolite and granulite facies. This could be a reason why, the IDs can survive the high-grade stages of PRM. The trapping of IDs by protecting minerals is possible even within the minimum field of their metastable existence at PRM conditions of greenschist and amphibolite facies (Fig. 5).

Here and below, the volume and P-T parameters of PRM facies are accepted following the scheme proposed by (Dobretsov et al. 1972).

Fig. 5. (opposite) Facies of metamorphism (after Dobretsov et al. 1972) and the possible field of resistance for non-protected impact diamonds under conditions of progressive regional metamorphism. Facies of metamorphism: contact (A_o - A_4), regional (B_1 - B_4) and high-pressure (C_1 - C_4). B_1 – granulite facies (800 – 1000°C, 0.4 – 1.5 GPa); B_2 –amphibolite facies (650 – 800°C, 0.3 – 1.1 GPa); B_3 –andalusite-muscovite schist facies (500 – 650°C, 0.2 – 0.8 GPa); B_4 – greenschist facies (350 – 500°C, 0.2 – 1 GPa). 1 – boundaries of facies; 2 – field boundaries of important minerals at P_{H2O} ~ 0.3 (low-temperature part of the diagram) to 0.9 (high-temperature part of the diagram) of P_{total}; 3 – melting boundaries for granite (P_{H2O} = 0.6 P_{total}) and basalt (lower line for usual compositions, higher line – for basic compositions); 4 –supposed kinetic threshold of metamorphism; 5 – P-T resistance field for non-protected impact diamonds and its assumed upper limit.

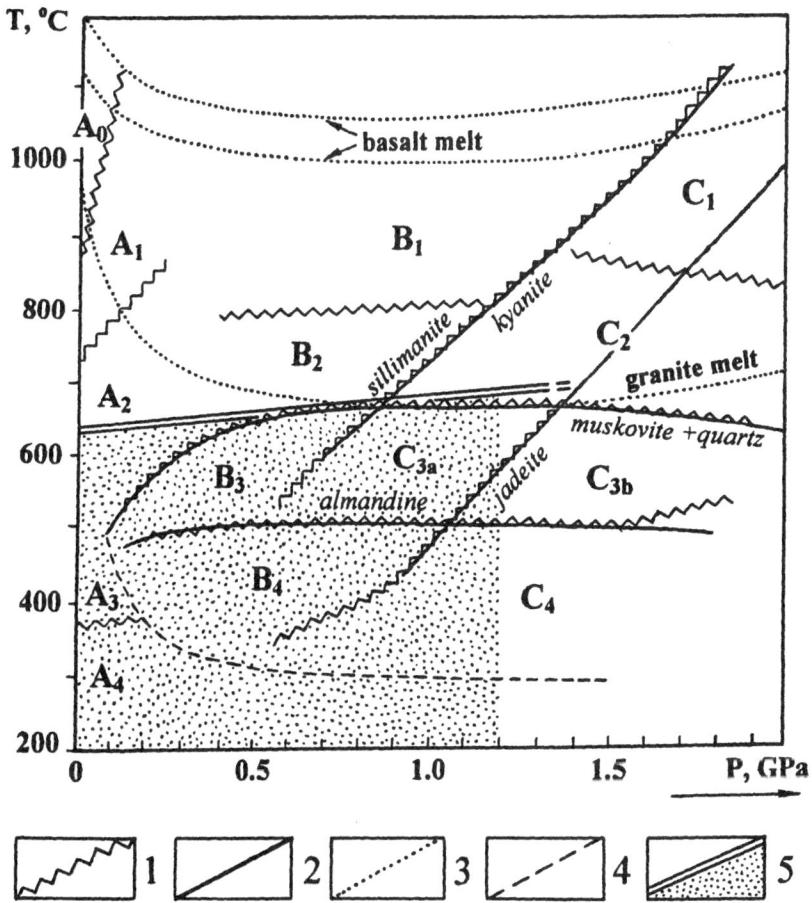

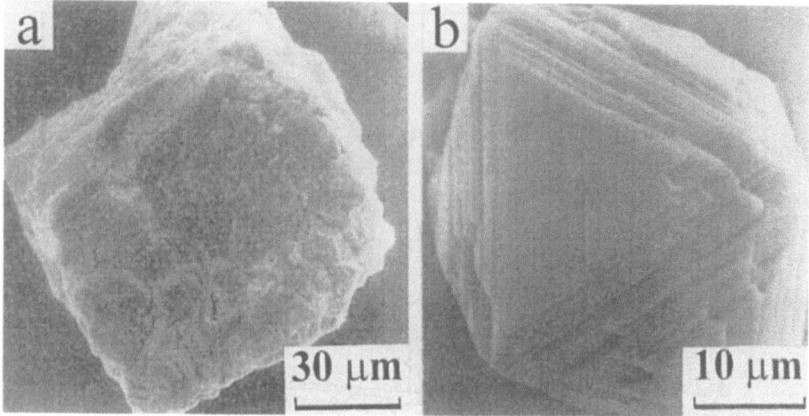

Fig. 6. Metamorphic diamonds, extracted from the rocks of the Kokchetav massif, Northern Kazakhstan: a – cube-like crystal; b – octahedron. SEM image (after Shatsky et al. 1998).

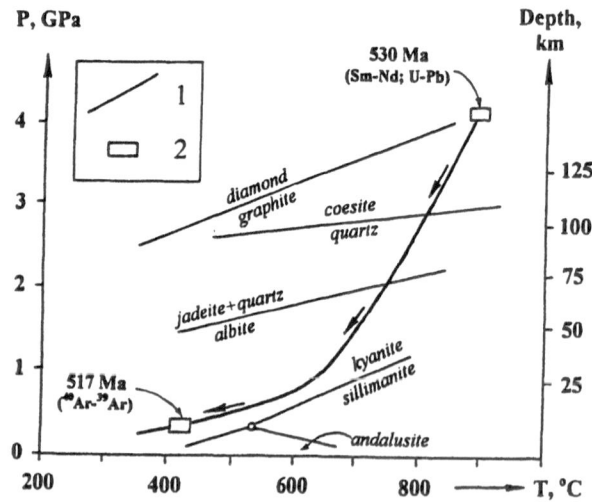

Fig. 7. Pressure – temperature – time evolution of the Kokchetav diamond-bearing rocks at the stage of regressive regional metamorphism (after Shatsky et al. 1995). 1 – equilibrium boundaries between fields of important minerals; 2 – pressure-temperature estimations and age determinations for high-pressure and greenschist metamorphic stages of diamond-bearing rocks.

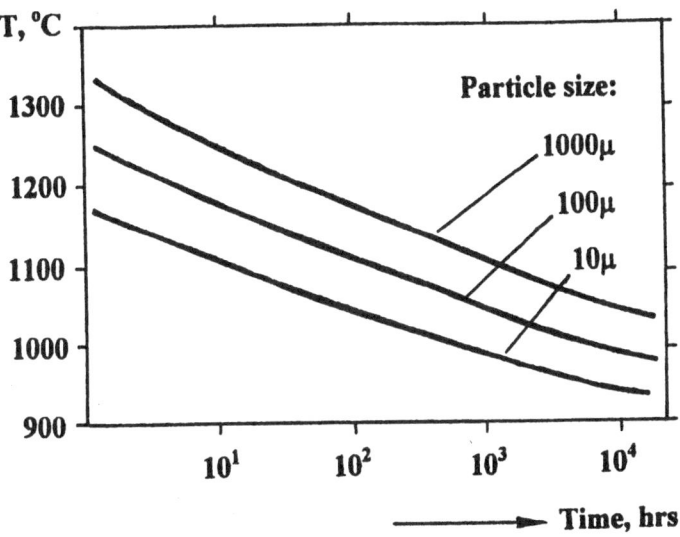

Fig. 8. Rate of 5 wt. % destruction of coesite vs. time at "dry" conditions for particles of various sizes (after Babitch et al. 1989).

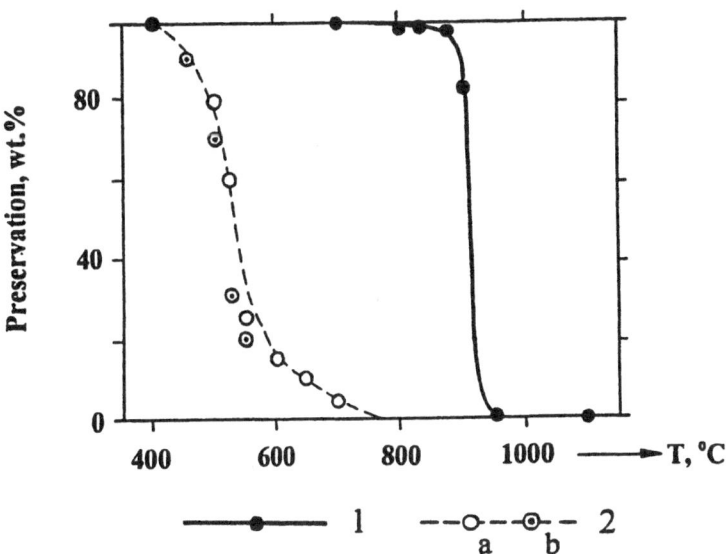

Fig. 9. Rate of coesite destruction vs. temperature for 1 hour of treatment at "wet" conditions (after Babitch et al. 1998). 1 - "dry" regime of transformation; 2 – "wet" regime of transformation at 1 GPa (a) and 2 GPa (b) confining pressure.

Preservation of other Evidence of Shock Metamorphism During Progressive Regional Metamorphism

Except for impact diamonds, other widespread mineralogical and petrographic indicators of shock metamorphism can be found with high-pressure polymorphs of silica (stishovite and coesite) and planar microdeformation features in shocked minerals being the most common.

Stishovite is quickly destroyed at temperatures > 400°C (Gigl and Dachille 1968), and in most cases it disappears at the stage of cooling of the impactites. Its preservation in impactites affected by PRM of even greenschist facies seems impossible.

Coesite, following experimental data, exhibits a variation in its resistance to temperature depending on the water content in the enclosing rocks. Under "dry" conditions it can be preserved at temperatures of 800 to 1300°C during a long time (Fig. 8). Under these conditions, the size of the mineral particles is also important for the level of preservation (Babitch et al. 1989). Coesite, preserved in relatively "dry" conditions and associated with quenched suevitic and other impact glasses, is the most widespread shock-indicator in rocks of terrestrial impact structures. However, the thermal resistance of the mineral decreases quickly in the presence of water, and at temperatures > 400°C it quickly decomposes out of the field of its pressure stability (Fig. 9). A common association of the mineral with diaplectic quartz glasses and lechatelierite in impactites is not favorable for its preservation either. These enclosing glasses, at first, are able to absorb the water from outside (Grieve et al. 1996). Then, secondly, they recrystallize easily into highly-permeable quartz-crystobalite aggregates with a "ballen" structure. Both processes occur practically in all cases at the stage of surface hydrothermal alteration of impactites and also at the initial stages of PRM. Hence, it is very doubtful that coesite can survive even the greenschist stage of PRM, unless it is trapped in water-protected minerals. One can point out that coesite of metamorphic origin, occurred in deep-seated rocks, and is able to survive combinations of pressure and temperature outside its stability field at conditions of regressive regional metamorphism. With metamorphic diamond, some trapping minerals (pyroxene, zircon, and others) provided such a preservation of coesite, which protects it from the action of water.

The preservation of diagnostic features of shock metamorphism, such as planar microdeformation structures in shocked minerals, becomes impossible under conditions of high-grade PRM, accompanied by total recrystallization of the rocks and by the development of grano- and lepidiblastic textures.

Conclusions

Among the only common mineralogical and petrographic signs of shock metamorphism, impact diamonds are able to survive high-grade PRM. They may be the only evidence of the shock metamorphism in strongly-reworked impactites from old geosynclinal belts of the pre-Cambrian Earth. The discovery of IDs in eclogites of the Kola Peninsula (Holovnya et al. 1977) may be a good example. The quest for IDs in old deeply metamorphosed rocks of possible impact origin has a definite scientific interest. The preservation of IDs at high-grade PRM conditions shows that these diamonds, originating from impact processes on the Earth's surface or entering the Earth from outer space within meteorites, occur in various horizons of the crust and the upper mantle, and thus serve as possible nucleation centers for the stable and metastable growth of endogenic diamond.

Acknowledgements

Reviews of the manuscript were provided by R.M. Hough and an anonymous reviewer, for which we are much appreciative. The improvement of the English language by R.M. Hough is also gratefully acknowledged. V.S. Shatsky kindly provided the photos of the Kokchetav microdiamonds.

References

Alvarez W, Clayes P, Kieffer SW (1995) Emplacement of KT boundary shocked quartz from Chicxulub crater. Science 269: 930-935

Ash RD, Arden JW, Grady MM, Wright IP, Pillinger CT (1988) Isotopically-light carbon in the Allende meteorite. Meteoritics 23: 255-256

Babitch YuV, Doroshev AM, Malinovsky IYu (1989) Heat-activated transformation of coesite at standard pressure. Soviet Geology and Geophysics 30: 127-130.

Babitch YuV, Turkin AI, Gusak SN (1998) Pecularities of high-pressure coesite to quartz transformation in the presence of water and carbon dioxide: some geological implications. Russian Geology and Geophysics 39: 694-698

Burki PR, Leutwyler S, Matsui Y, Sato Y (1996) Homogenous nucleation of diamond particles at atmospheric pressure. Meteoritics Planet Sci 31: A25-A26

Buseck PR, Barry JC (1988) Twinned diamonds in the Orgueil carbonaceous chondrite. Meteoritics 23 (3): 261-262

Carlisle DB, Braman DR (1991) Nanometer-size diamonds in the Cretaceous/Tertiary boundary clays of Alberta. Nature 352: 708-709

Clarke RS, Appelman DE, Ross PB (1981) An Antarctic iron meteorite contains pre-terrestrial impact-produced diamonds and lonsdaleite. Nature 291: 396-398

Daulton TL, Eisenhour DD, Bernatowicz TJ, Lewis RS, Buseck PR (1996) Genesis of presolar diamonds: comparative high-resolution transmission electron microscopy study of meteoritic and terrestrial nanodiamonds. Geochim CosmochimActa 60: 4853-4857

Dobretsov NL, Khlestov VV, Reverdatto VV, Sobolev NV, Sobolev VS (1972) .The Facies of Metamorphism. Sobolev VS (ed). Australian National University, Department of Geology, Canberra, Publication No. 214: 417 p.p

Dolgov YuA, Shugurova NA, Vishnevsky SA (1975) Gas inclusions in impactites. In: Dolgov YuA (ed.), Thermobarogeochemistry and genetic mineralogy Novosibirsk, Inst. of Geology and Geophysics Press: pp 129-140 (in Russian)

Evans T (1979) Changes produced by high temperature treatment of diamond. In: Field JE (ed). The Properties of Diamond London, Academic Press Inc, pp 403-424

Evans T, Santer DH (1961) Etching of diamond surfaces by gases. Philadelphia Magazine 6: 429-440

Ezersky VA (1982) The shock-metamorphosed carbon matter in impactites. Meteoritika 41: 134-140 (in Russian)

Ezersky VA (1986) High-pressure polymorphs originated by shock transformation of coal. Zapiski Vsesouznogo Mineralogicheskogo Obstchestva 115 (1): 26-33 (in Russian)

Fedoseev DV, Galimov EM, Varnin VP, Prokhorov VS, Deryagin BV (1971) Carbon isotopes fractionation at physical-chemical synthesis of diamond from gas. Doklady Akademii Nauk SSSR 201: 1149-1150 (in Russian)

Feldman VI (1990) Petrology of impactites. Moscow, Moscow University Press: 299 pp (in Russian)

Foote AE (1891) Meteoritic iron of Canyon Diablo. American Journal of Science 42: 413-417

Gigl P, Dachille F (1968) Effects of pressure and temperature on the reversal transitions of stishovite. Meteoritics 4: 123-136

Gilmour I, Russell SS, Arden JW, Lee MR, Franchi IA, Pillinger CT (1992) Terrestrial carbon and nitrogen isotopic ratios for Cretaceous-Tertiary boundary nanodiamonds. Science 258: 1624-1626

Grieve RAF, Langenhorst F, Stöffler D (1996) Shock metamorphism of quartz in nature and experiments. II. Significance in geoscience. Meteoritics and Planetary Science 31: 6-35

Gurov EP, Gurova EP, Rakitskaya RB (1995) Impact diamonds in the craters of the Ukrainian Shield. Meteoritics 31: 515-516

Harris JW, Vance ER (1974) Studies to the reactions between diamond and heated kimberlite. Contributions to Mineralogy and Petrology 47: 237-244

Hildebrand AR, Boynton WV (1990) Proximal Cretaceous-Tertiary boundary impact deposits in the Caribbean. Science 248: 843-847

Holovnya SV, Khvostova VP, Makarov ES (1977) The hexagonal modification of diamond (lonsdaleite) in the eclogites of the metamorphic complexes. Geokhimiya 5: 790-793 (in Russian)

Hough RM, Gilmour I, Pillinger CT, Arden JW, Gilkes KWR, Yuan J, Milledge HJ (1995) Diamond and silicon carbide in impact melt rocks from the Ries crater. Nature 378: 41-44

Hough RM, Gilmour I, Pillinger CT, Langenhorst F, Montanari A (1997) Diamonds from the iridium-rich K-T boundary layer at Arroyo el Mimbral, Tamaulipas, Mexico. Geology 25: 1019-1022.

Huss GR, Lewis RS (1995) Presolar diamond, SiC, and graphite in primitive chondrites: abundances as a function of meteorite class and petrologic type. Geochim Cosmochim Acta 59: 115-160

Karklina MI, Maslakovets YuP (1968) On the problem of diamond etching. Doklady Akademii Nauk SSSR 183: 1311-1312 (in Russian)

Kashkarov IF, Polkanov YuA (1964) On the diamond findings in titanite-zircon sands. Doklady Akademii Nauk SSSR 157: 1129-1130 (in Russian)

Kashkarov IF, Polkanov YuA (1972) Some special features of diamonds from titaniferous placers of the Northern Kazakhstan. Novye dannye o mineralakh 21: 183-185 (in Russian)

Koeberl C (1993) Chicxulub crater, Yukatan: tektites, impact glasses and the geochemistry of target rocks and breccias. Geology 21: 211-214

Ksanda CJ, Genderson EP (1939) Identification of diamonds in the Canyon Diablo meteorite. American Mineralogist 24: 677-680

Kvasnitsa VN (1985) Small diamonds. Kiev, Naukova Dumka Press: 215 pp (in Russian)

Kvasnitsa VN, Sobotovich EV, Kovalukh NN, Litvak AL, Rybalko SI, Sharkin OP, Egorova LN (1979) High-pressure carbon polymorphs in mosses from the Tunguska impact site. Doklady Ukrainskoi Akademii Nauk, B, 12: 1000-1004 (in Russian)

Lewis RS, Bright D, Steel E (1987) Presolar diamonds ($\delta^{13}C$) in carbonaceous chondrites: size distribution. Meteoritics 22: 445

Lewis RS, Anders E, Draine BT (1989) Properties, detectability and origin of interstellar diamonds in meteorites. Nature 339: 117-121

Lewis RS, Anders E, Wright IP, Norris SJ, Pillinger CT (1983) Isotopically-anomalous nitrogen in primitive meteorites. Nature 305: 761-771

Lipschutz ME (1962) Diamonds in Dualpur meteorite. Science 138: 1266-1267

Lipschutz ME, Anders A (1961) The record in meteorites. 4. Origin of diamonds in iron meteorites. Geochimica et Cosmochimica Acta 24: 83-105

Mal'kov IYu (1994) Carbon coagulation under conditions of non-stable flow of detonation products. Fizika gorenia i vzryva 30 (5): 155-157 (in Russian)

Mal'kov IYu, Titov VM (1995) The regime of diamond formation under detonation. In Metallurgy and material applications of shock waves and high strain rate phenomena. Proceedings of 1995 International Conference "ExploMet-1995", El Paso, USA. Elsevier Press pp 669-676

Marakushev AA, Bogaturev OS, Fenogenov AD, Paneyakh NA, Fedosova SP (1993) Impactogenesis and volcanism. Petrologia 1: 571-595 (in Russian)

Masaitis VL, Futergendler SI, Gnevushev MA (1972) The diamonds in impactites of the Popigai meteoritic crater. Zapiski Vsesouznogo Mineralogicheskogo Obstchestva 101: 108-113 (in Russian)

Masaitis VL, Shafranovsky GI, Grieve RAF, Langenhorst F, Peredery WV, Balmasov EL Fedorova IG, Therriault A (1997) Discovery of impact diamonds at the Sudbury structure. In Conference on large impacts and planetary evolution (Sudbury-1997). Lunar and Planetary Institute Contribution No. 922. Houston, Lunar and Planetary Institute Press: pp 33

Masaitis VL, Mastchak MS, Raikhlin AI, Selivanovskaya TV, Shafranovsky GI (1998a) Diamondiferrous impactites of the Popigai crater. All- Russian Geological Institute Sankt-Petersburg (VSEGEI) Press: 179 p. (in Russian)

Masaitis VL, Shafranovsky GI, Fedorova IG, Koivisto M, Kohronen JV (1998b) Lappajarvi crater: the first find of impact diamonds on the Fennoscandian shield. In Lunar and Planetary Science Conference 29: Abstract # 1171

Montanari A, Asaro F, Kennet JP (1993) Iridium anomalies of Late Eocene age at Massignano (Italy), and ODB Site 689B (Maud Rise, Antarctica). Palaios 8: 420-437

Newton J, Bischoff A, Arden JW, Franchi IF, Geiger T, Greshake A, Pillinger CT (1995) Acfer 094, a uniquely primitive carbonaceous chondrite from the Sahara. Meteoritics 30: 47-56

Nininger HH (1956) Arizona's meteorite crater: its past, present and future. Denver, Colorado, World Press Inc.: 232 p

Pal'yanov YuN, Khokhryakov AF, Borzdov YuM, Sokol AG (1995) Diamond morphology in growth and dissolution processes. In Sixth International Kimberlite conference, Novosibirsk, August 1995. Extended Abstracts. Novosibirsk, United Institute of Geology, Geophysics and Mineralogy Press: 415-417 p

Polkanov YuA, Er'omenko GK, Sokhor MI (1973) Impact diamonds from the Ukrainian fine-grained placers. Doklady Ukrainskoi Akademii Nauk, B: 989-990 (in Russian)

Rost R, Dolgov YuA, Vishnevsky SA (1978) The gas inclusions in the impact glasses of the Ries crater and the discovery of high-pressure polymorphs of carbon. Doklady Akademii Nauk SSSR 241: 695-698 (in Russian)

Shatsky VS, Sobolev NV (1993) Some aspects of diamond origin in metamorphic rocks. Doklady Akademii Nauk 331: 217-219 (in Russian)

Shatsky VS, Sobolev NV, Vavilov MA (1995) Diamond-bearing metamorphic rocks of the Kokchetav massif (Northern Kazakhstan). In: Coleman RG and Xiaomin Wang (eds). Ultrahigh pressure metamorphism Cambridge, Cambridge University Press. pp 427-455

Shatsky VS, Rylov GM, Efimova ES, Corte K, and Sobolev NV (1998) The morphology and real structure of microdiamonds from the Kokchetav Massif metamorphic rocks, kimberlites and alluvial placers. Russian Geology and Geophysics 39: 949-961

Sobolev NV, Shatsky VS (1990) Diamond inclusions in garnet from metamorphic rocks. Nature 343: 742-746

Smith JV, Dawson JB (1985) Carbonados: diamond aggregates from Early impacts on crustal rocks. Geology 13: 342-343

Urey HC, Mele A, Mayeda T (1957) Diamonds in stone meteorites. Geochimica et Cosmochimica Acta 13: 1-14

Val'ter AA, Er'omenko GP (1996) Carbon minerals in rocks of astroblemes. Meteoritics and Planetary Science 31: A144

Val'ter AA, Er'omenko GK, Kvasnitsa VN, Polkanov YuA (1992) The shock-metamorphic minerals of carbon. Kiev, Naukova Dumka Press: 172 p.p (in Russian)

Vdovykin GP (1967) The carbon matter of the meteorites (Organic compounds, diamonds, graphite). Moscow, Nauka Press: 272 pp (in Russian)

Vdovykin GP (1970) Diamonds in meteorites. Moscow, Nauka Press: 128 pp (in Russian)

Vdovykin GP (1991) Diamonds in stone meteorites-ureilites and their origin. Bulletin of Moscow Society of Naturalists, Geological Series, 66: 87-93 (in Russian)

Vishnevsky AS, Balagansky IA, Vishnevsky SA (1996) Computer simulation of the Popigai impact event (compression and initial excavation stages) and some consequences on global dispersion of projectile and tektite glasses. In: Drobne K, Gorican S , and Kotnik B (eds) International workshop "The role of impact processes in the geological and biological evolution of Planet Earth", September 27 – October 2 1996, Postojna, Slovenia. Abstracts. Ljubljana, Scientific Research Center (SAZU) Press pp 95-96

Vishnevsky SA, Pal'chik NA (1975) Graphite in the rocks of the Popigai structure: its destruction and transformation into other phases of the carbon system. Soviet Geology and Geophysics 16: 55-61.

Vishnevsky SA, Pal'chik NA, Raitala J (1999) Impact diamonds in impactites of the Lappajarvi crater, Finland. Russian Geology and Geophysics 40: in press (in Russian)

Vishnevsky SA, Afanasiev VP, Argunov KP, Pal'chik NA (1997) Impact diamonds: their features, origin and significance. Novosibirsk, Russian Academy of Sciences Press: 110 pp (in Russian and English)

Yerofeev MV, Lachinov PA (1888) About the Novo Urei meteorite. Zhurnal Russkogo Fiziko-Chimicheskogo Obstchestva 20: 185-213 (in Russian)

Yurk YuYu, Kashkarov IF, Polkanov YuA, Er'omenko GK, Yalovenko IP (1973) The diamonds from the Ukrainian sandy sediments. Kiev, Naukova Dumka Press: 167 pp (in Russian)

10 Deeply Exhumed Impact Structures: A Case Study of the Vredefort Structure, South Africa

R.L. Gibson and W.U. Reimold

Department of Geology, University of the Witwatersrand, P O WITS, Johannesburg 2050, South Africa. (065RLG@cosmos.wits.ac.za)

Abstract. The main evidence in the Vredefort Dome and surrounding parts of the Witwatersrand Basin is presented, which indicates that this terrain represents the deeply exhumed root zone of a large, 2023 ± 4 Ma, complex impact structure. Shock metamorphic features such as impact melt, shatter cones, high-pressure quartz polymorphs and shock microdeformation features are restricted to the Vredefort Dome, which constitutes the central uplift of the impact structure. High strain rate brittle deformation features, including pseudotachylites and clastic breccias, are found over a larger radial distance of at least 100 km from the center of the structure. Low-temperature (~300 °C) hydrothermal effects occur up to a similar radial distance but, in the central uplift, metamorphic mineral parageneses overprinting the shock and brittle deformation features in pelites indicate a strong radially-inward increase in temperature to between 700 and 900 °C. Pressure estimates of 0.2-0.3 GPa obtained from these mineral parageneses indicate that the Vredefort structure has been eroded by between 7 and 10 km. The original depth of burial of these rocks after the impact and their elevated temperatures explain the anomalous appearance and restricted distribution of many of the shock features compared to other, less eroded, structures. The metamorphism and hydrothermal activity are attributed to the combined effects of exhumation of deep (hot) crustal levels by the cratering event and formation of the central uplift, and shock heating, possibly with a minor component of heating caused by an overlying impact melt volume. The structural, shock and thermal impact-related features in the Vredefort Dome and Witwatersrand Basin provide a case study that should assist in the identification of other deeply exhumed impact structures.

Introduction

During the latter half of the Twentieth Century, impact cratering has become widely acknowledged as the most ubiquitous process influencing the surfaces of the solid planetary bodies and their satellites in our Solar System. On most of these planets, the absence of geological activity for hundreds, or even thousands, of millions of years has ensured the preservation of a comprehensive record of cratering events and provided an opportunity to study the surface morphologies of craters of a wide range of diameters. These extraterrestrial studies have provided important insights into impact cratering mechanisms and the large-scale response of targets to bolide impacts of various sizes (e.g., Melosh 1989). However, the physical inaccessibility of these craters and the blanketing effect of impact debris within craters hampers more detailed analysis of, in particular, the three-dimensional structure of the underlying crater basement and its response to the impact event. In contrast, the cratering record on Earth is far less comprehensive, due to rapid modification and/or destruction of many craters by geological and erosional activity (e.g., Grieve 1998). Nonetheless, such exhumation provides an opportunity to examine directly the deep-level effects of bolide impacts in the target rocks and, thereby, to better understand the processes by which such structures form.

This paper describes the main features of the Vredefort Dome in South Africa - interpreted as the central uplift of the world's oldest, and possibly largest, known impact structure. Recent data indicate that the levels of erosion in this structure are significantly greater than for any other impact structure known on Earth. This, together with its location within the intensively studied, gold-rich Witwatersrand basin, provides a natural laboratory in which to investigate impact-related phenomena at deep crustal levels. These observations should prove useful in the future identification of other deeply eroded impact structures.

Geological Setting

The Vredefort Dome is an ~70 km wide domical feature located within Archean and Paleoproterozoic rocks of the Kaapvaal craton of southern Africa, some 120 km southwest of Johannesburg (Fig. 1). It lies within the central parts of the Witwatersrand Basin, the world's richest gold province. The dome comprises a 40 km wide core of poly-deformed, pre-3.1 Ga, Archean granitoid gneisses, with subsidiary fragments of upper amphibolite to granulite facies mafic and metasedimentary gneisses, which is surrounded by a 15 km wide collar of greenschist to mid-amphibolite facies, metasedimentary and metavolcanic, Late Archean to Paleoproterozoic, strata. These supracrustal rocks, which range in age from 3.1 to 2.2 Ga (Armstrong et al. 1991), belong to (from oldest to youngest) the Dominion Group and the Witwatersrand, Ventersdorp and Transvaal Supergroups. In the western and northern sectors, the collar strata are generally subvertical to overturned (Fig. 2). The southern parts of the dome are obscured beneath Phanerozoic sediments and dolerite intrusions of the Karoo Supergroup (Figs. 1, 2), but are relatively well constrained from geophysics and borehole drilling. Although moderately steep normal dips are recorded in Witwatersrand Supergroup quartzites beneath the Karoo Supergroup cover, the sequence here is sparse and is clearly disrupted by faults.

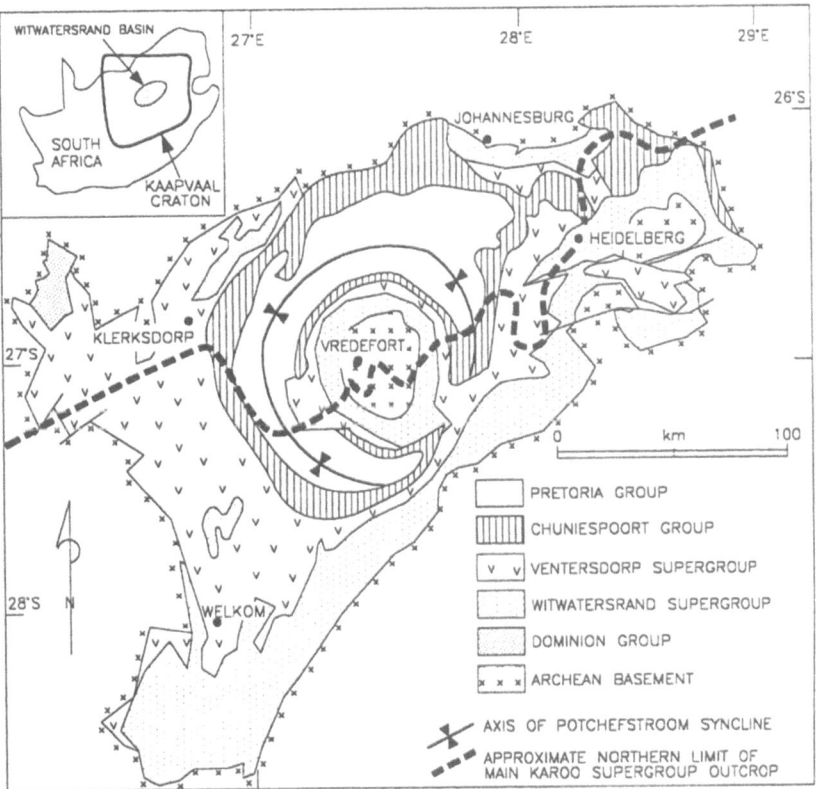

Fig. 1. Simplified geological map of the Archean-Paleoproterozoic geology of the central Kaapvaal craton, showing the distribution of strata within the Witwatersrand Basin. The long axis of the basin trends NE-SW, but it is dominated in its central parts by the Vredefort Dome and its rim syncline. Phanerozoic strata of the Karoo Supergroup cover much of the basin, particularly in the south.

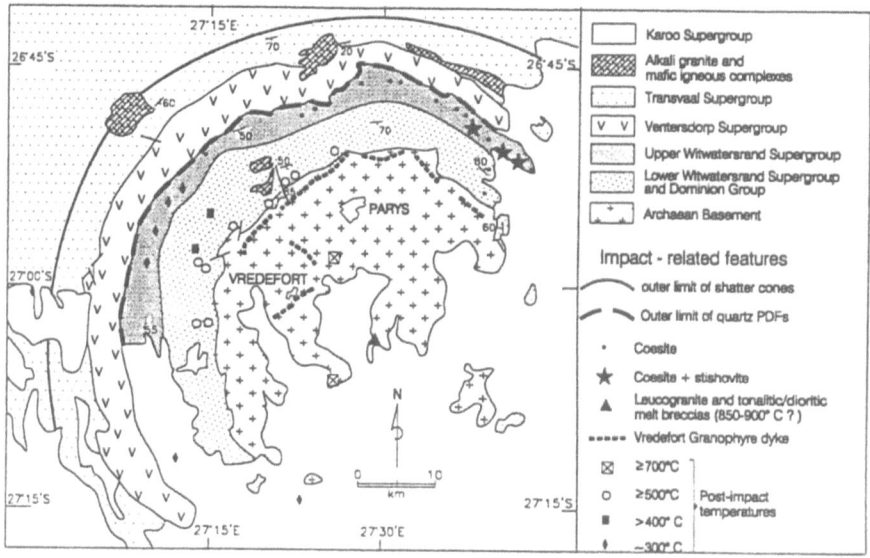

Fig. 2. Simplified geological map of the Vredefort dome, showing the distribution of the main impact-related features described in the text. Shock-related microdeformation features in the mineral zircon are restricted to rocks of the basement core and the inner collar of the dome.

The dome is surrounded by a 50-70 km wide rim syncline (Potchefstroom Syncline, Fig. 1), characterized by shallow dips in the Transvaal Supergroup and underlying strata. The Transvaal Supergroup strata are also affected by kilometer-scale open folds arranged tangentially around the dome (Simpson 1978). In addition, ductile shear zones with associated meter-scale folds and cleavage displaying centrifugal vergence with respect to the dome are found in the Transvaal Supergroup rocks along the northern margin of the Witwatersrand Basin up to 150-200 km from the dome (McCarthy et al. 1986; Gibson et al. 1999).

Intercalated within the strata in the collar of the dome are intrusions related to several magmatic events. Mafic sills in the Witwatersrand Supergroup are attributed to the ca. 2.7 Ga (Armstrong et al. 1991) Ventersdorp event (Pybus 1995). In addition, several small peralkaline and dioritic complexes in the collar (Fig. 2) have been attributed to a ca. 2.2 Ga event by Walraven and Elsenbroek (1991), but recently Moser (1997) obtained a U-Pb single-zircon age of 2.078 ± 0.012 Ga for one of these complexes, the alkali granite and nepheline syenite Schurwedraai body, situated in strata of the inner collar to the northwest of Parys. Mafic sills in the Transvaal Supergroup in the dome and rim syncline are attributed to the 2.06 Ga (Walraven 1997) Bushveld magmatic event. The folds in the rim syncline affect these sills, indicating that the dome is younger than 2.06 Ga.

Morphology of Large Impact Structures

The process of impact cratering has been extensively discussed in various books and papers (e.g., Grieve 1987; Melosh 1989; French 1998; Melosh and Ivanov 1999) and will not, thus, be described in detail here. However, it is important to note that the formation of large craters differs fundamentally from that of small craters - whereas small craters retain a bowl-shaped morphology due to limited slumping of the crater walls following impact, large craters are characterized by complex differential uplift of the crater basement, producing a *central structural uplift* simultaneous with inward slumping of the crater walls (Fig. 3). On Earth, the transition from simple, bowl-shaped to complex crater forms occurs where crater diameters exceed ~3 km.

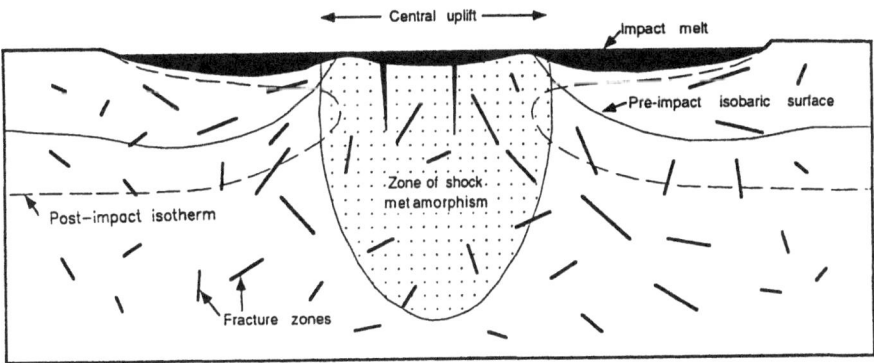

Fig. 3. Simplified schematic cross-section through a complex impact structure, showing the distribution of impact-related features as suggested by the Vredefort structure. Evidence from the Vredefort structure (see Fig. 2) suggests that the shock-diagnostic features are likely to be confined largely to the central structural uplift at deep levels (stippled area). This also applies to the post-impact thermal effects, although a narrow zone of contact metamorphism is likely beneath the impact melt body. Brittle deformation features, including occurrences of pseudotachylite and pseudotachylitic breccias, occur over a much wider area and may even develop outside the geographic limits of the crater.

From Fig. 3, it is apparent that the impact crater is only the surface manifestation of a large volume of rock that displays a wide variety of features related to the impact event. These include:

- *Impact melts and breccias*, located mainly within the crater, but also occurring outside the crater as impact ejecta, and as dykes intruded downwards along fractures in the crater basement;
- *Shock metamorphic features* in rocks and minerals, the intensity of which will diminish radially outwards and downwards, and which, in a large (complex) crater, are likely to extend over a significantly smaller radius than that of the final crater;
- *Brittle deformation features* related to the high strain rate adjustment of the crust adjacent to and below the zone of shock metamorphism during the initial shock compression and to readjustment of the crater basement during the subsequent modification of the crater – the full radial extent of these features is unlikely to be well-constrained, mainly because the lack of associated shock-diagnostic features is likely to hamper efforts to separate impact-related deformation from "normal" tectonic deformation (cf. also Reimold 1995 1998);
- *Thermal metamorphic and/or hydrothermal features* due to the combined effects of cooling of the superheated impact melts, shock and friction heating, and exhumation of rocks from deep (i.e., hotter) structural levels.

The three-dimensional distribution of each of these sets of features is shown schematically in Fig. 3. From this, it should be apparent that the term *impact structure* can have different meanings, depending on which features are being described. This is particularly the case for eroded impact structures where easily-recognizable features such as the crater and its impact melt and breccia fill have been removed, but it may also apply to well-preserved complex craters such as Chicxulub, where uncertainty currently exists about the significance of concentric geophysical and topographic rings beyond the accepted crater limits (e.g., Morgan and Warner 1999; Snyder and Hobbs 1999). A second point of note is that the understanding of an impact *event* should also be broadened to include metamorphic-hydrothermal processes, which may continue in the crater basement and crater fill for millions of years after the actual shock event.

Impact-Related Features in the Vredefort Dome

Although the Vredefort Dome is only one of several large domical structures cored by Archean basement gneisses in the region of the Witwatersrand Basin, it differs from the other domes, both in its gross structure and in the unusual structural and metamorphic features found in its rocks. In contrast to the other domes that have been produced by poly-phase tectonics, the Vredefort Dome displays both a more pronounced circular geometry and more extreme uptilting of the Late Archean to Paleoproterozoic supracrustal succession. Regional seismic sections (e.g., Durrheim et al. 1991) indicate that the base of the supracrustal sequence is raised by more than 10 km between the rim syncline and the dome. It is, however, the unusual deformation and metamorphic phenomena observed in rocks in the dome and its immediate vicinity that point to its origin as the central uplift of a large impact structure. These features also provide clues to the level of exhumation of the impact structure.

Melosh (1989) established that the diameter of the central uplift in complex impact structures varies between 0.22 and 0.5 of the final crater diameter. Interpretation of the 70-km-wide Vredefort Dome as the eroded root of the central uplift of a large impact structure would suggest that the dimensions of the Vredefort impact structure are similar to the ca. 200-250 km diameter Sudbury and ca. 180 km diameter Chicxulub structures (e.g., Deutsch et al. 1995; Sharpton et al. 1996a; Morgan and Warner 1999; Snyder and Hobbs 1999). However, Vredefort differs in at least one significant way from these structures - it does not contain any evidence of a coherent impact melt sheet and breccia crater fill such as is observed in the largely uneroded Chicxulub crater, and in the central parts of the Sudbury structure. As Sudbury appears to have experienced between 2 and 3 km of erosion (Rondot 1994), the level of erosion of the Vredefort structure must be deeper. Published estimates of the depth of erosion of the Vredefort Dome range from a few hundred meters (Martini 1991) to nearly 18 km (Schreyer and Abraham 1978).

The absence of a coherent impact melt sheet and fallback breccias suggests that the former is unlikely, and the latter, too, seems unrealistic given that stratigraphic and intrusive units that originally lay close to the surface prior to the formation of the dome, such as the upper Transvaal Supergroup and the Bushveld Complex, are still preserved with only gentle dips over large parts of the region. Recent work by our group (outlined below) has, however, provided new constraints on the depth of erosion of the Vredefort structure.

Impact Melt Rock

In the Vredefort Dome, occurrences of impact melt rock are restricted to several vertical dykes of so-called Vredefort Granophyre (Fig. 2). The dykes are largely confined to the basement core of the dome and display vertical NE and NW trends, although they are locally sinuous and may display short offshoots. They reach lengths of up to 9 km and widths of up to 65 m. The Granophyre occurs as both spherulitic

(Fig. 4) and granular varieties, sometimes even within the same dyke. Therriault et al. (1996) interpreted the mineral textures as indicative of supercooling from temperatures in excess of 1100 °C. The Granophyre displays a very uniform mineralogical composition comprising hypersthene, plagioclase, orthoclase, quartz, biotite, magnetite and ilmenite, with rare augite and pigeonite. Its chemical composition is both unusual (high Si and Ca contents, with a high Mg/Fe ratio) and remarkably homogeneous between dykes, indicating that the dykes are offshoots of a single, well-homogenized melt body (Reimold et al. 1990; Therriault et al. 1997a). A further clue to the origin of the Granophyre is the abundance of angular to rounded rock and mineral clasts in the dykes, many of which appear to have been derived from the collar strata, and which suggest that the parent melt body must originally have overlain the present level of exposure.

Several attempts have been made to model the chemical composition of the Granophyre by wholesale melting of the rock types in the dome (e.g., French and Nielsen 1990; Reimold et al. 1990; Therriault et al. 1997a). Early attempts to resolve the origin of the Granophyre by identifying unusual siderophile element contents, such as anomalous nickel or iridium values, were inconclusive, largely due to the anomalously high siderophile element contents of lower Witwatersrand Supergroup shales, which significantly contributed to the impact melt composition (French et al. 1989; Reimold et al. 1990). The impact origin of the Granophyre was only recently confirmed through Re-Os isotopic analysis: the Os content of the Granophyre is considerably higher than that of the rocks from which the bulk of the Granophyre is likely to have been derived, and the Granophyre displays very low, near-meteoritic, $^{187}Re/^{188}Os$ and $^{187}Os/^{188}Os$ ratios which indicate the presence of a very small (~0.2 %) meteoritic component (Koeberl et al. 1996). Radiometric age studies on the Granophyre provided the first constraints on the age of the impact event [a 2016 ± 24 Ma K-Ar isochron age from biotite and a 2002 ± 52 Ma U-Pb zircon age (Walraven et al. 1990), as well as a 2006 ± 9 Ma ^{40}Ar-^{39}Ar biotite age (Allsopp et al. 1991)].

Shock Deformation Features

A megascopic manifestation of shock deformation in impact structures is the presence of shatter cones - striated conical joints believed to form at shock pressures of between 4 and 30-45 GPa (Roddy and Davis 1977). The exact mechanism by which these features form is a matter of ongoing debate (e.g., Sharpton et al. 1996; Nicolaysen and Reimold 1999). Dietz (1961) and Hargraves (1961) first identified shatter cones in the Vredefort Dome. These cones are typically only partially developed, and are mainly (but not exclusively!) located along pre-existing planes of anisotropy such as bedding or joint surfaces (Fig. 5). For this reason they are largely restricted to the collar strata in the dome, occurring up to a radial distance of 40 km from its center (Fig. 2). Several studies in the dome have suggested that not all striated joint surfaces formed under shock compression (e.g., Nicolaysen and Reimold 1999). Given the complex structural movements that accompanied the modification of the Vredefort impact crater and formation of the dome, this may not be surprising.

Coesite and rare stishovite were identified by Martini (1978 1991) in quartzites of the upper Witwatersrand Supergroup at several localities in the northeastern collar of the dome (Fig. 2). Experimental shock data indicate minimum pressures of 5-10 GPa for coesite formation (White 1993) and 12-15 GPa for that of stishovite (Stöffler and Langenhorst 1994), although coesite has been synthesized at room temperature at pressures as low as ~2 GPa, and stishovite is stable at pressures as low as ~8 GPa under static conditions (Akaogi and Navrotsky 1984). Martini (1991) attributed the rather restricted distribution of the coesite and stishovite samples in the dome to post-shock thermal effects over much of the dome (see below) and noted that the coesite and stishovite are spatially associated with thin melt veinlets. He described these melts as so-called "Type A" pseudotachylites and proposed that they be distinguished from the more voluminous "Type B" pseudotachylites found elsewhere in the dome (see below), suggesting that the "A-type" pseudotachylites formed as a result of friction melting along slip surfaces generated during the shock compression phase of cratering, whereas the "B-type" pseudotachylite formed during crater modification - after decay of the shock wave. Martini also tentatively related his "Λ-type" melt phase to thin glass films occasionally observed on shatter cone surfaces. Reimold et al. (1992), however, questioned whether the coesite ± stishovite – bearing veins should not be considered as shock (impact) melt. Thus far, although glass has been documented along shatter cone surfaces, no evidence has been found that high-pressure silica polymorphs developed in association with it (Nicolaysen and Reimold 1999).

The most ubiquitous shock metamorphic feature in the rocks of the Vredefort Dome is found in quartz in the quartzites of the Witwatersrand Supergroup and in the Archean granitoid gneisses and pelitic xenoliths in the basement core. These remnants of planar microdeformation features (PDFs) typically involve one or more crystallographically-oriented sets of planar trails of fluid inclusions (Fig. 6a), or planar zones defined by fine-grained quartz mosaics (Fig. 6b). The outer limit of these planar features shown in Fig. 2 corresponds to the base of the quartz-free Ventersdorp Supergroup lavas. Although the similarity of the crystallographic orientations of these planes with those of true PDFs found in quartz crystals shocked to pressures of between 8 and 25 GPa was first established by Carter (1965) and further supported by Grieve et al. (1990), conclusive proof of their shock origin was only obtained relatively recently from transmission electron microscopy (TEM) by Leroux et al. (1994), who were able to ascertain that many of the planar features in the basal crystallographic orientation are mechanically-induced Brazil twins. Such twins, resulting from deformation, have only ever been described elsewhere as a product of shock deformation in rocks from confirmed impact structures and from the impact-related Cretaceous-Tertiary (K-T) Boundary Layer (Leroux et al. 1994) [they are, however, known as a formation from hydrothermal quartz growth – a process clearly not relevant when discussing the planar microdeformation of quartz in Archean basement rocks].

Radial traverses across the collar and core of the dome (e.g., Grieve et al. 1990; Hart et al. 1990) indicate that (a) the intensity of development of PDFs in quartz is irregular, and (b) the (0001) PDF orientation, indicative of relatively low shock pressures, dominates throughout, even close to the center of the dome, where other PDF orientations consistent with higher shock pressures might have been expected to occur in higher abundance than actually observed. Similar anomalies in the PDF distribution pattern in the Slate Islands impact structure have been attributed to differential uplift of the crater basement (B.O. Dressler, pers. comm. 1999; see also Sharpton et al. 1996b and Geology Forum 1997). However, Grieve et al. (1990) postulated that the preponderance of the (0001) orientation in the Vredefort Dome might reflect the greater robustness of this orientation during post-shock annealing.

The rocks in the dome lack conclusive evidence of shock metamorphism of other major rock-forming minerals. Intense kinkbanding is ubiquitous in biotite throughout the dome and is clearly related to the impact event (Wilshire 1971; Gibson et al. 1997a) but it is not, in itself, a shock-diagnostic feature. Gibson et al. (1997a) described multiple cleavage orientations in garnet from the lower Witwatersrand Supergroup as an impact-related feature, however, it is also not shock-diagnostic. Schreyer (1983) described fine-grained plagioclase aggregates from the central parts of the core of the dome which he suggested represent recrystallized diaplectic feldspar glass; however, recent work on rocks from the central core has suggested that these aggregates are part of polymineralic veins of pseudotachylitic or partial (anatectic) melt derived from the surrounding gneisses during or immediately after the impact event (Ashley et al. 1999). Ashley et al. (1999) have also identified up to three sets of subplanar lamellar structures in plagioclase from the centre of the dome which may be recrystallized PDFs; however, the level of post-shock annealing in these rocks is too high to preserve glass within these lamellae. In addition, re-appraisal of samples from the core lithologies indicates some evidence for partial breakdown of coarse feldspar crystals to very fine-grained aggregates. It remains to be demonstrated, however, whether these aggregates could represent diaplectic glass recrystallized during the post-impact thermal metamorphism (see below).

In recent years, Kamo et al. (1996) and Gibson et al. (1997b) identified multiple sets of planar microdeformation features in zircon crystals in rocks from several localities in the core of the dome (Fig. 6c). Kamo et al. (1996) also identified zircons in rocks from the dome that display a "strawberry texture" produced by shock melting of zircon and subsequent *in situ* crystallization of this melt as an aggregate of tiny, euhedral zircon crystals. These distinctive features have also been found in samples from the Sudbury, Manicouagan and Chicxulub impact structures, and from the K-T boundary layer (e.g., Krogh et al. 1984; Bohor et al. 1993). While not much work has been carried out on the nature and formation conditions of these deformation textures, a recent experimental study by Leroux et al. (1999) strongly suggests that planar microcleavage, as well as glassy planar deformation features, form in zircon in the shock pressure interval between 20 and 40 GPa.

High Strain Rate Brittle Deformation Features

The Vredefort Dome is the type locality for pseudotachylite (original spelling pseudotachylyte; Shand 1916), a relatively rare clast-laden melt rock comprising a generally glassy or microcrystalline matrix containing angular to rounded clasts of the wallrock lithology (-gies). Although it is most commonly found as small, millimeter- to centimeter-thick veins in fault zones, it also occurs in massive form in the Vredefort and Sudbury impact structures where individual veins may reach widths of tens to hundreds of meters and lengths of hundreds to thousands of meters. Pseudotachylites (breccia veinlets with *bona fide* evidence of melting) have been described from many more impact structures, however these occurrences are generally small and hardly ever exceed a few veinlets of mm to cm width.

Undoubtedly, melt-bearing breccias (here referred to as "pseudotachylite") that are different from impact melt rock do occur in both the Vredefort and Sudbury impact structures. However, in both these structures purely clastic breccias of general pseudotachylite-resemblance have also been described. In order to distinguish between proper pseudotachylite (friction melt) and other, similar appearing breccias, Reimold (1995 1998) introduced the term "pseudotachylitic breccia" for those breccias, for which no genetic information was available, but

Fig. 4. Spherulitic variety of Vredefort Granophyre containing clasts of quartzite, from a locality just south of Vredefort town. Spherules are defined by radiating acicular orthopyroxene crystals set in a feldspar-quartz matrix and indicate moderate to high supercooling of the melt after intrusion (Therriault et al. 1996).

Fig. 5. Shatter cones on joint surface in upper Witwatersrand Supergroup quartzite, western collar, showing dense horsetailing pattern.

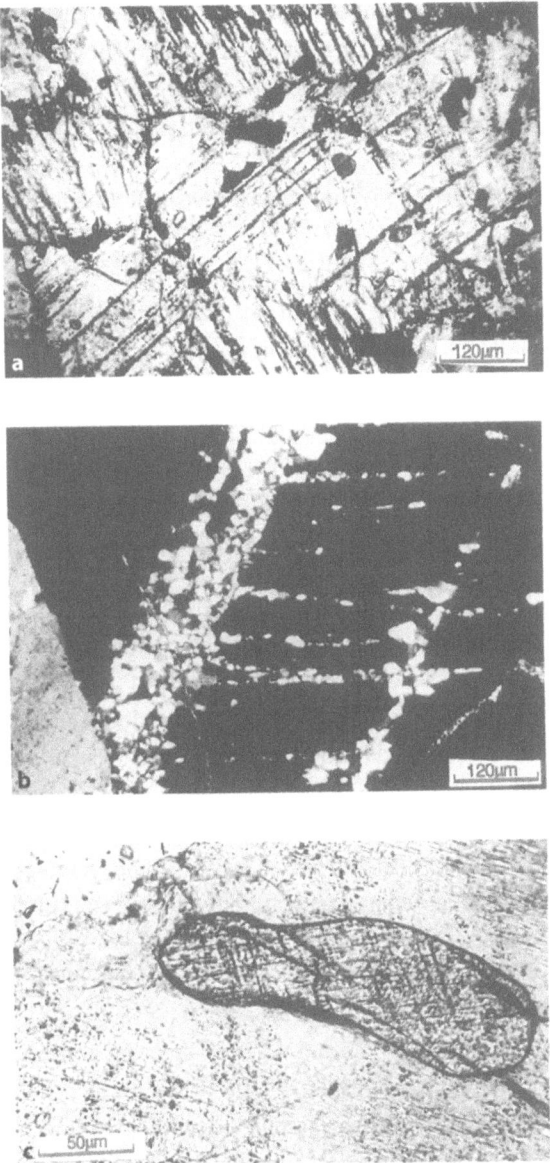

Fig. 6. (a) Decorated PDFs in quartz defined by trails of fluid inclusions, in lower Witwatersrand Supergroup quartzite. Plane polarized light. Width of field of view: 800 μm. (b) Mosaic recrystallization of PDFs in quartz from metapelitic granulite from the core of the dome south of Parys. Crossed polarizers. Width of field of view: 800 μm. (c) Planar microdeformation features in zircon inclusion in alkali feldspar from metapelitic granulite from the core of the dome SE of Vredefort town. Plane polarized light. Width of field of view: 300 μm.

which resembled tectonic or impact-produced pseudotachylite. In particular, when referring to the abundant breccias occurring in the Witwatersrand Basin in the environs of the Vredefort Dome, pseudotachylite, mylonitic, and cataclastic breccias need to be distinguished (e.g., Reimold and Colliston 1994; Reimold et al. 1999).

Current models (Spray 1995) suggest that the small volumes of pseudotachylite in tectonic fault zones may be generated as a result of a combination of cataclasis followed by frictional melting during high strain rate ($>10^{-1}$/s) coseismic slip in the elastico-frictional regime. Spray (1997) attributed the considerably more voluminous pseudotachylites in large impact structures to a similar process operating along so-called "superfaults" during cratering; however, in the absence of evidence of such large-displacement faults in association with most of the large pseudotachylite dykes in the Sudbury impact structure, Dressler (1984) suggested that explosive brecciation during impact cratering may play a role.

In addition to several large dykes of pseudotachylite up to 1 km in length in the Archean basement gneisses (Fig. 7a), smaller pseudotachylite veins ranging in length from millimeters up to tens to hundreds of centimeters in length occur ubiquitously in the core and collar rocks of the Vredefort Dome and in fault zones in the gold-fields along the northwestern margin of the Witwatersrand Basin (Fletcher and Reimold 1989; Killick 1992; Reimold and Colliston 1994). The pseudotachylites in the gold-fields show a clear spatial relation to bedding-parallel faults (Fletcher and Reimold 1989) which may represent ring-thrusts and/or normal-slip ring-faults related to the impact event (Killick 1992; Ellis and Reimold 1999). More rarely, veins are found in subvertical north-trending normal faults, many of which show evidence of pre-impact activity, but which were also active during the impact event (Reimold and Colliston 1994). Recent geochronological data have confirmed the syn-Vredefort age of pseudotachylites in the dome and in the gold-fields (Trieloff et al. 1994; Spray et al. 1995).

The large pseudotachylite dykes and network breccias in the core of the dome are frequently aligned along pre-impact ductile structures but, as in the case of the Sudbury dykes, slip displacements associated with pseudotachylite formation rarely exceed a few centimeters to a meter (Reimold and Colliston 1994). However, in the larger occurrences at Vredefort, observations of exotic clasts may indicate transport over distances of at least several tens of meters and, therefore, at least similar amounts of movement of the melt away from its site of generation. Pseudotachylites in the collar rocks of the dome occur either along bedding contacts or along closely spaced microfaults displaying, at maximum, millimeter to centimeter scale offsets (Fig. 7b). Although individual veinlets and pods of pseudotachylite associated with the microfaults seldom exceed a few millimeters in width, the close spacing of the microfaults (usually centimeters) means that the total volume of pseudotachylite in the collar rocks is also considerable (Gibson et al. 1997a). No consistent slip sense appears to exist along these microfaults.

Glass has only been rarely documented in the Vredefort pseudotachylites (White 1993). This may reflect either post-impact metamorphic recrystallization of glass (see below) or, as in the case of some of the larger dykes, formation of sufficiently large melt volumes, or sufficiently high host rock temperatures, to prevent quenching (Reimold and Colliston 1994). The generally more mafic chemical compositions of vein matrices relative to their host rock compositions, and the preponderance of quartz clasts, confirm that the pseudotachylites formed by selective melting of ferromagnesian and hydrous ferromagnesian minerals, followed by feldspar and some quartz, in a similar manner to tectonic pseudotachylites (Reimold 1991). However, the unusually high proportion of pseudotachylite in all rocks in the dome, and the lack of accompanying wide, high-strain mylonitic or fault zones typical of tectonic pseudotachylite occurrences, points to the unusual conditions under which the Vredefort pseudotachylites formed.

Thermal Metamorphic Features

The longest-lived consequence of an impact event will be the thermal perturbation in the crater basement caused by the combined effects of shock heating, exhumation of deep crustal rocks in the central uplift, frictional heating along faults, and cooling of superheated impact melt volumes lying within the crater and any melts generated during decompression of the crater basement. Whilst contact metamorphic effects have been observed adjacent to impact melt bodies such as the Sudbury Complex (Dressler 1984), it is only recently that workers have started to acknowledge the role of exhumation of pre-heated (due to the pre-impact geotherm and to shock heating) basement in the central uplifts of large impact structures in causing post-impact metamorphism and hydrothermal alteration (e.g., McCarville and Crossey 1996).

Qualitative evidence that the rocks in the Vredefort Dome experienced significant, and variable, post-impact thermal metamorphism was provided by studies of PDFs in quartz (Carter 1965; Grieve et al. 1990) and pseudotachylites in quartzites (Martini 1992). Both of these features show an increasing intensity of recrystallization towards the core of the dome (see Figs. 6a, b). Fricke et al. (1990) used fluid inclusions decorating PDFs to loosely constrain post-impact metamorphic conditions in the central core of the dome at P ~0.2-0.5 GPa, T > 700 °C.

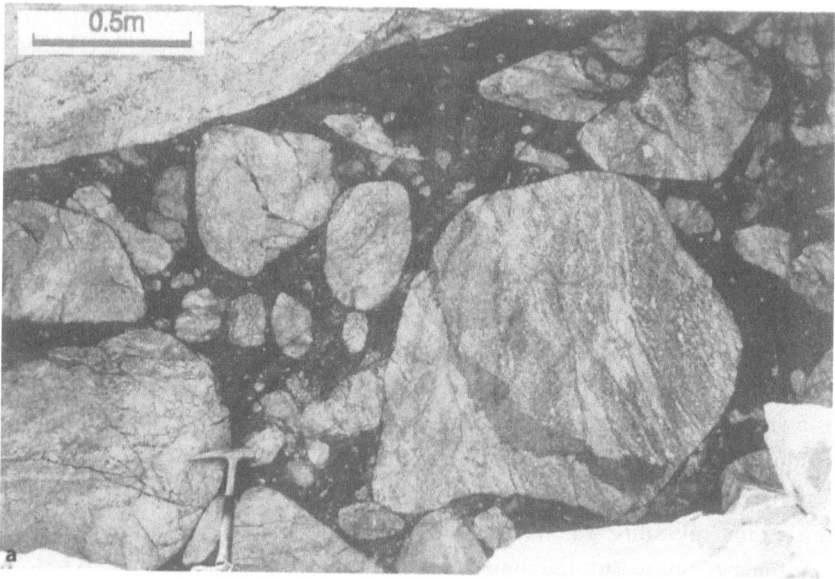

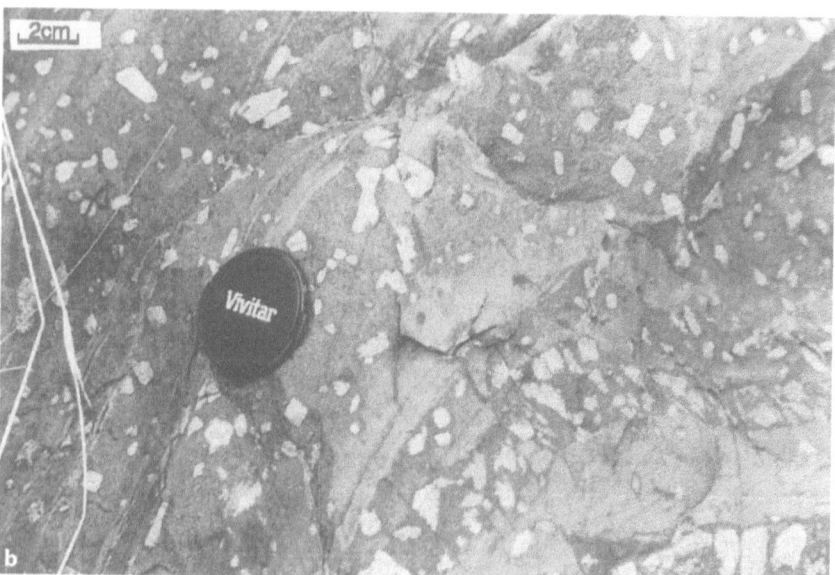

Fig. 7. (a) Pseudotachylitic breccia dyke from N of Parys, showing large clasts of Archean basement gneiss set in a dark, fine-grained matrix. (b) Intense microfaulting in lower Witwatersrand Supergroup metapelite in the northwestern collar. The microfaults truncate and offset bedding and andalusite porphyroblasts (right of lens cap), which grew during the earlier, pre-impact, metamorphic event, but are themselves overgrown by the post-impact paragenesis (see Figs. 8b,c). Discrete lenses and veins of pseudotachylite, commonly <1 mm wide, occur along the microfaults.

However, no attempts were made to use this technique to quantify the metamorphic conditions in the rest of the dome. In contrast to Grieve et al. (1990), Stepto (1990) and Hart et al. (1991) suggested that the post-impact thermal effects were highly localized in specific areas within the core of the dome. These authors envisaged either small mafic intrusions or highly variable impact shock intensities as the cause of this localization of thermal effects. However, studies of pseudotachylites and metapelitic mineral assemblages, both in the core and collar of the dome and in the gold-fields along the margins of the Witwatersrand Basin, enabled Gibson et al. (1997a 1998) and Stevens et al. (1997) to distinguish lower-pressure post-impact metamorphic parageneses from the older, higher-pressure, pre-impact parageneses and to use these to constrain the post-impact P-T conditions (Figs. 8a-c; Table 1).

Based on the variation in post-impact parageneses, Gibson et al. (1998) showed that the post-impact temperature decreases radially outwards, from >700 °C in the center of the dome to ~300 °C some 25 km from the center (Table 1). Temperatures were sufficiently high in the center of the dome, and were sustained for long enough, that a small body of coarse-grained post-impact granite, dated at 2017 ± 5 Ma (SHRIMP U-Pb single-grain zircon), was able to crystallize (Gibson et al. 1997b). The elevated temperatures also allowed the growth of authigenic zircon in the matrix of a breccia which Kamo et al. (1996) were able to date at 2023 ± 4 Ma (by the U-Pb single-zircon TIMS technique). Although Kamo et al. (1996) described the breccia as a pseudotachylite vein that crystallized under granulite facies conditions without quenching, Ashley et al. (1999) have not been able to rule out the possibility that it is a partial melt of tonalitic-dioritic composition produced by exhumation-induced decompression of rocks under granulite facies metamorphic conditions during formation of the central uplift. Such melting is well known in high-grade metamorphic terrains and should not be confused with shock decompression. The granulite and amphibolite facies textures shown in Fig. 8 are fully consistent with this decompressive P-T path following the shock event.

In addition to providing temperature estimates, the granulite-to-greenschist-facies assemblages allowed the estimation of post-impact metamorphic pressures and, consequently, the level of exhumation of the Vredefort impact structure. These pressures, of ~0.2-0.3 GPa, indicate that the structure has been exhumed by between 7 and 10 km since it formed, a considerably higher estimate than the previous maximum of 2.5 km postulated by Rondot (1994). Comparison of the post-impact parageneses with the higher-pressure pre-impact assemblages for the rocks in the innermost collar of the dome has also placed minimum constraints on the amount of uplift that occurred during formation of the impact structure. This figure of ~4 km is lower than the estimate obtained from displacement of the base of the supracrustal succession between the Potchefstroom Syncline and the dome, however, it is not certain that this was an isobaric surface at the time of impact.

Table 1. Post-impact metamorphic parageneses and P-T conditions in the Vredefort Dome and Witwatersrand Basin

Distance from center (km)	Post-impact paragenesis	P (GPa)	T (°C)	Reference
0	Cpx + hbl +pl + ksp ± op ± bt	-	± 850	Schreyer (1983)
±12	CO_2-H_2O fluid inclusions	~0.2-0.5	± 700	Fricke et al. (1990)
12	Crd + opx ± sp	0.27 ± 0.04	± 700	Gibson et al. (1998)
22	Crd + bt	0.30 ± 0.04	500-525	Gibson et al. (1998)
25	Bt + chl + ms	-	± 400	(Gibson et al. (1998)
30	Chl + ms	-	300 ± 10	S. Foya (pers. comm. 1999)
80-100	Chl + ms	-	300 ± 10	Frimmel (1997)

Bt – biotite; chl – chlorite; cpx – clinopyroxene; crd – cordierite; hbl –hornblende; ksp – alkali feldspar; ms – muscovite; op – opaque oxide; pl – plagioclase; sp – spinel

Fig. 8. (opposite) Photomicrographs indicating the variation in the post-impact metamorphic temperatures between the core and collar of the Vredefort Dome, as reflected in the silicate mineral parageneses. **a.** Pre-impact garnet porphyroblast partially replaced by orthopyroxene-cordierite symplectite and cut by pseudotachylite vein (from side to side, in about the middle of the photomicrograph) in Archean pelitic granulite from ca. 12 km from the center of the dome. Note the reaction between the pseudotachylite matrix and quartz clasts, which has produced blocky orthopyroxene (arrowed) similar to that observed between the garnet porphyroblast and matrix quartz. Plane polarized light. Scale bar = 200 μm. **b.** Backscatter electron image of pseudotachylite vein from same sample as (a) showing recrystallization of the matrix to orthopyroxene (light grey spots) + cordierite (light grey mass, left centre) + plagioclase (dark grey vein centre), with a K-feldspar rim (medium grey) against the wallrock quartz (black). Note development of blocky orthopyroxene around quartz clasts (arrowed) and against wallrock quartz (bottom). **c.** Microfault in lower Witwatersrand Supergroup metapelite (22km from the center of the dome) dextrally displacing pre-impact staurolite porphyroblast (high relief mineral, arrows) but which has been overgrown and garnet (grt) from the pre-impact metamorphic assemblage. Plane polarized light. Scale bar = 500 μm.

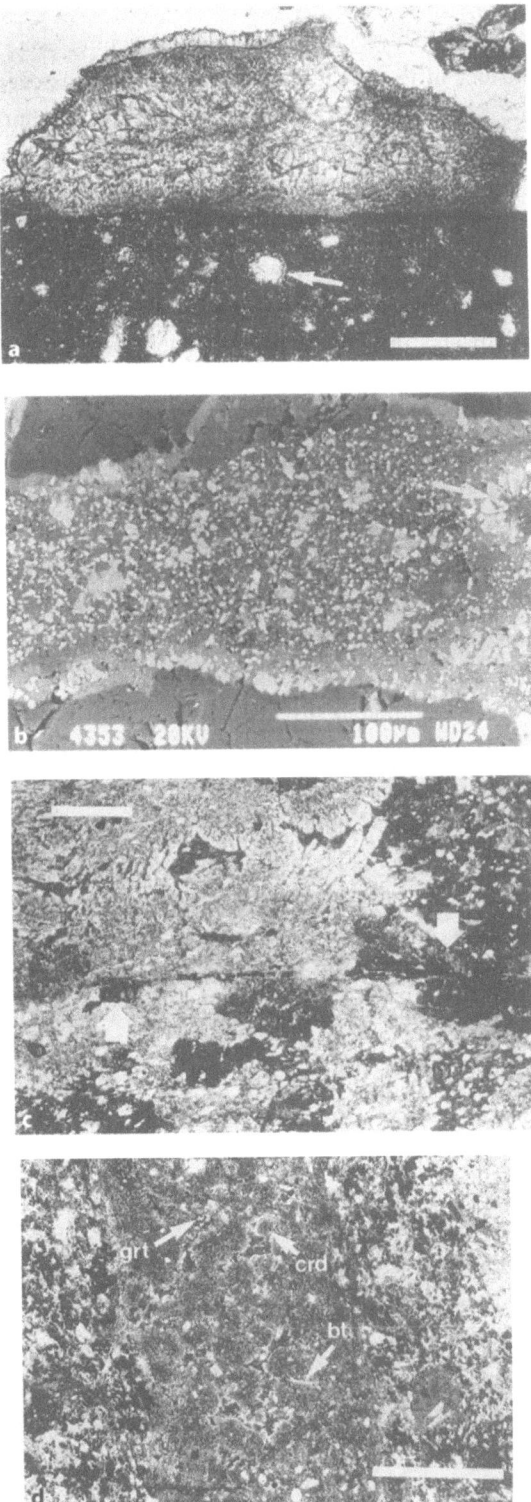

The use of pseudotachylites and related brittle structures as a time-line in the P-T-deformation history of the Vredefort impact structure has been extended to the margins of the Witwatersrand Basin. Several recent mineralogical and geochronological studies have established that these rocks experienced a lower greenschist facies hydrothermal overprint (T ~ 300 °C) slightly after, but otherwise penecontemporaneous with, the pseudotachylite-forming brecciation event (e.g., Trieloff et al. 1994; Frimmel 1997; Zhao et al. 1999; Reimold et al. 1999). Evidence that gold was mobile within the reef horizons during this event (Frimmel 1997) indicates the importance of understanding the precise nature of these post-impact thermal effects.

Based on the experimental results of Raikes and Ahrens (1979) and the maximum shock pressure estimates inferred by Grieve et al. (1990) from PDF orientations in quartz in the Vredefort Dome, it seems likely that rocks in the core and inner parts of the collar of the dome experienced a post-impact rise in temperature, due to shock heating, of between 150 and 250 °C. If, as suggested by metamorphic assemblages in the lower Witwatersrand Supergroup, the base of the supracrustal sequence was buried at a depth of ~14 km immediately prior to impact, the post-impact metamorphism could have been achieved by adding this shock heat to rocks pre-heated along a normal to slightly elevated continental geotherm (~20-25 °C/km). The roughly uniform ~300 °C temperature experienced by rocks between 25 and 100 km from the center of the dome reflects both the absence of a significant shock heating effect and the lower amounts of uplift in the outer parts of the impact structure. It is possible that a small component of heat for the post-impact metamorphism may have been derived from the now-eroded impact melt body (Fig. 3). However, as shown by Dressler (1984) for the Sudbury structure, significant contact metamorphic effects beneath such bodies may be extremely restricted due to rapid cooling of the clast-laden melt breccia and because most heat is lost upwards.

Recognition of Deeply Eroded Impact Structures – The Way Forward

Given the thousands of craters on the Moon, it is obvious that the current record of ~160 known impact structures on Earth cannot possibly reflect the number of bolide impacts experienced by our planet, even if allowance is made for the destruction of many of them by geological processes. In this regard, the accelerated rate of discovery of new impact structures in the last two decades is encouraging (Grieve 1998), but it is also clear that almost all craters discovered thus far have experienced comparatively little erosion and that initial identification has been based primarily on circular surface morphological features and/or the presence of impact melt rock and breccias. The prolonged debate concerning the origin of the Vredefort Dome and, by extension, many of the features within its rocks and those of the surrounding region (see review in Reimold and Gibson 1996), provides some indication of the problems likely to be faced in future attempts to identify other deeply eroded impact structures. However, together with Vredefort, such exhumed impact structures have a crucial role to play in expanding our understanding of the deep-level processes operating during and immediately after an impact event, as well as the geological structure of root zones of large, complex impact structures.

Perhaps the most striking aspect of the Vredefort impact structure is that, despite between 7 and 10 km of erosion, a strongly circular geological feature – the Vredefort Dome – remains (Figs. 1, 2). Although several workers have attempted to explain the dome in terms of polyphase tectonics or ballooning intrusions (see review in Reimold and Gibson 1996), its geometry and the scale of structural uplift documented by seismic studies (Durrheim et al. 1991) cannot be explained by such conventional geological processes, particularly in the light of the regional geological history. Nonetheless, the irregular shape of the core-collar contact in the southeast of the dome (Fig. 1) and the anomalous overturned dips of in turn by radiating aggregates of cordierite (white pseudohexagonal crystals with biotite-rich cores) related to the post-impact thermal metamorphism. Plane polarized light. Width of field of view: 700 μm. (D) Metapelite-hosted pseudotachylite (pt) in lower Witwatersrand Supergroup recrystallized to a cordierite ɪ biotitc paragenesis. The cordierite "porphyroblasts" comprise 100-250 μm diameter aggregates of radiating crystals, with dark cores rich in <10 μm long biotite inclusions, and inclusion-free rims. The breccia contains clasts of quartz, biotite (bt) the Witwatersrand Supergroup strata in large parts of the collar (Fig. 2) provide a timely reminder that a perfect circular geological geometry will result only where the target layering was originally horizontal and undisturbed by faults (not even considering the likely effects of post-impact tectonic overprinting). It remains unclear whether the main NE-SW trending bulk of the Witwatersrand Basin is, itself, a distorted circular feature (Henkel and Reimold 1998) or whether the elliptical shape is, in part, due to an elongate pre-impact basin onto which the Vredefort structure has been imposed. Certainly, the paucity of pseudotachylite in the northeastern (Johannesburg) and southwestern (Welkom) gold-fields (Fig. 1) may suggest that, at the time of impact, these areas were further from the impact center than the northern and northwestern (Carletonville and Klerksdorp) gold-fields.

A second encouraging aspect of the geology of the Vredefort Dome is that several dykes of impact melt penetrated at least several kilometers beneath the crater floor. This reinforces the patterns observed in the Sudbury Structure where dyke offshoots from the remnant of the main melt body are common (Dressler 1984); however, the smaller size of the Vredefort dykes indicates that such features may not extend much further below the present levels of exposure.

Ongoing studies of the regional structure of the Witwatersrand Basin using the wealth of seismic data available from mining companies have recently begun to identify major impact-related faults in the rim syncline around the Vredefort Dome and to relate these to discrete phases of movement during the cratering event (Brink et al. 1997; Henkel and Reimold 1998; Ellis and Reimold 1999). In contrast, although some large faults do disrupt the strata in the dome, the absence of evidence of *major* (kilometer-scale) syn-doming movement and the strong lateral continuity of the strata in the collar of the dome suggest a remarkable coherence of the rocks in the central uplift. This is supported by the absence of evidence of any significant structural disruption in the core of the dome. At the same time, the intensity of small pseudotachylite-bearing microfaults in the collar strata (Fig. 7b) and dome-wide abundance of brittle deformation in the form of multiply-striated joint sets (Nicolaysen and Reimold 1999) may explain how such large-scale coherence can be retained during such a high strain rate event. It remains to be seen whether evidence can be found of Spray's (1997) postulated "superfaults" or whether some other mechanism is required to explain the voluminous pseudotachylites in the core, as well as the abundant pseudotachylitic breccias in the wider environs of the dome. In short, the central uplift of the Vredefort impact structure shows, as yet, no evidence of being a megabreccia, despite the obvious space problems that must have accompanied its formation.

From Fig. 3, it is apparent that one of the biggest challenges presented by deeply eroded impact structures is the estimation of their original size. In the case of Vredefort, estimates of the original crater diameter have ranged from 60 to 300 km, with most current estimates in the 200-300 km range (see review in Therriault et al. 1997b; Henkel and Reimold 1998). In the Vredefort structure, shock features are largely confined to the central uplift which, according to Melosh (1989), typically accounts for less than half the final crater diameter (Fig. 3). The restricted distribution of high-pressure quartz polymorphs shown in Fig. 2 likely reflects the fact that, with increasing depth, the volume of rock affected by sufficiently high shock pressures to produce the polymorphs decreases, whereas the volume of rock affected by significant post-impact thermal metamorphism increases (Fig. 3). On the evidence from the dome, it also appears that planar deformation features in quartz – the most widely used feature for constraining shock pressures – are less reliable at deeper structural levels due to post-impact annealing (see also Huffman and Reimold 1996). This is well illustrated by the decreased abundance of PDFs in quartz towards the center of the dome (Grieve et al. 1990), a pattern which can be directly linked to the post-impact temperature distribution pattern (Gibson et al. 1998). A second feature of these PDFs in quartz from rocks that experienced subsequent moderate to high post-impact temperatures is that their use as a barometer consistently underestimates shock pressures compared with quartz polymorphs and PDFs in minerals such as zircon. A similar annealing mechanism may explain the lack of diaplectic glass in minerals such as feldspar which, according to experimental predictions, should be present in the rocks in the dome.

Shatter cones are the shock feature distributed farthest from the center of the Vredefort Dome, but the varied geology exposed in the dome clearly illustrates the lithological control on their distribution – they are not as extensively developed in the generally medium- to coarse-grained crystalline rocks in the core of the dome as in the fine-grained argillitic strata of the collar. As large parts of the continental crust are crystalline, the most obvious outcrop- or larger-scale evidence for an impact structure, in the form of shatter cones or even concentrically arranged strata, may be lacking in many impact structures.

In contrast to the restricted distribution of the shock-diagnostic features, non-diagnostic features such as pseudotachylites, cataclasites, and brittle faults may be abundant and occur over an area even wider than the extent of the original crater (Fig. 3). However, being brittle features, they may show no consistent concentric or radial pattern that indicates a circular feature of possible impact origin. Exceptionally large volumes of pseudotachylite or pseudotachylitic breccias – both individually and cumulatively – may, nonetheless, be a good preliminary sign that an impact structure may be present.

In contrast to the problems that the post-impact thermal effects pose for identification of shock microdeformation features, the facts that the rocks in the central uplift experience (a) an instantaneous increase in temperature by up to several hundred degrees, (b) virtually instantaneous uplift, and (c) an ensuing sufficiently slow cooling period to allow something approaching chemical equilibrium to be reached under the new P-T conditions will create distinctive medium- to high-grade decompression textures in rocks of appropriate (pelitic or mafic) bulk compositions. These parageneses should indicate growth in a hydrostatic stress field but, significantly, they should overprint the brittle deformation features (faults, fractures, and pseudotachylite) caused by the cratering event. While extremely voluminous pseudotachylites might themselves be an indicator of an unusual origin for the uplift structure, their overprinting by a high-temperature static thermal metamorphic event that decreases in intensity from the center to the margins of the uplift could be regarded as conclusive if the metamorphism cannot be directly linked to an intrusive heat source. The post-impact metamorphism in the crater basement beyond the central uplift will not produce decompression textures and is likely to be restricted to low-temperature hydrothermal effects.

The extensive growth of new minerals during the post-impact metamorphism and associated hydrothermal activity may contribute towards overcoming one of the most common problems of impact research – establishing the age of impact. In the case of the Vredefort structure, rocks over a 100 km radius have produced remarkably similar ages, with the greatest precision being attained in the central parts of the structure where post-impact metamorphic conditions were conducive to the production and slow crystallization of melts. On a cautionary note, however, the slower cooling rate experienced by the deep levels of a large impact structure such as Vredefort means that geochronological approaches that utilize minerals with low closure temperatures may produce ages that are several millions of years younger than the actual impact event.

Conclusions

The Vredefort impact structure is the only currently known extremely large impact structure in which erosion has removed the crater fill, allowing investigation of the deep levels of the crater basement. This investigation has shown that, at such deep levels (7-10 km), shock metamorphic features are restricted to the central structural uplift, but that even within the central uplift their distribution may be irregular. This is due, in part, to the variable nature of the target lithologies, but is also a consequence of the increased importance of the post-impact thermal history of the crater basement at deep structural levels. The Vredefort rocks show that post-impact temperatures of up to 700-900 °C can be achieved through shock heating of rocks pre-heated along a crustal geotherm. While these high-temperature metamorphic effects fall off drastically away from the center of the structure, relatively low-temperature hydrothermal effects extend over much, if not all, of the crater basement. Despite the complications which the thermal effects introduce into studies of shock microdeformation features, they add a new dimension to cratering studies by providing critical information, through new metamorphic mineral parageneses, of the amount of impact-induced uplift, the amount of subsequent exhumation, and the thermal budget of the deep crust during such events. The sustained high temperatures also facilitate the crystallization of minerals suitable for precise dating of the impact event.

Acknowledgements

Financial support from the National Research Foundation of South Africa (Grant Number GUN2034547) and the University of the Witwatersrand is gratefully acknowledged. Burkhard Dressler and Alex Deutsch provided helpful reviews. Lyn Whitfield drafted the diagrams and Henja Czekanowska assisted with photography. University of the Witwatersrand Impact Cratering Research Group Contribution # 10.

References

Akaogi M, Navrotsky A (1984) The quartz-coesite-stishovite transformations: New calorimetric measurements and calculation of phase diagrams. J Phys Earth Plan Int 36: 124-134

Allsopp HL, Fitch FJ, Miller JA, Reimold WU (1991) $^{40}Ar/^{39}Ar$ stepheating age determinations relevant to the formation of the Vredefort Dome, South Africa. S Afr J Sci 87: 431-442

Armstrong, RA, Compston W, Retief EA, Williams LS, Welke HJ (1991) Zircon ion microprobe studies bearing on the age and evolution of the Witwatersrand basin. Precambr Res 53: 243-266

Ashley AJ, Gibson RL, Koeberl C, Reimold WU, Greshake A (1999) A new type of melt rock and first evidence of shock deformation in plagioclase from the Vredefort impact structure, South Africa. Meteor Planet Sci 34 (Suppl): A9-A10

Bohor BF, Betterton WJ, Krogh TE (1993) Impact-shocked zircon: discussion of shock-induced textures reflecting increasing degrees of shock metamorphism. Earth Plan Sci Lett 119: 419-424

Brink MC, Waanders FC, Bisschoff AA (1997) Vredefort: A model for the anatomy of an astrobleme. Tectonophys 270: 83-114

Carter NL (1965) Basal quartz deformation lamellae - a criterion for recognition of impactites. Am J Sci 263: 786-806

Deutsch A, Grieve RAF, Avermann M, Bischoff L, Brockmeyer P, Buhl D, Lakomy R, Müller-Mohr V, Ostermann M, Stöffler D (1995) The Sudbury Structure (Ontario, Canada): a tectonically deformed multi-ring impact basin. Geol Rdsch 84: 697-709

Dietz RS (1961) Vredefort Ring structure: meteorite impact scar? J Geol 69: 499-516

Dressler BO (1984) The effects of the Sudbury event and the intrusion of the Sudbury Igneous Complex on the footwall rocks of the Sudbury Structure. In: Pye EG, Naldrett AJ, Giblin PE (eds) The geology and ore deposits of the Sudbury Structure. Ontario Geol Survey, Sudbury, Spec Vol 1, pp 97-136

Durrheim RB, Nicolaysen LO, Corner B (1991) A deep seismic reflection profile across the Archean-Proterozoic Witwatersrand Basin, South Africa. Am Geophys Union Geodyn Ser 22: 213-224

Fletcher P, Reimold WU (1989) Some notes and speculations on the pseudotachylites in the Witwatersrand basin and the Vredefort dome. S Afr J Geol 92: 223-234

French BM (1998) Traces of Catastrophe: A handbook of shock-metamorphic effects in terrestrial meteorite impact structures. LPI Contribution No. 954, Lunar and Planetary Institute, Houston. 120 pp

French BM, Nielsen RL (1968) Vredefort Bronzite Granophyre: chemical evidence for origin as a meteorite impact melt. Tectonophys 171: 119-138

French BM, Orth CJ, Quintana LR (1989) Iridium in the Vredefort Bronzite Granophyre: impact melting and limits on a possible extraterrestrial component. Proc 19th Lunar Planet Sci Conf, Cambridge Univ. Press, pp 733-744

Fricke A, Medenbach O, Schreyer W (1990) Fluid inclusions, planar elements and pseudotachylites in the basement rocks of the Vredefort structure, South Africa. Tectonophys 171: 169-183

Frimmel HE (1997) Chlorite thermometry in the Witwatersrand Basin: Constraints on the Paleoproterozoic geotherm in the Kaapvaal Craton, South Africa. J Geol 105: 601-615

Geology Forum (1997) Discussion of paper by Sharpton VL and Dressler BO (1996) New constraints on the Slate Islands impact structure: Comments and Reply. Geology 26: 666-669

Gibson RL, Armstrong RA, Reimold WU (1997b) The age and thermal evolution of the Vredefort impact structure: a single-grain U-Pb zircon study: Geochim Cosmochim Acta 61: 1531-1540

Gibson RL, Reimold WU, Wallmach T (1997a) Origin of pseudotachylite in the lower Witwatersrand Supergroup, Vredefort Dome (South Africa): Constraints from metamorphic studies. Tectonophys 283: 241-262

Gibson RL, Reimold WU, Stevens G (1998) Thermal-metamorphic signature of an impact event in the Vredefort Dome, South Africa. Geology 26: 787-790

Gibson RL, Courtnage PM, Charlesworth EG (1999) Bedding-parallel shearing and related deformation in the lower Transvaal Supergroup north of the Johannesburg Dome, South Africa. S Afr J Geol 102, in press *(Editors: available in November 1999)*

Grieve RAF (1987) Terrestrial impact structures. Ann Rev Earth Plan Sci 15: 245-270

Grieve RAF (1998) Extraterrestrial impacts on earth: the evidence and the consequences. In: Grady MM, Hutchison R, McCall GJH, Rothery DA (eds) Meteorites: Flux with Time and Impact Effects. Geol Soc, London, Spec Publ 140, pp 105-131

Grieve RAF, Coderre JM, Robertson PB, Alexopoulos J (1990) Microscopic planar deformation features in quartz of the Vredefort structure: Anomalous but still suggestive of an impact origin. Tectonophys 171: 185-200

Hargraves RB (1961) Shatter cones in the rocks of the Vredefort Ring. Trans Geol Soc S Afr 64: 147-154

Hart RJ, Andreoli MAG, Reimold WU, Tredoux M (1991) Aspects of the dynamic and thermal metamorphic history of the Vredefort cryptoexplosion structure: implications for its origin. Tectonophys 192: 313-331

Henkel H, Reimold WU (1998) Integrated geophysical modelling of a giant, complex impact structure: anatomy of the Vredefort Structure, South Africa. Tectonophys 287: 1-20

Huffman AR, Reimold WU (1996) Experimental constraints on shock-induced microstructures in naturally deformed silicates. Tectonophys (N.L. Carter Vol) 256: 165-217

Kamo SL, Reimold WU, Krogh TE, Colliston WP (1996) A 2.023 Ga age for the Vredefort impact event and a first report of shock metamorphosed zircons in pseudotachylitic breccias and Granophyre. Earth Planet Sci Lett 144: 369-388

Killick AM (1992) Pseudotachylytes of the West Rand Goldfield, Witwatersrand Basin, South Africa. Unpubl PhD thesis, Rand Afrikaans Univ, Johannesburg, 273 pp

Koeberl C, Reimold WU, Shirey SB (1996) A Re-Os isotope study of the Vredefort granophyre: Clues to the origin of the Vredefort structure, South Africa. Geology 24: 913-916

Krogh TE, Davis DW, Corfu F (1984) Precise U-Pb zircon and baddeleyite ages of the Sudbury Structure. In: Pye EG, Naldrett AJ, Giblin PE (eds) The geology and ore deposits of the Sudbury Structure: Ontario Geol Survey, Sudbury, Spec Vol 1, pp 431-446

Leroux H, Reimold WU, Doukhan JC (1994) A TEM investigation of shock metamorphism in quartz from the Vredefort dome, South Africa. Tectonophys 230: 223-239

Leroux H, Reimold WU, Koeberl C, Hornemann U (1999) Experimental shock deformation in zircon: a transmission electron microscopic study. Earth Plan Sci Lett 169: 291-301

Martini JEJ (1978) Coesite and stishovite in the Vredefort Dome, South Africa. Nature 272: 715-717

Martini JEJ (1991) The nature, distribution and genesis of the coesite and stishovite associated with pseudotachylite of the Vredefort Dome, South Africa. Earth Planet Sci Lett 103: 285-300

Martini JEJ (1992) The metamorphic history of the Vredefort dome at approximately 2 Ga as revealed by coesite-stishovite-bearing pseudotachylites. J Metam Geol 10: 517-527

McCarthy TS, Charlesworth EG, Stanistreet IG (1986). Post-Transvaal structural features of the northern portion of the Witwatersrand Basin. Trans Geol Soc S Afr 89: 311-324

McCarville P, Crossey LJ (1996) Post-impact hydrothermal alteration of the Manson impact structure, In: Koeberl C, Anderson RR (eds) The Manson Impact Structure, Iowa: Anatomy of an Impact Crater. Geol Soc Amer, Boulder, CO, Spec Pap 302, pp 347-376

Melosh HJ (1989) Impact Cratering: A geologic process. Oxford Mon Geol Geophys 11, 245 pp

Melosh HJ, Ivanov B (1999) Impact crater collapse. Ann Rev Earth Planet Sci 27: 385-415.

Morgan J, Warner M (1999) Chicxulub: The third dimension of a multi-ring impact basin. Geology 27: 407-410.

Moser DE, (1997) Dating the shock wave and thermal imprint of the giant Vredefort impact, South Africa. Geology 25: 7-10

Nicolaysen LO, Reimold WU (1999) Vredefort shatter cones revisited. J Geophys Res 104: 4911-4930

Raikes SA, Ahrens TS (1979) Post-shock temperatures in minerals. Geophys J R Astron Soc 58: 717-747

Reimold WU (1991) The geochemistry of pseudotachylites from the Vredefort dome, South Africa. N Jahrb Mineral Abh 162: 151-184

Reimold WU (1995) Pseudotachylite – Generation by friction melting and shock brecciation? – A review and discussion. Earth-Science Rev 39: 247-264

Reimold WU (1998) Exogenic and endogenic breccias: a discussion of major problematics. Earth-Science Rev 43: 25-47

Reimold WU, Colliston WP, (1994) Pseudotachylites of the Vredefort Dome and the surrounding Witwatersrand Basin, South Africa. In: Dressler BO, Grieve RAF, Sharpton VL (eds) Large Meteorite Impacts and Planetary Evolution. Geol Soc Amer, Boulder, CO, Spec Pap 293, pp 177-196

Reimold WU, Gibson RL (1996) Geology and evolution of the Vredefort Impact Structure, South Africa. J Afr Earth Sci 23: 125-162

Reimold WU, Reid AM, Horsch M, Durrheim RJ (1990) The 'Bronzite'-Granophyre from the Vredefort Structure - a detailed analytical study and reflections on the origin of one of Vredefort's enigmas. Proc 20th Lunar Planet Sci Conf Lunar Planet Inst, Houston, pp 433-450

Reimold WU, Colliston WP, Wallmach T (1992) Comment on "Nature, provenance and distribution of coesite and stishovite in the Vredefort structure" by JEJ Martini. Earth Plan Sci Lett 112: 213-217

Reimold WU, Koeberl C, Fletcher P, Killick AM, Wilson JD (1999) Pseudotachylitic breccias from fault zones in the Witwatersrand Basin, South Africa: Evidence of autometasomatism and post-brecciation alteration processes. Miner Petrol 66: 25-53

Roddy DJ, Davis LK (1977) Shatter cones formed in large-scale experimental explosion craters. In: Roddy DJ Pepin RO, Merrill RB (eds) Impact and Explosion Cratering. Pergamon Press, New York, pp 715-750

Rondot J (1994) Recognition of eroded astroblemes. Earth-Science Rev 35: 331-365

Schreyer W (1983) Metamorphism and fluid inclusions in the basement of the Vredefort Dome, South Africa: guidelines to the origin of the structure. J Petrol 24: 26-47

Schreyer W, Abraham K (1978) Symplectitic cordierite-orthopyroxene-garnet assemblages as products of contact metamorphism of pre-existing basement granulites in the Vredefort structure, South Africa. Contrib Mineral Petrol 68: 53-62

Sharpton VL, Marin LE, Carney C, Lee S, Ryder G., Schuraytz BC, Sikora P, Spudis PD (1996a) A model of the Chicxulub impact basin based on evaluation of geophysical data, well logs and drill core samples. In: Sharpton VL, Ward PD (eds), Global catastrophes in Earth history: An interdisciplinary conference on impacts, volcanism, and mass mortality. Geol Soc Amer, Boulder, CO, Spec Pap 247, pp 55-74

Sharpton VL, Dressler BO, Herrick RR, Schnieders B, Scott J (1996b) New constraints on the Slate Islands impact structure, Ontario, Canada. Geology 24: 851-854

Simpson C (1978) The structure of the rim synclinorium of the Vredefort Dome. Trans Geol Soc S Afr 81: 115-121

Snyder DB, Hobbs RW (1999) Ringed structural zones with deep roots formed by the Chicxulub impact. J Geophys Res 104:10743-10755

Spray JG (1995) Pseudotachylyte controversy: Fact or friction? Geology 23: 1119-1122

Spray JG (1997) Superfaults. Geology 25: 579-582

Spray JG, Kelley SP, Reimold WU (1995) Laser-probe ^{40}Ar-^{39}Ar dating of pseudotachylytes and the age of the Vredefort impact event. Meteoritics 30: 335-343

Stepto D (1990) The geology and gravity field in the central core of the Vredefort structure. Tectonophys 171: 75-103

Stevens G, Gibson RL, Droop GTR (1997) Mid-crustal granulite facies metamorphism in the Central Kaapvaal Craton: The Bushveld Complex connection. Precambr Res 61: 113-132

Stöffler D, Langenhorst F (1994) Shock metamorphism of quartz in nature and experiment: I. Basic observation and theory. Meteoritics 29: 155-181

Therriault AM, Reimold WU, Reid AM (1996) The Vredefort granophyre: Part 1. Field studies. S Afr J Geol 99: 1-21

Therriault AM, Reimold WU, Reid AM (1997a) Geochemistry and impact origin of the Vredefort Granophyre. S Afr J Geol 100: 115-122

Therriault AM, Grieve RF, Reimold WU (1997b) Original size of the Vredefort Structure: implications for the geological evolution of the Witwatersrand Basin. Meteor Planet Sci 32: 71-77

Trieloff M, Reimold WU, Kunz J, Boer RH, Jessberger EK (1994) ^{40}Ar-^{39}Ar thermochronology of pseudotachylites at the Ventersdorp Contact Reef, Witwatersrand Basin. S Afr J Geol 97: 365-384

Walraven F (1997) Geochronology of the Rooiberg Group, Transvaal Supergroup, South Africa. Econ Geol Res Unit Inf Circ 316, Univ of the Witwatersrand, Johannesburg, 27 pp

Walraven F, Elsenbroek JH (1991) Geochronology of the Schurwedraai Alkali Granite and associated nepheline syenite and implications for the origin of the Vredefort Structure. S Afr J Geol 94: 228-235

Walraven F, Armstrong RA, Kruger FJ (1990) A chronostratigraphic framework for the north-central Kaapvaal Craton, the Bushveld Complex and the Vredefort structure. Tectonophys 171: 23-48

White JC (1993) Shock-induced melting and silica polymorph formation, Vredefort Structure, South Africa. In: Boland JN, FitzGerald JG (eds) Defects and Processes in the Solid State: Geoscience Applications. The McLaren Volume. Elsevier Sci Publ, Amsterdam, pp 69-84

Wilshire HG (1971) Pseudotachylite from the Vredefort Ring, South Africa. J Geol 79: 195-206

Zhao B, Caluer N, Robb LJ, Zwingmann H, Toulkeridis T, Meyer FM (1999) K-Ar dating of white micas from the Ventersdorp Contact Reef of the Witwatersrand Bsin, South Africa: Timing of post-depositional alteration. Miner Petrol 66: 149-170

11 Identification of Ancient Impact Structures: Low-Angle Faults and Related Geological Features of Crater Basements

T. Kenkmann[1], B. A. Ivanov[2] and D. Stöffler[1]

[1]Museum für Naturkunde, Institut für Mineralogie, Humboldt-Universität Berlin, Invalidenstraße 43, D-10115 Berlin, Germany.
[2]Institute for Dynamics of Geospheres, Russian Academy of Science, Moscow, Russia 117939.

Abstract. Ancient impact craters are commonly deeply eroded, metamorphosed and/or deformed by later tectonics. The identification of such impact structures using microstructural or mineralogical criteria are very difficult to apply under such conditions. It is proposed that fault patterns in the crater basement can be used diagnostically in eroded structures. On the basis of field observations in the Rochechouart crater, France, and numerical computations, we suggest the existence of low-angle faults dipping gently towards the crater centre. In craters with final diameters of 20-30 km, low-angle normal faults might have a lateral extent of approximately 5 km and predominantly occur at a distance of approximately 3-8 km away from the crater centre. Our numerical model suggests that they formed at a depth of 0.5 to 4 km below the original crater surface. Low-angle normal faulting occurs during crater modification when the steep rim of the transient crater cavity starts to collapse. Gravity-driven sliding occurs in acoustically-fluidized rocks with low cohesion and reduced friction coefficients. It is suggested that the degree of fluidization changes with depth and distance from the impact centre. Faulting along gently dipping shear zones may be the result of an impact-induced rheological stratification of the crater floor and a passive rotation of the shear zones due to the uplift of the central peak. Rapid and unconstrained single slip events locally lead to frictional melting and the formation of pseudotachylites. Field observations are presented that strongly support the idea of block oscillation during crater modification.

Introduction

The identification of early Proterozoic and Archean impact craters is important in assessing the role of impact processes during the earth's early evolution. Because of their metastability, the commonly applied diagnostic criteria for impact structures such as shock effects in minerals, impact glasses etc., will no longer be applicable for identifying ancient impact sites. In addition, the originally circular crater morphology is expected to be lost due to the effects of erosive, tectonic, and metamorphic processes in the earth's crust. It is therefore most likely that the upper parts of an ancient impact structure are lost. In such cases, the diagnosis of an impact site has to rely on rocks and structural features characteristic of the crater basement. Among these are pseudotachylites, dykes with intrusive melts, breccia dykes and brecciated rock units (Stöffler et al. 1988). The structural features related to these rock types are fault systems originating from the modification of the transient crater cavity.

The formation of complex craters on the earth and other terrestrial planets is induced by a collapse of a parabolically-shaped transient crater cavity that formed after the hypervelocity impact of an asteroid or a comet. Interplanetary comparisons of the simple-to-complex crater transition prove that gravity is the principal force driving the collapse (Melosh 1989). It is achieved mainly by the uplift of the rocks underlying the crater's centre, while rock units near the rim slump downward and inward. Strength properties of conventional rock and debris cannot explain the collapse of the transient crater cavity. It is well-known that the strength properties of rocks beneath a collapsing crater are strongly modified by the shock wave with respect to unaffected rocks. Estimates of rock cohesion and angle of internal friction are in the order of 3 MPa and 0-5°, respectively (Melosh 1989; McKinnon 1978). However, rocks do not deform like a plastic continuum. Deformation is strongly localized and occurs along discrete fault zones. The distribution, geometry and kinematic history of fault zones exposed in the crater floor of complex impact craters is not understood very well and only a few, recent investigations have addressed this problem (e.g., Spray 1997 1998). In extraterrestrial craters fault zones and slumps are indicated by headscarps separating individual terraces. The morphology of headscarps suggests that they represent the surface of listric normal faults dipping crater inward (Spray 1997). The morphometrical analysis of terraced complex craters allows us to estimate the amount of horizontal slip of slumpings (Leith and McKinnon 1991). Terraces within an individual crater are narrow close to the crater centre, but increase in width as the rim is approached. The widest, best-defined, and last-formed terrace normally occurs just below the crater rim (Pearce and Melosh 1986; Leith and

McKinnon 1991). Morphological characteristics of extraterrestrial craters are helpful for deducing the mechanisms of near-surface faulting. However, the investigation of deeper structures rests upon observations and interpretations of terrestrial craters since geological cross-sections are not available for extraterrestrial craters.

Because of their specific origin, fault patterns in impact craters may be unique when they are compared to tectonically-induced fault patterns. We believe, therefore, that structural studies of faulting in the basement of well-preserved craters could provide a key information for recognizing ancient impact sites located in Archean and Proterozoic terranes. Considering the very high impact flux during the Archean it is to be expected that the ancient fault patterns, which are induced by the large and frequent Archean and early Proterozoic impacts (multiring basins), will be preserved in the ancient cratons of the earth.

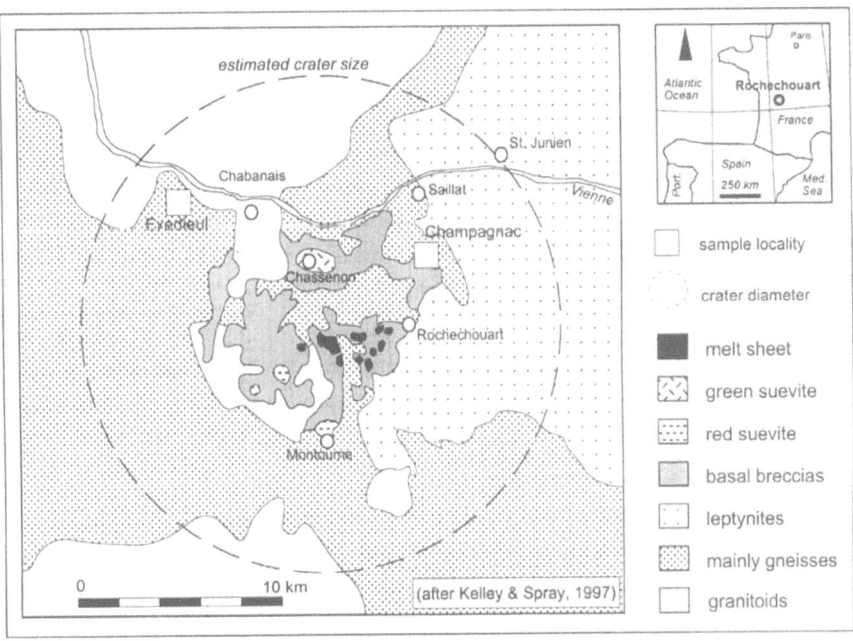

Fig. 1. Simplified geological map of the impact crater of Rochechouart, France, modified after Kelley and Spray (1997).

Geological Setting

We have initiated a case study of the relatively young but deeply eroded Rochechouart crater, France, which allows unique insights into the fault structure of a crater basement (Fig. 1). Results of this study are expected to be generally applicable to other complex terrestrial impact craters with similar target lithologies (crystalline rocks) and comparable diameter (20-30 km).

The Rochechouart impact crater is located at the north-western border of the Massif Central, France, about 40 km SW of Limoges (latitude N 45°50′, longitude E 0°56′)(Fig. 1). An impact origin of the Rochechouart structure was first postulated by Kraut (1969) and later confirmed by Kraut and French (1971), and Lambert (1974 1977a 1977b). They identified remnants of a thin melt sheet and suevite breccias. Indicators like shatter cones, planar shock features in quartz and isotropic plagioclase proved the impact hypothesis. Morphologically, a deep erosion of more than one kilometer of crater floor and crater fill has destroyed the crater. Continuous strata recording the impact are therefore absent. From the distribution of impact breccias and from a shock zoning study a crater size of approximately 20-23 km diameter was derived (Lambert 1974 1977a). However, recent geophysical data compiled by Pohl and Ernston (1994) reveal a 26-28 km diameter negative gravity anomaly fitting the circular Rochechouart structure. This indicates that the original crater diameter also exceeds the estimate of 25 km by Stöffler et al. (1988).

Because of its size, the crater falls into to the category of complex craters on earth. However, no direct structural proof for the existence of a central uplift has been established in the crater floor so far. Oskierski (1983) noted that a central peak may occupy only a small portion of the crater floor.

Target rocks include mica-rich paragneisses, leptynites, granites and granodiorites. North-south trending dykes with porphyric leucogranites intruded into this rock sequence before the rocks were affected by the impact. The age of the rock suite is predominantly upper Paleozoic (Reimold et al. 1983b). A main fracture system of submeridional orientation (striking 175°) was formed in Hercynian times. Satellite images show various linear fracture systems, but a concentric fracture system, which one would expect for an impact structure, cannot be detected (Lambert 1974b). In a regional fracture analysis performed by Bischoff and Oskierski (1987) no major differences in the pattern of joints inside and outside the impact structure were discovered. However, in the same paper these authors describe impact-related faults that dip towards the crater centre.

Detailed petrographic and geochemical investigations have been performed on different impact-induced rocks including the impact melt (Reimold et al. 1984c), impact breccias (Kraut 1969), impact breccia dykes (Oskierski 1983; Oskierski and Bischoff 1983; Bischoff and Oskierski 1987), and pseudotachylites (Oskierski 1983; Reimold et al. 1983a; Reimold et al. 1984a, b; Reimold et al. 1987, Bischoff and Oskierski 1987).

The most recent dating attempts of the Rochechouart impact reveal ages of 186 ± 8 Ma using the whole rock Rb-Sr technique on melt sheet rocks (Reimold and Oskierski 1987) and 214 ± 8 Ma using $^{40}Ar/^{39}Ar$ laser spot fusion dating of pseudotachylites (Kelley and Spray 1997). A late Triassic age for the impact is in accordance with an undisturbed Jurassic sedimentary sequence of the near Aquitaine basin that shows no evidence of ejecta deposits. Spray et al. (1998) have recently suggested that the Rochechouart impact is part of a multiple impact event that formed the Manicouagan and Saint Martin (Canada), the Obolon (Ukraine) and Red Wing (USA) craters. However, this conclusion is not firmly established as there are still uncertainties of the age determinations. The projectile of the Rochechouart structure may have been of chondritic composition (Horn and El Goresy 1981).

Fault patterns in the basement of the Rochechouart crater

Fault Geometries

Impact-induced faults have been studied predominantly in a quarry near the village of Champagnac north of Rochechouart (Fig. 1). This quarry is located about 6 km NE of the ancient impact centre and allows unique insights into the fracture morphology and fault geometry of the autochthonous crater floor. A second quarry is located about 9 km NW of the crater centre near Exedieul, west of Chabanais (Fig. 1). Although the area of exposure is restricted, we believe that these localities expose fault patterns that are very characteristic of the deformed crater basement and may exist in large parts of the crater floor. To distinguish faults induced by the impact event from pre-impact target faults we use, as our main criterion, the presence of pseudotachylites.

The structural analysis of the crater basement in Champagnac reveals the existence of several normal faults dipping with different angles towards the impact centre (Fig. 2 - 5). (*Normal faults* are inclined dip-slip faults along which the hanging wall block has moved down with respect to the footwall block.) The observed fault planes have complex geometries with bowl-shaped depressions and undulations and oblique dip-slip trajectories.

Fig. 2. Photographs from the quarry near the village Champagnac, north of Rochechouart. (a) The detachment dips gently towards the south. The fault has a ramp and flat geometry. In the footwall below the flat a pegmatite lens is transected into stacks of parallel faults forming an extensional duplex. (b) The dark plane dipping to the left is an exposure of a subsidiary fault surface. Black pseudotachylite masses of 20 cm width cover the fault.

The present outcrop situation exposes a prominent low-angle normal fault that dips gently with 5-35° SSW towards the crater centre (Figs. 2-4) (Kenkmann and Ivanov 1999). This fault crosscuts all lithologies. Striations on fault planes indicate normal faulting with an average slip vector of 210° (SSW). This major fault partly consists of an anastomosing network of sub-faults (Fig. 3). The fault surfaces are not plane, but contain ramp and flat geometries with different degree of inclination. To compensate these anomalies the faults cut into either the hanging wall or the footwall as faulting proceeds. Progressive cutting of active faults back into the footwall block during deformation leads to the formation of stacks of parallel faults (horses), which are connected by a roof and a floor fault. The resulting structure, having deca-metre size in the investigated quarry, can be denoted as an extensional duplex. In addition to this, other faults with antithetic (opposite) dip of 20-40° NE developed locally to accommodate the deformation of the hangingwall and footwall block. In two-dimensional outcrop sections faults occur as a network of fault trace-lines (Fig. 3). The trace of the major low-angle fault zone can be studied over approximately 500 m in an acute and obtuse angle to the fault zone strike.

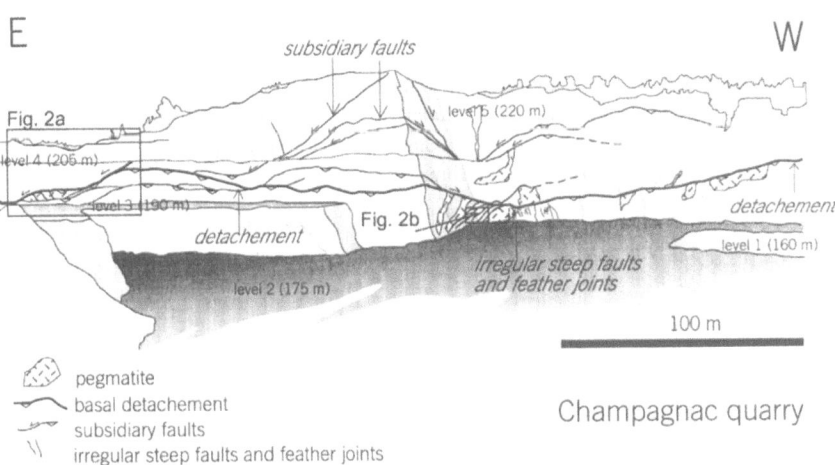

Fig. 3.   The figure shows the outcrop situation of July 1998 at the Champagnac quarry. An anastomozing network of fault trace lines that branch off from the major low-angle detachment can be observed along the south-facing quarry wall. Fault trace lines mostly cut the fault striking direction in an acute angle.

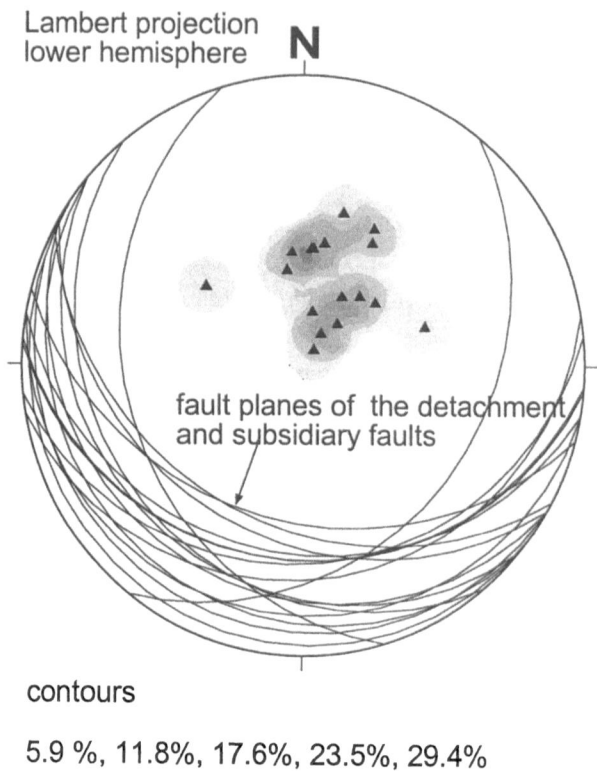

Lambert projection
lower hemisphere **N**

fault planes of the detachment
and subsidiary faults

contours

5.9 %, 11.8%, 17.6%, 23.5%, 29.4%

Fig. 4. Lambert projection of the detachment and subsidiary faults of the Champagnac quarry. Great circles and the related poles indicate the orientation of the detachment and the subsidiary faults. They all show a gentle dipping towards S or SSW. Note that the assumed impact centre is 6 km SW of the quarry.

More steeply-inclined subsidiary faults that branch off from the main low-angle fault form a set of normal faults, which can be denoted as an extensional imbricate fan (Figs. 3, 5). These faults are weakly listric (*listric* normal faults are concave-upward faults – i.e., faults whose dip decreases with increasing depth) and join the basal low-angle detachment at a depth of 30-100 m below the transition from the parautochthonous breccias to the autochthonous country rock. They are synthetic to the basal shear plane – i.e., they dip in the same direction. At the deca-metre scale they also show extensional duplexes like the main detachment. Some of the subsidiary faults are interconnected (Fig. 5). Steeper subsidiary faults partly transect parautochthonuous monomict breccias that lie on the top of the basement. These faults may have undergone unconstrained (free surface) dip-slip.

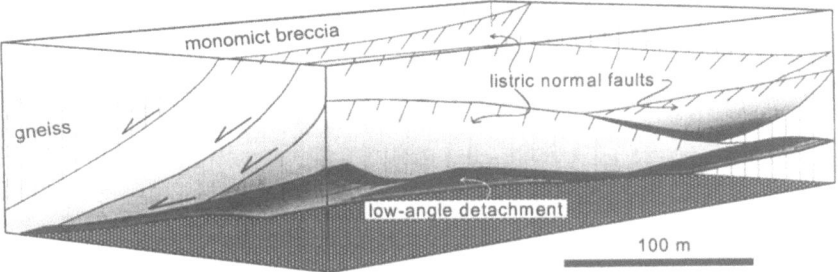

Fig. 5.   Simplified block model showing the geometry and kinematic interpretation of the fault zones observed in the Champagnac quarry. The basal detachment is a gently dipping normal fault that has a wavy surface. More steeply-inclined normal faults branch off from the basal detachment to form an extensional imbricate fan.

The quarry near Exedieul is located approximately 9 km away from the crater centre. Faults and fractures are most commonly steeply inclined and form conjugate systems (sets of faults with comparable strike and dip-angle but opposite dip orientation) that can be related to a vertically oriented maximum principal stress direction. These fractures may have formed due to overburden pressure and may not correlate to the impact. In contrast to these uniform fault patterns we observed some gently-dipping fault zones in the southeastern part of the quarry which are lubricated with chlorite and small relics of pseudotachylites. Like in the Champagnac quarry they form an imbricated set of faults. But unlike those in the former locality these faults dip away from the impact centre (W to WNW) with 20-45°. The sense of shear is ambiguous. Because their absolute ages are unknown, it is not possible to determine whether these faults are really linked to the impact If they are indeed related to the impact, we may conclude that faulting in the deeper parts of the crater floor is very distinct from near-surface faulting because morphological expression of such faults has never been found in lunar or other extraterrestrial craters. The distinct fault plane orientation may be linked to a more distal position of the locality with respect to the impact centre.

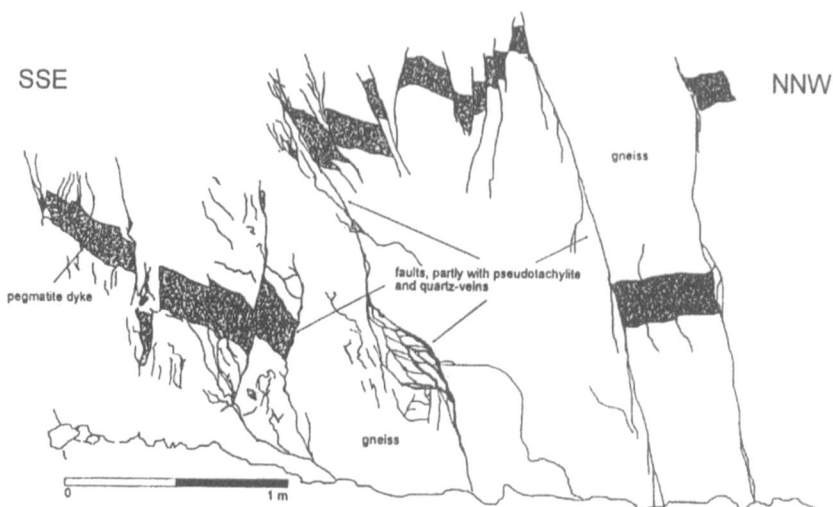

Fig. 6.   Distribution of subordinate faults which are exposed below the detachment at the Champagnac quarry. Feather joints, veins and faults are partly covered with quartz and pseudotachylites. A pegmatite dyke (dark-shaded) was used to infer the sense of shear along the irregular faults. Some fault zones were apparently activated as steep normal faults, others as steep reverse faults. It is believed that these faults formed while the shock wave passed the target rock and lead to an oscillation of blocks.

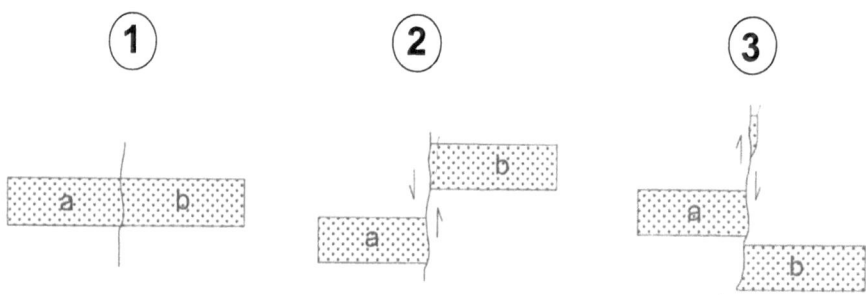

Fig. 7.   Diagrammatic sketch of the movement pattern that occurred along some of the small subordinate faults at Champagnac. An isolated slice of the transferred pegmatite dyke is imbricated into the fault zone. The apparent displacement may indicate a bimodal, oscillating movement along the fault plane that could evolve during acoustic fluidization.

Pseudotachylites and Subordinate Faults

The main and subsidiary faults of the Champagnac quarry are covered with grey-black colored rocks, which resemble pseudotachylites. These dyke-like masses (Fig. 2b) of centimetres to several decimetres width and tenth of metres length are strongly aligned along the discrete fault zones and crosscut all lithologies. Injections from the fault planes into the surrounding wall rock are rare. The dyke-like masses do not completely consist of pseudotachylites. Rather they form an interconnected network of veins and wraps around lenses of wall rock, breccias, or hydrothermal phases. It is believed that these pseudotachylites formed by frictional heating as a result of large amounts of unconstrained slip along fault surfaces activated during the modification of the transient crater cavity. They represent analogues to tectonic pseudotachylites (e.g., Magloughlin and Spray 1992) and hyalomylonites of large landslides (e.g., Masch et al. 1985).

Thin pseudotachylite veins (less than 1mm up to 1 cm) also cover subordinate faults, veinlets and fracture joints of the footwall block below the detachment. These pseudotachylites can be distinguished morphologically from those that cover the large fault zones. They formed more pervasively along an irregular or anastomozing fracture network of veins and feather joints. A preferred alignment occurs in a steeply-dipping orientation but, due to curvature in orientation, scattering is large (Fig. 6). A marker horizon (pegmatite dyke) was used to reconstruct the kinematics of these pseudotachylite-bearing faults and veins. Displacements are mainly vertical and range from 0.1 to 3.0 m. The spacing between adjacent faults with visible displacement is 0.1 to 1.0 m (Fig. 6). The shear sense is non-uniform: some being activated as steep normal faults, others as reverse faults. A few faults preserve evidence for multiple oscillating movements, as demonstrated in Fig. 7. Previous studies (Bischoff and Oskierski 1987; Reimold and Oskierski 1987) have shown that these pseudotachylites differ in time of formation with respect to the larger pseudotachylites. They predate the larger pseudotachylites and may have formed when the shock wave passed through the target rocks. The observed up and down movement along these subordinate faults is possibly related to an oscillation mechanism of blocks (Ivanov and Kostuchenko 1997) which is proposed to occur during acoustic fluidization of the target.

Recent dating of the impact performed by Reimold and Oskierski (1987) and Spray and Kelley (1997) was based on the analysis of pseudotachylite veins and pseudotachylite-bearing faults. A pre- or post-impact origin of those fractures and faults containing pseudotachylites can therefore be excluded. All pseudotachylites contain considerable amounts of hydrothermal phases (quartz, pyrite, calcite, and dolomite) which make microstructural investigations difficult. The hydrothermal component of vein fillings was generated simultaneously and shortly after the frictional melt (Reimold et al. 1987). A detailed microstructural investigation of pseudotachylites is beyond the subject of this paper. For a petrographic and geochemical description of breccia dykes, pseudotachylites, veins and veinlets we refer to Oskierski (1983), Bischoff and Oskierski (1987) and Reimold et al. (1987).

Numerical Modelling of Crater Modification

Model Design

In an attempt to understand the occurrence of low-angle normal faults in the crater floor of complex craters with final diameters of 20-30 km similar to the Rochechouart crater, a numerical model was designed (Kenkmann and Ivanov 1999). The modified SALE-2D hydrocode (Amsden et al. 1980; Ivanov 1994; Ivanov and Kostuchenko 1997, Ivanov and Deutsch 1999) was used in Lagrangian mode to compute material flow, temperature and stress distribution during crater modification. The generated mesh of the model has two parabolic and two straight edges and consists of 59 x 39 cells in the case of model A (Fig. 8a) and of 40 x 39 cells in the case of model B (Fig. 10). The inner parabolic edge represents the shape of the transient cavity while the outer parabola remains fixed presenting the assumed boundary of the fractured crater basement zone. The top horizontal edge represents the free surface. Elements along the symmetry axis are allowed to move in vertical direction only. The mesh refines towards the transient cavity. We used the Tillotson equation of state for basalt to describe the thermodynamics of the target rocks. Computation does not simulate the whole impact process from the very beginning but starts when the transient cavity stops growing. The inner parabolic edge in the model represents this transient crater cavity. We assume that the transient cavity was excavated to a depth of 4 km (model A) and 6 km (model B). The diameter of the transient cavity is set to 12 km (model A) and 18 km (model B). The depth/diameter ratio is 0.33 for both models. The chosen values mount the range of possible transient crater cavity sizes of the Rochechouart impact structure. The zone of rock disruption in the crater basement has a radial extension of 26 km at the surface in both models.

Before the model is described in more detail, a few remarks must be made concerning the rheology of the target rocks after the shock wave has passed the rocks: The collapse of a transient crater cannot be treated as a static process. It occurs immediately after excavation when large amounts of acoustic and seismic energy are still present in the rock surrounding the crater (Melosh 1989). Therefore, the mechanical behaviour of rocks during subsequent crater modification strongly deviates from usual rock rheology. Impact crater slumping, sliding and central peak uplift during crater modification requires rock failure under stresses much smaller than expected from conventional rock mechanics (Melosh 1979). The effective strength of these rocks lies around 3 MPa (Melosh 1977) and the effective angle of internal friction is less than 5° (McKinnon 1978). Although some controversy about the physical nature of strength reduction still exist, the theory of "acoustic fluidization" has been the most reliable theory so far that provides an adequate explanation of crater collapse. Acoustic fluidization of rock material was proposed by Melosh (1979, 1989), Melosh and Goetz (1982), Melosh and Gaffney (1983). In their model oscillation of blocks creates a sinusoidal variation of the normal load along a fault zone. During unloading traction exceeds friction and shear occurs. When the normal load decreases back to the friction limit, blocks stop moving. Ivanov and Kostuchenko (1997) implemented the equations for block oscillation into a numerical code that successfully reproduced the main features of impact crater collapse. Analogue experiments also demonstrated the drastic reduction in strength due to acoustic fluidization (Goetz and Melosh 1980).

We did not implement equations of block oscillation in our preliminary model. However, to take the effects of fluidization into account we artificially decreased internal friction and assumed a cohesion of 0.1 MPa. Wc also assumed that the effects of acoustic fluidization are not equally distributed throughout the whole crater floor. They are strongest in the central part of the crater basement and gradually decrease downward, sideward and upward (in case of a vertical impact). Iso-strength domains may have spherical-like shapes, although a flattening and distortion occur due to an increasing overburden pressure at depth and the existence of the transient crater cavity. The surface of the transient crater cavity (crater floor) which is covered with monomict breccias should not be affected by acoustic fluidization, since rocks do not form a coherent mass and cannot transport acoustic energy. A function describing the decay of oscillation in time and space is not available yet. To simulate a variable acoustic fluidization as a first approach we changed the friction coefficient linearly with depth. The friction, μ, of the rock mass is assumed to decrease linearly with depth from normal values of 0.43 (model A) and 0.58 (model B) at the free surface up to a horizontal plane which passes through the lowest point of the transient cavity (below the point of impact). At that level the friction coefficient μ is assumed to have the value of 0.058

(model A) and 0.12 (model B), and it remains constant from that level down to the end of the model mesh which is assumed to correlate with the bottom of the disrupted crater basement. The choice of the μ values are determined by fitting the model's predictions to observations of the studied collapsed crater. Model A has a noticeable softer rheology than model B.

In this model a comment about the definition of the friction coefficient is needed. The stress state is described with the principal stresses σ_1, σ_2 and σ_3 (positive for compression). In the numerical model shear failure occurs when the value of the second invariant of a stress tensor, $J_2 = 0.5 [((\sigma_1-\sigma_2)^2+(\sigma_2-\sigma_3)^2+(\sigma_1-\sigma_3)^2)]$ reaches its critical value. The dry friction is obtained by assuming that $J_2^{1/2}$ is proportional to the pressure, $p = (\sigma_1 + \sigma_2 + \sigma_3)/3$:

$$J_2^{1/2} = k_p\, p \tag{1}$$

where k_p is the coefficient of proportionality. In normal surface-to-surface friction measurements the friction coefficient μ is defined as the proportionality coefficient between the tangential friction force and the normal force. For complex stress states with an intermediate principal stress, σ, between the maximum, σ_1, and minimum, σ_3, stresses, the relation between μ and k_p may be a complex one. However, for the vast majority of computational cells in the numerical model the simple rule $\mu = 3^{-1/2} k_p = k_p / 1.73$ determines k_p for each μ, when μ is defined as

$$\tau = \mu\, \sigma_n \tag{2}$$

where $\tau = (\sigma_1 - \sigma_3)/2$ is the shear stress and $\sigma_n = (\sigma_1 + \sigma_3)/2$ is the normal stress at the sliding surface. The factor of $3^{-1/2}$ explains the above mentioned values of μ: $k_p = 1$ yields μ = 0.58, $k_p = 0.75$ yields μ = 0.43 etc.

Model Results

Model A (Figs. 8, 9)

A strong uplift of the central crater floor occurs from the very beginning. Sliding of the crater rim starts at the rim crest. After 20 sec (Fig. 8b) the first potential slip planes have formed. Their inclination is 60° near the surface but the dipping angle decreases with depth. The number of slip planes separating individual terraces grows in time and their spacing increases away from the crater centre. Shear zones in the periphery remain steeply inclined, but in the central part they bend into a subhorizontal orientation. Uplift of the central area reverses the dip orientation of the slip zone by passive rotation. After 40 sec high strain zones evolve (Fig. 8c).

Less-localized slip planes, dipping outwards with 30-45°, develop in response to the central uplift in deeper parts (3 - 4 km depth) of the model. After 60 sec the central uplift starts to collapse and slip trajectories indicate reverse motion (outward flow). Because inward flow continues in the more distal parts, collision occurs in a transition zone, which ultimately leads to some elements squeezing-out to the surface after 80 sec. Deformation ceased shortly after that deformation increment. A main characteristic of the collapsed crater model are weakly-inclined shear zones occurring in a distance of approximately 3 to 8 km away from the crater centre. They have a lateral extent of approximately 5 km at a depth of 0.5 to 2 km. Additionally, less localized shear zones develop with opposite dip orientation. The innermost one is strongly bent and overturned in the upper part, near the surface. A very complex style of extensional and compressional deformation dominates the uppermost region of the central peak. Fig. 9 represents a geological interpretation of the model results.

Model B (Figs. 10, 11)

Model B deviates from model A with respect to the initial transient crater size and the frictional properties (Fig. 10a). Despite this, results of model B are in principle similar to those of model A. At the beginning, domains of increased shear form parallel to the transient crater cavity. The dip-angle of the slip planes decreases with depth due to the central uplift and the reduced rock strength. Like in model A the first terraces form after 20 sec at the edge of the transient crater (Fig. 10b). Strain localization along the outer parabola is an artificial effect due to a fixed displacement of the edge and is not taken into account in the following discussion. The uplift of the central area causes passive rotation and listric geometries of shear planes (Fig. 10c). The motion along the uppermost shear plane changes with time. Sliding toward the center dominates the first 50 sec. Opposite sliding (flow outward), starting after 55 sec in the central parts while inward sliding still continues in the more distal parts, ultimately leads to collision and a squeezing-out of blocks to the crater surface. A lower high strain zone occurs at greater depth (2 to 4 km). It has a listric shape and shows a consistent dipping (20-30°) towards the crater centre. It extends from 4 to 9 km away from the crater centre. Below the central uplift a steeply-dipping thrust zone develops, which ends in the hinge of an antiform. In comparison to model A outward flow is subordinate and the central uplift of model B does not fail completely. Fig. 11 represents a geological interpretation of model B.

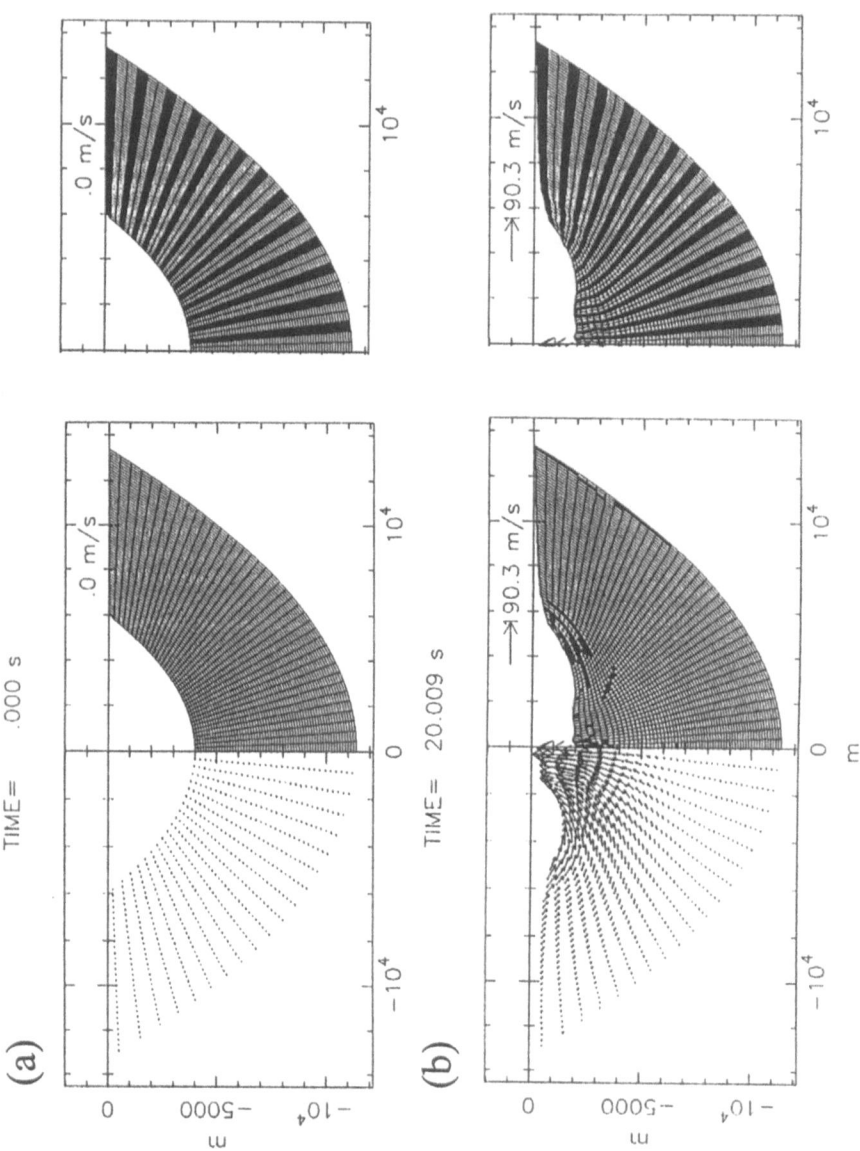

Fig. 8. (above and opposite) Results of the numerical model A as described in the text at different time increments: **(a)** 0 sec (initial conditions), **(b)** 20 sec, **(c)** 40 sec, **(d)** 80 sec. Deformation ceased shortly after deformation increment. Figures on the left side show the incremental velocity vectors of all cells during the selected time increments. Black shaded cells in the figures of the middle column are deformed more strongly than surrounding cells (by a factor of 1.5). They indicate zones of enhanced and localized shear. In the figures on the right side cells are shaded black and white in order to visualize the amount of displacement and slip along the shear zones. For model parameters and a discussion of the deformation history, see text.

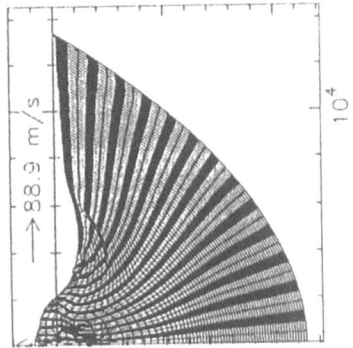

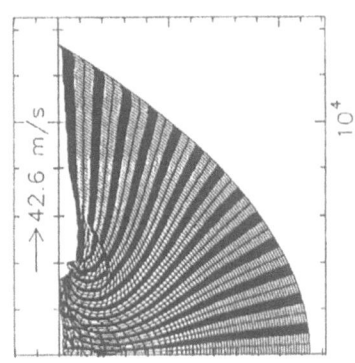

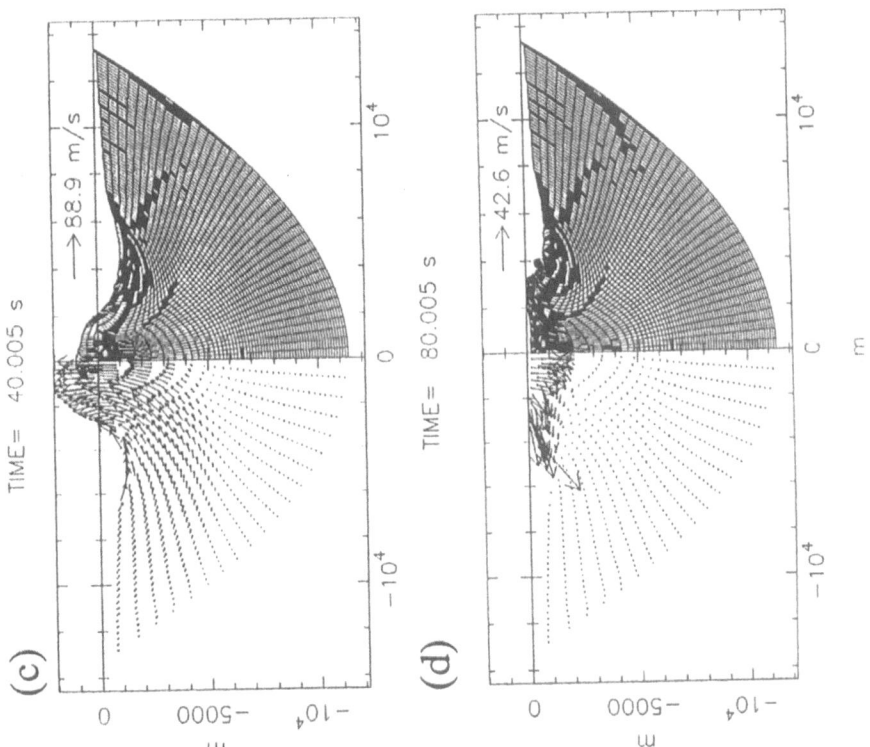

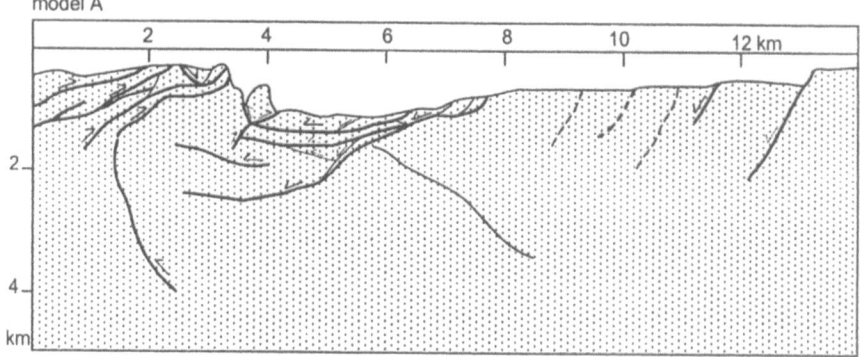

Fig. 9.   Geological interpretation and fault distribution inferred from the numerical model A after movement has stopped. Some of the faults in the central area may be activated in a multiple fashion. For explanation of the kinematic history, see text (after Kenkmann and Ivanov 1999).

Shear Heating

The computer code allows to calculate the values of friction work and corresponding shear heating of material in each computational cell, even though the computed shear heating describes the average heating of a material element with a size of 100 m. Corresponding shear heating was found to be in the range of 5 to 20 kJ kg^{-1} in most shear zones described above. For a typical rock heat capacity of about 1 kJ kg^{-1} K^{-1} the shear heating results in an *average* temperature rise of 5 to 20 K only. However, if one assumes that the shear is localized in a narrow zone, the shear heating should be ascribed to the this zone only. For example, if this zone has a width of 1 m, the shear heating in the shear zone would be 100 times larger in comparison with cell-averaged estimates. Consequently, the localized temperature rise due to shear may produce melting temperatures. The investigation of thermal softening effects in strongly-localized shear zones requires more complex models with a higher mesh resolution than the present one.

Discussion

Comparisons Between Field Data and the Numerical Model

The occurrence of low-angle normal faults in the crater floor of complex craters is documented by our structural investigations of the Rochechouart impact crater. However, the restricted area of field exposure does not allow us to fully assess the significance of these faults during crater modification. At this point it is important to notice that low-angle faults were also observed in the crater floor of other deeply eroded impact craters like the Upheaval Dome structure, U.S. (Shoemaker and Herkenhoff 1984) or the Gosses Bluff structure, Australia (Milton et al. 1972). Our observations are also in agreement with reflection seismic investigations. Subhorizontal seismic reflectors in the crater floor of the Siljan impact structure, Sweden (Juhlin 1988), occurring at a depth of 1.5 to 7.5 km at the rim of the central uplift, are induced by dolerite sills and extremely fractured granites plus pseudotachylites. The latter two lithologies indicate the existence of flat lying shear zones. On all deep reflection profiles of the inner part of the Chicxulub impact structure, Mexico, reported in Morgan et al., (1997), distinctive bands of gently dipping reflections (20° - 30°), called reflector G, can be observed at mid-crustal depth. These reflections dip towards the crater centre and are also interpreted to be fault traces. Unlike this, seismic profiles through distal parts of complex impact structures (Keiswetter et al. 1996; Morgan et al. 1997) show dominantly steeply inclined faults (60°).

In combining our field observations with the numerical computation of crater modification, the significance of low-angle normal faulting can be generalized. The investigated low-angle normal fault system of the Champagnac quarry is exposed at a distance of approximately 6 km from the calculated crater centre. This observation is in agreement with both numerical models, and suggests that low-angle normal faults may really extend at a distance of 4-8 km away from the impact centre, as indicated by the models. The models are also in agreement with the observation of more steeply-inclined subsidiary faults that branch off from the main detachments. The inclination of faults at the surface is normally 60°. These fault-surfaces may be equivalents to crater inward dipping headscarps which separate terrace-like features as observed in many extraterrestrial complex craters (Melosh 1977; Settle and Head 1979). The fault inclination reveals information about cohesion and frictional properties of these near surface rocks, which are assumed to be more intact than the underlying ones.

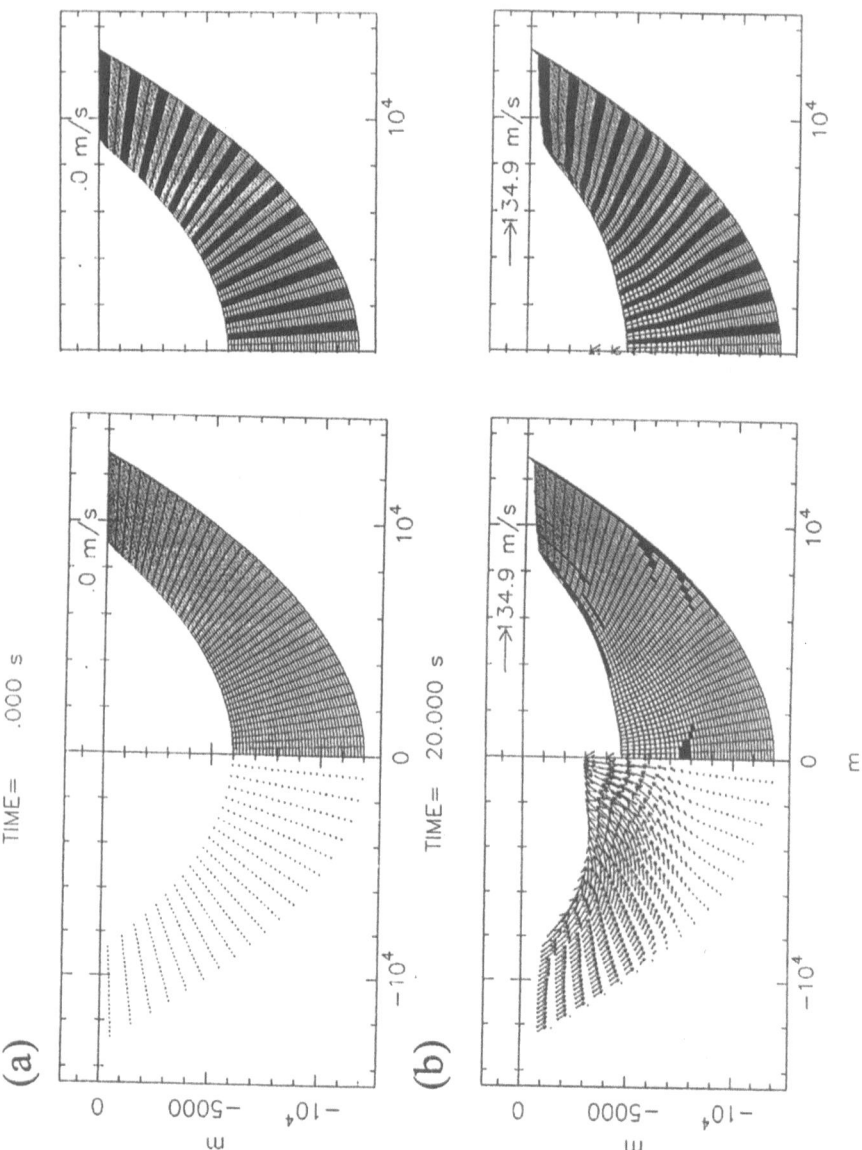

Fig.10.   (above and opposite) Results of the numerical model B as described in the text at different time increments: **(a)** 0 sec (initial conditions), **(b)** 20 sec, **(c)** 30 sec, **(d)** 60 sec. **(d)** represents the final shape of the collapsed crater. Figures on the left side show the incremental velocity vectors of all cells during the selected time increments. Black-shaded cells in the figures of the middle column are deformed more strongly than the surrounding cells (by a factor of 1.5). They indicate zones of enhanced and localized shear. In the figures on the right side cells are shaded black and white in order to visualize the amount of displacement and slip along the shear zones. For model parameters and a discussion of the deformation history see text.

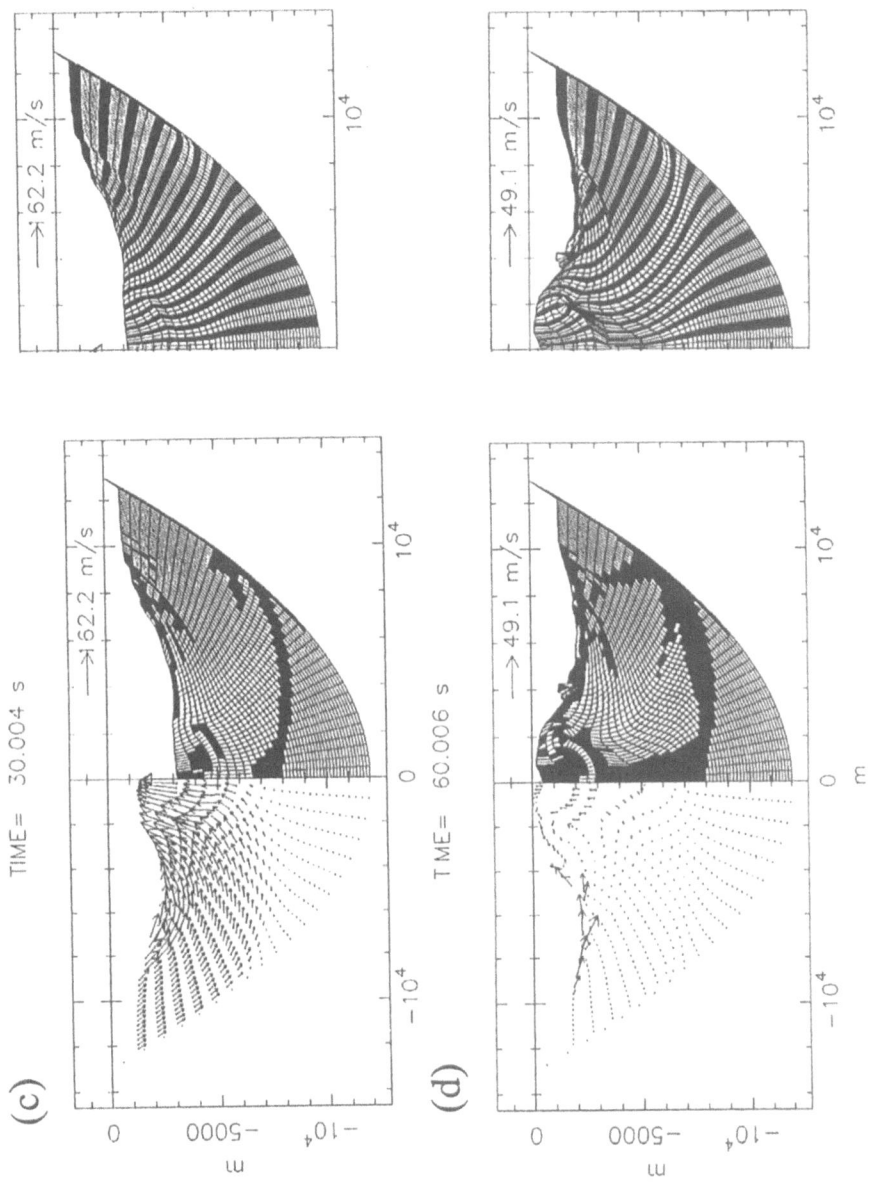

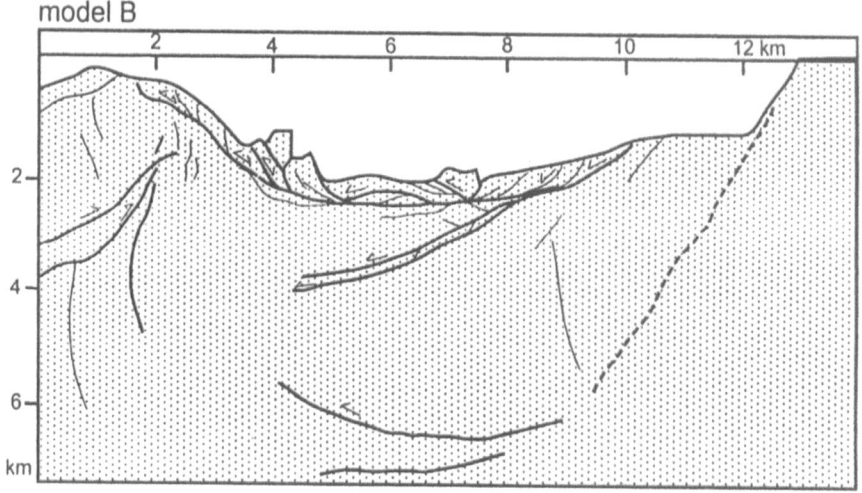

model B

Fig. 11. Geological interpretation and fault distribution inferred from the numerical model B, after movement has stopped. Some of the faults in the central area have been activated in a multiple fashion. For explanation of the kinematic history see text.

The low-angle normal faults are interpreted as being the base of downward and inward moving slides and slumps of rock units located primarily in the walls and near the rim of the transient crater. It is a geometric necessity of gravity-driven systems that extensional faults pass their displacement into a downslope translation. This requires the development of listric fault geometry, in which steeply dipping faults flatten markedly with depth. The deformation is dominantly concentrated in the sliding hanging wall block, but also the footwall block is expected to be affected by pervasive deformation as described above. Our modelling suggests that the degree of inclination of the detachment during crater modification changes due to the uplift and collapse of the central peak. In addition to passive rotation of the fault plane due to central uplift, a subhorizontal position is also favoured by the proposed impact-induced rheological stratification. A mechanically weak zone of low cohesion and friction may underlie a zone which is less affected by acoustic fluidization and which consequently has a higher strength. This layering produces a reorientation of the stress field and favours simple shear in the weak zone.

A minimum displacement of approximately 1 km for the single slip event along the detachments is calculated with the model. Estimates of single-slip displacements based on the width of pseudotachylites (Spray 1997) lead to values of about 1.2 km (width: 0.3 m, coefficient of friction μ: 0.1, latent heat fusion:

4.10^5 Jkg^{-1}, depth of burial: 100 m). Similar displacements were calculated by Spray (1997) for superfaults of the Sudbury multiring structure using a normal value of $\mu = 0.8$. However, if one decreases the friction coefficient to $\mu = 0.1$ due to effects of acoustic fluidization, displacements along the Sudbury superfaults are in the order of 20 km.

Tectonically-Induced Low-Angle Normal Faults

Tectonically-induced low-angle normal faults commonly occur in regions affected by crustal extension (core complexes, reactivated thrusts). Wernicke (1981) proposed a model, in which it is envisaged that lithospheric extension is accommodated on major, gently dipping zones of simple shear, which formed at a depth corresponding to the brittle-ductile transition. A feature characteristic of the so-called low-angle detachment faults is the presence of a strong contrast in metamorphic grade between hangingwall and footwall blocks. The hangingwalls typically expose low-grade upper crustal rocks, while the footwalls expose rocks at greenschist to amphibolite facies. The geometry of detachment faults was first identified in the Basin and Range Province, U.S., and later confirmed from other several metamorphic core complexes (e.g., Dinter and Royden 1993; Froitzheim et al. 1997). The problem with low-angle normal faults is that shear fractures commonly form at orientations of 30° from the maximum compressional stress axis. In extensional terrains the principal compressional stress axis is oriented perpendicular to the surface, thus the inclination of extensional faults is 60°. To explain the genesis of low-angle faults dipping with 15° on average one must postulate a 45° rotation of principal stress directions at mid-crustal depth. An explanation of stress rotation in the Basin and Range Province can be derived from the rheological structure and stratification of the crust in this high heat-flow region (Melosh 1990). The upper elastic crust is displaced horizontally with respect to a lower elastic layer and the intermediate viscous layer is subjected to simple shear.

The proposed rheological stratification, which is responsible for tectonic detachment faulting, may also occur during detachment faulting when the transient crater cavity fails. Numerical models have demonstrated (Melosh 1977; McKinnon 1978) that the uplift and collapse of the central peak of complex craters is only possible if friction coefficient and cohesion are reduced to very low values in this region. The effects of acoustic fluidization are strongest in the central crater floor and may gradually decrease towards the crater rim where the Hugoniot elastic limit of rocks is finally reached. This results in a mechanically weak zone that is sandwiched between two regions of relatively higher strength. In this region simple shear may dominate along horizontal flow lines to form the detachments. In addition, passive rotation of high-angle faults contribute to the present situation which is characterized by low-angle faults.

Low-angle faults of crater basements differ from low-angle faults of extensional terrains in many aspects. First, the observed structures must be put in the context of broad scale tectonic information and regional geology. We expect that impact-induced fault-patterns are disconnected from the regional tectonic context. Second, areas of tectonically-induced detachment faulting are much larger and reach several hundred kilometers of lateral extent (e.g., Wernicke 1981). As suggested from our numerical modelling, individual low-angle normal faults of impact craters with 20-30 km diameter are not expected to exceed a lateral extent of 5 - 6 km. However, if we extrapolate similar kinematic processes to larger impact craters, low-angle faults could be much more extended. For instance, recent seismic reflection data of the Chicxulub impact crater in Mexico (Morgan et al. 1997) suggest that gently dipping faults which are related to an outer ring of the multiring structure transect the whole continental crust and may even continue into the mantle. Rings of pseudotachylites, observed in the Sudbury-crater, Canada (Spray and Thompson 1996), may also represent analogues to such gently-dipping ring faults. In order to distinguish between tectonically and impact-induced detachments a very important criterion of low-angle faults formed during crater modification is their spatial ring-like distribution.

Furthermore, the strong interaction of structures indicating extension with those indicating collision is diagnostically important for identifying crater structures. Specifically, in the central parts of collapsed impact structures where deformation is very intense, faults activated originally as normal faults are reactivated as thrust faults and vice versa.

In an attempt to distinguish between tectonically and impact-induced faults, the presence, morphology and thickness of pseudotachylites have to be considered. Impact-induced faults commonly form in a single-slip event, lasting a few minutes at most, and are characterized by large displacements (hundreds of meters up to kilometer) (see model results). Spray (1997) proposed the term "superfault" for unconstrained faults with this type of behavior. In contrast to superfaults, many slip-events accumulate along tectonic detachment faults through time. Deformation increments can sometimes be distinguished on the basis of different slip vectors at different P-T conditions. Tectonic detachment faulting is most commonly an aseismic process, which occurs in the brittle-ductile transition zone of the crust. Due to the lack of ruptural slip events in tectonic detachment faulting, pseudotachylites have been rarely observed along low-angle detachments.

However, pseudotachylites can form along gently-dipping slopes during landslides (Masch et al. 1985; Maddock 1986).

A careful structural investigation linked to broad scale tectonic information may be able to unravel the different genetic scenarios. Under the assumption that an early heavy bombardment affected the earth similarly to the moon, then metamorphosed relics of impact fault patterns with pseudotachylites should be frequent in Proterozoic and Archean terranes.

Limitations of the Numerical Model

The hydrocode SALE-2D has been successfully applied to model different aspects of cratering (e.g., Ivanov and Deutsch 1999). However, some precautionary notes have to be considered in relation to our preliminary model. First of all, it does not calculate the whole process of cratering but starts at an intermediate point when the transient cavity has formed dynamically. Second, the equations of acoustic fluidization are not implemented in the model; rock fluidization is introduced artificially by a reduction in strength and internal friction. Melosh and Gaffney (1983) show that acoustically fluidized debris may be described as power-law, non-Newtonian fluids. They also emphasize that the stress sensitivity to strain rate changes with the intensity of the acoustic field. However, in our model we ignore any strain rate effects and assume a Coulomb failure. Fluidization of rocks abates in time and in space. However, equations for acoustic diffusion have not been available so far. Therefore, we applied a very simple model and assumed a linear decrease in friction with depth, whereas a change in time was not considered.

Finally, the effect of thermal softening, which may be quite effective in narrow shear zones, is underestimated in our model because the thermal state is calculated for whole elements. The resolution of the mesh does not allow any localization of deformation below the grid size.

Conclusions

For the identification of Archean and Proterozoic impact craters, morphological, microstructural, and mineralogical criteria are difficult to apply, because of their tendency to change in time. We propose that the distribution and kinematics of faults in the crater floor of complex craters provide a diagnostically powerful criterion for the identification of ancient impact sites. Based on our field observation in the relatively young but deeply eroded Rochechouart crater, France, in combination with numerical computations, we arrived to the following:

1. Low-angle normal faults dipping gently towards the crater centre may widely occur in the crater basements of impact craters with final diameters of 20-30 km. They may have a lateral extent of approximately 5 km and predominantly occur at a distance of approximately 3-8 km away from the crater centre. The depth of formation may range from 0.5 to 4 km below the original crater surface. Thus, they are not detectable by remote sensing of extraterrestrial crater. They have flat and ramp structures. Because of the circular shape of impact craters these faults should be arranged in a ring-like fashion.

2. More steeply-inclined subsidiary faults branch off from the detachment to form extensional imbricate fans. At the surface of the crater floor these faults form headscarps and separate terrace-like features that are well-known from extraterrestrial craters.

3. Low-angle normal faulting occurs during crater modification, when the transient crater cavity fails. Gravity-driven sliding occurs in acoustically fluidized rocks with low cohesion and reduced friction coefficients. It is suggested that the degree of fluidization changes with depth and distance from the impact centre. Faulting along weakly-dipping shear zones may be induced by passive rotation due to the uplift of the central peak and the impact-induced rheological stratification of the crater floor with a specifically weak zone underlying less fluidized rocks.

4. Due to rapid and unconstrained single slip events, impact-related fault zones are partly covered with pseudotachylites.

Acknowledgements

T. Kenkmann wishes to thank C. Marchat for logistical support during the field work. A part of this work was done during a scientific stay of B. A. Ivanov at the Museum für Naturkunde, Berlin. We are grateful to E. Pierazzo and S. Kelley for review of the manuscript and constructive suggestions for improvements.

References

Amsden A A, Ruppel H M, Hirt C W (1980) SALE: A simplified ALE computer program for fluid flow at all speeds. Los Alamos National Laboratory LA-8095 Los Alamos NM. 101 pp

Bischoff L, Oskierski W (1987) Fractures pseudotachylite veins and breccia dykes in the crater floor of the Rochechouart impact structure SW-France as indicators of crater forming processes. In: Pohl J (ed) Research in Terrestrial Impact Structures, pp 5-29

Dinter D A, Royden L (1993) Late Cenozoic extension in northeastern Greece: Strymon valley detachment system and Rhodope metamorphic core complex. Geology 21: 45-48

Froitzheim N, Conti P, Van Daalen M (1997) Late Cretaceous synorogenic low-angle normal faulting along the Schlinig fault (Switzerland Italy Austria) and its significance for the tectonics of the Eastern Alps. Tectonophysics 280: 267-293

Goetz P, Melosh H J (1980) Experimental observation of acoustic fluidization in sand. EOS Trans AGU 61: 373

Horn W, El Goresy A (1981) Discovery of metallic residues of the Rochechouart meteorite in basement rocks. Bull Minéral 104: 587-593

Ivanov B A (1994) Geomechanical models of impact cratering: Puchezh-Katunki structure. In: Dressler B O, Grieve R A F, Sharpton V L (eds) Large meteorite impacts and planetary evolution: Geol Soc America Spec Paper 293: pp 81-91

Ivanov B A, Deutsch A (1999) Sudbury impact event: Cratering mechanics and thermal history. In: Bressler BO, Sharpton VL (eds) Large meteorite impacts and planetary evolution II. Geol Soc America Spec Paper 339, In Press

Ivanov B A, Kostuchenko V N (1997) Block oscillation model for impact crater collapse, Lunar Planet Sci 28: 631-632

Juhlin C (1988) Interpretation of the seismic reflectors in the Gravberg-1 well. In: Boden A, Erikson K G (eds) Deep Drilling in Crystalline Bedrock Volume 1: The Deep Gas Drilling in the Siljan Impact Structure Sweden and Astroblemes. Springer Verlag, Berlin, pp 113-121

Keiswetter D, Black R, Steeples D (1996) Structure of the terrace terrane Manson impact structure Iowa interpreted from high-resolution seismic reflection data, In: Koeberl C, Anderson RR (eds) the Manson impact structure: Anatomy of an impact crater. Geol Soc America Spec Paper 302: pp 105-114

Kelley S P, Spray J G (1997) A late Triassic age for the Rochechouart impact structure, France. Meteorit Planet Sci 32: 629-636

Kenkmann T, Ivanov B A (1999) Low-angle faulting in the basement of complex impact craters: numerical modelling and field observations in the Rochechouart structure, France. Lunar Planet Sci 30: 1544

Kraut F (1969) Über ein neues Impaktitvorkommen im Gebiet von Rochechouart-Chassenon (Department Haute-Vienne und Chartres), Frankreich. Geol Bavarica 61: 428-460

Kraut F, French B M (1971) The Rochechouart meteorite impact structure France: Preliminary geological results. J Geophys Res 76: 5407-5413

Lambert P (1974) La structure impactitique de Rochechouart (Limousin) et son contexte structural régional par interprétation de photo-satellite: image ERST. Bulletin du BRGM Sec I 4: 177-188

Lambert P (1977a) The Rochechouart Crater: Shock Zoning Study. Earth Planet Sci Lett 35: 258-268

Lambert P (1977b) Rochechouart impact crater: Statistical geochemical investigations and meteoritic contamination. In: Roddy D J, Pepin R O, Merrill R B (eds) Impact and Explosion Cratering. Pergamon Press, New York, pp 449-460

Leith A C, McKinnon W B (1991) Terrace width variations in complex Mercurian craters and the transient strength of cratered Mercurian and Lunar crust. J Geophys Res 96: 20923-20931

Maddock R H (1986) Frictional melting in landslide-generated frictionites (hyalomylonites) and fault-generated pseudotachylites – discussion. Tectonophysics 128: 151-153

Magloughlin J F, Spray J G (eds) (1992) Frictional melting processes and products in geological materials. Tectonophysics 204: 197-337

Masch L, Wenk H R, Preuss E (1985) Electron microscopy study of hyalomylonites – evidence for frictional melting in landslides. Tectonophysics 115: 131-160

McKinnon W B (1978) An investigation into the role of plastic failure in crater modification. Proc Lunar Planet Sci 9: 3965-3973

Melosh H J (1977) Crater modification by gravity: a mechanical analysis of slumping. In: Roddy J, Pepin R O, Merill R B (eds) Impact and explosion cratering. Pergamon Press, New York, pp 1245-1260

Melosh H J (1979) Acoustic fluidization: A new geologic process? J Geophys Res 84: 7513-7520

Melosh H J (1989) Impact cratering. A geologic process. Oxford Monographs on Geology and Geophysics, Oxford University Press Oxford, 11: pp 245

Melosh H J (1990) Mechanical basis for low-angle normal faulting in the Basin and Range Province. Nature 343: 331-335

Melosh H J, Gaffney E S (1983) Acoustic fluidization and the scale dependence of impact crater morphology. J Geophys Res 88: A830-A834

Melosh H J, Goetz P (1982) The rheology of acoustically fluidized debris: experiments and application to crater slumping. Lunar Planet Sci 13: 511-512

Milton D J, Barlow B C, Brett R, Brown A R, Glikson A Y, Manwaring E A, Moss F J, Sedmik E C E, Van Son J, Young G A (1972) Gosses Bluff impact structure, Australia. Science 175: 1199-1207

Morgan J, Warner M, Brittan J, Buffler R, Camargo A, Christeson G, Denton P, Hildebrand A, Hobbs R, Macintyre H, Mackenzie G, Maguire P, Marin L, Nakamura Y, Pilkington M, Sharpton V, Snyder D, Suarez G, Trejo A (1997) Size and morphology of the Chixculub impact crater. Nature 390: 472-476

Oskierski W (1983) Geologisch-petrographische Untersuchungen im Zentralbereich der Impaktstruktur von Rochechouart, SW Frankreich. Diploma Thesis, Univ Münster

Oskierski W, Bischoff L (1983) Petrographic geochemical and structral studies on impact breccia dikes of the Rochechouart impact structure, SW France, LPS Lett 14: 584-585

Pearce S J, Melosh H J (1986) Terrace width variations in complex lunar craters. Geophys Res Lett 13: 1419-1422

Pohl J, Ernstson K (1994) Gravity investigation in the Rochechouart impact structure. Third International Workshop "Shock wave behaviour of solids in nature and experiment, Limoges/Rochechouart, France, European Science Foundation: p 53

Reimold W U, Oskierski W (1987) The Rb-Sr age of the Rochechouart impact structure France and geochemical constraints on impact melt-target rock-meteorite compositions. In: Pohl J (ed) Research in Terrestrial Impact Structures, Vieweg Verlag, Wiesbaden, pp 94-114

Reimold W U, Bischoff L, Nieber-Reimold J, Oskierski W, Rehfeldt A (1983a) Petrographic and geochemical studies on the basement rocks of the Rochechouart meteorite crater, France and pseudotachylite therein. Lunar Planet Sci 14: 636-637

Reimold W U, Nieber-Reimold J, Oskierski W, Rehfeldt A (1983b) A geochemical studies and chronological study on amphibolite and granitic rocks from the Haute Limousin, Massif Central. Fortsch Mineral 61: 178-180

Reimold W U, Bischoff L, Oskierski W, Schäfer H (1984a) Genesis of pseudotachylite veins in the basement of the Rochechouart impact crater France. I. Geological and Petrographical Evidence. Lunar Planet Sci 15: 683-684

Reimold W U, Bischoff L, Oskierski W, Rehfeld A, Schmidt A (1984b) Genesis of Pseudotachylite Veins in the Basement of the Rochechouart Impact Crater France. II. Geochemical Evidence and a Genetic Model. Lunar Planet Sci 15: 681-682

Reimold W U, Oskierski W, Schäfer H (1984c) The Rochechouart Impact Melt: Geochemical Implications and Rb-Sr Chonology. Lunar Planet Sci 15: 685-686

Reimold W U, Oskierski W, Huth J (1987) The pseudotachylite from Champagnac in the Rochechouart meteorite crater, France. J Geophys Res 92: E737-E748

Settle M, Head J W (1979) The role of slumping in the modification of lunar impact craters. J Geophys Res 84: 3081-3096

Shoemaker E M, Herkenhoff K E (1984) Upheaval Dome impact structure. Lunar Planet Sci 15: 778-779

Spray J G (1997) Superfaults. Geology 25: 579-582

Spray J G (1998) Localized shock- and friction-induced melting in response to hypervelocity impact. In: Grady M M, Hutchinson R, McCall G J H, Rothery D A (eds) Meteorites: Flux with Time and Impact Effects. Geol Soc London Spec Pub 140: pp 195-204

Spray J G, Thompson L M (1996) Friction melt distribution in a multi-ring impact basin. Nature 373: 130-132

Spray J G, Kelley S P, Rowley D B (1998) Evidence for a late Triassic multiple impact event on earth. Nature 392: 171-173

Stöffler D, Bischoff L, Oskierski W, Wiest B (1988) Structural deformation breccai formation and shock metamorphism in the basement of complex terestrial impact craters: implications for the cratering process. In: Boden A, Eriksson K G (eds) Deep Drilling in Crystalline Bedrock 1: pp 277-297

Wernicke B (1981) Low-angle normal faults in the Basin and Range Province: nappe tectonics in an extending orogen. Nature 291: 645-648

12 Basic Remote Sensing Signatures of Large, Deeply Eroded Impact Structures

Andreas Abels[1,2], Heiko Zumsprekel[1,2] and Lutz Bischoff[2]

[1]Institut für Planetologie, D-48149 Münster, Germany.
[2]Geologisch-Paläontologisches Institut, D-48149 Münster, Germany.
(abels@uni-muenster.de).

Abstract. Orbital remote sensing has discovered a number of impact structures on Earth. Complex structures eroded below allochthonous breccias and sheets of impact melt typically reveal a multicircular bull's-eye pattern when the projectile struck a subhorizontally bedded sedimentary target. These structures are striking because of steeply dipping strata in the central uplift, sometimes encircling a crystalline plug. The central anomaly is usually surrounded by an annulus representing a shallow ring-shaped syncline that develops during the cratering process. If not exposed, vegetation, drainage, and weathering products can delineate the bedrock geology. In crystalline targets, the signatures of central uplift and ring syncline are much more subtle, and hard to recognize by satellite imagery. At large structures, however, traces of slumping faults at the crater rim may remain perceptible as curved drainage or scarps. In particular radar systems are potentially useful for detecting such features. The presently known cratering record indicates that more deeply eroded impact structures exist, and this especially in crystalline terrains.

Introduction

Of about 100 of the presently known complex impact structures exposed on Earth only some 15 are eroded below the true crater floor, i.e., below allochthonous crater-fill breccias and sheets of impact melt (Grieve et al. 1995, Glikson 1996, Pesonen 1996, Koeberl and Anderson 1996, and references therein). As the largest part of a vertical profile through the plastically deformed rock volume lies beneath the true crater floor (Fig. 1; Melosh 1989, Grieve and Pilkington 1996), this proportion does probably not reflect the actual record, but indicates that sub-crater floor portions are more difficult to identify. Thus, a potential population of deeply eroded complex impact structures exists which has not been discovered yet. Some of these 'missing' structures may be already known, either as suspected or as misinterpreted geological features, or their signature is simply less distinct than for better preserved craters and hence their first recognition still pending.

Complex impact structures are large physiographic features, of some kilometers to several tens of kilometers, and hence they are hard to recognize within the view field of ground studies. For this task, orbital remote sensing (RS) is of great utility, because large areas of virtually the whole globe can be investigated in a synoptic way.

In order to ascertain general superficial expressions of complex, heavily eroded impact sites, and to evaluate the possibilities of RS for detecting them, we have compiled and analyzed information about the surface properties of some known structures eroded below or almost below the true crater floor. In addition, some published information on the unexposed sub-crater floor properties of those and better-preserved complex impact sites have been taken into account.

Although many impact craters have been initially discovered by RS, detailed studies of individual structures are still rare (e.g., Garvin et al. 1992, Prinz 1996). The same holds for RS analyses of occasionally similar appearing geological objects, for instance concentrically stratified ultrabasic-alkaline intrusions and ring dikes. This makes it virtually impossible to define RS signatures that may be exclusively associated with deeply eroded complex impact structures, but in consideration of our own experience and the published information few features seem to be typical at least. The perception of these impact-typical features depends somewhat on the type of RS data; therefore, important characteristics of those are briefly outlined as well.

This contribution is not a comprehensive treatise of the phenomenology of every individual structure mentioned, neither in RS images, nor in other data. We intend to provide some general clues on how complex heavily eroded impact structures appear or may appear in RS images, by taking into account the suggested construction of craters, the associated appearances in plane view, and the factors that influence the signature of these configurations in RS data.

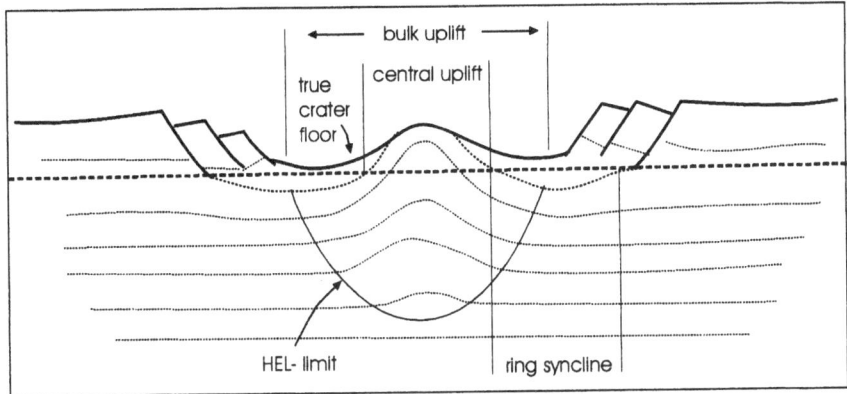

Fig. 1. Schematic cross-section through complex central-peak impact structure. Horizontal broken line marks level at which material was placed during uplift above and below its pre-impact position, respectively. Lateral Hugoniot Elastic Limit (HEL)-limit may vary within extent of ring syncline.

General items

Internal Geometry of Complex Impact Craters

First, it is necessary to briefly address the cratering process as far as it concerns the outlined purpose. This process is still not well understood, and hence some of the following statements are poorly constrained. However, for attempts to assign the observed configuration of deeply eroded impact structures and to predict possible yet unobserved configurations some reasonable assumptions are justified.

The final picture of a pristine complex impact structure is established during the collapse of a transient cavity under the influence of gravity and probably elastic rebound of the compressed target rocks (Grieve et al. 1981, Melosh 1989). The crater rim slumps down into zones of stepped terraces that are bounded by inward dipping faults, while the crater floor lifts up with an increasing amplitude towards the center of the structure. Only rocks in the central part of this bulk uplift raise above their pre-impact position defining a central uplift (Fig. 1). A peripheral annulus that reaches up to below the final terraced rim remains beneath this level.

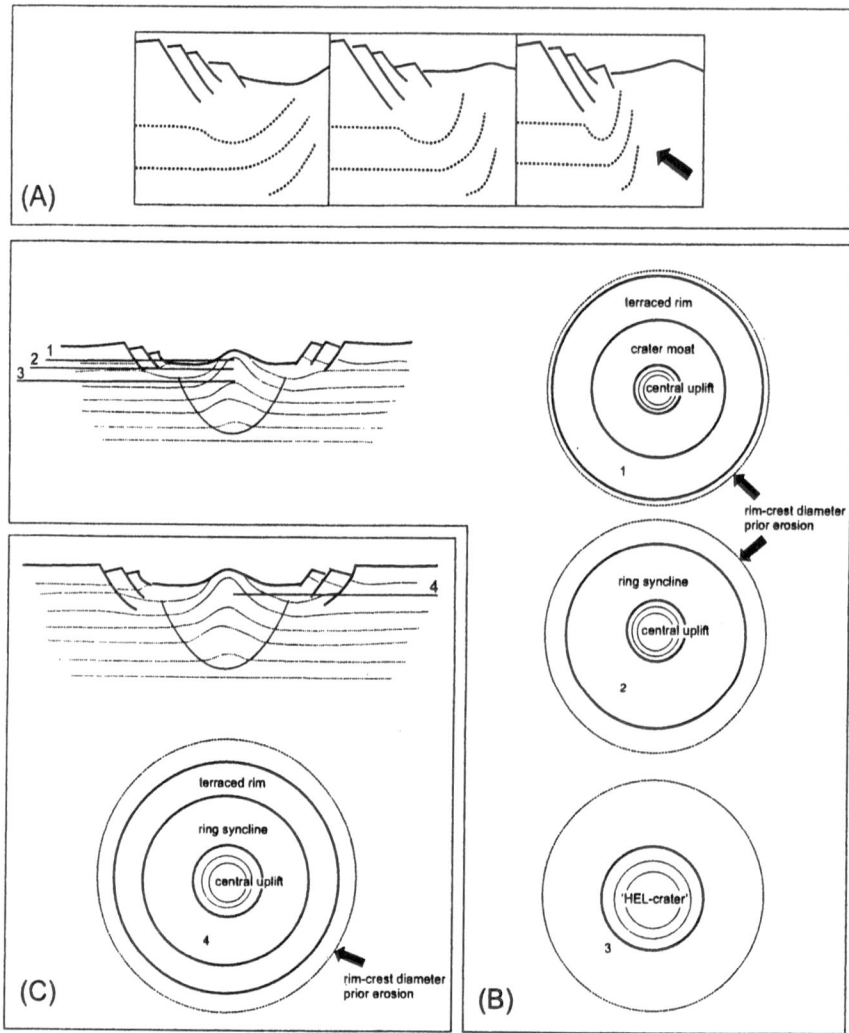

Fig. 2. (a) Sketch showing hypothetically decreasing width of ring syncline relative to central uplift which transforms from central-peak to peak-ring morphology. (b, c) sketches of profile and plane views for different erosion levels; approximated erosion levels: 1 = Araguainha, 2 = Gosses Bluff, 3 (no indubitable structure known), 4 = Siljan. Note that Siljan and Gosses Bluff, both eroded below true crater floor, appear principally different in plane view.

Taking hypothetical horizontal planes as reference horizons, this annulus represents geometrically a shallow ring-shaped syncline, both in sedimentary and crystalline targets, although only in horizontally stratified rocks an actual syncline in terms of structure develops. The upper limit of the bulk uplift, including the ring syncline up to the innermost slumping fault, is the true crater floor. Cross-sectional shape and width of ring syncline and central uplift presumably change with the size of the impact structure. Especially the diameter-dependent transition from central-peak, via central-peak-basin (a central peak surrounded by a peak ring) to peak-ring-basin structures (a basin surrounded by a peak ring) could be combined with a decreasing width ratio of ring syncline and central uplift (Fig. 2a). The relative depth extension, distance to the crater center, and curvature of slumping faults in the terraced rim are also variable and probably strongly influenced by target stratification and intensity of bulk uplift in the crater interior. In most cases, however, the position of the outermost slumping fault and the outer limit of the ring syncline are presumably interdependent. Whereas the transient cavity almost disappears during collapse, a plastically deformed rock volume, i.e., in which the Hugoniot Elastic Limit (HEL) was exceeded, remains preserved below the true crater floor and roughly keeps its maximum extent throughout the cratering process (Fig. 1). Its diameter is distinctly smaller than that of the final uneroded crater, and may correspond to the limit of bulk structural uplift.

Data Types and Effectiveness

Two principally different orbital data types have been used for the investigation of terrestrial impact structures so far. First, multispectral and panchromatic data of passive sensor systems (Landsat-TM and MSS, SPOT) and second, data of active imaging radar systems (ERS-1/2, RADARSAT, SIR-C/X-SAR, SIR-A).

Passive systems detect the reflected radiation of the earth's surface derived from its solar illumination. Reflection properties are measured either in the whole visible part of the electromagnetic spectrum (panchromatic mode) or in various, distinct spectral bands (multispectral mode).

The discrimination of lithotypes, soils or vegetation in multispectral imagery is mainly based on the characteristic reflectance and absorption features that the surface of these materials show due to their chemical composition and structure. By means of digital image processing, it is possible to improve the visualization of specific spectral variations of surface materials. Thus, multispectral imagery is especially useful when the impact-induced distribution of lithologies is exposed or if secondary factors, like vegetation, reflects this distribution. In addition, the geological analysis of optical sensor data, both panchromatic and multispectral, relies on interpretation techniques known from studies of aerial photographs, like lineament detection or determinations of texture variations and drainage pattern etc.

Radar systems actively emit microwaves of different length and subsequently detect the backscatter that is characteristically modified and reflected by the different surface materials.

Radar sensors differ from passive sensors in the way that the surface is artificially illuminated with microwaves. The reflection received from a radar sensor with specific characteristics (wavelength and polarization emitted, received polarization, side looking direction, incidence angle etc.) crucially depends on the topography, dielectric properties of the surface and its roughness. Radar is therefore particularly useful for detecting fault traces and scarps. Further advantages of RS radar imagery are the potential capability to penetrate specific vegetation, clouds and dry sand veneers. Furthermore, it might be possible to detect specific weathering patterns established in the area damaged by the impact event (Blumberg et al. 1995). Potentially indicative fracture density changes associated with a deeply eroded impact structure may be detectable by radar studies, but this has not been demonstrated yet.

Examples from the Known Cratering Record

Apart from the erosion level and initial size of an impact structure, further factors influence its appearance in RS data. Of importance are the climatic region in which the structure was located since its formation, and the target rock, especially whether it is of primary sedimentary or crystalline type. Both factors combined determine erosional effects and the kind of vegetation and weathering products that may cover the impact structure. In the following some examples in primary sedimentary and crystalline terrains are described, each group in the supposed order of increasing erosion level. If no raw data are available, we add sketches and refer to publications in which RS images are shown and/or interpreted. Most comprehensive is still the compilation of Grieve et al. (1988), which is in the following not additionally cited, but suggested throughout.

Primary Sedimentary Targets

Most terrestrial impact structures known to be eroded below or close to the true crater floor have been initially discovered because of a distinct central uplift. This uplift is typically visible in RS images as a highly circular ‚bull's-eye‘ when the projectile impacted a primary sedimentary target.

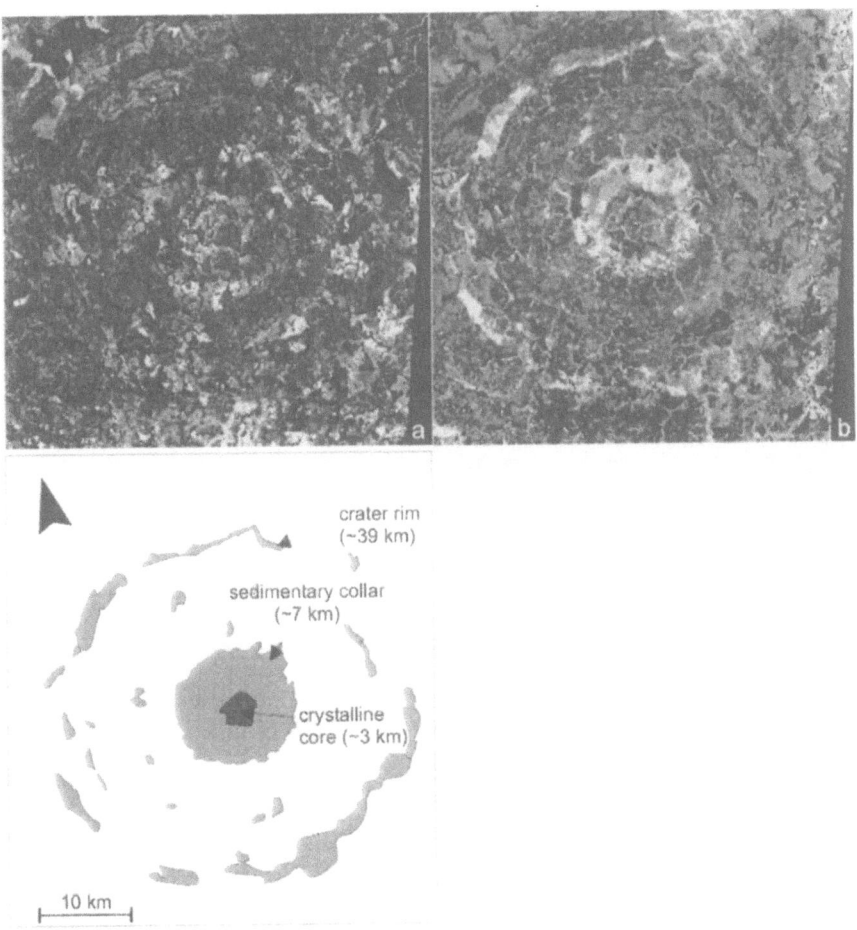

Fig. 3. Araguainha, Mato Grosso/Goiás, Brazil. (a) Structure appears indistinct in panchromatic image composed of Landsat TM bands 1, 2, 3. Circular arrangement of lithologies is excellently reflected in ratio 4/3 by changing vegetation (b). The granitic center of uplift is surrounded by steeply dipping sandstone ridges. Present outer limit of Araguainha is marked by remnants of down-faulted sedimentary rocks. Sketch map simplified after Bischoff and Prinz (1994).

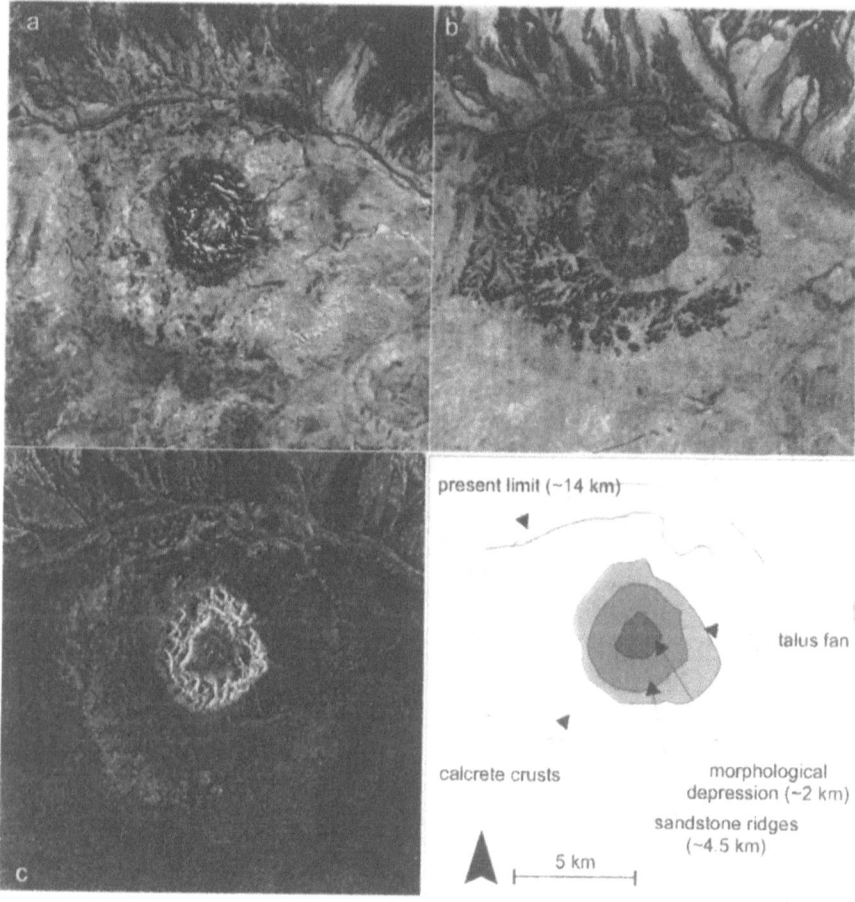

Fig. 4. Gosses Bluff, Northern Territory, Australia. Transforming six Landsat TM bands (1-5, 7) into six principal components (PC) maximizes spectral differences between single bands and hence leads to strong decorrelation and reduction of data. (a) First PC mainly contains overall albedo of all TM bands. Steeply dipping sandstones surrounding depressed center of uplift appear distinct due to their morphology. Present outline of structure can be suspected. (b) Second PC emphasizes spectral differences between visible and infrared spectra whereas morphological features are suppressed. Calcrete deposits are well recognizable as black patches, indicating increased porosity (fracturing) of underground. Light grey ring surrounding central ridges is interpreted as talus fan. (c) Sandstone ridges in uplift are well detectable in Radarsat image, as westward side-looking radar illumination produces topographic distortion effects of foreslopes. Note also well recognizable present outer limit of the structure outlined by a ring of whitish grainy texture, indicating increased surface roughness.

An example is *Araguainha*, which is situated in the densely vegetated boundary region between Goiás and Matto Grosso, Brazil (Fig. 3). Dietz and French (1973) first suggested that the structure was of impact origin after the analysis of a corresponding Landsat-MSS scene. The target rock is a thick clastic sedimentary sequence underlain by crystalline basement of granites and subordinate phyllites (Theilen-Willige 1982, Crosta 1987, Bischoff and Prinz 1994). In the central uplift the crystalline rocks are now exposed and encircled by a 8 km-diameter ridge of steeply dipping sandstone and siltstone, this, in turn, is surrounded by a depressed, 2-4 km wide belt of siltstones. The centrally uplifted crystalline plug is in its periphery covered by impact breccias suggesting an erosion level slightly above the true crater floor. Outwards on the central uplift follows a 8-10 km wide zone of variable relief that was not known until the RS study of Dietz and French (1973). It structurally represents a deep erosion level of the crater moat and terraced rim. The largely removed terraced rim is still indicated by a segmented graben system in which impure limestones and siltstones are preserved that have been totally denuded in the surrounding area. At Araguainha the circular arrangement of rock types can be indirectly mapped by multispectral RS imagery through a discrimination of the vegetation, because the lithology influences its kind and density either directly by the nutrients or indirectly by the topography (Bischoff and Prinz 1994). The depressed crystalline rocks in the center roughly correspond to dense grassland interspersed with small bushes, whereas the ring-ridge is characterized by loose palm tree stands. Variable bush vegetation and small trees dominate the outer crater. Theilen-Willige (1987) demonstrated the utility of airborne radar for detecting ridges and scarps at Araguainha, as the applied wavelength tends to suppress local vegetative details and to emphasize topography that reflects the underlying geological structure.

Another example for a distinct central uplift signature is that of *Gosses Bluff*, central Australia (Fig. 4; Milton et al. 1996). Here a purely sedimentary target was impacted and subsequently eroded for the most part just below the true crater floor, although allochthonous breccias are still preserved in small patches. Due to the well exposure in an arid region, the various lithologies of the structure are directly detectable by RS (Prinz et al. 1996). The central uplift with an average diameter of 4.5 km is outlined by a prominent ring-ridge, mainly consisting of steeply dipping sandstone and well detectable in radar and optical imagery. A sharply limited zone of homogeneous spectral behaviour, probably representing an annular talus fan, surrounds the exposed ridge. No clear relics of peripheral graben structures are present, although the partly annular drainage pattern that surrounds the structure may have developed because of subcropping slumping faults. This drainage defines roughly the presently visible (by RS) diameter of the structure to a maximum of ~14 km. It probably marks the outer limit of the ring syncline, which shows a distinct signature in radar as well as multispectral imagery. The actual present diameter is assumed to be 24 km as suggested by reflection seismics (Milton et al. 1996). Increased fracturing and hence porosity is indicated by calcrete evaporites, which do not occur outside the structure's limit (Prinz 1996). The phenomenology of Gosses Bluff resembles that of the 12 km-diameter *Serra da Cangalha* impact structure in Goiás, Brazil (McHone 1979).

The *Carswell* structure in Saskatchewan, Canada, has been distinctly deeper eroded than Gosses Bluff (Fig. 5). No traces of allochthonous impactites have been preserved (e.g., Grieve and Masaitis 1994). The structure is situated in a glaciated terrain and the target consists of a thick sedimentary sequence that overlays crystalline basement. Carswell appears inconspicuous in Landsat-MSS images, but it is still perceptible due to curvilinear dolomite ridges, approximately ~39 km in outer diameter (Grieve 1984). The ridges surround less resistant clastic rocks and a centrally exposed crystalline plug 18 km across. The dolomite annulus, forming the outermost part of the present structure, is structurally part of the outer hinge of a strongly faulted overturned syncline, and might be the remnant of a larger ring syncline surrounding the central uplift (Baudemont and Fedorowich 1996; Fig. 4). However, as noted in section 1.1 the width of the ring syncline may change relative to the degree of central uplift. Garvin and Schnetzler (1993) note that Carswell is in ERS-1 radar data no more easily discerned than in optical RS data.

The 30-km *Shoemaker* impact structure in western Australia (Fig. 6; formerly Teague Ring) is apparently more deeply denuded but structurally similar to Carswell. Metasediments, including a banded iron formation, surround a central plug of crystalline rocks about 11 km across, which is almost entirely covered by alluvium (Shoemaker and Shoemaker 1996). As at Carswell, the metasediments structurally form a disturbed syncline that may be the last vestige of a formerly wider depression (Bunting et al. 1982). This syncline is broader in the NE than in the SW resulting from post-impact NE-up-SW-down tilting and subsequent erosional truncation to a horizontal level (Bunting et al. 1982, Shoemaker and Shoemaker 1996). Depressed areas adjacent to the more resistant, elevated banded-iron formation are filled with salt lakes, well distinguishable in processed Landsat images.

A special case for an eroded impact site in a sedimentary target is the ~12-km *Spider* structure, northwest Australia. This impact site is characterized by a conspicuous radially splayed fault system that encloses a small structural dome (Shoemaker and Shoemaker 1996). In comparison with other impact structures of similar size, the central uplift of Spider is with ~0.5 km in diameter unusual small, either resulting from an oblique impact or from very deep erosion (Shoemaker and Shoemaker 1996), possibly deeper than in all of the aforementioned examples.

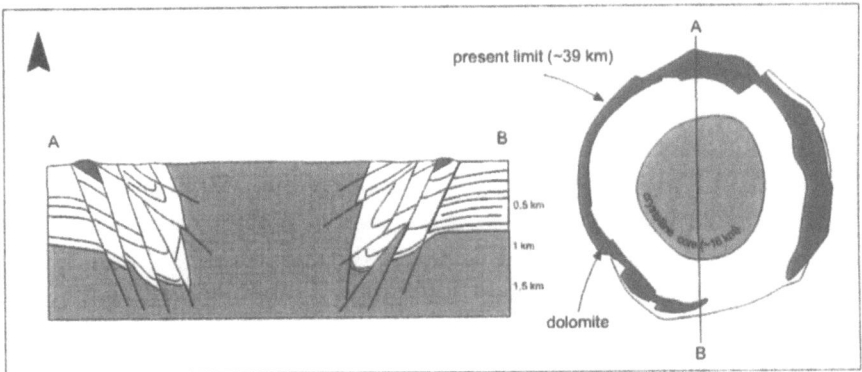

Fig. 5. Outline map and cross-section of Carswell structure, Saskatchewan (simplified after Baudemont and Fedorowich 1996).

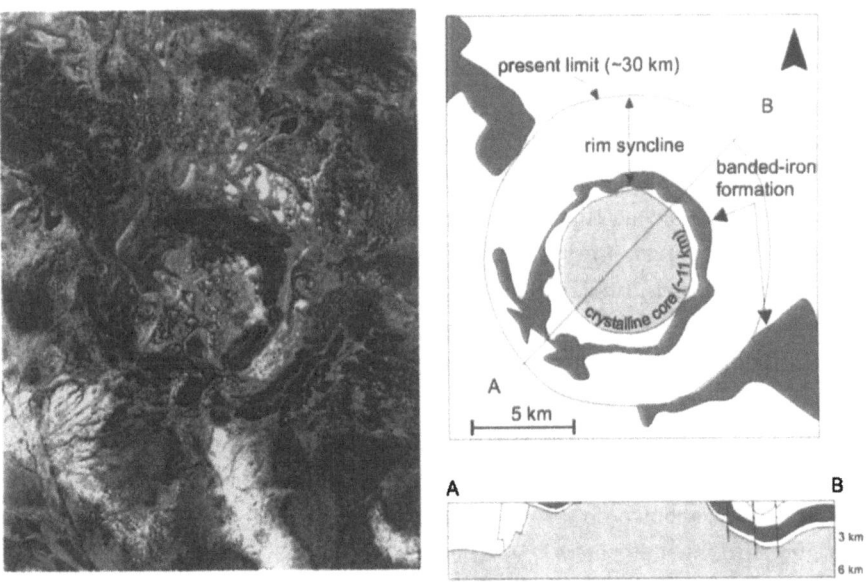

Fig. 6. Shoemaker, Western Australia. The deeply eroded impact structure is a prominent feature in this ratio 5/7 of Landsat TM data due to circular arrangement of banded iron formations (dark grey) encircling granitic basement rocks. Quaternary alluvial deposits and lake sediments cover most of the crater. Sketch map and profile simplified after Shoemaker and Shoemaker (1996) and Bunting et al. (1982).

However, even at this structure a subtle annular feature around the central uplift is still preserved. McHone et al. (1995) note that the size and internal complexity are clearly delineated in processed X-SAR/SIR-C radar images.

Further impact structures exposed in sedimentary targets and probably eroded below the true crater floor are *Sierra Madera* (Ø 12 km; United States), *Tin Bider* (Ø 6-10 km, Algeria), *Serpent Mound* (Ø 8 km, United States), *Ramgarh* (Ø 5.5 km, India), *Upheaval Dome* (Ø 5 km; United States), and *Goyder* (Ø 3 km, Australia). All these structures are striking, well exposed physiographic features. Finally, it should be noted that the *Vredefort* structure, S-Africa, with 200-250 km across much larger than all aforementioned examples, appears in orbital view similar to, for instance, Gosses Bluff. The Vredefort dome, i.e., the central uplift, is partly enclosed by the sub-annular Witwatersrand basin, which resembles a ring syncline (Koeberl 1994).

Primary Crystalline Targets

At the *Siljan* impact structure, central Sweden, the only prominent features in RS data are circumferential half-graben structures of the terraced rim that outline the periphery of the former crater (Fig. 7; e.g., Henkel 1992). In this half-graben ring, which is about 45 km in outer diameter and partly filled with lakes, down-faulted Paleozoic sediments have been preserved which covered the crystalline basement during the impact event. The existence of centrally uplifted rocks approximately 11 km across is proven, for instance, by geophysical investigations (Henkel 1992, Juhlin and Pedersen 1987, 1993), but it has not been described in 'satellite imagery data' (Vattenfall 1991). The distinct annular feature that surrounds central uplifts in sedimentary terrains, is at Siljan only detectable by geophysics as a zone of increased fracturing (Henkel 1992).

The *Lac de la Presqu'île* structure in Québec, Canada, might be deeper eroded than Siljan (Higgins and Tait 1990). In this glaciated terrain the roughly annular Presqui'île lake, about 7 km across, surrounds a peninsular on which shatter cones occur, but also few kilometers beyond the eastern lake shore. This location might be the only known case of a sub-crater floor impact site in crystalline terrain that has been discovered because of its central uplift. However, this discovery was based on shatter cones and not on the physiography of the structure.

Potential Features Observable by RS

The described examples indicate that complex impact structures eroded below the true crater floor are much more difficult to identify by RS studies in crystalline than in sedimentary terrains. In sedimentary terrains, the variable spectral properties and erosional resistance of steeply dipping strata in the central uplift are in the first line responsible for the eye-catching nature of the impact structures. The well ordered annular distribution of different lithologies also determines secondary indicators, like the pattern of vegetation, drainage and deposition of weathered material. The annular appearance remains striking when uplifted crystalline basement is exposed in the central uplift, but it may become increasingly less distinct the more shallow centrally uplifted strata dip away from the crystalline core (Fig. 2). The latter depends on the relative erosion level and on the original thickness of the sedimentary cover relative to the transient cavity size. If another annular feature is present in some distance surrounding a distinguishable central part, the putative central uplift, it may represent the outer edge of the shallow ring syncline or of the uplifted, plastically deformed region underneath. A decision on what this ring feature is due to, is only convincing by considering additional constraints, e.g., dip of bedding, preservation of structural features of the terraced rim, or distribution of shock features. Correspondingly, the width ratio between this ring feature and the central uplift can not be used in isolation to estimate the relative erosion level. In general, multicircularity is typical for sub-crater floor impact structures in primary sedimentary terrains.

Crystalline targets are usually not well stratified in the horizontal dimension (at least if only the upper crust is considered). The rheological contrast of uplifted rocks among themselves and of uplifted and surrounding non-uplifted rocks is usually not sufficient in order to facilitate significant differences of spectral properties and erosional resistance (and hence give a RS signature). If sufficient rheological contrast is present yet, possibly impact-induced, it would be at least in highly metamorphosed terrains spatially more irregular than in sedimentary targets due to pre-existing structural grains. Thus, distinct circular features caused by an eroded central uplift are unlikely to be observable by RS in crystalline terrains. The described case of the Siljan impact structure, however, points to another possibility. The structural half-grabens in the periphery of Siljan are bounded by listric slumping faults that grade presumably a few kilometers below the present surface, as indicated by reflection seismics (Juhlin and Pedersen 1987, 1993). These faults may be detectable by RS as fracture valleys even after the distinct graben structures would have been eroded. The drainage pattern could be influenced by such curvilinear fault traces. A deep continuation of inward dipping ring faults has also been proposed for the buried *Chicxulub* structure, Mexico, here even through the whole crust (Morgan et al. 1997). However, because of its diameter (> 170 km) it is suggested that Chicxulub is a multiring basin, and the formation of peripheral concentric faults may be different from those at complex impact structures (discussion in Melosh 1989). The only known case for the deeper-level expression of a multiring basin in a primary crystalline terrain could be *Sudbury* in Ontario, Canada (e.g., Deutsch et al. 1995). Based on Landsat-MSS image interpretations, Dressler (1984) and Butler (1994) proposed that around the main mass of impact-melt, the Sudbury Igneous Complex (SIC), arrays of concentrically arranged lineaments are present, which they interpret as the traces of ring faults. Parts of the inner lineament arrays are in fact associated with thick pseudotachylites and dikes of injected impact-melt (Spray and Thompson 1995, Wood and Spray 1998). The downward extension and curvature of these putative faults is not known, but they may be recognizable in RS imagery even when Sudbury would have been considerably deeper eroded, perhaps even to below the SIC. A diagnostic value of the often quoted $\sqrt{2}$ ring spacing for multiring basins is nevertheless questionable, as it is unknown whether potentially fault-generating mechanisms would be independent of inhomogeneous lithospheric properties on Earth. In this respect, it is of interest that the $\sqrt{2}$ ring spacing may also be present around the main intrusion of the Kola alkaline province, Russia (Beliaev et al. 1976).

Conclusion and Discussion

All known impact structures eroded below the true crater floor in primary sedimentary terrains are conspicuous features and are hard to overlook with the synoptic view capabilities of RS. They are distinct because of variable lithologies in stratified targets, which leads to well ordered annular spectral and morphological patterns for the central uplift. In most cases, an annular feature surrounds the central uplift. It represents a horizontal section through the ring syncline or the plastically deformed region underneath (Fig. 2). In crystalline terrains the recognition of impact structures eroded below the crater floor is more difficult by RS. These rocks are usually not stratified and hence the transient cavity collapse does not produce a distinct annular distribution of different lithologies. In contrast to sedimentary targets, the terraced rim potentially leaves a signature in RS data in the form of fault-induced valleys or moats. However, it is possibly only at large structures that these features are of sufficient depth that they remain preserved at deep erosion levels. The small number of presently known structures eroded below the true crater floor, when compared to the entire terrestrial record, clearly indicates that more such structures exist, especially in crystalline terrains.

Large uncertainties about the surface appearance remain largely due to poorly understood nature of the cratering process. This is certainly true for complex impact structures and in particular for multiring basins. In addition, modifications by post-impact magmatic and tectonic alteration always needs to be considered. The effectiveness of the kind of RS data depends essentially on the outcrop conditions, but different data types are always useful. Comparisons of the same crater with different sensors can potentially provide useful information for the structure itself, but also for impact structure search strategies in different regions. Instructive examples are presented by McHone (1979), Theilen-Willige (1987), McHone and Greeley (1987), and Blom et al. (1998).

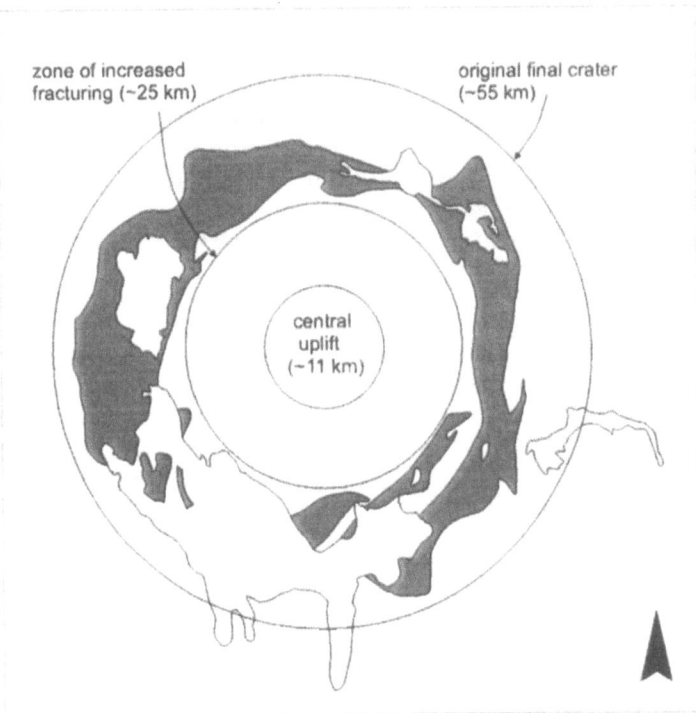

Fig. 7. Outline map of Siljan structure, Sweden (simplified after Juhlin and Pedersen 1987, and Henkel 1992)

Acknowledgements

We gratefully acknowledge Richard Grieve for placing Radarsat data of Gosses Bluff at our disposal.

References

Baudemont D, Fedorowich J (1996) Structural control of uranium mineralization at the Dominique-Peter deposit, Saskatchewan. Econ Geol 91: 855-874

Beliaev KD, Uvadiev LI, Shulga TF (1976) Regularities of the allocation of central-type massifs of the Kola Peninsula. Akademiya Nauk SSSR Doklady 226: 163-165 (in Russian.)

Bischoff L, Prinz T (1994) Der Araguainha-Krater in Brasilien. Geowissenschaften 12: 354-360

Blom RG, McHone JF, Crippen RE (1998) Satellite detection of possible 770 m diameter impact crater at 18 degrees 9 minutes north 50 degrees 4 minutes east Yemen Arab Republic. Lunar Planet Sci,29: #1559

Blumberg DG, McHone JF, Kuzmin R, Greeley R (1995) Radar imaging of impact craters by SIR-C/X-SAR.. Lunar Planet Sci, 26: 139-140

Bunting JA, Brakel AT, Commander DP (1982) Nabberu, Western Australia. 1: 250 000 Geol Series map and explanatory notes. Geol Surv Western Australia

Butler HR (1994) Lineament analysis of the Sudbury multiring impact structure. In: Dressler BO GSA Spec Pap 293: 319-329

Crosta AP (1987) Impact structures in Brazil. In: Pohl J (ed) Research in terrestrial impact structures, Springer-Verlag, Berlin, pp 30-38

Deutsch A, Grieve RAF, Averman M, Bischoff L, Brockmeyer P, Buhl D, Lakomy R, Müller-Mohr V, Ostermann M, Stöffler D (1995) The Sudbury Structure (Ontario, Canada): a tectonically deformed multi-ring impact basin. Geol Rundsch 84: 697-709

Dietz RS, French B (1973) Two probable astroblemes in Brazil. Nature 244: 561-562

Dressler BO (1984) The effects of the Sudbury event and the intrusion of the Sudbury Igneous Complex on the footwall rocks of the Sudbury structure. In: Pye EG, Naldrett AJ, Giblin PE (eds), The geology and ore deposits of the Sudbury structure; Ontario Geol Surv, Spec Vol 1, pp 97-136

Garvin JB, Schnetzler CC, Grieve RAF (1992) Characteristics of large terrestrial impact structures as revealed by remote sensing studies. Tectonophys 216: 45-62

Garvin JB, Schnetzler CC (1993) Remote sensing signatures of terrestrial impact features. EOS, Suppl, Transactions, AGU 74: 387

Glikson AY (1996) A compendium of Australian impact structures, possible impact structures, and ejecta occurrences. AGSO J Austral Geol Geophys 16: 373-375

Grieve RAF, Robertson PB, Dence MR, (1981) Constraints on the formation of ring impact structures, based on terrestrial data. In: Schultz PH, Merrill RB (eds) Multiring basins. Proc Lunar Planet Sci 12A: 37-57

Grieve RAF (1984) The impact cratering rate in recent time. Proc Lunar Planet Sci, Part 2, J Geophys Res Suppl 89: B403-B408

Grieve RAF, Wood CA, Garvin JB, McLaughlin G, McHone JF (1988) Astronaut's guide to terrestrial impact craters. LPI Tech Rep 88-03, Lunar Planet Inst, Houston, Texas

Grieve RAF, Masaitis VL (1994) The economic potential of terrestrial impact craters. Intern Geol Rev 36: 105-151

Grieve RAF, Rupert J, Smith J, Therriault A (1995) The record of terrestrial impact cratering. GSA Today 5: 189-196

Grieve RAF, Pesonen LJ (1996) Terrestrial impact craters: Their spatial and temporal distribution and impacting bodies. Earth Moon Planets 72: 357-376

Grieve RAF, Pilkington M (1996) The signature of terrestrial impacts. AGSO J Austral Geol Geophys 16: 399-420

Henkel H (1992) Geophysical aspects of meteorite impact craters in eroded shield environment, with special emphasis on electrical resistivity. Tectonophys 216: 63-89

Higgins M, Tait L (1990) A possible new impact structure near Lac de la Presqu'île, Québec, Canada. Meteoritics 25: 235-236

Juhlin C, Pedersen LB (1987) Reflection seismic investigations of the Siljan impact structure, Sweden. J Geophys Res 92: 14113-14122

Juhlin C, Pedersen LB (1993) Further constraints on the formation of the Siljan impact crater from seismic reflection studies. Geol Fören Stockholm Förhand 115: 151-158

Koeberl C (1994) African meteorite impact craters: characteristics and geological importance. J African Earth Sci 18: 263-295

Koeberl C and Anderson RR (1996) Manson and company: Impact structures in the United States. In: Koeberl C, Anderson RR (eds) the Manson impact structure: Anatomy of an impact crater. Geol Soc America Spec Paper 302, pp 1-29

McHone JF (1979) Riachão Ring, Brazil: A possible meteorite crater discovered by the Apollo astronauts. NASA SP-412: 193-202

McHone JF, Greeley R (1987) Talemzane: Algerian impact crater detected on SIR-A orbital imaging radar. Meteoritics 22: 253-264

McHone JF, Blumberg DG, Greeley R, Underwood JR (1995) Space shuttle radar images of terrestrial impact structures: SIR-C/X-SAR. Meteoritics 30: 543

Melosh HJ (1989) Impact Cratering. A geologic process. Oxford University Press, Oxford

Milton DJ, Glikson AY, Brett R (1996) Gosses Bluff – a latest Jurassic impact structure, central Australia. Part 1: geological structure, stratigraphy, and origin. AGSO J Austral Geol Geophys 16: 453-486

Morgan J, Warner M, Brittan J, Buffler R, Camargo A, Christeson G, Denton P, Hildebrand A, Hobbs R, Macintyre H, Mackenzie G, Maguire P, Marin L, Nakamura Y, Pilikington M, Sharpton V, Snyder D, Suarez G, Trejo A (1997) Size and morphology of the Chicxulub impact crater. Nature 390: 472-476

Pesonen LJ (1996) The impact cratering record of Fennoscandia. Earth Moon Planets 72: 377-393

Prinz T (1996) Multispectral remote sensing of the Gosses Bluff impact crater, central Australia (N.T.) by using Landsat-TM and ERS-1 data. ISPRS J Photogram Remote Sens 51: 137-149

Shoemaker EM, Shoemaker CS (1996) The Proterozoic impact record of Australia. AGSO J Austral Geol Geophys 16: 379-398

Spray JG, Thompson LM (1995) Friction melt distribution in a multi-ring impact basin. Nature 373: 130-132

Theilen-Willige B (1982) The Araguainha astrobleme/Central Brazil. Geol Rundsch 71: 318-327

Theilen-Willige B (1987) The use of airborne and spaceborne radar images for the detection and investigation of impact structures. In: Pohl J (ed) Research in terrestrial impact structures. Springer, Berlin, pp 115-130

Vattenfall (1991) Scientific summary report of the deep gas drilling project in the Siljan Ring impact structure. Report, Vattenfall, Swedish State Power Board

Wood CR, Spray JG (1998) Origin and emplacement of Offset Dykes in the Sudbury impact structure: Constraints from Hess. Meteor Planet Sci 33: 337-347

13 The Terrestrial Cratering Rate over the Last 125 Million Years

David W. Hughes

Department of Physics and Astronomy, The University, Sheffield S3 7RH, United Kingdom.

Abstract. The rate at which the surface of Earth is being cratered can be measured by analyzing the sizes and ages of the craters that are found on certain stable areas of the Earth's landmass. It is shown, over the range 2.5 < D < 100 km, that the number $N(D)$ of craters that are formed per unit area per unit time larger than diameter D does not (as previously thought) follow a simple relationship of the form $N(D) \propto D^{\alpha}$, where the power α is about -2.0. It is also found that the present day cratering rate for Earth is such that craters of diameters greater than say 3, 8, 20, 40 and 100 km are formed somewhere on the surface every 160 ka, 300 ka, 740 ka, 2.8 Ma and 17 Ma, respectively.

Introduction

During its annual journey around the Sun, at a speed of 30 km s^{-1} the Earth can, and does, hit asteroids and comets. Because of this high velocity, impacting objects with equivalent diameters larger than around 100 m punch through the atmosphere with little retardation and will produce craters larger than about 2 km on the Earth's land surfaces. (All the impactors discussed in this paper are irregular in shape. The equivalent diameter is the diameter of a sphere that has the same surface area).

Researchers into cratering rates have realized that certain parts of the Earth's oldest continents are not only unaffected by continental drift and volcanism, but are also sufficiently stable that it takes on the order of 125 million years for wind erosion and weathering to completely erase craters that are larger than a few kilometers in diameter, and, consequently, were initially over 1 km deep. The findings of this paper suggest that large majority of the craters larger than 2.5 km in diameter that have formed on the stable shield areas of North America, northern Europe and north-western Australia during the last 125 Ma, are not only still detectable, but are also so clear that, with care, their age and initial diameter can be estimated with reasonable accuracy. In fact craters larger than 25 km seem to take longer than 400 Ma to disappear. That said, 400 Ma is still less than 10 % of the total age of Earth and this period is not long enough, nor are the 'stable' regions large enough, to provide an accurate record of the formation rate of craters larger than 100 km.

The number of known terrestrial impact structures has increased rapidly over the last half century. About 20 were known in 1940, 50 in 1970, 100 in 1980 and the present day count is about 160 (see Grieve 1988). Their spatial distribution on the Earth's surface has often been plotted and a recent version of this map is given in Grieve (1998). The geological stability and thus 'crater recording potential' of the North American Craton, the Northern European craton and the north Australian desert stand out clearly. It must be stressed, however, that even though these regions have been carefully explored, there is every expectation that some more craters in the age and size range being discussed by this paper will be found there in the future. With this in mind the cratering rates quoted in this paper should be taken as lower limits.

The estimation of the recent, average, terrestrial cratering rate is relatively simple. The words recent and average are significant. It is widely accepted that the number of asteroids and comets passing through the inner solar system, and thus the crater formation rate, has been decreasing in an approximately exponential fashion since the origin of the Earth. Every now and again, however, a large asteroid or comet breaks up and the consequential production of a new asteroidal family or a new comet shower can lead to a short-lived increase in the crater production rate in certain size ranges.

In the idealized situation where craters are not eroded, and the incident flux of projectiles is constant, the number, $N(D)$, of craters larger than a specific diameter D, that are found on a freshly exposed area, A, of Earth would increase monotonically as a function of the exposure time of the area. If the same conditions apply to the recent past as do to the near future, this is equivalent to saying that a plot of the number of craters $N(D)$, larger than a specific diameter D, that have been found on an uneroded surface, should increase monotonically as a function of age, T, of the surface. Under these circumstances the rate, $\Phi(D)$, at which craters larger than diameter D are produced is given by:

$$\Phi(D) = \frac{N(D)}{T} \cdot A^{-1} \tag{1}$$

Calculating the Cratering Rate

Let us take the North American Craton as an example (see Figure 1). The known craters on this area are listed in Table 1. The diameters of these craters are known to an accuracy of between 5 and 10 %. Unfortunately, with rare exceptions, the crater ages are only known to an accuracy of between 10 and 20 %. Some ages are upper limits, these being the stratigraphic ages of the target rocks.

The area of the North American Craton is about $(1.24 \pm 0.10) \times 10^7$ km^2, this being obtained by adding the areas of the latitude-longitude rectangles contained within the boundary shown in Figure 1 (see Hughes 1999). The error in the area comes from the uncertainty in defining the craton boundary.

Equation 1 can be applied to the data in Table 1 by plotting $N(D)$ as a function of crater age, T. This has been done for four representative crater diameters, 2.4, 8 19 and 35 km, and these plots are shown in Fig. 2. As erosion progressively makes craters less easy to recognize, the $N(D)$ versus T plots are expected to depart from linearity for craters that are older than a certain age. Fig. 2 shows that the plots of crater numbers larger than 2.4, 8 and 19 km are linear up to ages of about 125 ± 15 Ma. This can be interpreted (bearing in mind the provisos mentioned above) as meaning that the North American Craton is so stable that *nearly all* craters larger than 2.4 km, formed during the last 125 ± 15 Ma are still visible and measurable. Erosion is occurring all the time. Take the upper curve in Figure 2 as an example. Its deviation from linearity indicates that after 200 Ma of exposure, the area in question has 16 known craters, whereas the extrapolation of the linear relationship indicates that there should be just over 22. So only about 70 % of the $D > 2.4$ km craters formed on it during that time period are still detectable. After an exposure of 150 Ma, about 85 % would still be detectable; and an exposure for 125 ± 15 Ma ensures that the vast majority (nearly 100 %) of the $D > 2.4$ km craters formed on it during that time period are still detectable.

The fact that $[N(D) / T]$ in equation 1 is the gradient of the lines in Fig. 2 means that the errors in individual diameter and age measurements are subsumed into the gradient error. Also the accurate positioning of the break-point from linearity does not affect the final results.

As is to be expected, all the lines in Fig. 2 pass nearly through the origin. The gradients $[N(2.4) / T]$, $[N(8) / T]$, $[N(19) / T]$ and $[N(35) / T]$ are (0.113 ± 0.005), (0.0827 ± 0.004), (0.0528 ± 0.0009) and (0.0171 ± 0.0010) Ma^{-1} respectively. After 125 Ma the area of $(1.24 \pm 0.10) \times 10^7$ km^2 contains about 2, 7, 10 and 14 craters larger than 35 19, 8 and 2.4 km respectively. If, however, $N(D)$ was proportional to D^{-2}, as previously suggested, the numbers would scale as 2, 8, 43, and 476.

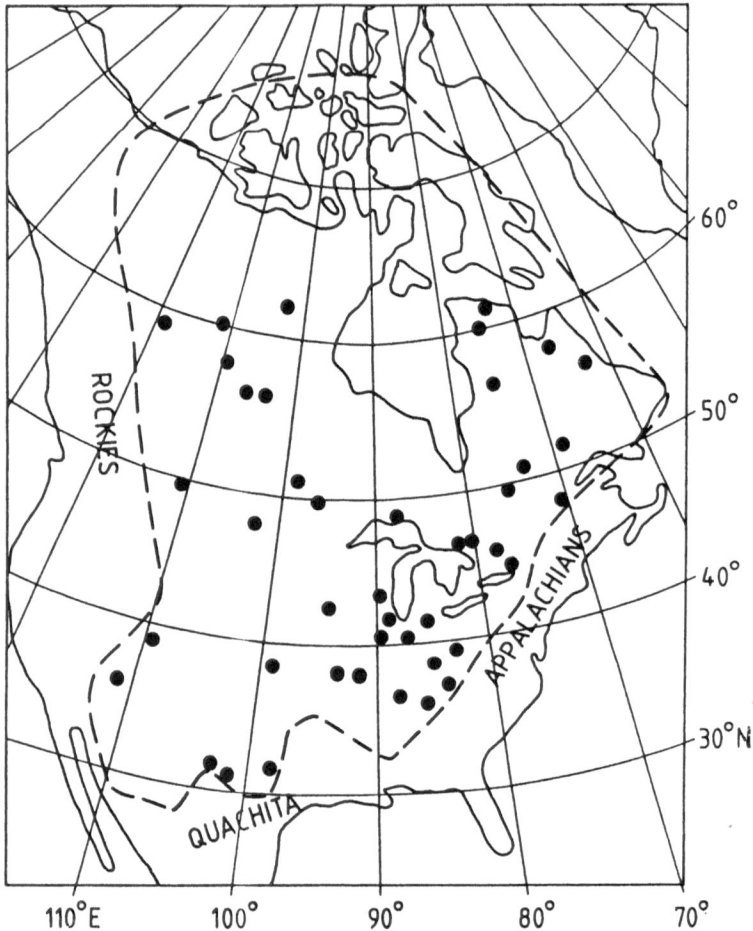

Fig. 1. The stable North American Craton is bounded in the west by the Rockies, in the south and east by the Quachitas and Appalachians and in the north by the Arctic. The black dots represent the craters listed in Table 1.

In Hughes (1999) figures similar to those of Fig. 2 have been produced for Europe and Australia. These show very similar trends and indicate that it is valid to add the data of these three regions together, for craters that occurred in the last 125 Ma.

The combined area of the North American Craton, the European planes craton and the north Australian desert is $(22 \pm 4) \times 10^6$ km^2, a value that is 4.3% of the total area of the Earth's surface (which is 5.11×10^8 km^2). Mindful of the fact that many researchers in the past (e.g., Hughes 1981, and papers referenced therein) have represented certain diameter ranges of the data using power laws such as N (D) μ $D^{-\alpha}$, Fig. 3 shows how the logarithm of the cumulative number, N (D), varies as a function of log D for the craters that have been produced in the last 125 Ma on the North American, European and Australian areas.

Previously, researchers [e.g., Morrison et al. (1994), Grieve (1989), Shoemaker (1983), Hughes (1981) and Grieve and Dence (1979)] would have assumed that only the 'linear' region, $20 < D < 100$ km was valid, and that the deviation from this linearity in the $D < 20$ km region was due to the smaller craters being erased by erosion. To correct for this supposed erosion, and to calculate the expected rate at which small craters are formed, these researchers would then simply extrapolate the linear portion of Fig. 3 to lower and lower diameters. If this procedure was valid one would, however, expect the mean age of the craters to decrease as one considered small and smaller (and thus more easily erased) craters. To test this supposition, the average age of the craters in specific diameter ranges has been calculated and is shown in the small graph at the top of Fig. 3. Notice that (discounting statistical errors) this mean age only deviates from a constant value of 125 / 2, i.e. 62.5 Ma for craters with $D < 2.4$ km. This underlines the conclusion that the data for *all* the $D \geq 2.4$ km craters shown in Fig. 3 is 'true', and that the specific area of $(22 \pm 4) \times 10^6$ km^2 provides a *near complete* record of the $D \geq 2.4$ km crater production that has occurred in the last 125 Ma.

An easy way of representing the data given in Fig. 3 is to note that the average time, T_i (D), between impacts that produce craters larger than diameter D, somewhere on the whole surface of Earth, is given by:

$$T_i = \frac{A_{AEA}T}{A_E N(D)}$$

where A_{AEA} is the cratered area of America, Europe and Australia, (i.e., 22×10^6 km^2), T is the exposure time of this area (i.e., 125 Ma for the main data set in Fig. 3 and 400 Ma for the inset), A_E is the surface area of Earth (5.11×10^8 km^2), and N (D) is the number of craters larger than D found on the area A_{AEA}.

Values of T_i (D) obtained from a smoothed curve through the data of Fig. 3 are given in Table 2. The third column of this table gives the values of T_i (D) quoted by Morrison et al. (1994). For the $D \geq 20$ km data the ratio between the T_i (D) values obtained in the present paper and in Morrison et al. (1994) is about 5.5. This ratio progressively increases as one goes down in size from 20 km to 3 km craters. The present paper concludes, for example, that craters bigger than 3 km are produced 40 times less frequently than predicted by Morrison et al. (1994). Their prediction that craters larger than say 3 km are produced every 4 ka is based on the invalid extrapolation of the $D \geq 20$ km data to lower crater sizes.

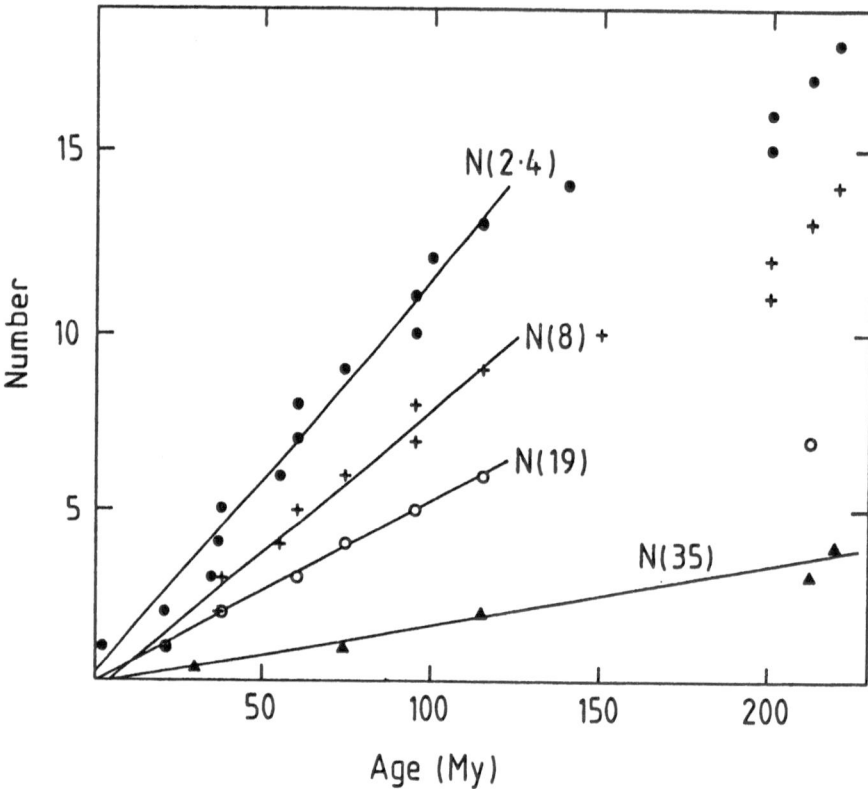

Fig. 2. The number of known craters, $N(D)$ on the North American Craton, having diameters greater than D, is plotted as a function of their age, for values of D equal to 2.4, 8, 19, and 35 km respectively. Notice that the $N(2.4)$ data fall away from linearity for craters older than about 120 My, indicating that craters older than this are under-represented on the Craton. Some of these older craters have been lost due to erosion. This figure indicates that *all* the craters that are both larger than 2.4 km and that have been formed on the area in the last 120 My, have been recognized and recorded.

Their prediction would indicate that the North American region inside the dashed curve in Figure 1, if uneroded, would be peppered with 400 craters, larger than 2.4 km and younger than 115 Ma. Only 13 are observed.

Authors of 'hazard' and 'Armageddon' papers (see for example Chapman and Morrison 1989, 1994) predict that the production of a crater on Earth larger than about 30 km in diameter (threshold energy dissipation 3×10^5 Megatons TNT equivalent yield, i.e., 10^{21} J) will lead to a "Global Catastrophe" or a "Civilization Destroying Event" (i.e., the deaths of more than a quarter of the world's population). According to the present paper (see Table 2) this momentous event occurs about every 1.6 Ma!

Many researchers (see below) have specifically given values for the present-day cumulative Earth surface cratering rate for $D \geq 20$ km craters. Fig 3 indicates that 8.2 ± 0.4, $D \geq 20$ km craters were formed on an area of $(22 \pm 4) \times 10^6$ km^2, in a time period of (125 ± 20) Ma, leading to a cumulative flux of $(3.0 \pm 0.7) \times 10^{-15}$ km^{-2} y^{-1}. Hughes (1981) gave $(2.6 \pm 0.9) \times 10^{-15}$ km^{-2} y^{-1}, Grieve and Dence, (1979) gave $(3.5 \pm 1.3) \times 10^{-15}$ km^{-2} y^{-1}, Grieve (1984) gave $(5.4 \pm 2.7) \times 10^{-15}$ km^{-2} y^{-1} and Grieve and Shoemaker (1994) gave $(5.6 \pm 2.8) \times 10^{-15}$ km^{-2} y^{-1}. The data in Morrison et al. (1994) yields 15×10^{-15} km^{-2} y^{-1}.

Returning to Table 1 it can be seen that the few listed $D \leq 2.4$ km craters are extremely young. Very little can be said about their production rate. I do not recommend extrapolating from the data given in Table 2.

There are two main reasons why so few of these small craters are found on Earth. First many of the potential causative impactors do not reach the surface of our planet. The mass loss and deceleration of an impactor increases as a function of its surface area to mass ratio, and this ratio increases as size decreases. It has been suggested that considerable deceleration occurs when the kinetic energy of the air molecules that hit the incoming object greatly exceeds the energy required to vaporize it.

Table 2. A comparison between average time interval, $T_i(D)$, between impacts producing craters larger than diameter D for the whole Earth's surface found in this paper, and those quoted by Morrison et al. (1994).

D (km)	$T_i(D)$ (years) This paper	$T_i(D)$ (years) Morrison et al.
100	17,000,000	2,700,000
80	11,000,000	2,000,000
60	6,300,000	1,000,000
40	2,800,000	500,000
30	1,600,000	290,000
25	1,100,000	210,000
20	740,000	140,000
15	490,000	80,000
10	340,000	38,000
8	300,000	25,000
6	250,000	15,000
4	180,000	7,000
3	160,000	4,000

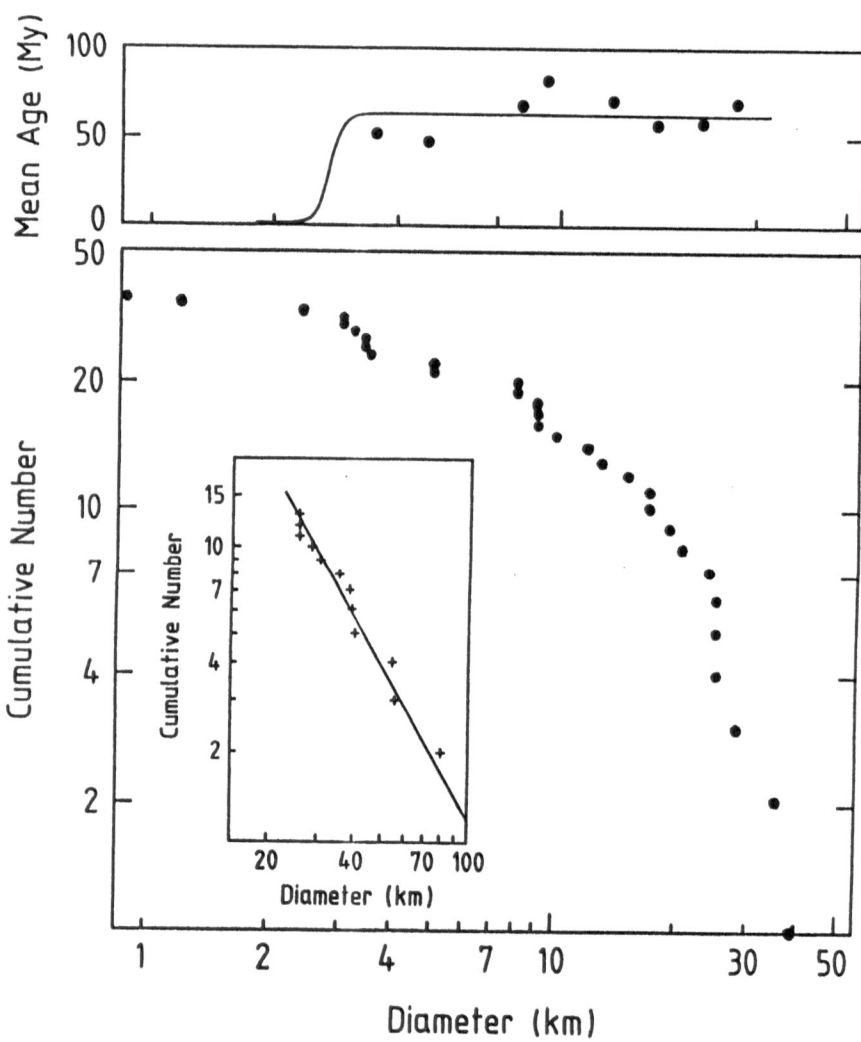

Fig. 3. A logarithmic plot showing the cumulative number, N, of craters larger than diameter D that have been produced in the last 125 My on the combined North American, European and Australian areas. The inset diagram shows the large crater data ($D > 25$ km), this data having been accumulated over 400 Ma. The data at the top of the figure shows the average age of the craters in specific diameter ranges. This only deviates from about 62.5 My for craters with $D < 2.4$ km.

Regarding the atmosphere as a gas layer 8500 m deep and of uniform density 1 kg m^{-3}, and giving an incident asteroid a density of 3650 kg m^{-3}, a specific heat of 1000 J kg^{-1} K^{-1}, a latent heat of sublimation of 6 x 10^6 J kg^{-1} and a trajectory that makes an angle of 45° to the ground, and then vaporizing the asteroid at a temperature of 2100 K after raising the temperature from the space value of 280 K, indicates that that an initial asteroid diameter of 100 m and a final crater diameter of 2 km is a crude limit. Changing the density of the impactor and the angle of incidence can alter this limit. So the 1.2 km diameter Barringer Meteorite Crater in Arizona (age 49,000 y) was produced because the impactor was iron and of high density (and low surface area to mass ratio.) In fact the vast majority of the known, small, terrestrial craters seem to have been formed by incident metallic as opposed to rocky asteroids, this being clear from the residual impactor fragments littering the crater scene. The thirteen authenticated explosive meteorite craters which have classified meteorite fragments in their near vicinity listed in Graham et al. (1985) have sizes in the $0.01 < D < 3$ km range. With the exception of Haviland ($D = 0.01$ km; H5 chondrite) and Brent ($D = 3$ km; L or LL chondrite) they have iron meteorite precursors. [There is even some doubt about Haviland being an explosive crater, and it might just be a large penetration hole that has formed in soft ground.] Remember that $D \sim 3$ km is a watershed. It can be seen from the small graph at the top of Fig. 3 that craters bigger than this size are difficult to erase, whereas smaller craters have a very much shorter lifetime, this explaining why so few of these small craters are known. Meteorite fragments are very rarely found associated with the larger (D > 5 km) impact structures.

Statistics from the meteorites that have been recorded falling to the Earth surface indicate that only about 5 % of the extraterrestrial impactors *are* metallic (Sears 1978). [This percentage probably varies as a function of impactor size]. So iron-metallic impactors penetrate the atmosphere easier than do the crumbly carbonaceous chondrites impactors (which probably have densities of only twice that of water). Small cometary impactors (with densities 0.5 to 0.05 that of water) find it even more difficult to penetrate the Earth's atmosphere.

The Power Law

The main graph of Fig. 3 indicates that it is completely inappropriate to force a single $N (D) \mu D^{-\alpha}$ power law through any part of the $D < 25$ km data. Considering the small numbers of craters available for analysis, it is somewhat difficult to even justify the use of a single power law for the $D > 25$ km data. A plot of number verses age for $25 < D < 100$ km craters is linear up to ages of about 400 Ma. A linear regression fit to the log N (D) verses log D data shown in the inset plot in Fig.3 gives:

$$\log N\ (D) = 3.5 - 1.7 \log D.$$

These data indicates that $\alpha = 1.7 \pm 0.4$, a value that is in reasonable agreement with the $\alpha = 1.9$ result given by Morrison et al. (1994) for the same size range. Similar power laws, close to $D^{-2.0}$, have been found for the Moon. Hughes (1993), for example, re-analyzed the size data for the large craters that are on the Earth-facing hemisphere of the Moon and found:

$$\log N(D) = (5.84 \pm 0.40) - (2.00 \pm 0.19) \log D(\text{km}) \qquad (2)$$

Martian craters have also been found to obey an N (D) μ D^{-2} law (see Hartmann 1973 and Neukum and Wise 1976), as have the $D > 35$ km Venusian craters (see Schaber et al. 1992).

In passing, notice that equation (2) indicates that the Moon's Earth-facing hemisphere contains about 1700 craters larger than 20 km in diameter. This hemisphere has an area of 1.90×10^7 km^2, and has been exposed to a flux of bombarding objects for about 4×10^9 y, so the $D \geq 20$ km craters have been produced at an average rate of 23×10^{-15} km^{-2} y^{-1} over that time period. This is much larger than the present-day flux to Earth, underlining the fact that, among other things, the flux has been decaying over this period.

The number of craters formed per unit time on the Earth's surface is about 20 times that formed per unit time on the Moon's. This is due to a combination of (a) the surface area of the Earth being $(6738.164 / 1738.2)^2 = 13.46$ times greater, (b) the gravitational field enhancement of the collisional cross-section being larger for Earth than for the Moon. [The collisional cross-section area, A_C, of a body of radius r and mass M, and escape velocity, v_E, is given by:

$$A_C = \pi r^2 \left(1 + \frac{GM}{ru^2} \right) = \pi r^2 \left(1 + \frac{V_E^2}{u^2} \right) \qquad (3)$$

where G the constant of gravity and u the relative velocity between the body and the impactor when they are "well apart" (see Encrenaz and Bibring 1990), and (c) the fact that asteroids hit the Earth at an average velocity of 20.8 km s^{-1}, whereas they hit the Moon at 17.7 km s^{-1} (Harris and Hughes 1994), so mass for mass they produce larger craters on Earth than they would on the Moon. [The difference between the acceleration of gravity at the surface of the two bodies also slightly affects the final crater dimensions because material is ejected further on the Moon than it is on Earth.]

The relationship between the diameter, d, and kinetic energy, E, of an incident body and the diameter D of the crater that it produces on a planetary or satellite surface is difficult to assess, especially in the size range being dealt with in this paper. Quoted examples rely heavily on extrapolation because craters in the $3 < D < 100$ km range, have *not* been produced experimentally using known amounts of energy. A handful of small $(0.066 < D < 0.368$ km), bowl-shaped craters have, however, resulted from surface nuclear bomb tests. Here it was found that:

$$\log E \text{ (erg)} = (23.02 \pm 0.03) + (3.18 \pm 0.03) \log D \text{ (km)} \tag{4}$$

where E is the energy of the nuclear explosion and D is the diameter of the crater that was formed (see Hughes 1998). This equation will be applied to impact phenomena, even though there is clearly a considerably difference between the type of explosion taking place. Typical impacting asteroids have velocities of 21000 m s^{-1}, and densities of around 2000 kg m^{-3} (this density, needless to say, is a rough estimate, but takes into account the fact that most asteroids seem to have some type of macro-porosity and that recent spacecraft observations indicate that the mean S-type asteroid has a density of 2800 kg m^{-3} and the mean C-type a density of 1300 kg m^{-3}). Replacing E by the "in space" kinetic energy of the impacting asteroid gives D / d values of 28.0, 24.5, 23.4, and 21.3 for typical 1, 10, 20, and 100 km craters.

Moon Craters

Table 1 indicates that the most commonly found terrestrial craters are about 10 km across, and equation 4 shows that these are produced by asteroids that have diameters of about 0.4 km. Bearing in mind the differing mean impact velocities, it can be seen from equation 4, and data given above, that a 20 km crater on Earth is formed by a 6.5×10^{11} kg impacting asteroid, whereas a 20 km crater on the Moon is formed by a 9.4×10^{11} kg asteroid. The size distribution of asteroids (see Hughes and Harris 1994) indicates that there are about 1.47 more of the former than the later. Secondly the collisional cross-section given by equation 2 indicates that unit area of Earth is $[1 + (11.2^2 / 20.8^2)] / [(1 + (2.38^2 / 17.7^2)]$, i.e., 1.27 times more effective at attracting impactors than unit area on the Moon.

Multiplying these two factors together indicates that the formation rate, per unit area, of 20 km craters is 1.86 times higher on Earth than on the Moon.

The Moon has no continents, no continental drift, no extant volcanoes, no atmosphere, no wind and is dry. So, unlike Earth, large craters on the Moon have lasted for at least three billion years (i.e., throughout the post volcanic Eratosthenian and Copernican period, see Cadogan 1981). Samples returned from the Apollo Missions have helped astronomers date different regions of the Moon, and the number of craters per unit area on those regions can easily be measured from Earth and from lunar orbit. It has been found that the lunar cratering rate has been decreasing approximately exponentially as the Solar System gets older. Neukum and Ivanov (1994) quote:

$$\phi\,(1) = 3.77 \text{ x } 10^{-13} \exp{(6.93\ t)} + 8.38 \text{ x } 10^{-4} \tag{5}$$

where $\phi\,(1)$ is the cumulative number of craters larger than 1 km in diameter that were produced on every km^2 of the lunar surface every year at a time t aeons in the past (an aeon is 10^9y). So even though the rate has been reasonably constant for the last 500 Ma or so, it was about 1.6 times greater 1500 Ma ago, and 2.6, 6, and 1900 times greater 2500, 3500 and 4500 Ma ago.

This exponential function has an important consequence. The Earth, with its erosion, continental drift and mountain building, has a very new surface, which has been recording the impact damage produced by the geologically recent (i.e., last 125 Ma) population of interplanetary bodies. The Moon, however, has a very old surface, which has, in the main, recorded craters produced by earlier populations of impacting bodies. Considering a region of lunar surface that is 4.2 x 10^9 y old one finds that half the craters on it were produced in the time period between 4.2 x 10^9 and 4.10 x 10^9 y before the present and the other half in the time interval between 4.10 x 10^9 y and now. Likewise on surfaces that are 4.1 x 10^9, 4.0 x 10^9, 3.5 x 10^9, 3.0 x 10^9, 2.0 x 10^9 and 1.0 x 10^9 y old the median age of the craters is 3.99 x 10^9, 3.89 x 10^9, 2.73 x 10^9, 1.60 x 10^9, 1.0 x 10^9, and 0.5 x 10^9 y old.

Needless to say, there is no reason why the mass distribution of the impacting population of asteroids should not have changed during this time interval. Hughes (1991) suggested that the solar system's family of asteroids initially had not only a total mass that was over a thousand times greater than it is today but also that the mass distribution index, s, of the early population was about 1.65 (the value obtained by an accreting family of bodies) whereas the population today has an index of about 2.00 (the value that is produced after a great deal of collisional fragmentation). [The mass distribution index is defined such that the number of particles with masses between m and $m + dm$ is proportional to m^{-s}, and the number of particles with masses greater than m is proportional to $m^{(1-s)}$.]

If the mass distribution index of the asteroidal impactors changes as a function of time, so does the size distribution of the resultant craters. Unfortunately the way in which the asteroidal s value has varied with time is not known empirically. The evolution of asteroidal size distribution has, however, been modelled by Wetherill (1989).

Discussion

An indirect route could potentially establish the form of Fig. 3. This would require both

(a) a detailed knowledge of the size distribution of Near Earth Objects (NEOs) with diameters between about 0.1 km and 5 km, coupled with a knowledge of their orbital statistics. This would enable us to calculate the rate at which these objects struck Earth. And

(b) an understanding of the relationship between the kinetic energy, density, strength and physical form (i.e., whether they are solid rocks or loose aggregates of smaller bodies) of the impacting object and the diameter of the crater that is produced when that object hits the surface of Earth.

Unfortunately neither of these are known with any degree of certainty. A NEO search program in the volume of space of radius of about 0.1 AU (i.e., about 40 Earth-Moon distances) will not be able to detect all the 0.1 to 5 km asteroids that pass through. And the relationship between crater diameter ($2.5 < D < 100$ km) and possible impactor kinetic energy ($10^{23} < E < 10^{30}$ ergs) is a matter of considerable debate, and way beyond the limits of experimental testing (see Hughes 1998).

The main conclusion of this paper is that the curve shown in Fig 3 is a true representation of the way in which the Earth's crater numbers vary with size in the $2.4 < D < 40$ km diameter range. This distribution is clearly not represented by a simple power law. Even though the $D > 20$ km results presented in this paper agree in general with the results present by many other researchers, we contend that the production rate of the smaller craters is much less than previously thought.

References

Cadogan PH (1981) The Moon - our sister planet. Cambridge University Press, Cambridge 362 pp

Chapman CR, Morrison D (1989) Cosmic Catastrophes. Plenum Press, New York, 302pp

Chapman CR, Morrison D (1994) Impacts on the Earth by asteroids and comets: assessing the hazard. Nature 367: 33 - 40

Encrenaz T, Bibring J-P (1990) The Solar System. Springer-Verlag, Berlin Heidelberg, New York, 349 pp

Graham AL, Bevan AWR, Hutchison R (1985) Catalogue of Meteorites. British Museum (Natural History), London, 421 pp

Grieve RAF (1984) The impact cratering record in recent time. J. Geophys Res Suppl 89: B403 - B407

Grieve RAF (1989) Hypervelocity impact cratering: a catastropic terrestrial geologic process. In: Clube SVM (ed) Catastrophes and Evolution: Astronomical Foundations, Cambridge University Press, Cambridge, pp 57-79

Grieve RAF (1998) Extraterrestrial impacts on earth: the evidence and the consequences. In: Grady MM, Hutchison R, McCall GHJ, Rothery DA (eds) Meteorites: flux with time and impact effects, Geol Soc Lond Spec Pub 140, pp 105 - 131

Grieve RAF, Dence MR (1979) The Terrestrial cratering record II. The crater production rate. Icarus 38: 230 - 242

Grieve RAF, Shoemaker E.M (1994) The record of past impacts on Earth. In: Gehrels T (ed) Hazards due to comets and asteroids, University of Arizona Press, Tucson, pp 417 - 462

Harris NW, Hughes DW (1994) Asteroid-Earth collision velocities. Planet Space Sci 42: 285 - 289

Hartmann WK (1973) Martian cratering 4. Mariner 9 initial analysis of cratering chronology. J. Geophys Res 78: 4096 - 4116

Hughes DW (1981) The Influx of Comets and Asteroids to the Earth. Phil Trans Roy Soc Lond A 303: 353 - 368

Hughes DW (1991) The largest asteroids ever. Q J Roy Astr Soc 32: 133 - 145

Hughes DW (1993) Meteorite Incidence Angles. J Br Astron Assoc 103: 123 -126

Hughes DW (1998) The mass distribution of crater-producing bodies In: Grady MM, Hutchison R, McCall GHJ, Rothery DA (eds) Meteorites: flux with time and impact effects, Geol Soc Lond Spec Pub 140, pp 31 - 42

Hughes DW (1999) The Cratering Rate of Planet Earth. J British Interplanetary Society 52: 83 - 94

Hughes DW, Harris NW (1994) On the distribution of asteroid sizes and its significance. Planet Space Sci 42: 291 - 295

Hodge P (1994) Meteorite Craters and Impact Structures of the Earth. Cambridge University Press, Cambridge, 124 pp

Morrison D, Chapman CR, Slovic P (1994) The Impact Hazard. In: Gehrels T (ed) Hazards due to Comets and Asteroids. University of Arizona Press, Tucson, pp 59 - 91

Neukum G, Wise DU (1976) Mars: A standard crater curve and possible new time scale. Science 194: 1381 - 1387

Neukum G, Ivanov B A (1994) Crater size distributions and impact probabilities on Earth from lunar, terrestrial-planet, and asteroid cratering data. In: Gehrels T (ed) Hazards due to Comets and Asteroids. University of Arizona Press, Tucson, pp 359 - 416

Schaber GG, Strom RG, Moore HJ, Soderblom LA, Kirk RL, Chadwick DJ, Dawson DD, Gaddis LR, Boyce JM, Russel J (1992) Geology and distribution of impact craters on Venus: What are they telling us? J Geophys Res 97: 13257 - 13301

Shoemaker EM (1983) Asteroid and comet bombardment of Earth. Ann Rev Earth Planet Sci 11: 461 - 494

Sears DW (1978) The nature and origin of meteorites. Oxford University Press, Oxford, New York, 40 pp

Wetherill GW (1989) Origin of the asteroid belt. In: Binzel RP, Gehrels T, Mathews MS (eds) Asteroids II. University of Arizona Press, Tuscon, pp 661-680

14 Impact Melting of Carbonates from the Chicxulub Crater

A. P. Jones[1], P. Claeys[2] and S. Heuschkel[2]

[1]Department of Geological Sciences, University College London, Gower Street, London, WC1E 6BT, United Kingdom. (adrian.jones@ucl.ac.uk)
[2]Institut fur Mineralogie, Museum fur Naturkunde, Invalidenstrasse 43, Berlin, Germany.

Abstract. We have recently interpreted distinctive feathery-textured "spinifex" carbonate in the upper part of the Chicxulub suevite breccia as quenched carbonate melts (Jones et al. 1998); these distinctive fragments make up to ~10 vol% of the breccia. Carbonate clasts and spherules occurring in the ejecta-rich basal part of the coarse clastic sequence, which marks the K/T boundary all around the Gulf of Mexico, may represent distal quenched droplets of carbonate liquids. In seeking to explain this widespread carbonate impact-melting phenomenon, we have re-examined the available experimental evidence. The important decarbonation reaction for calcite $CaCO_3 = CaO + CO_2$ is inhibited by very small pressures up to temperatures >2000 K. We conclude that massive decarbonation by direct shock pressure is unlikely without attainment of temperatures >~4000 K. Therefore, decarbonation generally can only occur during post-shock cooling for carbonates at low pressure (~< 10 bars). We assume that post-shock cooling is quasi-thermodynamic, and provide a general P-T model for carbonate spanning 11 orders of magnitude in pressure (atmosphere to core). Subtle differences in sample preconditioning can probably explain the wildly divergent experimental shock data. A major planetary implication for the formation of the Earth's early atmosphere is that impacts on limestone would be less likely to have contributed substantial CO_2 than has previously been assumed. Lastly, we note that carbonate melts at high pressures serve as excellent catalysts for diamond growth, and may have contributed to the widespread formation of some impact diamond.

Introduction

The shock degassing of impacted carbonate- and sulphate-bearing rocks is a very important feature of, for example, the Chicxulub impact event that has global atmospheric implications. However, two lines of evidence suggest to us that this phenomenon may have been seriously oversimplified in previous work based on assumptions of the behaviour of carbonates under shock conditions; (i) petrographic identification of abundant carbonate melt textures (ii) reassessment of experimental data.

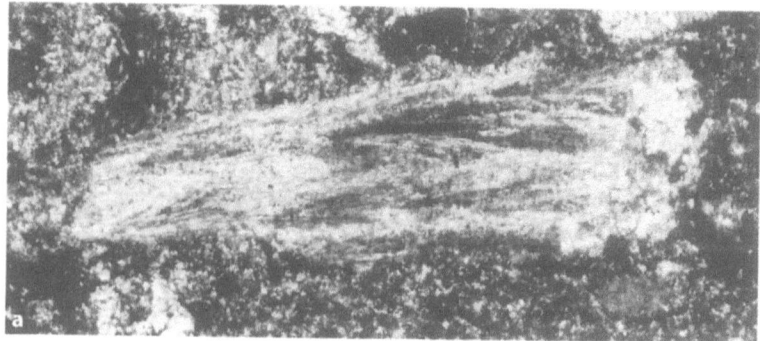

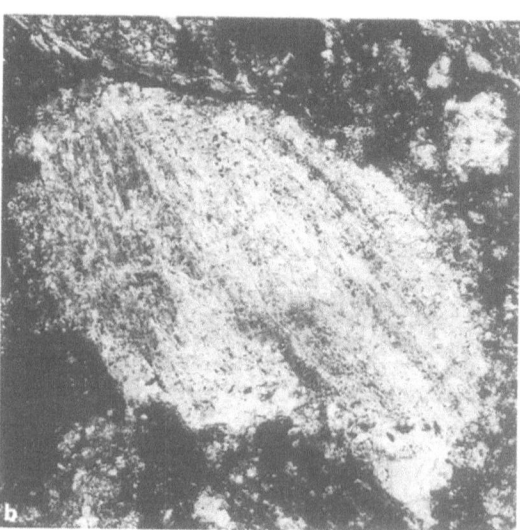

Fig. 1.   Petrographic views of N13 feathery carbonate textures (a and b) interpreted here as *quenched carbonate melts* (see text). Field of view ~ 3 mm; cross-polarized light.

Quenched Carbonate Melt Textures

Feathery, or Spinifex Textures

At Chicxulub we have recently interpreted feathery-textured carbonate in the sue-vite breccia as quenched carbonate melts (Jones et al. 1998). These distinctive carbonate fragments make up to ~10 vol.% of parts of the breccia, with skeletal branching and radiating aggregates of calcite crystals. They occur as rounded to subangular ~mm-sized fragments randomly dispersed in the carbonate breccia (Fig. 1). Preliminary cathodoluminescence (CL) data shows that all types of carbonate fragment in the suevite (micrite, fossiliferous limestone, secondary calcite, and matrix carbonate) show luminescence, except for the feathery calcite (Heuschkel et al. 1998). This could be due to *luminescence quenching* by dissolved iron in the quench carbonate, but awaits confirmation by electron microprobe analysis. Additional geological and petrographic details of the suevite rocks are provided in Appendix I. Texturally they are carbonate analogues of the well known silicate spinifex textures which characterize komatiites in Archean greenstone belts. They also resemble spinifex silicates and "barred" olivine seen in chondrules. These textures are a well understood quenching phenomenon, and have been carefully synthesized in laboratory experiments for olivine. They occur when quench rates are high, in liquids with low nucleation densities (Donaldson 1974) and may reflect thermal conduction paths (e.g., Shore and Fowler 1999). The diagnostic interpretation of radiating quenched crystal aggregates of carbonates as representing quenched carbonate liquids, is routine practice in petrologic experiments over a wide range of pressures and temperatures (e.g., Hamilton et al. 1979; Wyllie 1989; Jones et al. 1998). We present a schematic diagram to illustrate the changes in calcite morphology with degree of undercooling ($\Delta T^\circ C$) in Fig. 2. Typical laboratory quench rates of up to 400°C per second do not produce carbonate glass. This is because carbonate liquids are essentially ionic, and do not polymerize; thus carbonate glasses have not been found in nature, and are not generally expected (Genge et al. 1995). Carbonatite lavas formed from igneous carbonate melts, also show similar quench textures, and where vent-fountaining has produced spherical lapilli ejecta, these too are completely crystalline (Keller 1981; Church and Jones 1995), despite very fast quench rates. These volcanic carbonate lapilli quenched through ~500°C in a few seconds and are quite similar to the subangular, carbonate fragments seen in the Chicxulub suevite. They are also similar to spherical carbonates recently found in bedded spherulite deposits in Southern Mexico where carbonate spherulites are often cored by silicate materials (Grajales et al. 1996; Salge et al. 1998). Further work to establish the true nature of these distal ejected carbonate spherulites is planned (see below).

Carbonate Clasts, Spherules and Lapilli

Exotic carbonate material also occurs in the base of the coarse clastic sequence, which marks the K/T boundary all around the Gulf of Mexico. The basal unit is essentially composed of a mixture of solid, shocked and molten ejecta-curtain material expelled on ballistic trajectories from the Chicxulub crater (Smit et al. 1996, Claeys et al. 1996, Claeys et al. 1998). In this unit, closely associated with shocked quartz and clay spherules, some with a preserved impact glass core, are numerous irregular clasts of limestones with subangular outlines that show that the clasts were solid at the time of emplacement (see Fig. 3). These clasts are composed of tiny euhedral calcite crystals that are free of deformation and have a remaining texture of annealed carbonate marbles. Many limestone clasts contain subrounded to circular bodies of coarser calcite, which appears to be filling vesicles, or which could represent more complex melt-solid phenomena. Locally the limestone clasts are in close contact with altered impact silicate glass, suggesting that they might have pressed together when molten. The presence of dissolved carbonate in the Chicxulub ejecta curtain is implied by the CaO-rich (up to 29 wt%) end member composition of the impact glasses. Major and trace element compositions as well as isotopic data suggest that this yellow glass originates from the mixing of the "granitic" Chicxulub basement with overlying carbonates (Blum and Chamberlain 1992; Claeys et al. 1993).

The recently discovered one meter thick carbonate-rich accretionary lapilli bed overlying the K/T boundary breccia in Southern Mexico (Grajales et al. 1996) is an important, although yet poorly understood component of the Chicxulub ejecta. At this moment we only have a preliminary description of this unit: the cm-size lapilli are composed of a silicate core surrounded by concentric layers of carbonate. Shocked quartz grains are found in the same bed. The K/T age of the unit is well constrained by the stratigraphy and Ir analyses are underway.

Unfortunately, despite its abundance the significance of the carbonate component in the Chicxulub ejecta material has until now been overlooked, and in fact most material has been dissolved in order to search the frequently included silicate cores for shocked minerals. We are now in the process of re-evaluating the role of carbonate in the ejecta, but clearly our initial conclusion is that they may represent an important additional piece of evidence in support of the melted target carbonates ejected from the Chicxulub crater. The limestone clasts found all around the Gulf of Mexico at a distance of ~1000 km from the crater, and the carbonate lapilli found in Southern Mexico, some ~500 km from the crater may include distal carbonate melt ejecta (Fig. 3). We do not yet know their relationship with the spinifex carbonate melts found in the proximal Chicxulub suevites, as described above, but the bulk chemistry of the accretionary lapilli bed is rather similar to the upper part of the Chicxulub suevite.

Immiscible Melt Textures?

We have not yet found direct evidence for carbonate-silicate immiscibility in the Chicxulub suevite; however, we suggest that it should occur, for reasons given below. Liquid immiscibility has been recognized in the West Clearwater Lake impact structure (Dence et al. 1974) and most recently at the Ries crater (Graup 1999; e.g., Fig 4).

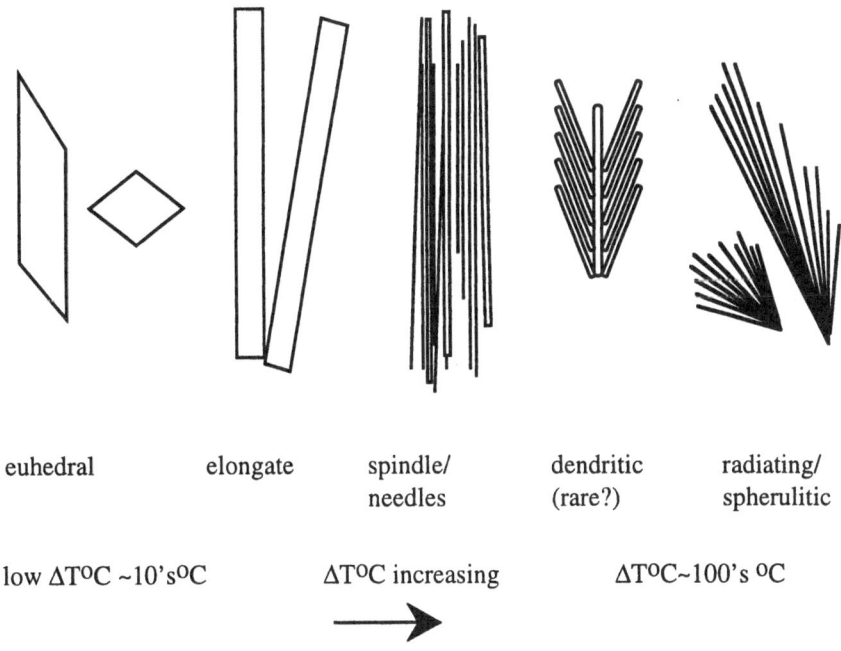

| euhedral | elongate | spindle/
needles | dendritic
(rare?) | radiating/
spherulitic |

low $\Delta T^{o}C$ ~10's^{o}C $\Delta T^{o}C$ increasing $\Delta T^{o}C$~100's ^{o}C

$\longrightarrow$

Fig. 2. Carbonate crystallization textures (stylized) as a function of cooling rate/degree of undercooling (DT °C).

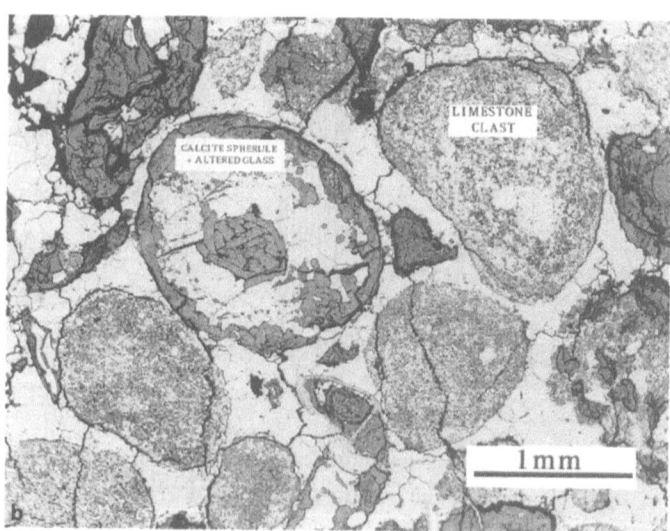

Fig. 3. (a) Field outcrop of lapilli bed. (b) Petrographic view of carbonate clasts from carbonate-rich accretionary lapilli bed overlying KT boundary breccia in southern Mexico: mm-sized fragments of limestone are abundant in the basal ejecta unit forming the KT boundary coarse clastic sequence all around the Gulf of Mexico (Claeys et al. 1998). They often display rounded or elongated morphologies. They are closely associated with impact glass, locally the rounded carbonate clasts, the silicate glass and the limestone clasts seem to be pressed into each other as if they had come in contact as molten material in flight. The carbonate clasts are composed of tiny (10 µm size) euhedral calcite crystals that are free of deformation and have a texture suggestive of annealing or resembling carbonate marbles. Dark shades are altered silicate glass, light coloured areas are predominantly carbonates.

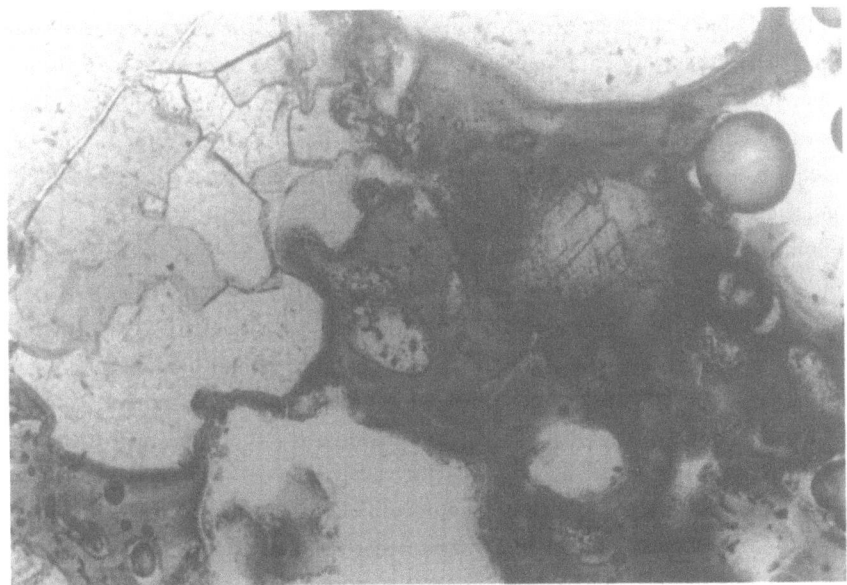

Fig. 4. Immiscible texture in vesicular-glass bearing suevite from Ries crater (APJ sample). Light coloured rounded blobs of carbonate in centre field of view are surrounded by dark silicate glass showing additional spherical objects at smaller scales. Coarse-grained calcite (limestone fragment) at top and left is clearly embayed by silicate melt and arrested in process of melting. Width of field ~1 mm.

We have observed abundant sometimes "coagulated" rounded textures, usually sub-mm scale (not shown), of fine-grained carbonates whose strongly curved surfaces are reminiscent of meniscus-controlled silicate-carbonate immiscible phenomena. The nature of the suevite however, means that both carbonate and silicate melts are now in juxtaposition with an array of variable lithic and impacted fragments, including recognizable fossil relics which are apparently unshocked. We are currently exploring cathodoluminescence methods to identify different generations of carbonate crystallization. This is important because in some geological situations, secondary carbonate textures can produce rather similar features (e.g., carbothermal alteration or calcretisation). We expect in due course to find immiscible phenomena, since they are particularly common in rapidly quenched experiments. The strongly curved surfaces at the interface between carbonate droplets and silicate melt are caused by radical differences in physical properties between the two liquids, including surface tension (Jones et al.

1995, Dobson et al. 1996, Genge et al. 1995). In fact with cooling from high temperatures, the carbonate droplet phase could well remain liquid to temperatures below the silicate glass transition (Genge et al. 1995). Under equilibrium conditions, the occurrence of two-liquid fields in mixed silicate-carbonate systems have been extensively explored as a genetic mechanism to explain the origin of carbonatite magmas (e.g., Hamilton et al. 1979; Wyllie 1989). It is also well known from static experiments that disequilibrium conditions favour the appearance of immiscible carbonate-silicate phenomena, so called "quench immiscibility".

Some evidence for immiscible carbonate melts may be present in the Haughton crater, formed almost entirely in carbonate target rocks. Here, Martinez et al. (1994a) illustrate rounded carbonate blebs in quenched silicate (pyroxene) which they have interpreted (their Fig. 2) as carbonate vapour condensates from recombined shocked oxides. We suspect that these textures, and others described by the same authors, are simply immiscible textures developed between coexisting silicate and carbonate melts. The partially vacated space in the carbonate blebs representing coexisting vapour. As mentioned above, such immiscible textures are quite common in association with natural carbonatites, and can be produced experimentally with ease; they are generally more common during situations of rapid cooling, or quenching. We suggest that because of carbonate melting and rapid quenching, silicate-carbonate immiscibility is also likely to occur in large carbonate sediment (~limestone) dominated impact craters; immiscible carbonate-silicate textures have recently been reported from the Ries crater (Graup 1999).

Decarbonation Experiments

We have attempted to combine experimental shock data, with static high P-T data; this also includes an assessment of calculated equation of state (EOS) data. Several groups have attempted to quantify degassed CO_2 from carbonate during laboratory shock experiments, with spectacular lack of agreement (Boslough et al. 1982, Lange and Ahrens 1983 1986, Martinez et al. 1995). Most of these experiments determined evolved CO_2 gas directly, but Martinez et al. (1995) also searched for the decarbonated oxides as reaction products (CaO and MgO). Direct shock methods for most of the data, appear to yield different results from the reverberation method of Martinez et al. (1995). The data is summarized below in Table 1.

The important decarbonation reaction for calcite $CaCO_3 = CO_2 + CaO$, is inhibited by even very small pressures (about 40-90 bars at 900-1200°C; Huang and Wyllie 1976) even up to temperatures of at least >2000 K. So, decarbonation can (generally) only occur during post-shock cooling for carbonate at near atmospheric pressure. Thus, target carbonate platforms at burial pressures equivalent to depths of just a few hundred metres, would not decarbonate, but would melt instead; in other words only exposed surficial limestone would be likely to decarbonate. In the case of the Chicxulub crater, where the sedimentary cover was ~2-3 km thick (Morgan et al. 1997), we will soon be able to make better estimates of the amount of decarbonation, based on further studies to determine the extent and distribution of melt within the crater (see below).

Water has two additional effects in promoting melting and inhibiting decarbonation; a vertical column of water ~ 1/2 km deep would contribute sufficient bottom pressure to prevent limestone decarbonating (at least for T< 3500 K), and water in limestone (porous) would enhance the degree of melting by greatly lowering melting points. The disagreements between the shock experiments are possibly explained by subtle variations in partial pressures of CO_2 and shock duration (Martinez et al. 1995, Tyburczy and Ahrens 1986), and/or water content and porosity of the pre-shock samples.

Table 1. Shock data for decarbonation.

	Shock Method	Carbonate	Pressure	CO_2-loss?
Boslough et al. 1982	Gun	calcite	18 GPa	0.03-0.3 mol%
Lange and Ahrens 1983 1986	Gun	calcite	10 GPa	some
			20 GPa	30-40%
			30 GPa	50%
Kotra et al. 1983	Gun	calcite	50-60 GPa	5-10%
Martinez et al. 1994 1995	Plate, reverb	dolomite, calcite	60 GPa	almost none, (but MgO)

Decarbonation P-T Diagram

Decarbonation curves were calculated for both calcite and dolomite using EOS data by Martinez et al. (1995); this included a calculation method extended to high temperatures approaching 4000K, where there are no experimental data. We have represented the P-T data for carbonate from static experiments together with the calculated decarbonation curves of Martinez et al. (1995) in Figure 5.

The two forms of data have not been rigorously combined, and this somewhat unconventional diagram is meant for illustrative purposes only. In particular, detailed representation of data precision has not been thoroughly assessed, and some of the slopes for phase boundaries (dT/dP) are approximate only. By plotting pressure on a log scale ($P = PCO_2$) we are able to consider more than eleven orders of magnitude pressure on one diagram, extending from atmospheric all the way to core pressures. The most important features are the decarbonation curves, and the extensive field(s) of carbonate liquid. The decarbonation curves for both calcite and dolomite are subparallel, and can be considered together. They show two distinct regions; a high temperature region (>3500K) that is insensitive to pressure, and a low temperature region with positive dT/dP. For a wide range of shock conditions direct decarbonation can only occur for temperatures in excess of ~3500 - 4000K. Decarbonation in shock experiments that do not achieve these conditions must occur during post-shock cooling, as recognized by Martinez et al. (1995). Under these conditions, decarbonation can only take place under very low ambient pressures. Experiments by Huang and Wyllie (1976) place limits of 40-90 bars pressure (at 1200-900°C), which is reasonably similar to calculated values of ~1250K for 10 bars by Martinez et al. (1995).

The liquidus curve for calcite (Huang and Wyllie 1976) is similar to that for dolomite (Irving and Wyllie 1973) and has only been determined over a small "crustal " pressure interval. The invariant point "q" where solids (calcite and CaO) liquid and vapour coexist is shown; in this region calcite melts incongruently. Further details, but not essential for the following arguments, can be found in Tyburczy and Ahrens (1986). With increasing temperature above the liquidus, or at lower pressures, carbonate melt would coexist with increasing amounts of· vapour (CO_2). Although there is very little data on measured solubility of CO_2 in calcite melt (~11 % CO_2 ; Irving and Wyllie 1973), it is known that carbonate melts can dissolve additional CO_2. This complicates the picture of CO_2 release (from the liquid plus vapour phase field in Fig 5) and clearly more solubility data are needed.

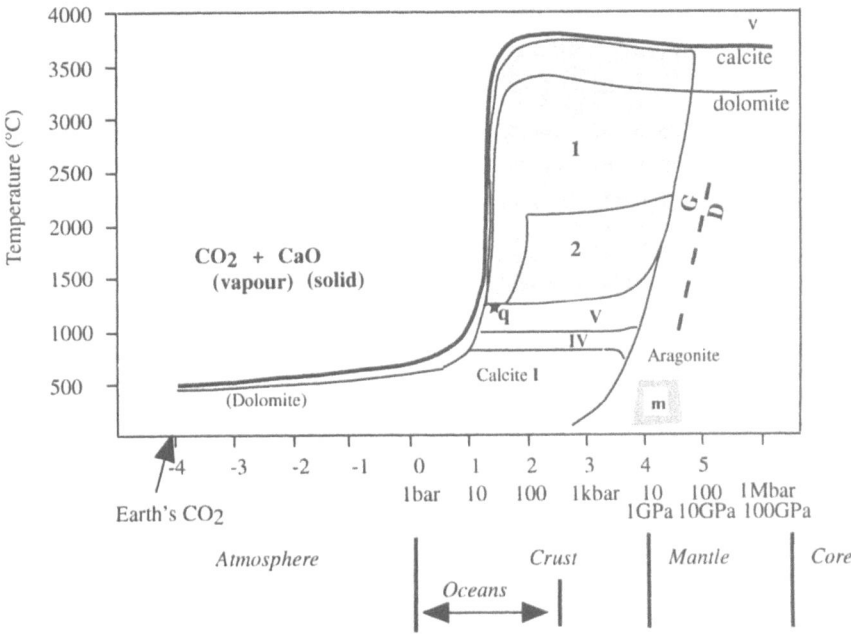

Fig. 5. Schematic P-T diagram for the decarbonation of calcite (heavy continuous curve). Mainly constructed using data from Tyburczy and Ahrens (1986), Martinez et al. (1995) and Pierazzo et al. (1998). The additional decarbonation curve for dolomite (light curve) is shown for comparison (from Martinez et al. 1995). Invariant point "q" was determined experimentally by Huang and Wyllie (1976) and Irving and Wyllie, (1973). Roman numerals refer to various high pressure polymorphs of calcite, and dashed curve G-D is graphite-diamond stability curve. Hatched fields "1" and "2" are different interpretations of the extent of calcite liquid. Both fields "1" and "2" coexist with some CO_2 vapour but the details are unknown; see text or Pierazzo et al. (1998) for discussion. In general, many post-shock peak P-T conditions will decompress through the fields for liquid carbonate; higher temperatures will produce more CO_2.

The two hatched fields in Fig 5 are the most important for this paper; they represent areas where carbonate liquid can potentially exist, coexisting with varying amounts of CO_2 vapour. Field "1" is bounded by the upper temperature limit for calcite decarbonation as calculated by Martinez et al. (1995). The decarbonation curve for dolomite is subparallel to calcite but displaced to slightly lower temperatures. In general we can expect the behaviour of calcite and dolomite to be rather similar. Field "2" is the field for calcite liquid (and calcite liquid plus CO_2 vapour) as presented by Tyburczy and Ahrens (1986). Note that the field for liquid and liquid plus vapour (labelled "1" in Fig. 5) estimated by Martinez et al. (1995) extends to significantly higher temperatures and over a wider range of pressure than the earlier estimate of Tyburczy and Ahrens (1986): labelled "2" in Fig. 5. This expanded liquid field is purely theoretical, but is consistent with a recent calcite phase diagram calculated from EOS data by Pierazzo et al. (1998). We take field "1" to be indicative of the P-T field for appearance of calcite liquid. It is clear that a wide variety of unloading paths (post-shock; not shown) will be likely to pass through field "1", leading to carbonate liquid, plus some amount of vapour (CO_2) as yet unquantified. Because of the very low decarbonation pressure, any carbonate liquids so generated would only decarbonate if they were to remain liquid to near atmospheric pressure. This is consistent with the known stability of natural carbonate melts over a range of crustal pressures. The Earth's oceans may be quite significant in determining carbonate melting since small pressures (tens of bars) occur in the critical region where the calcite decarbonation curve has very steep dT/dP. Also shown in Fig. 5 is the proximity of the graphite-diamond curve. The lack of overlap between the calcite liquid field ("1") and stability of diamond is largely compositional; dolomite and then magnesite ($MgCO_3$ not shown) replace calcite at higher pressures relevant to mantle conditions (Katsura and Ito 1990; Dobson et al. 1996), and could exist as a liquid at pressures within the diamond stability field (see next).

Carbonate Melts and Diamond Growth?

The magnesium carbonate, magnesite ($MgCO_3$) decarbonates at much higher pressure than calcite or dolomite, of around 22 kbar or 2.2 GPa (Irving and Wyllie 1973). Magnesite is generally only thought to occur as a very minor component in the Earth's mantle. Nonetheless, one shock experiment has successfully synthesized diamond by direct shock from magnesite, incorporating the reduction reaction $MgCO_3 + Fe = FeO + CO_2$ (Sekine 1987). Diamond is of course an important shock indicator mineral. Therefore, also shown for reference on Fig. 5 is the position of the diamond-graphite stability curve. We infer that under certain circumstances, it may be possible for diamond to form in the presence of shock-induced carbonate melt, though this would require a multi-stage process, changes in oxygen fugacity and persistence of high pressures in the carbonate melt for sufficient time. As shown in Fig. 5, carbonate melts of pure calcite composition do not overlap the diamond stability field; Mg-rich carbonate such as dolomite or magnesite ($MgCO_3$; not shown) however, could (Taniguchi et al. 1996). This seems reasonable in principle, because carbonate melt is known to be an excellent catalyst for diamond growth (Taniguchi et al. 1996). The diamond precursor form of carbon could be graphite, which follows conventional interpretations, or in a multistage process, the carbon itself may be derived from the carbonates. Recent interpretation of experiments involving one of us (APJ: Taniguchi et al. 1996) for synthetic diamond grown from carbonate-graphite mixtures shows stable isotopic signatures consistent with carbon contribution to the grown diamond both from the carbonate melt and from the starting graphite. This differs from our initial interpretation of the growth mechanism (Taniguchi et al. 1996) and offers a new potential mechanism for impact-related diamond growth. Diamond grown from carbonate melts is typically cubo-octahedral (Fig. 6) compared with diamond formed by shock transformation of graphite, which is irregular and elongate (e.g., Hough et al. 1995).

Fig. 6. Diamond grown from synthetic carbonate melts (system graphite-$K_2Mg(CO_3)_2$ at 9 GPa, 1650°C, acid washed to remove carbonate after experiment; note morphology is dominated by euhedral crystals with cubo-octahedral form showing some elongation (after Taniguchi et al. 1996).

Discussion

Studies of terrestrial craters suggest that impact melt volumes typically comprise 1-5 % equivalent of the total crater volume excavated. Our recent recognition of carbonate melting in large impact structures, such as Chicxulub, goes some way to reconcile the apparent anomaly between lower volumes of impact melts generated in sedimentary compared with crystalline targets (see Kieffer and Simonds 1980). With the exception of spherulites, most of the carbonate melt indicators we have described can only be diagnosed petrographically. At Chicxulub, carbonate melt textures account for about 10 % by volume of the carbonate suevite unit, which in borehole Y6 is nearly 300 m thick. But, the carbonate melt is not homogeneously distributed in the whole suevite; we assume, for arguments sake, that of 300 m suevite, one third is dominated by the carbonate melt-like sample N13, then approximately 3-4 % pure carbonate melt exists. In the most comprehensive seismic-derived model to date (Morgan et al. 1997), the combined thickness of impact breccia (including the carbonate suevite) and silicate-dominated melt breccia is interpreted to vary from ~500 m to 1.7 km, relative to a pre-impact Cretaceous stratigraphy of ~2-3 km sediments including lower interbedded anhydrites and carbonates. In principle, it will be possible to convert these numbers into ratios of silicate and carbonate melt volumes for the crater as a whole. This would require further petrographic examination of additional regional borehole samples. We note that even a small proportion (say, carbonate to silicate melt = 1:9) of total conservative melt estimates based on a Chicxulub excavation volume of 50,000 km^3 (Morgan et al. 1997) could have yielded a total carbonate melt volume (100 - 300 km^3) far in excess of any known carbonate igneous body or volcano.

Of course near the centre of the crater, much carbonate is expected to have decarbonated by direct shock. However, based on theoretical entropy calculations for pure calcite (Yang et al. 1996) this may require much higher pressures (say 100 GPa) than previously estimated (~45 GPa for calcite, Kieffer and Simonds 1980), unless the loading path creates high temperatures > 4000 K. At pressures >100 GPa, where direct decarbonation can occur, available shock data suggests that the gas produced is not the expected combination of simple oxides (M. Gerasimov pers. comm 1998; T. Sekine pers. comm. 1998). Using our own simplistic scaling based on a hemispherical Ries crater model (Stöffler 1971) or more sophisticated hydrocode models (Pierazzo et al. 1998), based on the 100 GPa contour, we might expect less than ~500 km^3 target carbonates to have decarbonated. This is even lower than previous lowermost estimates by Ivanov (1996) and Pierazzo et al. (1998). Our evidence for abundant carbonate melting serves to reinforce the recent appraisal of Pierazzo et al. (1998) that previous estimates of CO_2 released from the Chicxulub impact, and their effects on global climate have, until now, been consistently overestimated.

By analogy with the phase relations for carbonates, we also conjecture that the rock-forming evaporite minerals anhydrite and gypsum would not necessarily have volatilized, and may have melted, but their phase relations are less well known (Pistorius et al. 1969). Phase relations for the sulphate barite ($BaSO_4$) are quite well known, due to the occurrence of this mineral in a number of natural carbonate melt systems; in the broadest sense it has phase relations not too different to calcite. Once again, recent pressure-entropy modelling by Pierazzo et al. (1998) confirms our deductions that relatively high pressures (~100 GPa) are needed for shock volatilization of anhydrite. In this respect, it is interesting that specimens from the Haughton crater (estimated peak pressure 60 GPa) also preserve coherent gypsum (Martinez et al. 1994) which has not therefore volatilised, although it is also not clear that it melted. At Chicxulub, anhydrite is preserved in the suevites (see Appendix I), though in lesser amounts than previously expected (Heuschkel, in prep.). Just as for CO_2, we infer that previous estimates of sulphur species volatilised from target anhydrite at Chicxulub have also been overestimated (Pierazzo et al. 1998).

In conclusion, we have established some textural criteria for the recognition of quenched carbonate melts in impact rocks. Our evidence for the melting of carbonates during impact requires a reduction in the amount of CO_2 thought to have degassed during the Chicxulub K-T boundary event. This appears to be consistent with both the experimental and theoretical data for calcite, and may be extended to other "volatile" phases such as dolomite and anhydrite. A major planetary implication for the formation of the early Earth's atmosphere, is that impacts on limestone in general would be much less likely to have contributed substantial CO_2 than has previously been assumed (e.g., Kieffer and Simonds 1980, Lange and Ahrens 1986). Lastly, we conjecture that certain impact conditions for generation of carbonate melts might also favour growth of some forms of impact diamond.

Acknowledgements

We thank several colleagues including, in particular, both Paul DeCarli and Dieter Stöffler for stimulating discussions. We also thank Rob Hough and an anonymous reviewer for careful criticism of an earlier draft of this paper. The work of both P. Claeys and S. Heuschkel is supported by the DFG Priority Program "International

Continental Drilling Program" (ICDP)/Kontinentales Tiefbohr-progamm der BRD (KTB).

References

Blum JD, Chamberlain CP (1992) Oxygen isotope constrains on the origin of impact glasses from the Cretaceous-Tertiary boundary. Science 257: 1104-1107

Boslough MB, Ahrens TJ, Vizgirda J, Becker RH, Epstein S (1982) Shock induced devolatilization of calcite. Earth Planet Sci Lett 61: 166-170

Church AA, Jones AP (1995) Silicate-carbonate immiscibility at Oldoinyo Lengai. J Petrol 36: 869-889

Claeys P, Alvarez W, Smit J, Hildebrand AR, Montanari A. (1993) KT boundary impact glasses from the Gulf of Mexico region. Lunar Planet Sci 24: 297-298

Claeys P, Alvarez W, Smit J,, Montanari A. (1996) Shocked (?) limestone fragments - A new component of the proximal impact ejecta of the Chicxulub crater. Meteoritics Planet Sci 31: A29

Claeys P, Smit J, Montanari A, Alvarez W (1998) L'impact de Chicxulub et la limite Crétacé-Tertiaire dans la région du Golf du Mexique. Bull Soc Géol France 169: 3-9

Dence MR, Engelhardt WV, Plant AG, Walter L S (1974) Indications of fluid immiscibility in glass from West Clearwater Lake impact crater, Quebec, Canada. Contrib Mineral Petrol 46: 81-97

Dobson DP, Jones AP, Rabe R, Sekine T, Kurita K, Taniguchi T, Kondo T, Kato T, Shimomura O, Urakawa S (1996) In-situ measurement of viscosity and density of carbonate melts at high pressure. Earth Planet Sci Lett 143: 207-215

Donaldson CH (1974) Olivine crystal types in harrisitic rocks of the Rhum pluton and in Archean spinifex rocks. Geol Soc Am Bull 85:1721-1726

Genge MJ, Jones AP, Price GD (1995) An infrared and Raman study of carbonate glasses: Implications for the structure of carbonatite magmas. Geochim Cosmochim Acta 59: 927-937

Graup G (1999) Carbonate-silicate liquid immiscibility upon impact melting: Ries crater, Germany. Meteoritics Planet Sci 34: 425-438

Grajales M, Moran DJ, Padilla P, Sanchez MA, Cedillo E, Alvarez W (1996) The Lomas Tristes breccia: A KT related breccia from Southern Mexico. Geol Soc Am Abstracts with programs 28(7): A183

Hamilton DL, Freestone IC, Dawson JB, Donaldson CH (1979) Origin of carbonatites by liquid immiscibility. Nature 279: 52-54

Heuschkel S, Lounejeva Baturina E, Jones AP, Sanchez-Rubio G, Claeys P, Stöffler D (1998) Carbonate melt in the suevite breccia of the Chicxulub crater. Eos (Abs) 79(45): F554

Hough RM, Gilmour I, Pillinger CT, Arden JW, Gilkes HWR, Yuan J, Milledge HJ (1995) Diamonds and silicon carbide in impact melt rock from the Ries impact crater. Nature 378: 41-44

Huang W-L, Wyllie PJ (1976) Melting relations in the systems $CaO-CO_2$ and $MgO-CO_2$ to 33 kilobars. Geochim Cosmochim Acta 40: 129-132

Irving AJ, Wyllie PJ (1973) Melting relationships in $CaO-CO_2$ and $MgO-CO_2$ to 36 Kilobars with comments on CO_2 in the mantle. Earth Planet Sci Lett 20: 220-225

Jones AP, Dobson D, Genge M (1995) Comment on physical properties of carbonatite magmas inferred from molten salt data. Geol Mag 132: 121

Jones AP, Heuschkel S, Claeys P (1998) Comparison between carbonate textures in Chicxulub impact rocks and synthetic experiments. Meteoritics Planet Sci 33: 4 A79

Keller J (1981) Carbonatitic volcanism in the Kaiserstuhl alkaline complex: evidence for highly fluid carbonatite melts at the Earth's surface. J Volcanol Geother Res 9: 423-431

Kieffer SW, Simonds CH (1980) The role of volatiles and lithology in the impact cratering process. Rev Geophys Space Phys 18: 143-181

Kotra RK, See JH, Gibson EK, Horz F, Cintala MJ, Schmidt RS (1983) Carbon dioxide loss in experimentally shocked calcite and dolomite. Lunar Planet Sci 14: 401-402

Kraft SE, Knittle E, Williams Q (1991) Carbonate stability in the Earth's mantle: a vibrational spectroscopic study of aragonite and dolomite at high pressures and temperatures. J Geophys Res 96: 17,997-18,009

Lange MA, Ahrens TJ (1983) Shock-induced CO_2 production from carbonates and a proto-CO_2-atmosphere on the Earth. Lunar Planet Sci 14: 419-420

Lange MA, Ahrens TJ (1986) Shock-induced CO_2 loss from $CaCO_3$: implications for early planetary atmospheres. Earth Planet Sci Lett 77: 409-418

Martinez I, Agrinier P, Scharer U, Javoy M (1994a) A SEM-ATEM and stable isotope study of carbonates from the Haughton impact crater, Canada. Earth Planet Sci Lett 121: 559-574

Martinez I, Scharer U, Guyot F, Deutsch A, Horneman U (1994b) Experimental and theoretical investigation of shock induced outgassing of dolomite. Lunar Planet Sci 25: 839-840

Martinez I, Deutsch A, Scharer U, Ildefonse P, Guyot F, Agrinier P (1995) Shock recovery experiments on dolomite and thermodynamical calculations of impact induced decarbonation. J Geophys Res 100: 15,465-15,476

Morgan J, Warner M, Brittan J, Buffler R, Camargo A, Christeson G, Denton P, Hildebrand A, Hobbs R, Macintyre H, Kackenzie G, Maguire P, Marin L, Nakamura Y, Pilkington M, Sharpton V, Snyder D, Suarez G, Trejo A (1997) Size and morphology of the Chicxulub impact crater. Nature 390: 472-476

Pierazzo E, Kring DA, Melosh HJ (1998) Hydrocode simulation of the Chicxulub impact event and the production of climatically active gases. J Geophys Res 103: 28607-28625

Pistorius CWFT, Boeyens JCA, Clark JB (1969). Phase diagrams of $NaBF_4$ and $NaClO4$ to 40 kbar and the crystal-chemical relationship between structures of $CaSO_4$, $AgMnO_4$, $BaSO_4$ and high-$NaClO_4$. High Temp High Press 1: 41-52

Salge T, Claeys P, Grajales M, (1998) in press.

Sekine T (1987) Diamond from shocked magnesite. Naturwiss 75(9): 462-463

Shore M, Fowler AD (1999) The origin of spinifex texture in komatiites. Nature 397, 691-694

Smit J, Roep TB, Alvarez W, Montanari A, Claeys P, Grajales M, Nishimura JM, Bermudez J (1996) Coarse-grained, clastic sandstone complex at the KT boundary around the Gulf of Mexico: Deposition by tsunami waves induced by the Chicxulub impact, In: Ryder G, Fastovsky D, Gartner S (eds), The Cretaceous-Tertiary Boundary Event and other Catastrophes in Earth History, Boulder, Colorado, Geol Soc Am Sp Pap 307, pp 151-182

Stöffler D (1971) Coesite and stishovite in shocked crystalline rocks. J Geophys Res 76: 5474-5488

Taniguchi T, Dobson D, Jones AP, Rabe R (1996) Synthesis of cubic diamond in the graphite-magnesium carbonate and graphite-$K_2Mg(CO_3)_2$ systems at high pressure of 9-10 GPa region. J Mat Res 11: 2622-2632

Tyburzcy JA, Ahrens TJ (1986) Dynamic compression and volatile release of carbonates. J Geophys Res 91: 4730-4744

Wyllie PJ (1989) Origin of carbonatites; evidence from phase equilibrium studies. In (ed. K Bell) Carbonatites: Genesis and evolution. Unwin Hyman, London: pp 500-545

Yang W, Ahrens TJ, Chen G (1996) Shock vaporization of anhydrite and calcite and the effect on global climate from K/T impact crater at Chicxulub. Lunar Planet Sci 27: 1473-1474

Appendix

We distinguish 3 different parts of the suevite, drilled at Yucatan-6: i) a carbonate-dominated polymict breccia with shocked basement clasts and melt fragments (Nucleo (N) 13: 1100-1103 mbsl), ii) a carbonate-rich suevite-like breccia with shocked basement clasts and altered silicate glass fragments (N14: 1208-1211 mbsl) and iii) a typical, carbonate-poor suevite with shocked basement clasts and melt fragments (N15: 1253-1256 mbsl). At a depth of 1293 mbsl, the drill core encountered impact melt rock (N16).

Polymict Breccia Dominated by Carbonate

The carbonate-dominated polymict breccia (Y6-N13-3; Y6-N13-4, Y6-N13-5, Y6-N13-9) is mainly composed of carbonates (~ 75 vol%), < 3 vol% of dolomite clasts as well as anhydrite, feldspar and quartz fragments. The quartz fragments are recrystallized or show planar deformation features (PDFs) or mosaicism. The groundmass is rich in calcite (25 vol%) with feldspar (40 vol%, both K-feldspar and plagioclase), quartz (30 vol%) and chlorite (5 vol.%). The carbonate clasts can be divided into (i) bioclastic material with preserved foraminifera, (ii) clasts of pure fine dark micrite, (iii) limestone fragments and (iv) calcite grains with a feathery structure. Using cathodoluminescence, every carbonate phase displays luminescence, except for the calcite with the feathery structure (Heuschkel et al. 1998).

Carbonate-rich, Suevitic Breccia

Fragment sizes in thin section scale range from 2-5 mm (Y6-N114-1,Y6-N14-5a/5b, Y6-N14-10, Y6-N13-11a/11b, Y6-N14-15, Y6-N14-x1/x2). Lithic and mineral fragments are homogeneously distributed throughout the matrix, there are no specific fragment-poor or fragment-rich parts. Most fragments are highly altered and of gneiss, quartzite or limestone composition. Some of the limestone fragments are so pristine that fossils are still recognizable. The amount of limestone clasts can locally vary between 6 and 15 vol%. In addition, a large number of greenish subrounded fragments with pores which are now of clay mineral composition were identified (up to 30 vol%). They probably represent former silicate glass fragments, which are now altered. Furthermore, dark (brownish), subrounded fragments with a sub-microscopically fine matrix and partly with fluidal texture, were identified. Some melt fragments are observed with a granitic composition, containing recrystallized quartz crystals. The amount of silicate melt fragments lies < 5 vol%.

Scanning electron microscopic (SEM) and semi-quantitative energy-dispersive x-ray analyses (SEM-EDX) were carried out on several samples, especially to focus on the composition of the matrix. The matrix consists of small $CaCO_3$ grains (± 10 μm) surrounded by a mixture of K-feldspar, plagioclase, quartz and pyroxene crystals (< 5 μm). The pyroxene has a marked augitic composition. Locally the amount of calcite reaches up to 40 vol%. Additional EDX analyses show that all the $CaCO_3$ is in the groundmass is calcite and not aragonite. Altogether the amount of calcite in these samples varies between 20 and 24 vol%.

Carbonate-poor, Suevitic Breccia

Below 1253 mbsl, Y6-N15 represents another type of Chicxulub suevite material. Few basement clasts are observed, most of the clasts are of quartz, feldspar and anhydrite composition, partly digested into the matrix. Quartz and feldspar clasts are surrounded by a corona of clinopyroxene. Characteristic are aerodynamically shaped melt clasts, up to 0.5 mm in size, sometimes also with recrystallized quartz. The composition of these melt blebs is silica-rich (60-64 wt%), with moderately high CaO (3-4 wt%).

In comparison to N14, the amount of $CaCO_3$ in the matrix is less in N15, whereas the amount of melt fragments increases to 10 vol%. The groundmass consists mainly of feldspar (alkali feldspar and plagioclase), augitic pyroxene, quartz and minor amounts of calcite (< 5 vol%).

15 The Gallejaur Structure, Northern Sweden

Robert Lilljequist

North Atlantic Natural Resources, Kungsgatan 62, S-752 18 Uppsala, Sweden.
(r.lilljequist@nanr.se)

Abstract. Rocks, interpreted as impact generated lithologies, occur in a large area surrounding the Gallejaur magnetic structure at latitude $65°10'$/longitude $19°30'$ in northernmost Västerbotten County in northern Sweden. These rocks comprise a variety of different types of breccias: authigenic/autochthonous monomict breccias from the underlying rock units (monomictly brecciated basement), parautochthonous, monomict breccias, and polymict melt breccias. No convincing planar deformation features (PDFs) in shock metamorphic minerals have yet been identified.

The Gallejaur structure is located in the central part of the Skellefte mining district in the Precambrian Baltic Shield. The ca. 1.9 Ga old Skellefte district is an extensively mineralized, mainly felsic, submarine volcanic belt. The predominant lithologies are acid to mafic volcanic and volcanoclastic rocks, interbedded with, and overlain by, graphitic schists and greywackes. The rocks, which in the present article are described as impact-generated, have been called the Vargfors Group and overlie the Skellefte volcanics and sediments with an angular unconformity. In general, the older rocks are deformed and folded, which results in a more or less vertical position, whereas the younger rock sequence is flat-lying and undeformed outside of the regional shear zones. The youngest rocks in the area, intruding the Skellefte district supracrustals and probably the impact-generated rock, are A/I-type granitoids belonging to the Revsund-Adak granite suite, which have been dated at ca. 1.80 to 1.78 Ga.

The highly magnetic ring around the centre of the structure is interpreted as an impact melt body with varying amounts of more or less absorbed clasts of various basement lithologies. Density measurements indicate a mafic composition, implying that the original rocks were andesitic to basaltic. The gravity anomaly of the central rise region, above +15 mGal (150 gu), is high. This gravity high is surrounded by an encircling gravity low of about −20 mGal amplitude. The central uplift area is between 10 and 12 km in diameter according to the gravity and magnetic anomaly maps, which corresponds to a final crater diameter of 50-60 kilometers, and a transient cavity of about half that size.

In the center of the Gallejaur ring structure a fine-grained crystalline rock of monzonitic composition is found, which is loaded with rounded clasts of volcanic origin. The origin is not yet established, but the lack of deformation and location within the inferred central uplift indicate that the monzonitic rock could have an origin as crystallized impact melt. The Gallejaur monzonite has been dated at 1873 ± 10 Ma, which is taken as the age of the impact event. The monzonite and an underlying porphyritic rock seem to be differentiates of one melt body.

In a northwest-southeast zone across the structure occur water-deposited immature sediments (pelitic, arkosic, and conglomeratic), which are interpreted as fill.

The original impact structure has been affected by later deformation events, which are mainly represented by shearing, faulting and erosion.

Introduction

During exploration fieldwork in the Skellefte mining district in summer 1998, on behalf of North Atlantic Natural Resources, rocks were observed that indicated a possible formation by impact. Highly magnetic, fine-grained melt rocks with a polymict clast population were found to surround a central structure with fine- to medium-grained rocks of monzonitic composition. These rocks are concentrically surrounded by allochthonous formations with megablocks, polymict and monomict clastic breccias, and authigenic breccias.

The Gallejaur structure is situated in the northernmost part of the Province of Västerbotten, northern Sweden, at latitude 65°10′ and longitude 19° 30′ (Fig. 1).

The uppermost sequences consist of flat-lying sediments of low metamorphic grade, including conglomerates, slump breccias (with volcanic and sedimentary components), and slates. These rocks have previously been interpreted as supracrustal units of the Vargfors Formation (Lundberg 1980). They discordantly overlie higher-grade metamorphic, steeply dipping metavolcanic and -sedimentary rocks of the Skellefte Formation (Högbom 1937; Grip 1942; Kautsky 1957).

In earlier publications, the central Gallejaur structure has been described as a granitoid intrusion (e.g., Högbom 1937; Lundberg 1980). Gavelin (1955) named the central melt rock a monzonite – a classification followed by later authors (e.g., Skiöld 1988; Billström and Weihed 1996).

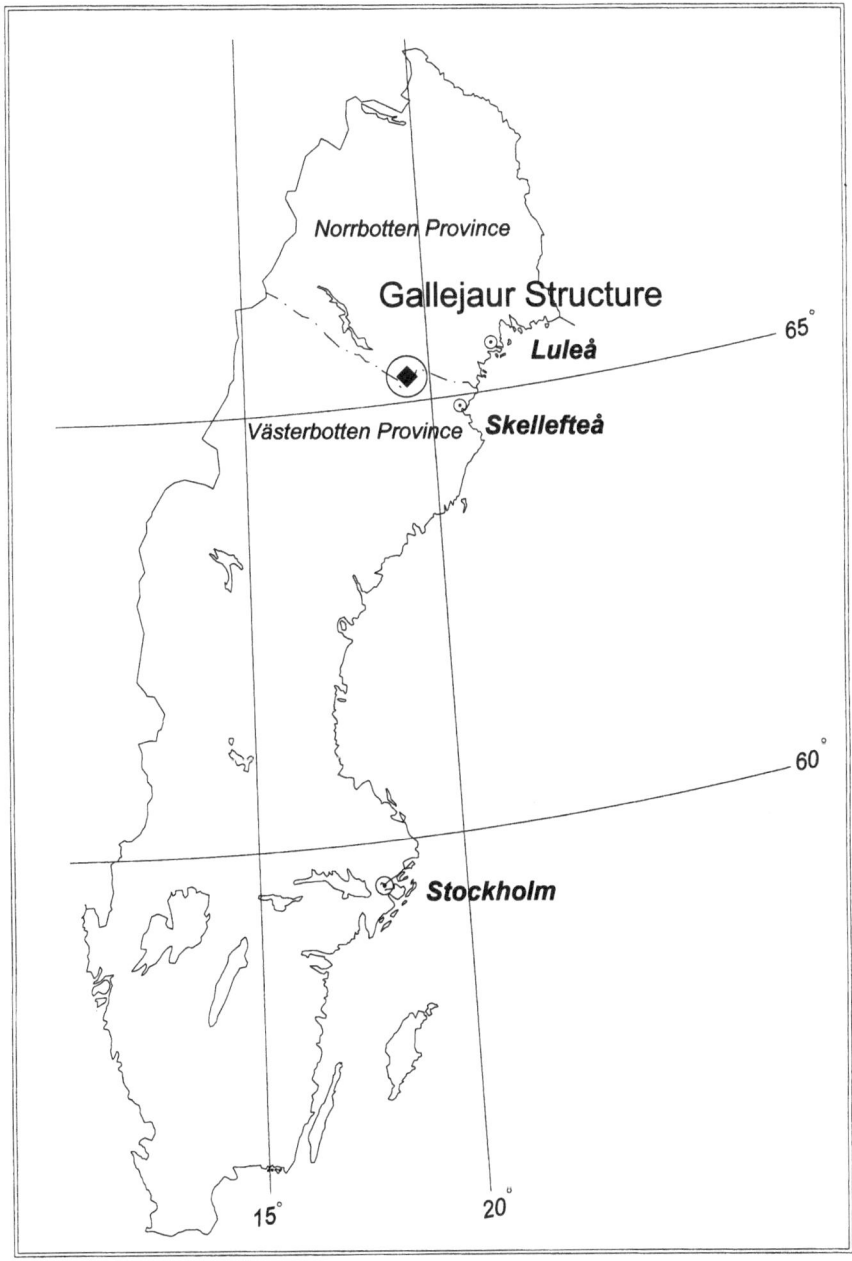

Fig. 1. Location map of the Gallejaur structure in northern Finland

Geological Framework and Previous Geological Interpretation

The Gallejaur Structure is situated in the central part of the Skellefte district, an area of approximately 150x50 km in northern Västerbotten and southern Norrbotten Provinces (in Swedish: län). The Skellefte district is rich in mineral occurrences and deposits. Thick piles of submarine felsic volcanic rocks hosting over 85 pyritic base metal deposits (Allen et al. 1996) dominate the area (Fig. 2). Currently, six mines are exploiting volcanic hosted massive sulfide ore bodies, and two gold mines are active.

To the north, the Skellefte district is bordered by the continental Arvidsjaur volcanic rocks and, to the south, by the Bothnian basin with metagreywackes, gneisses, and migmatites.

The lowermost volcanic unit is named the Skellefte Group (Allen et al. 1996). The main rock types are Paleoproterozoic (ca. 1.88-1.89 Ga) volcanoclastic rocks, porphyritic intrusions, and lavas with intercalated meta-sedimentary rocks (including graphitic black schists reflecting the waning stages of volcanic eruptions). The Skellefte Group rocks trend WNW and are generally regarded as formed in a volcanic arc setting. Synvolcanic intrusions, considered as comagmatic, are of similar age. Post-volcanic, feldspar-porphyritic granitoids are referred to as Revsund–type granites, with ages from 1.80 to 1.78 Ga (Billström and Weihed 1996).

Unconformably overlying the Skellefte Group lies the so-called Vargfors Group, which comprises autochthonous breccias, mostly flat-lying sedimentary formations, and polymict conglomerates. Earlier investigators (Högbom 1937; Gavelin 1939; Grip 1942 1951; Kautsky 1957;, Lundberg 1980; Claesson 1985; Rickard 1986) recognized a hiatus between the deformed and steeply dipping Skellefte Group rocks and the Vargfors sediments. Weihed et al. (1992) and Allen et al. (1996), however, have challenged this relationship in the last years. In their view both the Skellefte and Vargfors Groups contain the same tectonic fabrics and no major time difference between the two units.

Many stratigraphic schemes have been proposed for the Skellefte district. Despite the economic importance of the area, only a relatively few age determinations have been carried out. Table 1 gives a simplified stratigraphic scheme for the Skellefte district and adjacent areas with the approximate time span of published U-Pb zircon ages. For detailed descriptions the reader is referred to the original papers.

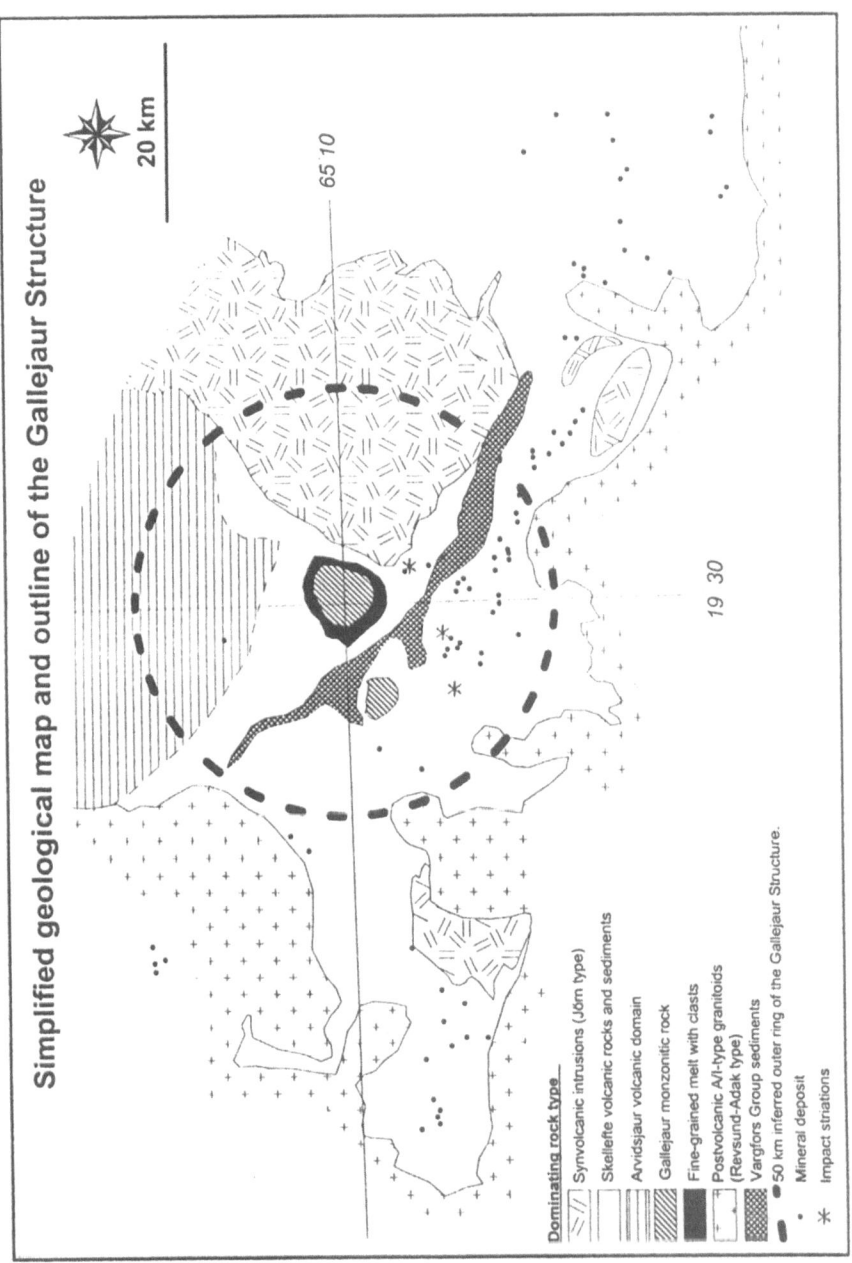

Fig. 2. Simplified geological map of the Skellefte district and the inferred outer limit of the Gallejaur structure.

Table 1. Simplified stratigraphic scheme for the Skellefte district and adjacent areas

Rock units	approximate age (Ga)	References
Revsund granitoids	1.78-1.80	Skiöld 1988
		Patchett et al. 1987
		Claesson and Lundqvist,1995
Regional metamorphism	1.80-1.82	Weihed and Mäki 1997
Gallejaur monzonite	1.87	Skiöld 1988
Vargfors Group*; "ignimbrite" with polymict clasts	1.88	Billström and Weihed 1996
Arvidsjaur Group	1.88	Skiöld et al. 1993
(mainly subaerial volcanic rocks)	1.87-1.88	Weihed and Mäki 1997 (Table 1)
Jörn Granitoids	1.87-1.89	Wilson et al. 1987
	1.86-1.95	Billström and Weihed 1996
Skellefte Group of supracrustal rocks		
Graphitic schists and greywackes**		
Acid to mafic volcanic and volcanoclastic	1.88-1.89	Weihed et al. 1992; Weihed and Mäki. 1997
rocks and subintrusives.		
	1.89	Billström and Weihed 1996
Meta-sedimentary rocks of the Bothnian Basin	1.88-1.95	Weihed and Mäki 1997

* The Vargfors Group is divided from stratigraphic low upwards into Mensträsk conglomerate, Aborrtjärn conglomerate, Gallejaur volcanics, and Dödmanberg conglomerate (Allen et al. 1996)
**The Elvaberg Formation (Kautsky 1957, Allen et al. 1996)

The Skellefte Group rocks exhibit low greenschist to amphibolite facies metamorphism and, locally, intense hydrothermal alteration. The Vargfors Group rocks generally display a very low-grade metamorphic grade and are only deformed within large-scale shear-zones.

Geophysics and Petrophysics

Regional geophysical maps were produced in the 1960s and 1970s by the Swedish Geological Survey (SGU). A positive gravity anomaly of about 15 mGal (150 gu) over the central Gallejaur Structure was observed (Lundberg 1980; Enmark and Nisca 1983) instead of a negative gravity anomaly as would be expected for a granitic intrusive (Fig. 3). Combining geological, petrophysical and geophysical data, Enmark and Nisca (1983) came to the conclusion that the Gallejaur Structure was formed as a large, roundish, mafic laccolith with a shallow cap of monzonite derived by melt differentiation. The maximum gravity anomaly is about 18 mGal (180 gu) and occurs 2 km west of the centre of the structure.

The Geological Survey of Sweden collected 262 rock samples from the area, including 150 from the Gallejaur Structure. Density, magnetic susceptibility and the natural remnant magnetization (NRM) of the samples were measured and reported by Enmark and Nisca (1983). In-situ measurements show a susceptibility distribution with peak values of 2.4×10^{-2} SI for the monzonite and 4.0×10^{-2} SI for the surrounding fine-grained melt rocks, including large clasts of intermediate to mafic rocks (Enmark and Nisca 1983: Fig. 5). The Q-values (ratio of remnant to induced magnetization) vary from about 0.1 to 30, demonstrating a wide spread of both susceptibility and q-values in the monzonite, the Gallejaur "gabbro", and the "acid-intermediate volcanics" inside the magnetic structure" (Enmark and Nisca 1983: Fig. 8). The density distribution for the samples from the magnetic annular ring shows a frequency peak at 2 900 kg/m^3, whereas that of the samples from the center structure shows a frequency peak at 2 690 kg/m^3.

The Gallejaur structure is prominent on both the regional Bouguer gravity (Fig. 3) and aeromagnetic total field maps (Fig. 4). Both anomalies have a circular configuration and are 10-12 km wide. On the magnetic map, a ring of very high magnetization surrounds a more moderate magnetic anomaly in the center of the roundish structure (Fig. 11). Apart from the inner, almost circular structures, two partly continuous rings can be discerned on the magnetic map, at a distance of 8 and 17 km from the inner ring (Fig. 4a).

Magnetic modelling by Enmark and Nisca (1983) indicates an antiformal shape of the source structure. The central gravity high was an enigmatic feature, as it

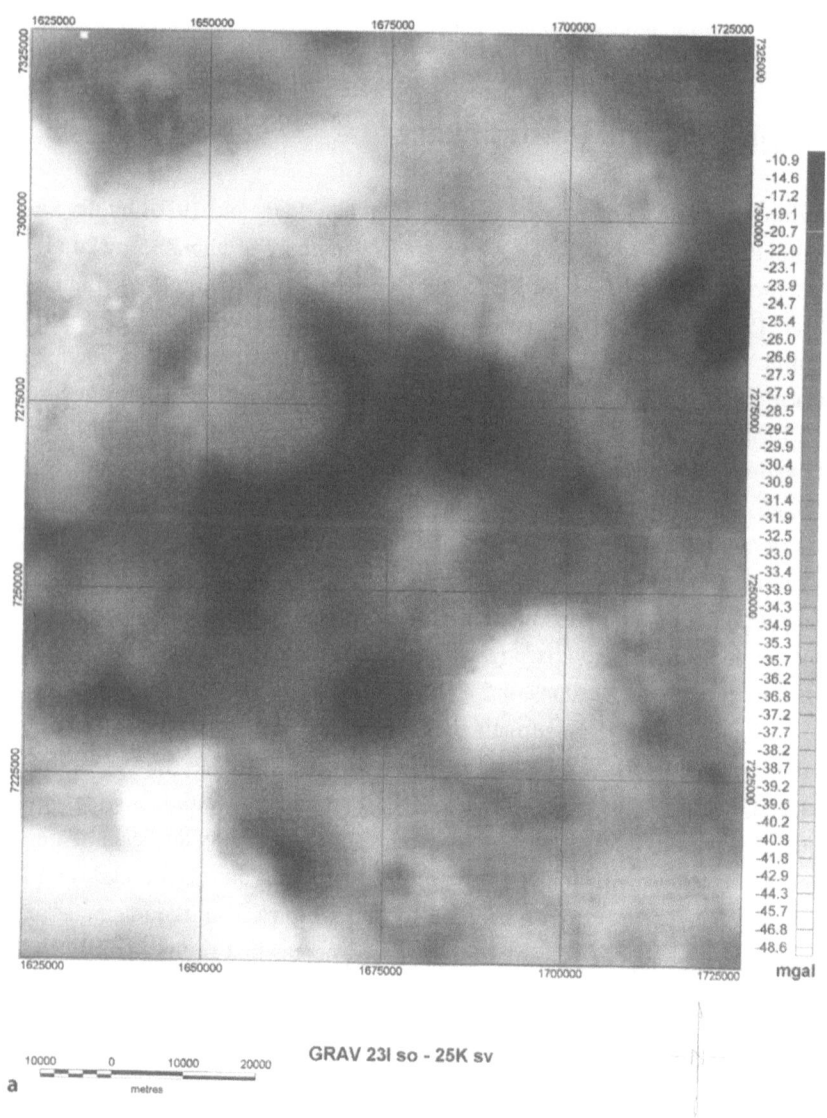

GRAV 23I so - 25K sv

a

Fig. 3. a. Gravity anomaly map with grey-scale values from –48.6 to –10.9 mGal. The average station spacing is 0.84 points/km. Prepared by Johan Söderman and published with permission from the gravity data base, © Sveriges geologiska undersökning, Dnr 00-216/99.

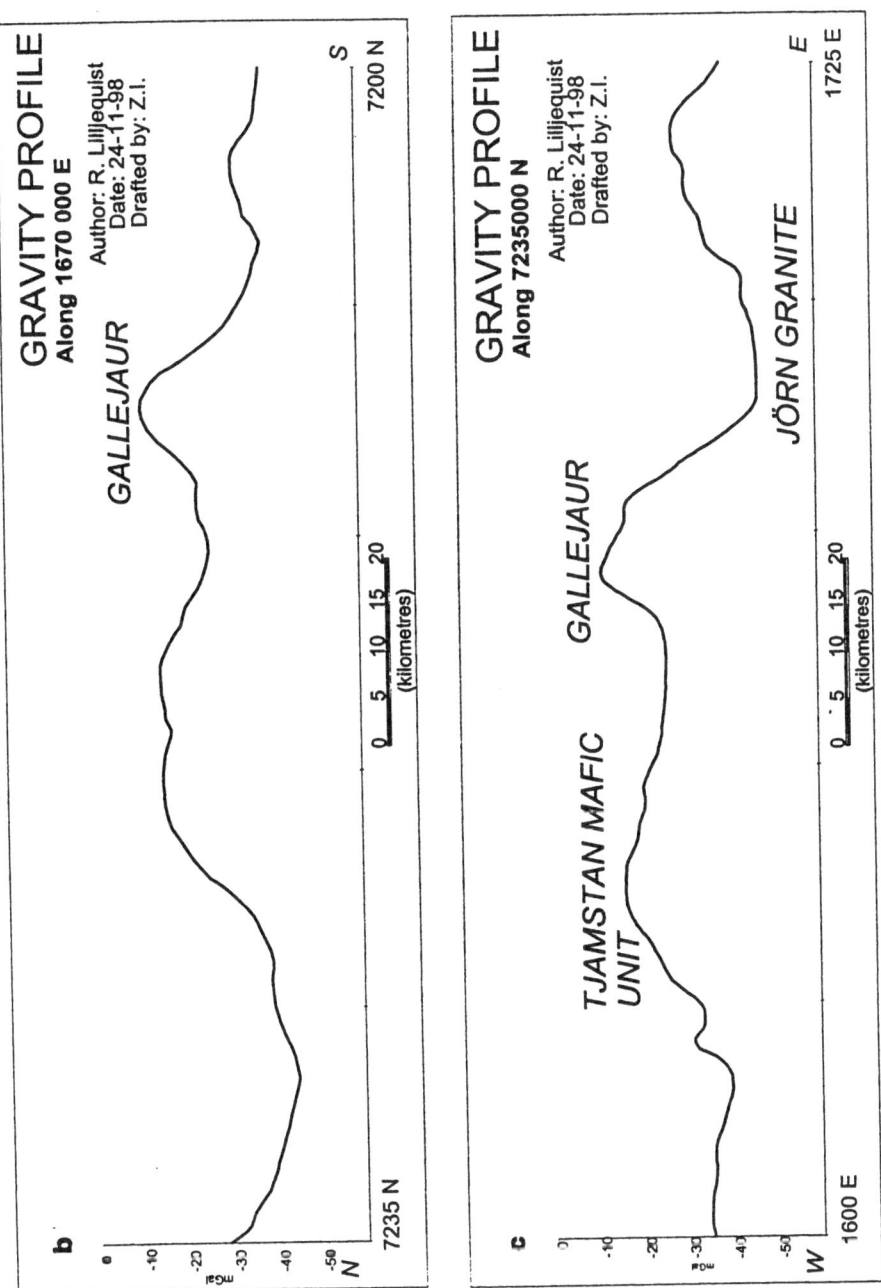

Fig. 3. b. Gravity profile in N-S direction (along E-1670000) **c.** Gravity profile in E-W direction (along N-7235000).

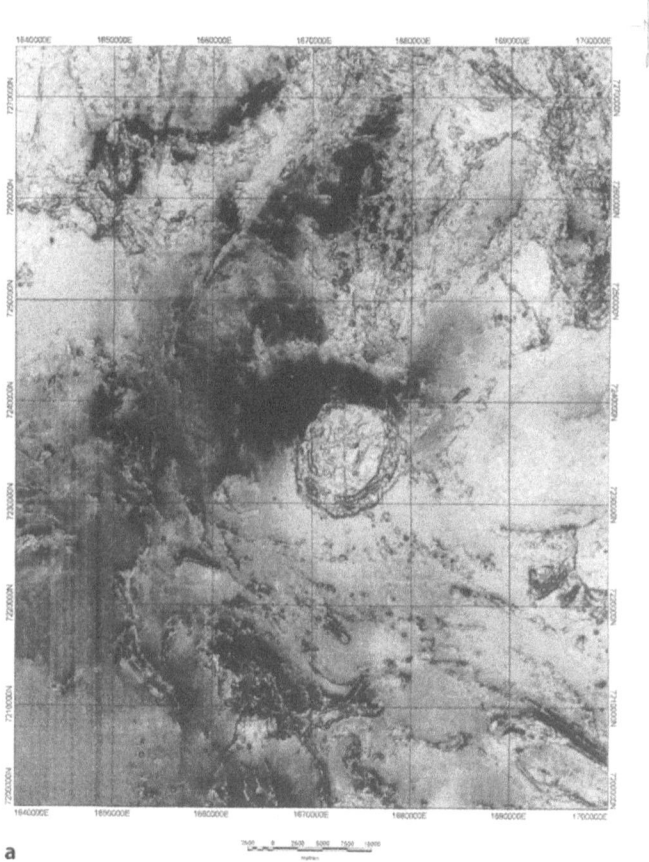

a

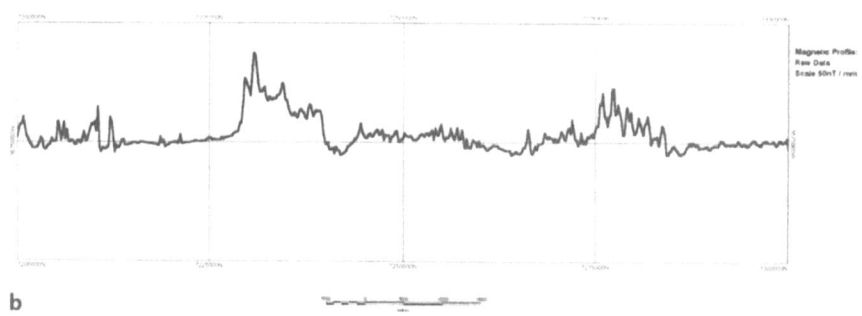

b

Fig. 4. a. Magnetic anomaly map. The magnetic declination is +3 to +4° and the inclination are 75 degrees (Eriksson and Henkel 1994). The map is based on airborne magnetic surveys at 30 m altitude and 200 m line spacing. **b.** Magnetic profile in N-S direction (along E-1675000)

Fig. 5. "Mensträsk breccia". Autochthonous breccia in acid volcanic target rocks (N-7233950, E-1657600). The hammer shaft is 60 cm long.

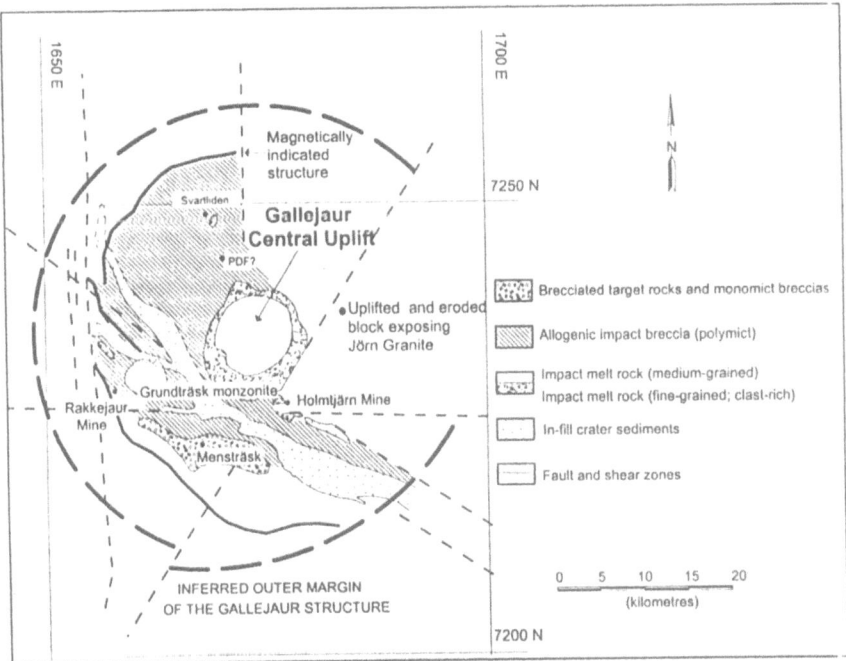

Fig. 6. Simplified geological map of the Gallejaur Structure according to the impact hypothesis. The outermost ring denotes the inferred structure. Major deformation zones and impact-generated formations are shown.

seemingly was associated with low-density rocks (monzonite). The modelling has clearly indicated that a high-density source below the monzonite is needed to explain the anomaly (Figures 4 and 11). Enmark and Nisca (1983) concluded, based on gravity and magnetic modelling, that the structure causing the anomaly must be a high-density body with a depth extent of at least 2-3 km, indicating that the "monzonite" is a shallow layer at the top of this high-density rock.

Breccia Occurrences

Authigenic and parautochthonous breccias and shatter cones.

The autochthonous breccias (Fig. 5) have earlier been named the Mensträsk breccia and occur most extensively in an approximately 15x3 km area at Mensträsk (Fig. 6).

Autochthonous breccias are also found to the north of the Rakkejaur Mine in the eastern part of the structure, close to the Holmtjärn Mine southwest of the inferred central uplift, and at Svartliden in the northern part of the structure (Fig. 6). Most of these occurrences, with exception of the Mensträsk area, are thought to form part of allogenic megablocks. In some of the megablocks (e.g., southeast of Holmtjärn) the contact between relatively undisturbed basement rocks and breccia is preserved with what is interpreted as shatter cone structures in the less brecciated basement (Fig. 9).

At the Mensträsk Mine, Grip (1951) described a complete section from authigenic breccia to slates. Later shear movements affected the rocks and caused mineralization along a NNW trend. The brecciated target rocks (quartz porphyries) are overlain by a 20 to 30 m thick sequence of polymict breccia. Upwards the quartz porphyry is completely broken up into angular fragments contained in a more fine-grained matrix. The breccia grades upwards into a coarser breccia with one-meter sized clasts. Higher up, clasts become smaller and more rounded with decreasing depth. Splinter-like fragments are frequently found, and the variation in facies is great. Uppermost there are layers of finer sediments varying from psammite to pelite. The low metamorphic sedimentary rocks that overlie the polymict breccia consist of pelites, which often show a well-preserved banding. The metamorphism is so slight that these rocks are named slates. An abundance of graphite and pyrrhotite is a characteristic feature of the finest sediments. Abrupt breaks in the sequence are common. Large grains always consist of quartz, but there are also occasional grains of albite, quartzite and slate.

Sericite, fine-grained quartz, small amounts of light coloured biotite, powdered graphite and occasional crystals of pyrrhotite are additional components. Several lenses of massive sulfide are confined to the breccia.

What is interpreted as impact-generated striations or shatter cones have been found at three localities (Fig. 2), within the autochthonous brecciated target rocks. The shatter cones are not perfectly developed, as the lithology of the target rocks is inhomogeneous (Fig. 9)

Monomict breccias with felsic volcanic clasts are common in the area to the north of Rakkejaur and in an area to the ENE of Mensträsk (Fig. 2). Monomict breccias with clasts of amygdaloidal andesitic lava have been found SE of the Holmtjärn Mine (Figs. 2, 6 and 7).

Allochthonous breccias

Polymict breccias are found along the old Skellefte River course and in the NW sector of the inferred impact structure (Fig. 2). Some of the polymict breccias contain melt-fragments. At Elvaberget, south of Lake Mensträsk, a polymict breccia with meter-sized clasts is preserved.

Suspected PDFs in quartz have been observed in a sample from a polymict breccia with granite clasts, 4 km north of the inferred central uplift (Fig. 6). This observation has, however, to be confirmed.

The size and frequency of outcrops limits a detailed interpretation of breccia formations of the Gallejaur structure. The size of megablocks, for example, exceeds the dimensions of outcrops. Erosional effects and later deformation further complicate the picture.

In the magnetic ring around the central Gallejaur Structure, gabbroic to dioritic rocks are seen in contact with a fine-grained clast-bearing melt rock, indicating a complex interstratification of megablocks and melt.

Melt rocks

Two types of melt rocks are found within the Gallejaur Structure. A >600 m thick medium-grained body of monzonitic composition occupies the inner part. The SiO_2 content of the medium-grained melt varies from 64.4 wt% to 68.2 wt% with an average of 66.9 wt%. The variation in weight% of the other major elements is almost insignificant, indicating a strongly homogenized melt. Its lower part is a slightly more mafic porphyritic melt of latite andesite composition (L Widenfalk, pers. comm.).

Fig. 7. Monomict andesite breccia (N-7226364, E-1677833). The GPS instrument is 14 cm long and 5 cm wide.

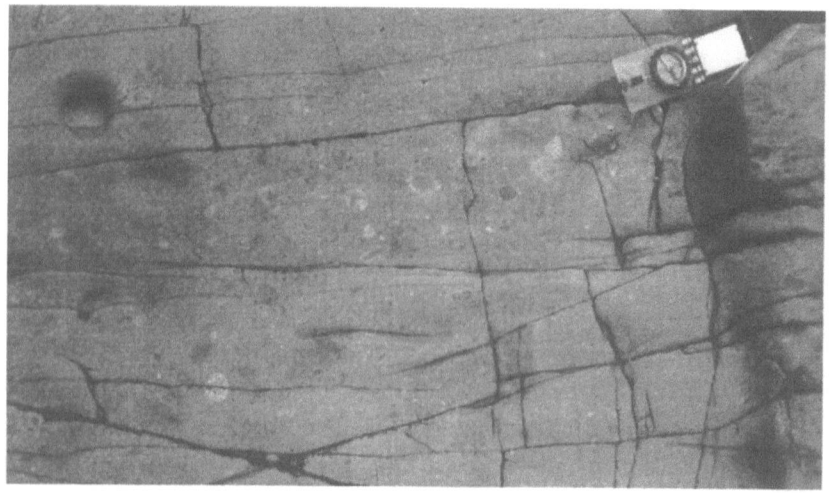

Fig. 8. Impact melt rock with polymict, mainly granitic clast content (N-7230266, E-1669028).

Fig. 9. Impact striations indicating shatter cone structures. Photo by Manne Lindwall. The scale bar is 10 cm.

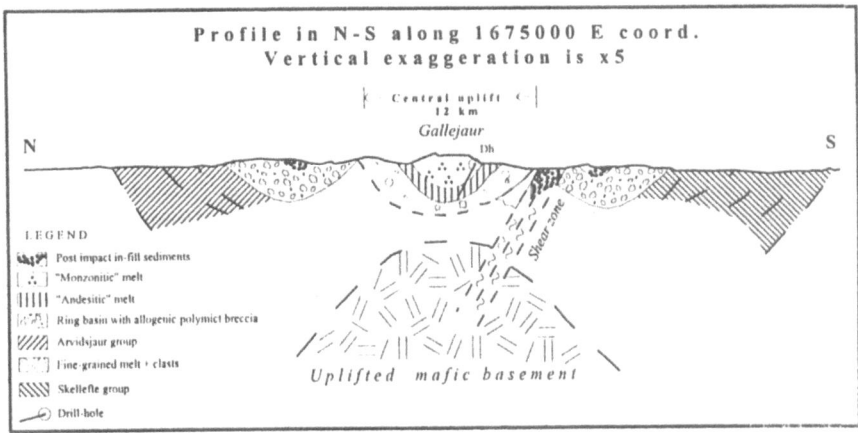

Fig. 10. Sketch section through the Gallejaur impact structure. The sedimentary fill north of the Gallejaur central uplift is inferred. The impact melt might have extended over part of the allogenic polymict breccia. The target rocks to the north (Arvidsjaur Group) and to the south (Skellefte Group) have been downslided along the impact crater walls.

Table 2. Chemical analyses of the medium-grained melt rock form the central part of the Gallejaur Structure. KBS drillhole: 7233800N/1672600E (Swedish Grid System)

monzonite

	SiO_2	TiO_2	Al_2O_3	Fe_2O_3	MnO	MgO	CaO	Na_2O	K_2O	P_2O_5	Ba	Sr
78814	67.5	0.56	14.60	4.40	0.10	0.76	1.70	5.20	3.30	0.13		
78843	68.2	0.53	14.30	4.00	0.10	0.47	1.60	4.90	3.40	0.11		
78848	66.4	0.64	14.60	4.50	0.10	0.89	2.30	5.00	3.20	0.16		
78844	66.5	0.71	14.40	4.39	0.08	0.68	2.04	4.91	3.36	0.28	1200	340
78847	67.3	0.62	14.40	4.74	0.10	0.51	1.99	4.99	3.25	0.26	1500	310
KBS-035 m	67.9	0.51	14.00	4.69	0.19	0.53	1.94	5.15	3.00		1490	295
KBS-064 m	67.0	0.55	14.40	4.47	0.08	0.62	1.49	5.06	3.01		1570	225
KBS-127 m	65.8	0.50	14.20	4.56	0.09	0.55	2.38	4.64	2.92		1060	235
KBS-257 m	65.7	0.63	14.10	4.80	0.10	0.81	2.04	4.54	2.92		1420	280
KBS-380 m	66.6	0.62	14.10	4.84	0.10	0.73	1.96	5.37	2.78		1450	274
KBS-417 m	66.9	0.60	14.20	4.64	0.12	0.73	1.81	5.36	3.01		1370	303
average:	**66.9**	**0.59**	**14.27**	**4.55**	**0.11**	**0.66**	**1.93**	**5.01**	**3.09**	**0.18**		

porphyritic latite andesite

	SiO_2	TiO_2	Al_2O_3	Fe_2O_3	MnO	MgO	CaO	Na_2O	K_2O	P_2O_5	Ba	Sr
KBS-439 m	66.3	0.61	14.50	4.77	0.08	1.10	2.33	5.58	2.49		1320	228
KBS-479 m	64.4	0.56	13.90	4.70	0.10	1.12	2.86	4.31	2.75		1110	207
KBS-552 m	65.0	0.60	14.40	4.61	0.08	1.37	3.28	4.72	2.20		1090	207
KBS-594 m	65.7	0.54	14.30	4.55	0.07	1.54	3.25	4.14	2.50		1130	210
average:	**65.4**	**0.58**	**14.30**	**4.66**	**0.08**	**1.28**	**2.93**	**4.68**	**2.49**			

Data for major elements in weight%, Ba and Sr in ppm. The KBS samples are analysed with ICP, and the samples numbered 78 with XRF at the SGAB laboratory in Luleå.

The lower porphyritic rock contains a little less silica, twice as much MgO, and a little increased content of CaO, K_2O, and Na_2O than the monzonite (Table 2).

A drill-hole has been sunk into this central part of the structure (Fig. 10) by the Swedish Nuclear Fuel and Waste Management Company (SKB). The hole penetrated 434 meters of a reddish-brown monzonite, mainly composed of feldspar and small aggregates of mafic minerals. The rock contains a varying amount of round or elongated greyish black mafic inclusions, normally about 2 cm long. From 434 to 600 meters the rock is described as a latite andesite with porphyritic texture, composed of a fine-grained matrix with 1-2 mm sized reddish feldspar phenocrysts. The type and abundance of inclusions are similar in both lithologies.

Outside of the monzonite, a clast-laden fine-grained greenish black melt rock (Fig. 8) forms a ring of magnetic anomalies (Enmark and Nisca 1983). Locally perlitic texture is preserved. This melt rock contains varying amounts of clasts of a polymict population. These features are best seen along the former, now mostly dry, course of the Skellefte River.

Table 3. Chemical analyses (ICP) of melt rocks from the magnetic rim of the Gallejaur Structure (after Allen et al. 1995). Major element data in wt%, trace element data in ppm.

N-coord	E-coord	Code	Description
7229000	1670750	Ni 4b	Pyroxene-porphyritic "basalt"
7230000	1669420	Ni 8	"intrusive", "basalt"
7229950	1668850	Ni 10	Feldspar porphyric "dacite"
7232000	1667980	Ni 28b	Feldspar porphyric "basalt"
7227910	1671940	Ni 58	"andesite"
7227960	1671950	Ni 59	High Mg "basalt"

	"basalt"	"basalt"	"dacite"	"basalt"	"andesite"	"basalt"
	Ni 4b	Ni 8	Ni 10	Ni 28b	Ni 58	Ni 59
SiO_2	50.20	49.40	66.60	51.30	61.10	48.10
TiO_2	0.56	0.64	0.48	0.57	0.46	0.58
Al_2O_3	14.30	16.80	16.30	19.00	16.80	13.60
Fe_2O_3	10.40	10.70	3.23	7.97	5.68	11.40
MnO	0.17	0.18	0.07	0.14	0.09	0.20
MgO	8.80	6.96	1.01	4.09	2.79	11.30
CaO	9.94	8.82	2.97	12.60	4.03	8.04
Na_2O	1.96	2.70	4.41	2.99	5.47	1.73
K_2O	1.22	0.89	3.67	0.65	0.57	1.24
P_2O_5	0.21	0.23	0.13	0.07	0.17	0.17
LOI	0.80	2.75	0.95	1.05	1.60	2.95
Total	98.60	100.10	99.80	100.40	98.80	99.40
Ba	386	345	1 490	174	352	815
Rb	39	23	58	32	19	26
Sr	386	525	478	272	468	356
Ga	-		22	24		
Ta	-		13	17		
Nb	5	5	12	4	8	4
Zr	80	114	203	62	112	53
Y	17	23	17	17	14	15
Th	-		7	4		
U	-		5	4		
Ni	96	87	-	-	13	97
Cu	52	57	13	34	35	31
Pb	8	31	18	2	6	7
Zn	75	94	60	89	76	99
As	3	47	16	23	9	7

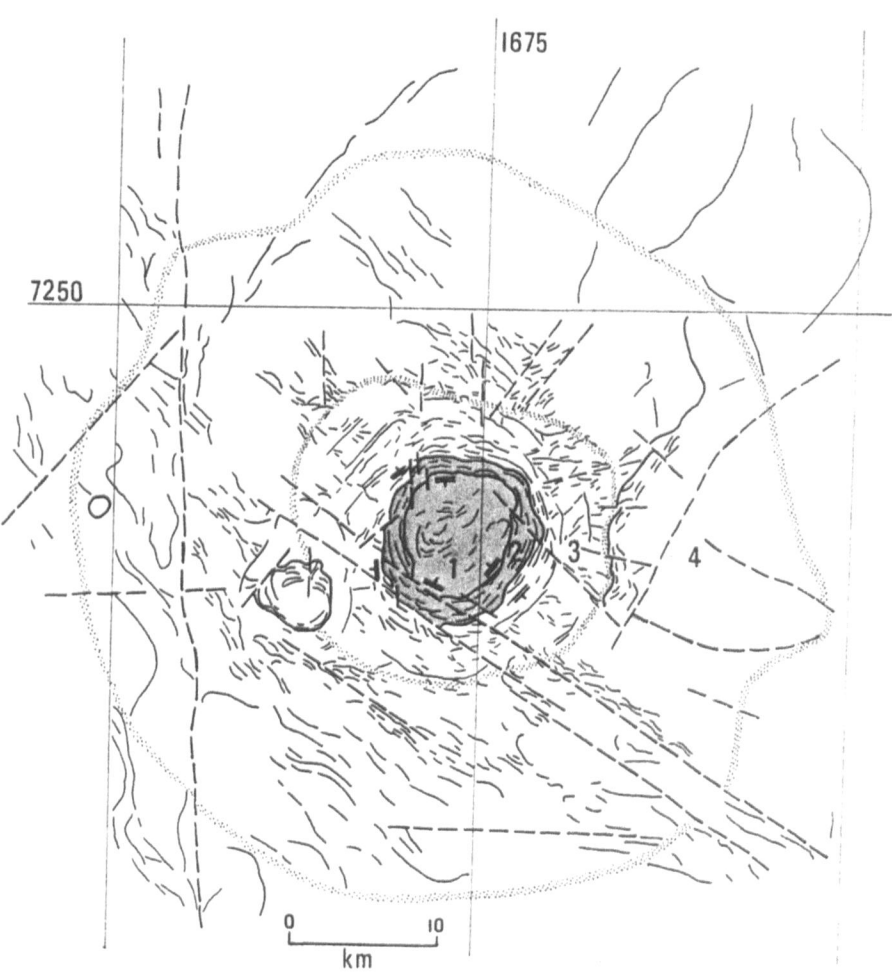

Fig. 11. Interpretation map of aeromagnetic structures. Modified and extended from Enmark and Nisca (1983). The finer lines indicate magnetic lineations, the coarser lines magnetic contacts, and the dashed lines represent faults. Magnetic contact dips are shown as thick lines with dip indication. **1.** Central rise region indicated by gravity and magnetic high, **2.** Highly magnetic central annular ring, **3.** Low magnetic annular ring limiting the concentric magnetic structures, and **4.** Inferred outer limit of impact crater to seemingly undisturbed target rocks. The grid system is the Swedish National grid.

Chemical analyses for samples from the outer magnetic ring are given in Table 3. The SiO_2 content of the outer magnetic rock ranges from 48 to 66.6 wt%. The chemical variations are probably reflecting the percentage and composition of the clasts. However, the glassy to fine-grained melt seems to have an intermediate to basaltic composition, indicating the composition of the dominating parent (target) rocks.

A melt body is also suspected near Svanfors, 21 km SW of the centre of the Gallejaur Structure (Fig. 2). In the literature (Kautsky 1957) it has been described as a 20-50 m thick green andesitic lava, between conglomerate with granite clasts and a polymict conglomerate. Kautsky (1957) observed abundant granite conglomerates with an andesitic matrix, and (unexpectedly) granite pebbles in the middle of the "effusive andesitic lava". The description is almost identical to the description of the "Vargfors andesite" (Lundberg 1980, and field book annotations by Lundberg and Claesson) in the magnetic ring around the central part of the Gallejaur Structure.

An Impact Model for the Gallejaur Structure

An origin by impact for the Gallejaur Structure was first considered in the course of geological recognizance work in June 1998. However, sometime earlier, F.E. Wickman had discussed the possibility of an impact origin of the Vargfors Formation (oral communication). The indications of an impact origin of the rocks around the Gallejaur central uplift area are based on occurrences of extensive autochthonous brecciation, of monomict clastic breccias, a melt rock with exotic clasts along the former course of the Skellefte River, and some locations with shatter cones. The recognition of PDFs from a polymict melt breccia is still not confirmed but would support the field observations.

The occurrence of a granitoid (monzonitic) rock in the central part of the topographic highs around the Gallejaur Mountain is enigmatic, but this could also represent the well crystallized inner part of a once more extensive intrusive body.

During fieldwork rocks previously described as formed through volcanic and fluvial activity were remapped and reinterpreted. Extensive areas (about 15x5 km in the southern parts) with monomict and polymict breccias were identified. The so-called Vargfors andesite has been reinterpreted as impact melt with abundant clasts of a polymict clast population, including granite fragments.

Erosion Level and Major Post-impact Structures.

Erosion has not reached the basement under the impact melt body in the central parts of the structure. It can be assumed that the impact melt body originally was over 600 m thick within an inferred central peak or peaking ring. Above the melt sheet, suevite and crater-fill sediments were deposited. Erosion could have removed about 0.5 – 1 km of material. Low metamorphic grade of the in-fill sediments indicates modest erosion.

A prominent shear zone transverses the structure in a NW-SE direction, down-faulting and preserving the post-impact Vargfors conglomerates and sediments (Fig. 6). A block of Jörn Granite has been uplifted in the north-eastern sector of the structure; destroying evidence of impact formations in that region. Another large shear zone runs N-S in the western part of the structure (Figures 6 and 12).

Age of the Gallejaur Impact Structure.

Zircons from the monzonitic part of the Gallejaur impact melt body indicate a crystallization age of 1873 ± 10 Ma (Skiöld 1988). Zircons from an "ignimbrite" – a polymict melt breccia from the outer magnetic ring - indicate an age of 1875 ± 4 Ma (Billström and Weihed 1996). It is, however, not clear whether these data represent ages of zircons from older rocks or represent the true age of the impact event. The zircons from the monzonitic part of the Gallejaur Structure are, in general, small, transparent and colourless. They are only faintly zoned, stubby and exhibit no second order prismatic faces (Skiöld 1988). They are likely to represent the age of the crystallization of the monzonite. The Vargfors "ignimbrite" contains both rock and broken crystal fragments. The zircons are clear and transparent with concentric zoning observed. A few grains are of gem quality. Both stubby, short prismatic grains (2:1 ratio) and needle-shaped crystals (6:1 ratio) exist (Billström and Weihed 1996). This could indicate a mixture of new and inherited zircons.

Clasts of Revsund granite have not been observed in the allogenic breccias. This indicates that the crater was formed before the intrusion of the coarse-grained feldspar porphyritic Revsund granite, yielding ages of 1.7 –1.8 Ga (Patchett et al. 1987; Skiöld 1988; Claesson and Lundqvist 1995) or that such granite did not exist within the area affected by the impact. Additionally, there are no observations of migmatites or gneisses (highly metamorphic parts of the Bothnian Basin metasediments) in the Vargfors conglomerates (in-fill or post-crater sediments). The impact event is definitely younger than the Jörn granite suite of 1.89 to 1.88 Ga age (Wilson et al. 1987) from which clasts are frequently found in the polymict melt breccia. The relation to Arvidsjaur supracrustal rocks is, at present, uncertain.

Inferred Crater Size.

The final crater edge is estimated to be situated about 25 km from the centre of the structure, giving an original diameter of about 50 km (Figures 2, 6 and 11). An inner annular structure can be distinguished on the geophysical and geological/structural maps about 15-20 km from the centre. The northeastern sector of the Gallejaur Structure is uplifted and eroded. Jörn granite rocks occupy the terrain, and exhibit only minor signs of impact deformation. The present estimate of the collapsed crater size is based on the diameter of the central uplift, which is 10-12 km in diameter, as interpreted from gravity and aeromagnetic data. In many craters this is about 20% of the final crater and thus a diameter of 50-60 km is envisaged (compare models of the Vredefort Structure, presented in Henkel and Reimold 1988). This dimension fits the geological structure as interpreted from aeromagnetic maps and is not contradicted by the configuration of the gravity and magnetic anomaly maps. It could be realistic to assume a transient cavity of about half that size

A schematic N-S cross-section through the structure is shown in Fig. 10, based on field and the drill-hole located near the center of the Gallejaur central rise. In the profile the topographical expression of the central rise can be noted.

According to existing estimates of the relation between impact melt thickness (Cintala and Grieve 1992) and volume (Grieve 1993), the Gallejaur Structure could have been larger then assumed here. Degradation caused by 1.8 Ga of erosion and post-impact tectonic events has probably removed much of the evidence of the original shape and dimension.

Calculations based on a 300 km³ volume of impact melt indicate an original transient crater of 23-24 km and an original crater diameter of 46-56 km (Melosh 1989, and references therein)

Concluding Remarks

The thickness of the melt body is at least 600 meters, as can be inferred from a drill-hole in the central uplift. The present melt rock volume can be estimated of about 180 km³ on the basis of drill-hole, gravity and magnetic data, and outcrop observations. Including a possible melt rock outlier (Grundträsk – 14 km from the center of the Gallejaur Structure), the original melt body might have comprised > 300 km^3.

The central rise in large craters evolves into a peak ring. The thickness of melt in the central part of the Gallejaur structure can best be explained as an accumulation of large volumes of impact melt within a central peak ring structure. Much of the impact-induced heat remained for a long period in the uplifted region, permitting the melt to crystallize into a medium-grained plutonic rock, with most clasts absorbed or annealed. The gravity high is best explained by a central rise, bringing up deeper situated mafic to ultramafic rocks. Expected structural uplift is almost 5 km.

Table 4. New simplified stratigraphic scheme for the Skellefte district

Rock units[a]	approximate age (Ga)	References
Revsund granitoids	1.78-1.80	Claesson and Lundqvist 1995
Rocks related to the Gallejaur Structure		
Meta-sedimentary fill		
Gallejaur monzonite and impact melt	1.87	Skiöld 1988
Polymict melt breccia		
Monomict clastic breccia, autochthonous breccia		
Arvidsjaur Group	1.88	Skiöld et al. 1993
Could be contemporaneous with the Skellefte Group of supracrustal rocks	1.87-1.88	Weihed and Mäki 1997 (Table 1)
Skellefte Group of supracrustal rocks and comagmatic Jörn granitoids	1.88-1.89	Weihed and Mäki 1997

[a]It has not been possible in the field to observe the age relationship between the Revsund-Adak granites and the rocks related to the Gallejaur Structure. The age of the Gallejaur monzonite (1 873 Ma) is in this paper regarded as a maximum age.

The magnetic structural pattern indicates three annular rings (Fig. 11). The inner central rise is obvious from the magnetic structures, the high magnetic anomaly and the gravity high. The inner ring is surrounded by a magnetic low. Such patterns are observed in the Lake Mien and Dellen impact structures (Henkel

1992). An outermost irregular ring structure is inferred where the magnetic anomalies outside the crater appear undisturbed.

The outer margin of the central part of the structure displays a highly magnetic ring, caused by the presence of magnetite in a fine-grained to glassy melt rock. This magnetic ring is likely to be caused by additional remnant magnetism as indicated by high Q-values (Enmark and Nisca 1983).

The effects of the Gallejaur impact structure on the geologic evolution in the Skellefte district have not previously been recognized. The new model explains satisfactory the presence of low-metamorphic rocks in flat lying to moderately dipping strata discordantly imposed on the Skellefte Group of supracrustals. A new stratigraphic scheme is presented in Table 4.

Several ore deposits are located within the affected area and both the impact-generated rock formations and the post-impact crater fill sediments have local sulfide mineralization and alteration zones indicative of post-impact hydrothermal activities. Whether they are the result of immediate post-impact hydrothermal activity or whether the complexly deformed volume of the impact structure represented a trap for later mineralization is, however, not clear.

Acknowledgments

North Atlantic Natural Resources AB (NAN) and its Managing Director Edward T. Posey have in every aspect supported this study and made the publication possible. Herbert Henkel has dedicated much time to discuss several of the aspects of impact cratering and the Gallejaur Structure in a creative and constructive way. The manuscript has been considerably improved with valuable suggestions of Uwe Reimold, Christian Koeberl, and Jüri Plado. Lars-Åke Claesson, Lennart Widenfalk and Per Weihed have kindly contributed analytical results and thin sections of rocks from the area. SKB gave permission to inspect the Gallejaur drill-hole. Richard F. Horsnail kindly improved the manuscript style, and Zdenka Ivanic and Jon Rasmusen skillfully prepared the drawings.

References

Allen RA, Weihed P, Svenson S-Å (1995) Physical volcanology of the Skellefte mining district: The relationship between ore deposits and regional volcanic setting. PIM project 92-1207, NUTEK, Sweden

Allen RA, Weihed P, Svenson S-Å (1996) Setting of Zn-Cu-Au-Ag massive sulfide deposits in the evolution and facies architexture of a 1.9 Ga marine volcanic arc, Skellefte District, Sweden. Economic Geology 91: 1022-1053

Billström K, Weihed P (1996) Age and provenance of host rocks and ores in the Paleoproterozoic Skellefte District, northern Sweden. Economic Geology 91: 1054-1072

Cintala MJ, Grieve RAF (1992) Melt production in large-scale impact events:calculations of impact-melt volumes and crater scaling (abstract). In: Large Meteorite Impacts and Planetery Evolution. Lunar and Planetary Institute Contribution 790 pp 14

Claesson LÅ (1985) The geochemistry of early Proterozoic metavolcanic rocks hosting massive sulfide deposits in the Skellefte district, northern Sweden. Journal of the Geological Society 142: 899-909

Claesson S, Lundqvist T (1995) Origins and ages of Proterozoic granitoids in the Bothnian basin, central Sweden; isotopic and geochemical constraints. Lithos 46:115-140

Enmark T, Nisca D (1983) The Gallejaur intrusion in northern Sweden – a geophysical study. Geologiska Föreningens i Stockholm Förhandlingar 105: 287-300

Eriksson L, Henkel H (1994): Geophysics. In: Fredén, C (ed.) Geology. National Atlas of Sweden. SNA Publishing, Stockholm, pp. 76-101

Gavelin S (1939) Geology and ores of the Malånäs District, Västerbotten, Sweden. SGU C 524

Gavelin S (1955) Beskrivning till berggrundskarta över Västerbottens län. (in Swedish with English summary). Sveriges geologiska undersökning Ca 37: 1-99

Grieve RAF (1993) An impact model of the Sudbury structure. In: Sudbury-Norill'sk OGS Special Volume 5, Geological Survey of Canada contribution 29092: 119-132

Grip E (1942) Die Tektonik und Stratigraphie der Zentralen und Östlichen Teile des Skelleftefeldes (in German). Bulletin of the Geological Institution of the University of Uppsala 30: 67-90

Grip E (1951) Geology of the sulfide deposits at Mensträsk and a comparison with other deposits in the Skellefte District. Sveriges geologiska undersökning C 15: 1-52

Henkel H (1992) Geophysical aspects of meteorite impact craters in eroded shield environment, with special emphasis on electric resistivity. Tectonophysics 216:63-89

Henkel H, Reimold WU (1998) Integrated geophysical modelling of a giant, complex impact structure: anatomy of the Vredefort Structure, South Africa. Tectonophysics 287:1-20

Högbom A (1937) Skelleftefältet (in Swedish with English summary). Sveriges Geologiska Undersökning C 389: 1-122

Kautsky G (1957) Ein Beitrag zur Stratigraphie und dem Bau des Skelleftesfeldes, Nordschweden (in German with English summary). Sveriges geologiska undersökning C 543: 1-65

Lundberg B (1980) Aspects of the geology of the Skellefte field, northern Sweden. Geologiska Föreningens i Stockholm Förhandlingar 102: 156-166

Melosh HJ (1989) Impact Cratering – a geological process. Oxford Monographs on Geology and Geophysics 11, pp 1- 245

Patchett PJ, Todt W, Gorbatschev R (1987) Origin of continental crust of 1.9-1.7 Ga age: Nd isotopes in the Svecofennian orogenic terrains of Sweden. Precambrian Research 35:145-160

Rickard D (ed) (1986) The Skellefte Field. Sveriges Geologiska Undersökning Ser C 62:1-54

Skiöld T (1988) Implications of new U-Pb zircon chronology to Early Proterozoic crustal accretion in northern Sweden. Precambrian Research 38: 147-164

Weihed P, Bergman J, Bergström U (1992) Metallogeny and tectonic evolution of the Early Proterozoic Skellefte district, northern Sweden. Precambrian Research 58:143-167

Weihed P, Mäki T (eds.) (1997): Volcanic hosted massive sulfide and gold deposits in the Skellefte district, Sweden and western Finland. Geological Survey of Finland, Guide 41: 1-81

Wilson MR, Claesson L-Å, Sehlstedt S, Smellie JAT, Aftalion M, Hamilton PJ, Fallick AE (1987) Jörn: an Early Proterozoic intrusive complex in a volcanic arc environment. Precambrian Research 36: 201-225

16 Neugrund Structure – the Newly Discovered Submarine Early Cambrian Impact Crater

Kalle Suuroja and Sten Suuroja

Geological Survey of Estonia, Kadaka tee 80/82, EE12618, Tallinn, Estonia.
(s.suuroja@egk.ee)

Abstract. The Neugrund Bank is situated on the southern side of the entrance of the Gulf of Finland (59°20'N; 23°31'E) between Osmussaar and Krass islands. It is a shoal of a very peculiar multi-ring shape. In the coastal and offshore area of North-Western Estonia, numerous erratic boulders, consisting of rocks resembling impact breccias, have been found. The investigations proved that under Neugrund Bank and in its surrounding is located a classic buried and partially newly exposed impact crater. The studies revealed the general morphology of the structure. In the summer of 1998, during three expeditions, the submarine impact structure was investigated in detail by diving and sidescan sonar profiling. As a result, the hypothesis, which had been indirectly indicated by remote sensing, was finally verified.

The Neugrund impact crater formed ca. 540 My ago in a shallow epicontinental sea as the result of the impact of an extraterrestrial body with a diameter of ca. 400 m. As a consequence of the impact a crater with the rim-to-rim diameter of 7 km was formed. The depth of the ca. 5 km-diameter crater has not yet been determined, but is assumed to range over 500 m. At the distance of about 10 km around the crater, the target rocks are strongly disturbed. After the impact the crater deep was filled with clastic deposits and was buried within a rather short time (some millions of years). The sedimentation conditions in the crater differed from those of the surrounding area until the Middle Ordovician. Since then, and up to the Tertiary, the crater remained buried and was partially uncovered during the Neogene.

In the impact-derived rocks (clast-supported mono- and polymict impact breccias), the following evidence for shock metamorphism has been observed: shatter cones, mosaicism of quartz and feldspar, planar fractures and planar deformation features in quartz, kink bands in biotite, diaplectic glass, partial melting of clasts, occurrence of recrystallized glassy grains, occurrence of thin veins of melt. The impact origin of the Neugrund structure may be considered verified. In the impact-affected Precambrian basement rocks of the Neugrund structure, we observed an enrichment of potassium, and a decrease in the sodium content, similar to the behavior of these elements in some rocks of the Kärdla impact crater.

Introduction

Neugrund Bank is situated on the southern side of the entrance of the Gulf of Finland (59°20' N; 23°31' E) between Osmussaar (Odensholm) and Krass islands. It is a shoal of a very peculiar multi-ring shape (Figs. 1, 2). The integrated large-scale (1:50 000) geological mapping of coastal and offshore area in North-Western Estonia has provided abundant interesting information on this geologically problematic area (Suuroja et al. 1997). In the coastal and offshore area of North-Western and Western Estonia, as well as on Osmussaar, Muhumaa and Saaremaa islands occur more than 600 singular erratic boulders, consisting of brecciated metamorphic rocks (amphibolites, migmatites, and gneisses) of the Precambrian crystalline basement (Figs. 1, 4, 5).

Firstly the gneiss-breccia boulders were described on Osmussaar Island by Öpik (1927), who named the rock type as gneiss-breccia. Somewhat later, Thamm (1933) investigated the mineralogy of these rocks more thoroughly. He tried to identify the origin and source area of these boulders, but without any success. Orviku (1935) and Viiding (1960), too, have described several finds of these boulders in Western Estonia.

In 1984–1988, the Geological Survey of Estonia carried out marine geological mapping of the North-West Estonian shelf at a scale of 1:200 000. The Neugrund structure was interpreted as a glacial formation – moraine wall (Raukas and Hyvärinen 1992; Lutt and Raukas 1993).

During the large-scale (1:50 000) geological mapping, geologists Suuroja and Saadre (1995), who had for several years participated in the investigations of the nearby (ca 60 km SW) Kärdla impact crater (Puura and Suuroja 1992) noted the obvious similarity of the gneissbreccias to the allo- and authigenic impact breccias of the Kärdla impact crater. Microscopic investigations of the fine-grained clast-supported impact breccias (Suuroja et al. 1997) showed that they belong to the low shock metamorphism stage (0–I). The distribution fan of the gneiss-breccia erratic boulders pointed to the sea area eastward of Osmussaar Island, where the submarine Neugrund Bank is located (Figs. 1, 2). Therefore, the hypothesis was proposed that the peculiar multi-ring-shaped Neugrund Bank and its surrounding could be a buried impact structure, partially exposed as a result of the Quaternary glaciation (Suuroja and Saadre 1995).

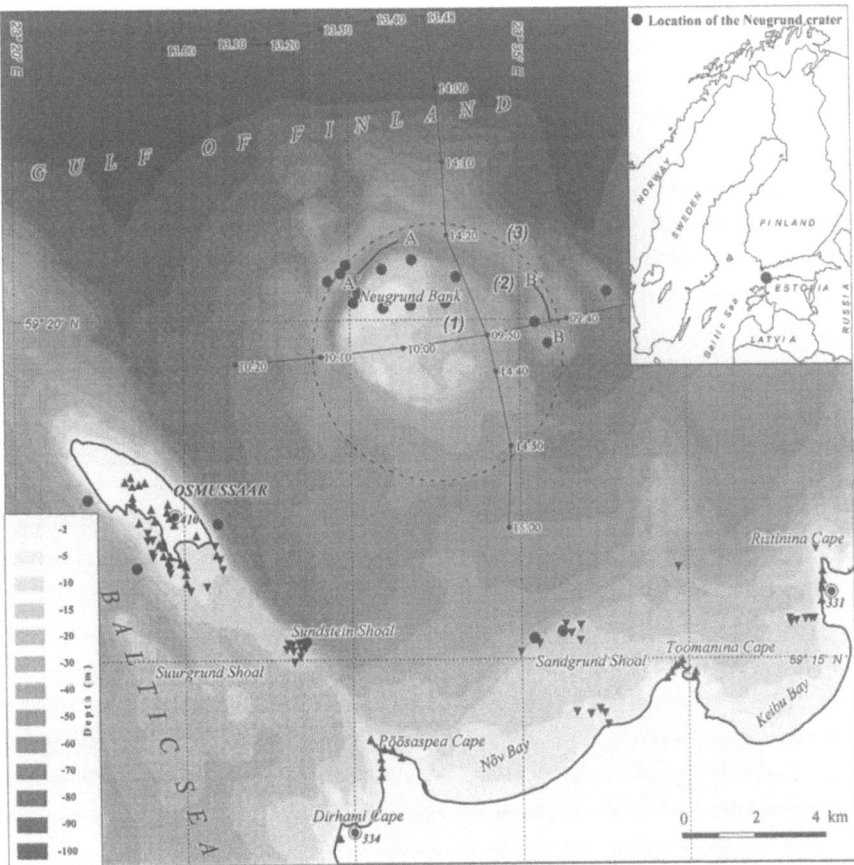

Fig. 1. The location and bathymetric maps of the Neugrund crater area . 1 = central plateau; 2 = circular erosional canyon; 3 = rim-wall. Triangles = erratic boulders made of Neugrund Breccia; dots = diving and sampling sites; dotted line = rim- wall ridge; circles = drill hole and its number; lines with numbers = seismo-acoustic profiles in Fig. 6 and 7.; lines A – A' and B- B' = side-scan sonar profiles.

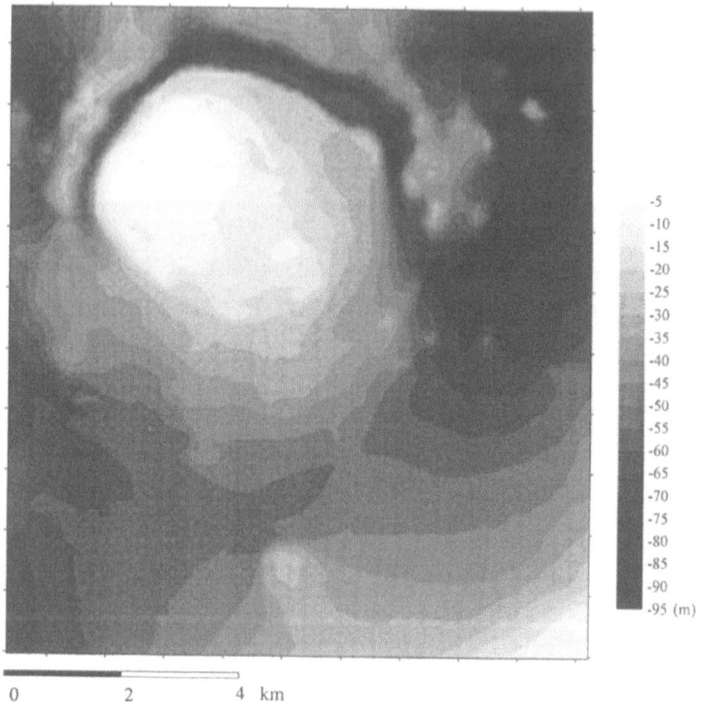

a

0   2   4 km

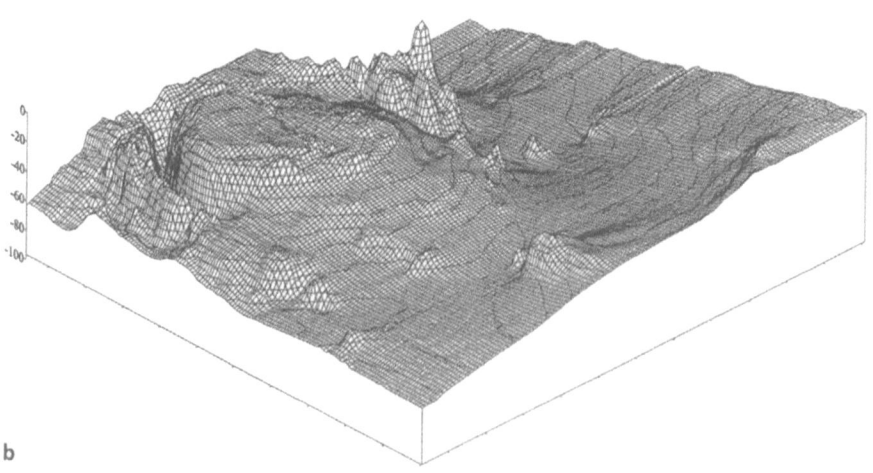

b

Fig. 2. Sea bottom relief of the Neugrund crater area. (a) bathymetric map (data of the Board of Waterways of Estonia); (b) bar chart of the same area.

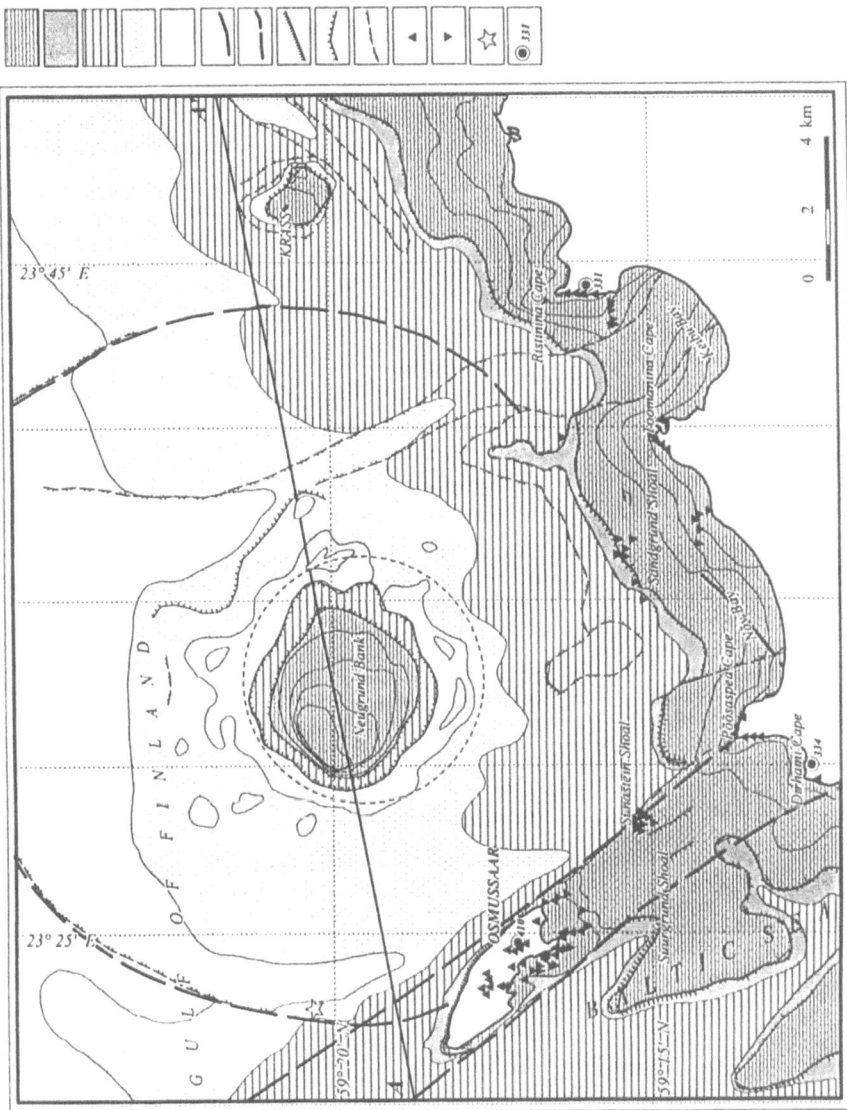

Fig. 3. Bedrock geological map of the Neugrund crater area. 1 = Precambrian basement metamorphic rocks; 2 = pre-impact terrigenous rocks (Vendian and Early Cambrian); 3 = post-impact terrigenous rocks (Early Cambrian); 4 = Early Ordovician terrigenous rocks; 5 = Middle Ordovician limestones; 6 = coastline; 7 = supposed fault; 8 = terrace on land; 9 = terrace on sea bottom; 10 = buried terrace on the sea bottom; 11 = erratic boulders made of Neugrund breccia on land; 12 = same on the sea bottom; 13 = inferred epicentre of the Osmussaar earthquake of 1976. 14 = drill hole and its number (cross-section A – A').

Fig. 4. Some erratic boulders of Neugrund breccia derived by glacial action from Neugrund crater rim-wall or ejected blocks on the North-Western Estonian coast: (a) on Toomanina Cape; (b) on Põõsaspea Cape.

Fig. 5. The impact-affected Precambrian metamorphic rocks (autochthonous impact breccia) from the Neugrund crater area: (a) amphibolite rock; (b) granitoid.

Considering the above-mentioned presumption, in the summer of 1996 Stockholm University and the Geological Survey of Estonia organized a joint expedition on r/v "Strombus", headed by Tom Floden. Neugrund Bank and the surrounding sea bottom were geophysically investigated using continuous seismo-acoustic and magnetometric profiling (Fig. 6; 7). These studies proved that under Neugrund Bank and in its vicinity lies a buried and partially newly exposed impact crater. The general morphology of the structure was revealed as well (Suuroja 1996; Suuroja et al. 1997), but direct verification – by studying samples of impact-derived rocks collected directly from the crater structure – was not obtained.

In the summer of 1998 the submarine impact structure was investigated in detail by diving and sidescan sonar. All reachable structures and rocks at the depth of 2–43 m were investigated, sampled and recorded by video camera. 100 miles sidescan sonar profiling was carried out. As a result, the hypothesis previously indirectly proposed from remote sensing was supported by direct evidence. Macroscopic and geochemical similarity of the rocks provide clear evidence that all so-called gneiss-breccia (or the Neugrund Breccia – as they could be named in the future) erratic boulders in Western and North-Western Estonia are derived from the Neugrund structure (crater) rim-wall and most likely are of impact origin. A lot of information about of the lithology of target rocks and impact and post impact processes has been obtained by investigating the Osmussaar (no. 410, 5 km SW of the crater rim), Ristinina (no. 331, 10 km SE of the crater rim) and Dirhami (no. 334, 10 km S of the crater rim) drill cores (Fig. 1; 3).

Geological and paleogeographical setting of the crater

The Neugrund impact event took place in the Early Cambrian, during the middle of the Lontova age (ca 540 Ma) in shallow epicontinental sea not far from the shoreline of that time (Fig. 8). In the Early Cambrian the impact site on the Baltic Continent was situated on the Southern Hemisphere at latitudes of 30–40° (Torsvik et al. 1996). At that time the Precambrian crystalline basement of the impact site was covered with Vendian (ca 60 m) and Early Cambrian (ca 40 m) terrigeneous rocks (sandstones, siltstones, clays) and water (ca 50 m). Prior to the impact, the sedimentary cover presented a horizontal monocline (Fig. 9a). A hypothetical cross-section of the target rocks has been reconstructed on the basis of some drill cores from the surrounding area (Osmussaar Island, Ristinina Cape, and Dirhami Cape).

The Early Cambrian Lontova age sedimentation in the basin was characterized by relatively smooth changes in facies environment (Mens and Pirrus 1997), considering the gradual decrease of sand content and increase in clay content from the west to the east (Fig. 8). Evidently the changes in facies were mainly due to the changes in the depth of the basin. In the impact site the deposits of the pre-impact Lontova Formation are represented by fine- to coarse-grained weakly lithified quartzose sandstones (ca. 70 %) with interlayer of clayey siltstone. Such association of sediments is typical for a shallow offshore sea, were the water depth may be 50-100 m (Mens and Pirrus 1997). Biostratigraphically the impact event is sufficiently characterized and marks the pre-trilobite Early Cambrian *Platysolenites antiguissimus* biozone of the East-European Platform (Mens et al. 1990). The above-mentioned skeletal fragments have been found in impact-related sedimentary breccias in surrounding area, as well as in the deposits above and below them (Fig. 8).

In the pre-impact section the ca 60-m thick Vendian complex (Fig. 10) consisted mainly of fine- to coarse-grained, weakly lithified and water-saturated sandstones (the Kotlin Stage). Along the base of the Vendian complex there is a 2–10-m thick bed of multicoloured unsorted clayey-sandy-gravelly deposits (mixtite). The clayey matrix of this rock consists mostly of kaolinite, in contrast to the Early Cambrian deposits, where illite is the dominant clay mineral.

The Svecofennian orogenic metamorphic rocks form the Precambrian crystalline basement of the impact site. Information about of the crystalline basement has been obtained by drilling in the surrounding area, investigation of numerous erratic boulders derived from the crater structures (rim-wall and ejected megablocks) and by the direct observing and sampling submarine outcrops.

Amphibolites are the most widely distributed rock type in this area, forming over 50 % of the crystalline basement. They occur mainly as large layer-like bodies containing varying amounts (5–30 %) of migmatite granite veins and lenses. In addition to granites, intercalated quartz-feldspar gneisses and biotite gneisses have been recorded as well.

The amphibolites of the Neugrund crater area are migmatized to a variable degree and are fine- to medium-grained (grain size 1–5 mm), of linear or massive texture and greyish-green or dark-green in colour. Mineralogical composition (vol. %) of amphibolites, which based on microscopic assessment 15 thin sections is the following (Table 1): plagioclase (An 40–50 %) = 40–45 %; amphibole (Hbl) = 50–60 %; quartz = 0–2 %; biotite 1–5 %; pyroxene (clinopyroxene) = 1–5 %; K-feldspar = 0–2 %. Accessory minerals are represented by apatite, orthite and zircon; ore minerals are mainly magnetite, pyrite and ilmenite. According to the chemical composition ($SiO_2 = 48$–50 %, $Na_2O+K_2O = 3$–6.5 wt%), the amphibolites of the Neugrund crater area correspond to the rocks of the basalt group. The above-mentioned rocks are very similar to the amphibolites of the crystalline basement of the Kärdla impact crater area (Koppelmaa et al. 1996)

The information obtained from the study of erratic boulders and submarine outcrops indicates that biotite gneisses in the basement of the Neugrund crater area are of more restricted distribution than amphibolites, but they form rather large layer-like bodies in the eastern part of the area. Biotite gneisses are relatively homogeneous, but fragmentary-banded reddish-grey medium-grained strongly migmatized rocks. Major minerals are plagioclase (An 30–35 %), quartz, K-feldspar and biotite, the ratios of which may vary to some extent. In places the biotite gneisses contain also some hornblende. Accessory minerals are apatite, magnetite, zircon and orthite. By chemical composition (SiO_2 = 63–68 wt%; Na_2O+K_2O = 5.5–7.5 wt%) the biotite gneisses of the Neugrund crater area correspond to the rocks of the andesite group and some amphibolites, and they greatly resemble analogous rocks of the Kärdla crater area (Koppelmaa et al. 1996). The crystalline basement rocks in this area are weathered below the contact with the Vendian sedimentary rocks (Fig. 10). Evidence of Precambrian (Vendian) weathering has been found up to ten meters below the peneplane surface of the crystalline basement.

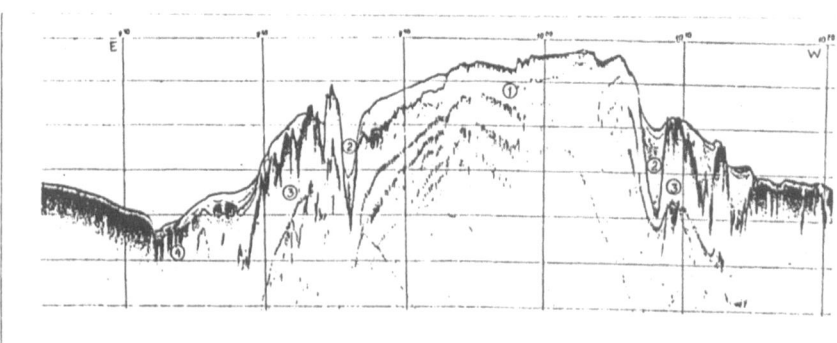

Fig. 6. E–W direction continuous seismo-acoustic profiling record across the Neugrund crater area. Record frequency 4 kHz, space between horizontal lines ca. 18 m in water; space between vertical lines 10 minutes or ca. 2.3 km. 1 = Neugrund Bank or central plateau; 2 = circular erosional canyon; 3 = rim-wall; 4 = surrounding area.

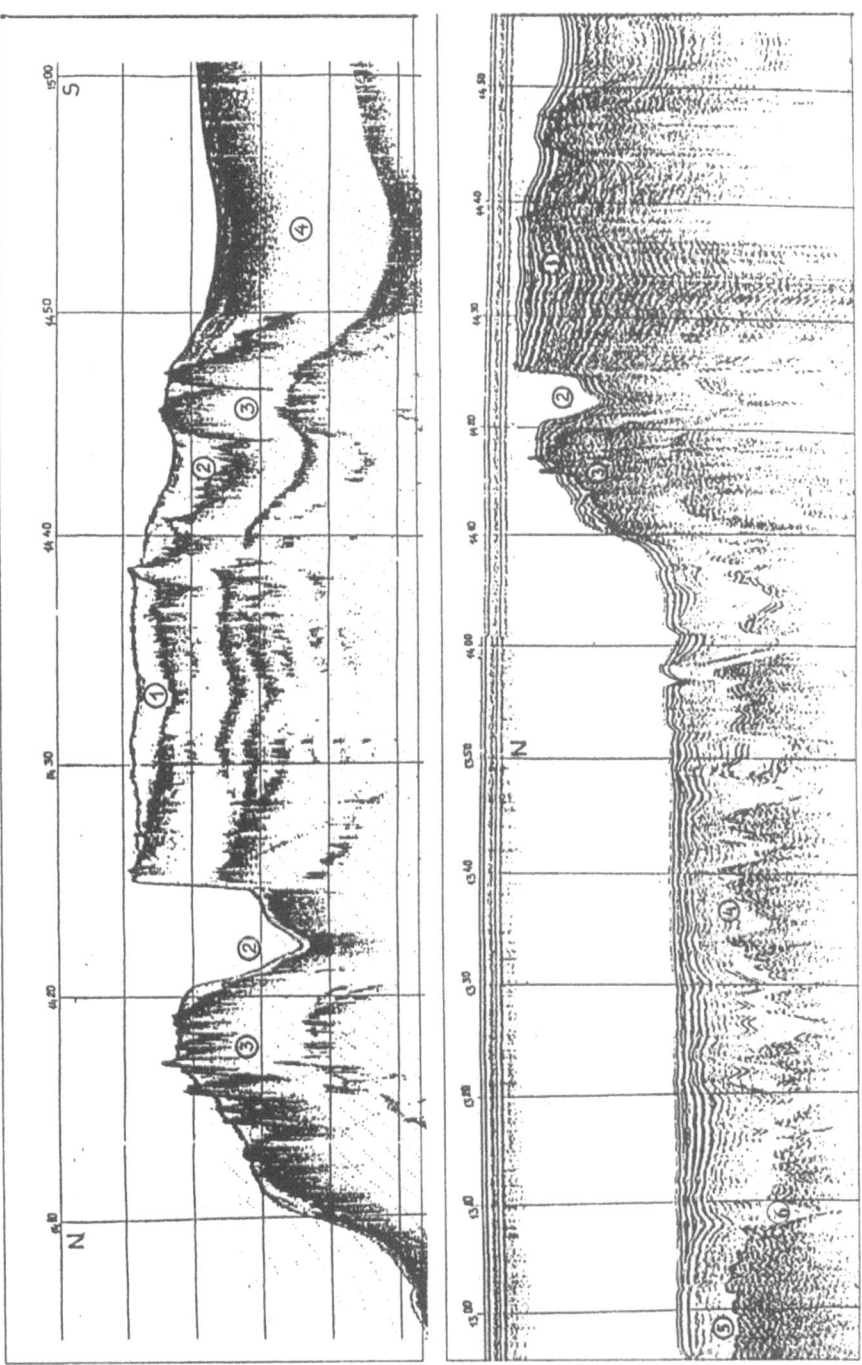

Fig. 7. N–S direction seismo-acoustical record across the Neugrund structure. Space between horizontal lines ca 18 m in water; space between vertical lines 10 minutes or ca. 1.8 km. (a) record frequency 4 kHz. (b) record frequency 250 – 500 Hz. 1 = Neugrund Bank or central plateau; 2 = circular erosional canyon; 3 = rim-wall; 4 = disturbed target rocks in the surrounding area; 5 = undisturbed target rocks; 6 = ring-fault.

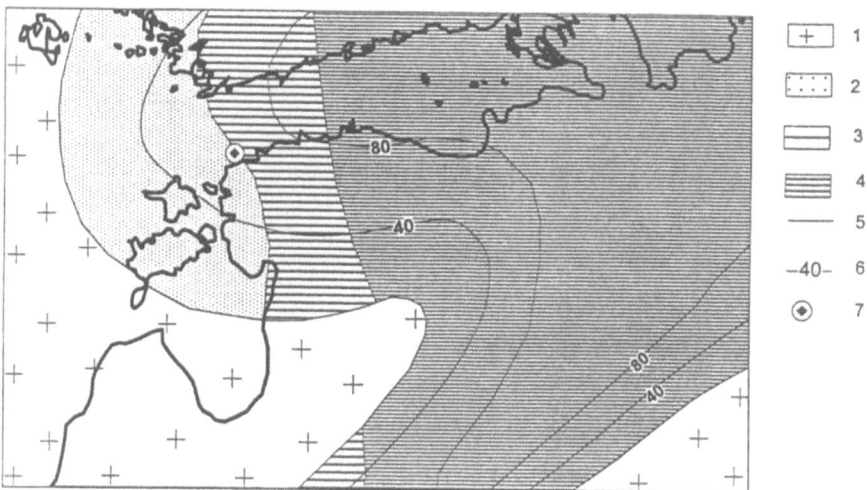

Fig. 8. Lithological sketch-map of the Neugrund crater area during Lontova Age (modified from Mens and Pirrus 1997): 1 = land, represented by igneous and metamorphic rocks; 2 = area of sandstones; 3 = area of ca 50 % argillaceous rocks; 4 = area of >75 % argillaceous rocks; 5 = boundary of facies belt; 6 = isopachs of Lontova Formation; 7 = Neugrund impact site.

Morphology of the crater

By now, the morphology of the crater has been relatively well investigated (Table 2). The major structural features of the submarine impact crater (rim-wall, crater proper, ejecta layer and ring fault) have been distinguished. The best overview of the preserved and exposed crater structure was obtained by diving and sidescan sonar profiling (Fig. 11). The crater rim-wall is made of brecciated crystalline basement metamorphic rocks (authigenic impact breccias) and is exposed on the sea bottom as a 25–50 m high, 100–500 m wide and 9–10 km long semicircular (southern part of the rim is buried under the Quaternary deposits) range of smooth hills resembling *roche moutonnée* features (Fig. 11 b). The external slope of the rim is gently sloping (10–20°), while the inner slope is steeper (30–50°) and in places penetrated by joints sub-parallel to the rim. Along these joints blocks of the rim have slid down towards the crater deep. The rim-wall, together with ejected blocks, has served as the main source of the Neugrund Breccia erratic boulders of Western Estonia. The preserved rim-wall is composed of fragments of up to 120 m high elevation (Fig. 6). The primary height of the structure may have been over

300 m, and total height of the rim-wall together with the talus slope, about 500 m. The rim-wall was partially eroded immediately after the impact by resurge wave and short-time (some millions of years) erosion before burying of the structure. It is as yet unknown, which parts of the rim-wall were eroded shortly after the impact in the Early Cambrian, and which during the Tertiary erosion and the Quaternary glaciation (or Paleocene).

Between the rim-wall and at the crater filling and overlying complexes (central plateau), there is up to 70 m deep and 200–500 m wide circular canyon-like depression (Fig. 2; 6; 7). The external slope of the canyon is the rim-wall and the internal slope the section of the central plateau. The internal slope of the circular canyon, where the filling and overlying complexes of the crater is exposed for ca. 50 m, is remarkably steeper (40–90°). It is especially well expressed in the northern part of the Neugrund structure (Fig. 7), where over a difference of ca. 100 m, a step-like lowering of the limestone plateau from the depth of 2–3 m to the depth of 18–20 m can be observed. A-6–8 m high precipice, exposing Middle and Late Ordovician limestones follows the "steps". The foot of the escarpment is presented by ca. 30–40 m high slope with an angle of 30–40°, where Early Ordovician deposits (glauconitic sandstone, Dictyonema shale, biodetritic sandstone) and Early Cambrian post-impact silt- and sandstones crop out. The exact time of final burial of the crater is unknown, but there is reason to assume that it occurred a short time after the impact at the end of the Early Cambrian. Therefore, the Early Cambrian terrigeneous deposits of the Tiskre and Lükati Formations are considered to belong to the crater filling complex, and Early and Middle Ordovician deposits to the overlying complex.

The relatively flat bottom of the circular canyon (Fig. 11 a) is covered by a 10–20 m thick layer of Holocene (Baltic Ice Lake) varved clay (Fig. 6). In the southern part of the structure the varved clays fill the canyon and the thickness of the layer reaches up to 40 m (Fig. 7). The canyon is obviously of erosional origin and was formed by flowing water at the contact of two rock complexes with very different physical-mechanical properties: the hard and durable complex of the brecciated crystalline basement metamorphic rocks, and friable complex of Cambrian terrigeneous rocks (silt- and sandstones). The time of formation of this erosional structure as well as the primal denudation of the buried crater, is related to the Tertiary (Neogene) and intensive tectonic uplifting of the area (Puura 1991; Puura and Floden 1997; Mozayew 1973) and is directly connected with the formation of the important erosional structure known as the Baltic Clint. During the Pleistocene glaciation the rim-wall, as well as the surrounding area, were additionally eroded by glaciation.

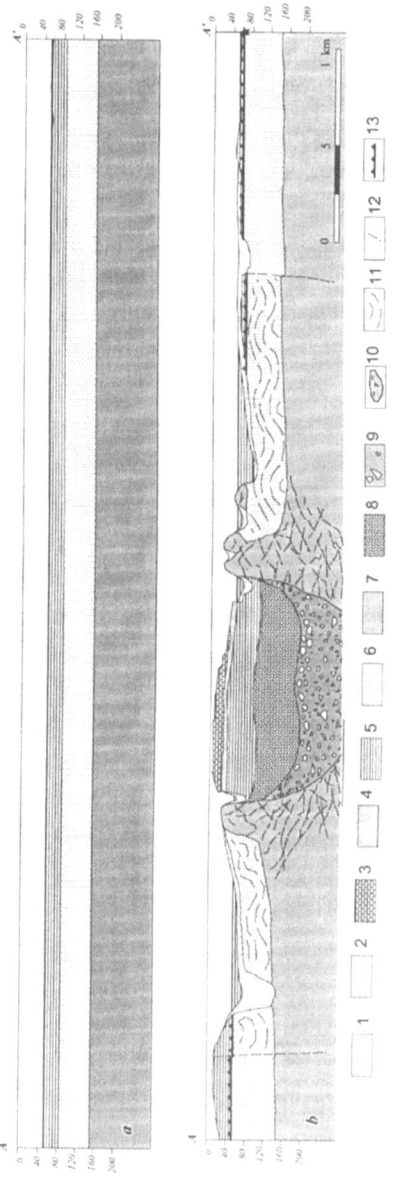

Fig. 9. Cross-section of the Neugrund crater area (for location of the cross-section, see Fig. 3.): (a) reconstruction of the impact target; (b) present configuration. 1 = sea water; 2 = Quaternary deposits; 3 = Middle and Early Ordovician limestones; 4 = Early Ordovician terrigeneous rocks; 5 = post-impact Early Cambrian terrigeneous rocks ; 6 = pre-impact Early Cambrian and Vendian terrigeneous rocks; 7 = metamorphic rocks of the Precambrian basement; 8 = crater filling complex; 9 = inferred impact breccias; 10 = ejected megablocks; 11 = disturbed sedimentary target rocks; 12 = faults; 13 = ejecta layer.

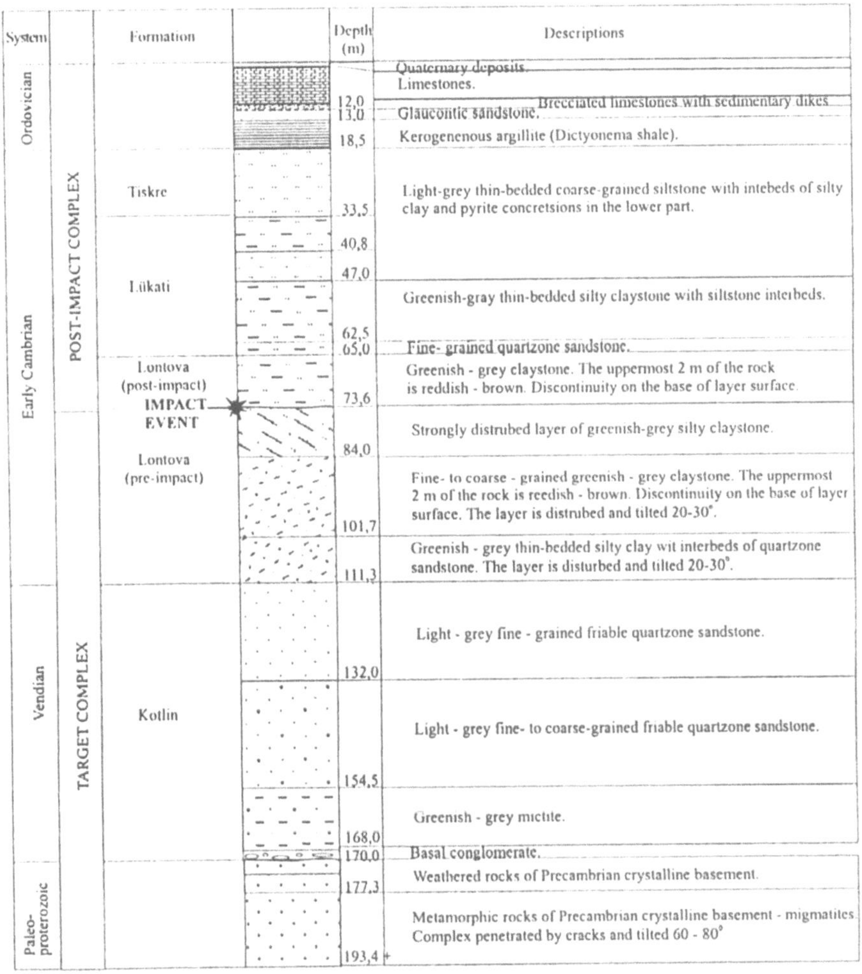

System	Formation		Depth (m)	Descriptions
Ordovician	POST-IMPACT COMPLEX			Quaternary deposits.
				Limestones.
			12,0 13,0	Glauconitic sandstone. Brecciated limestones with sedimentary dikes
			18,5	Kerogenenous argillite (Dictyonema shale).
Early Cambrian		Tiskre	33,5	Light-grey thin-bedded coarse-grained siltstone with intebeds of silty clay and pyrite concretsions in the lower part.
			40,8	
		Lükati	47,0	
			62,5	Greenish-gray thin-bedded silty claystone with siltstone interbeds.
			65,0	Fine- grained quartzone sandstone.
		Lontova (post-impact)		Greenish - grey claystone. The uppermost 2 m of the rock is reddish - brown. Discontinuity on the base of layer surface.
		IMPACT EVENT	73,6	
			84,0	Strongly distrubed layer of greenish-grey silty claystone.
		Lontova (pre-impact)	101,7	Fine- to coarse - grained greenish - grey claystone. The uppermost 2 m of the rock is reedish - brown. Discontinuity on the base of layer surface. The layer is distrubed and tilted 20-30°.
			111,3	Greenish - grey thin-bedded silty clay wit interbeds of quartzone sandstone. The layer is disturbed and tilted 20-30°.
Vendian	TARGET COMPLEX	Kotlin		Light - grey fine - grained friable quartzone sandstone.
			132,0	
			154,5	Light - grey fine- to coarse-grained friable quartzone sandstone.
			168,0	Greenish - grey mictite.
			170,0	Basal conglomerate.
Paleo-proterozoic			177,3	Weathered rocks of Precambrian crystalline basement.
			193,4	Metamorphic rocks of Precambrian crystalline basement - migmatites. Complex penetrated by cracks and tilted 60 - 80°

Fig. 10. The Osmussaar drill core (nr. 410) lithostratigraphic section. The drill hole is situated 9 km SW from the centre of the impact site.

In the centre of the Neugrund structure is an almost circular central plateau 4.5 km in diameter, the so-called Neugrund Bank. The plateau is covered with Middle Ordovician limestones and it is located directly at the crater (Fig. 1; 9b). Water depth on the Neugrund Bank varies between 1–15 m, increasing from NW to SE. Geomorphologically the central plateau may be considered a submerged erosional relict or a so-called klint island. On the plateau are exposed the Middle Ordovician limestones of the Viivikonna, Uhaku, Väo, Aseri and Pakri formations. The preliminary results of diving showed that the thickness and composition of limestones of these formations within the crater differ greatly from rocks of above-mentioned formations in the surrounding area. The thickness of the limestone cover on the plateau is approximately 20 m, decreasing from NW to SE. The plateau is from all sides bordering with 6–10 m high escarpment, which is buried under the Quaternary deposits (varved clays) in the south-eastern part of the crater.

Submarine investigation of the canyon section showed that the crater is filled mainly with Early Cambrian and Early Ordovician clastic rocks. Due to the discovery of Early Cambrian and Early Ordovician deposits in the crater, it became necessary to correct the previous assessment (474 Ma) of the age of the crater (Suuroja 1996). The latter was preliminary determined mainly by the evidence of catastrophic seafloor crushing which is widespread in North-Western Estonia and is recorded also in a section of Osmussaar Island. The new age determination is based on the presence of the Early Cambrian deposits in the crater proper and the occurrence of over 20 m thick layer of disturbed and brecciated (impact-influenced) rocks on the top of the Lontova Formation (the Early Cambrian, Lontova age, ca 540 Ma) in drillhole sections located not far from the impact centre (Osmussaar Island – 9 km SW, Ristinina –13 km SE).

Recognition of impact origin of the Neugrund structure

Various criteria have been developed for the recognition and confirmation of impact structures. The characteristics are the evidence of shock metamorphism, crater morphology, geophysical anomalies, and presence of meteorites or geochemical discovery of traces of the meteoritic projectile (Koeberl and Anderson 1996; Melosh 1989).

The morphology of the Neugrund structure is strongly suggestive of an impact origin. The negative magnetic anomaly at the crater deep and positive corresponding to the rim-wall are clearly expressed (Fig. 12). Up to now, gravimetric studies and detail geochemical investigations of impact-influenced rocks have not been carried out in the Neugrund crater area. Data about the existence of a central uplift in the crater are absent, but from experience based on a slight (some meters) uplift of the limestone beds in the central part of the crater (Fig. 6) indicate its possible presence. A ring fault with a diameter ca 20 km as an outer boundary of the structure, which concentrically surrounds the inner crater (Fig. 3), is also typical for impact craters of this type.

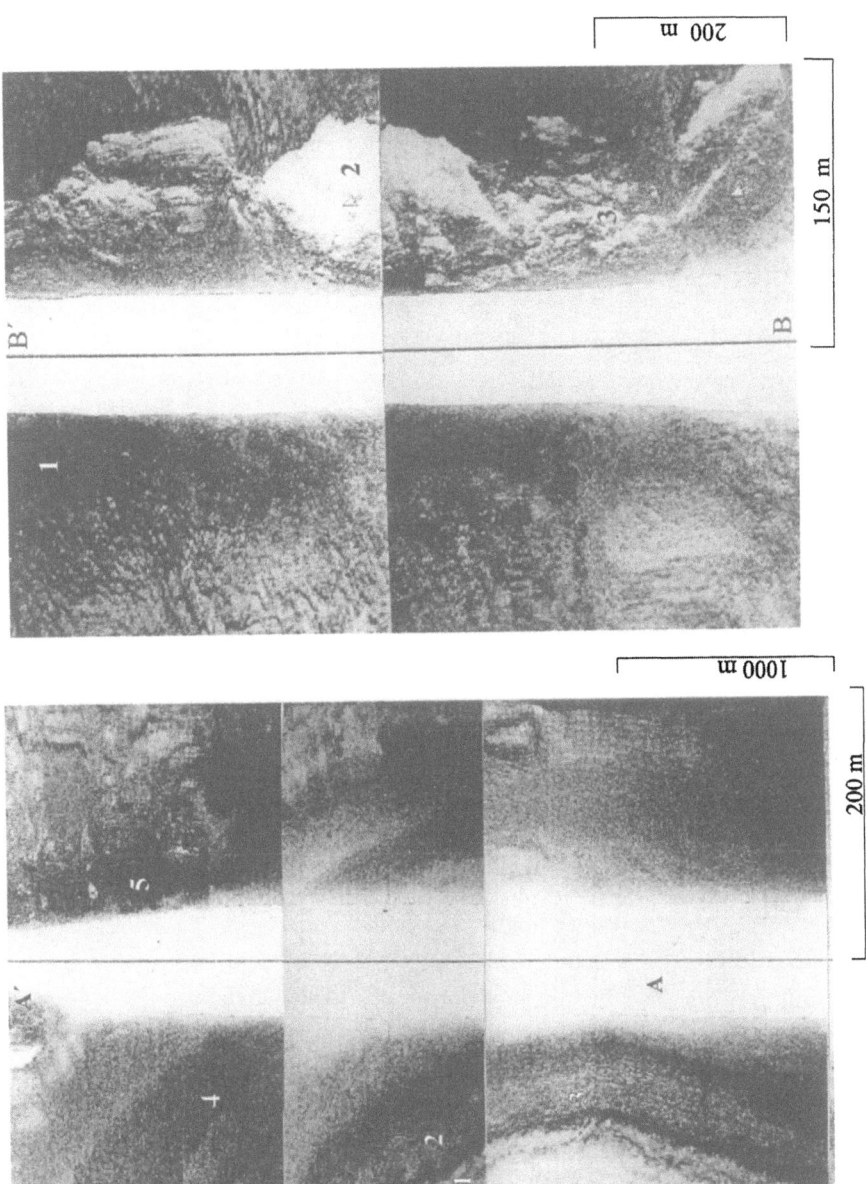

Fig. 11. Sidescan sonar profiles across the circular canyon of the Neugrund structure. A –
A' = across the circular erosional canyon in the western part of the structure. 1 = gradually
lowered limestone plateau; 2 = ca. 6 m high limestone precipice; 3 = ca. 30 m high
escarpment (from top to bottom) glauconitic sandstone, Dictyonema shale, biodentritic
sandstone and siltstone; 4 = flat bottom of the canyon covered with varved clay; 5 = rim-
wall were cropped out brecciated Precambrian metamorphic rocks. Width of the recording
strip 200 m. B – B' = along the rim-wall. 1 = inner slope of the rim-wall; 2 = outcrop of
brecciated Precambrian metamorphic rocks; 3 = glacially eroded part of the rim-wall (*roche
mountaunce* features).

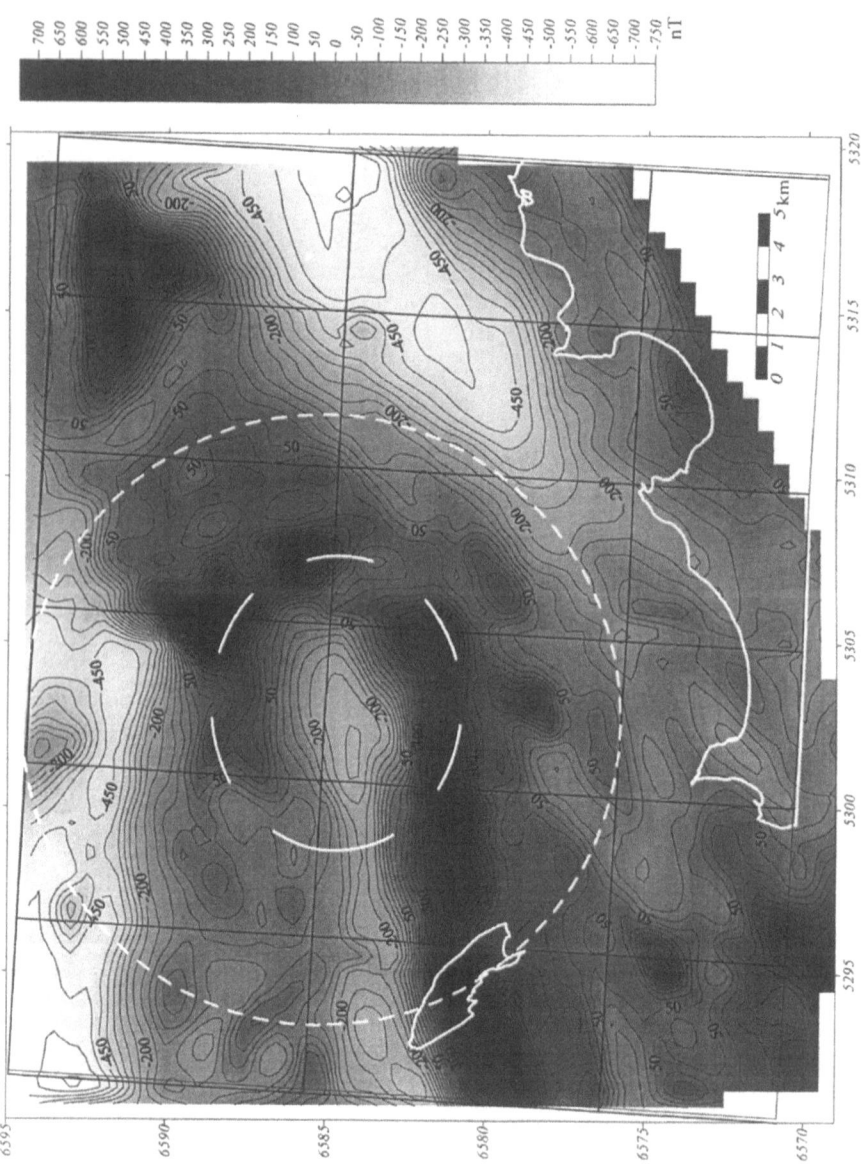

Fig. 12. DGRF-65 aeromagnetic anomaly (total field) map of the Neugrund crater area. The dashed lines denote the ridge of crater rim (inner) and the ring fault (outer). Compiled by Tarmo All (Geological Survey of Estonia).

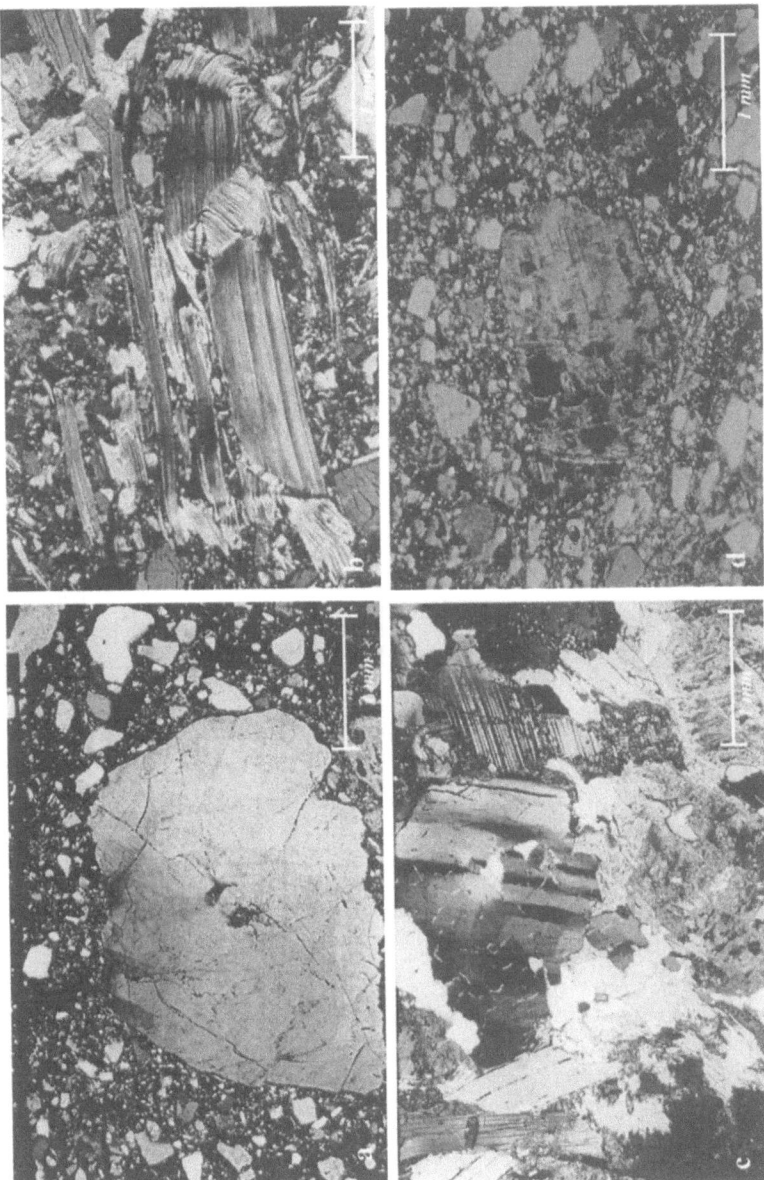

Fig. 13. Photomicrographs of matrix supported polymict impact breccia veins from the Neugrund crater rim-wall and erratic boulders. (a) Two weakly developed sets of nondecorated planar deformation features (PDF) in a quartz clast. Sample from rim-wall. Crossed polarizers (b) Biotite leaves near – horizontal cleavage and well-developed kink bands. Crossed polarizers, sample from matrix supported polymict impact breccia vein in crater rim. (c) Quartz grain with intensive band-like mosaicism.. Sample from erratic boulder. Crossed polarizers (d) Strongly altered and partially melted feldspar (plagioclase) clasts from matrix-supported polymict impact breccia (in the center), sample from ejected megablocks. Crossed polarizers.

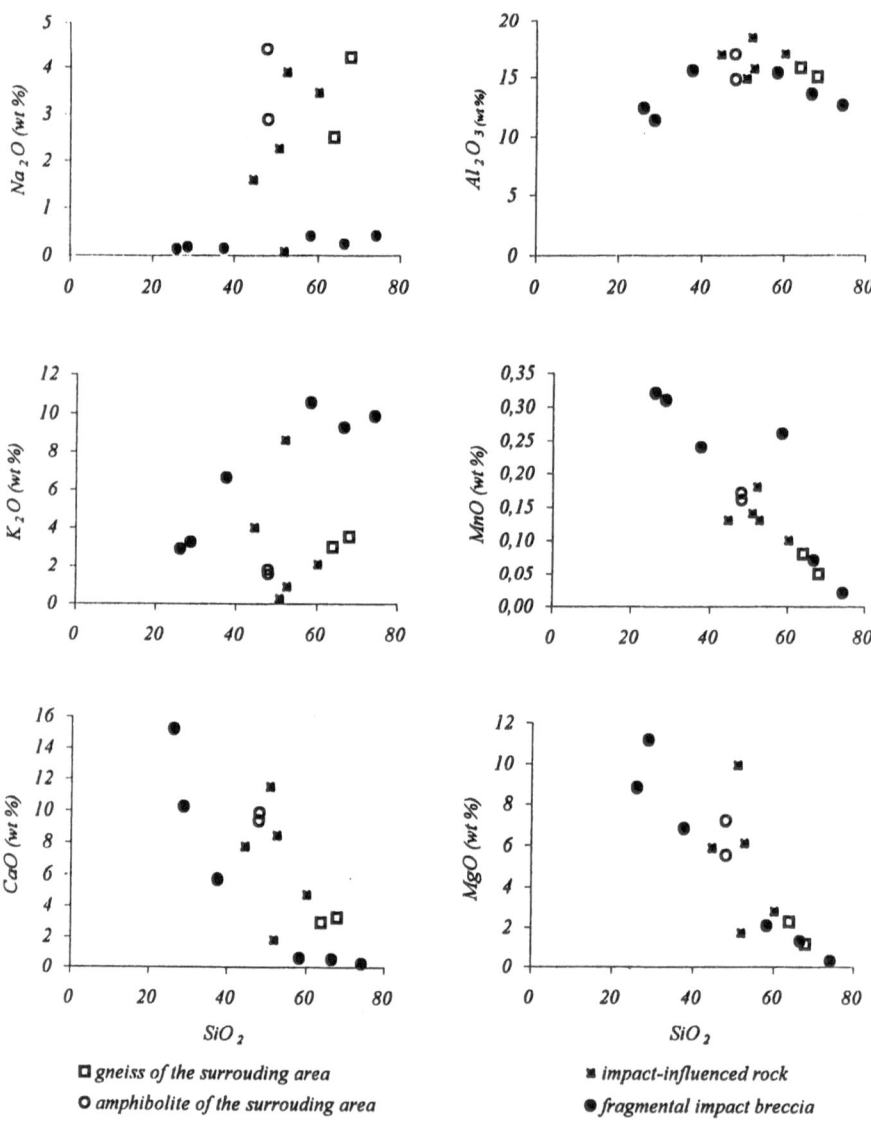

Fig. 14. Harker-diagrams of the variations of silica versus selected major element contents (wt %) in the Precambrian rocks and impact breccias of the Neugrund impact structure.

Table 1. Chemical composition (wt%) of Precambrian target rocks and impact breccias of the Neugrund area.

Sample nr.	Surrouding area				Crater area											
	gneisses		Amphibolites		Impact influenced target rocks					Matrix supported breccia						Melt
	3312460	3342536	3342030	3342336	31	63	8285-2	8285-3	8229-28	18	17	8285-1	3-26	8285-4	8229-29	20
SiO$_2$	63,92	68,02	47,97	47,84	50,70	60,14	51,88	44,54	52,50	74,28	66,64	26,06	58,30	37,64	28,70	50,90
TiO$_2$	0,40	0,34	0,95	1,11	0,51	1,22	1,12	0,58	0,70	0,09	0,43	0,45	0,67	1,24	0,50	0,05
Al$_2$O$_3$	15,72	14,98	14,74	16,94	14,83	16,99	18,42	16,91	15,71	12,56	13,53	12,36	15,31	15,52	11,36	10,52
Fe$_2$O$_3$	5,00	2,05	6,46	5,16	1,90	2,76	-	-		1,23	3,68	-	7,33	-		-
FeO	1,36	2,01	7,76	6,75	4,89	3,45	-	-		<0,20	1,80	-	0,22	-		-
Fe$_2$O$_3$ total	-		-	-	7,33	6,59	12,42	12,24	10,43	1,23	5,68	16,32	7,57	13,93	16,55	19,76
MnO	0,08	0,05	0,16	0,17	0,14	0,10	0,18	0,13	0,13	0,02	0,07	0,32	0,26	0,24	0,31	0,14
MgO	2,25	1,15	5,50	7,18	9,91	2,79	1,72	5,86	6,08	0,29	1,28	8,79	2,05	6,81	11,15	4,37
CaO	2,93	3,24	9,85	9,37	11,52	4,74	1,81	7,72	8,44	0,21	0,53	15,2	0,58	5,67	10,25	3,27
Na$_2$O	2,50	4,21	2,89	4,40	2,26	3,45	0,08	1,59	3,90	0,41	0,24	0,13	0,40	0,14	0,17	0,13
K$_2$O	3,00	3,53	1,56	1,81	0,26	2,10	8,60	4,00	0,88	9,84	9,26	2,88	10,52	6,59	3,25	1,57
P$_2$O$_5$	0,13	0,17	0,16	0,80	0,02	0,35	0,50	0,26	0,15	0,03	0,24	0,24	0,41	0,28	0,25	0,07
S	<0,10	0,05	0,36	0,19	<0,10	<0,10	<0,10	<0,10	0,08	0,10	<0,10	0,10	<0,10	0,10	0,04	0,10
CO$_2$	-		-	-		-		2,55		0,55	0,21	12,54	-	7,00	11,75	-
LOI	2,10	0,49	1,18	1,23	1,44	0,98	3,51	5,85	0,95	1,15	2,31	17,99	2,77	11,33	17,26	6,56
Total	99,39	100,29	99,54	99,95	99,02	99,56	100,34	99,78	100,06	100,22	100,22	100,84	98,95	99,50	99,79	97,44

Table 2. Characteristics of the Neugrund crater

Characteristics	Size
Pre-impact geological succession of the target:	
sea water	ca 50 m
Lower-Cambrian claystones, siltstones and sandstone's	40 m
Vendian sandstones	60 m
Sedimentary rocks cover in all	100 m
Precambrian crystalline basement metamorphic rocks	>3000 m
Lateral dimensions of the impact structure:	
Diameter (rim-to rim) of the crater	7 km
Diameter of the crater on the level of target surface	6 km
Diameter of the ring fault	20 km
Diameter of the area with impact-derived debris (ejecta) layer	>30 km
Diameter of the central uplift	ca 1 km
Diameter of the biggest ejected megablock	ca 500 m
Vertical dimensions:	
Depth of the crater proper below the target surface	>500 m?
Height of the eroded rim above the target surface	ca 120 m
Height of the central peak	ca 100 m?
Thickness of impact-related breccias:	
In the crater	>200 m
Outside the rim (ejecta)	0-100 m
Magnetic anomalies:	
Negative, correspondending to the crater proper	200 nT
Positive, over the crater rim	500 nT
Thickness of the post impact cover:	
Lower and Middle Ordovician limestone in the crater proper	24 m
Lover Ordovician and Lower Cambrian terrigenous rocks in the crater proper	>50 m

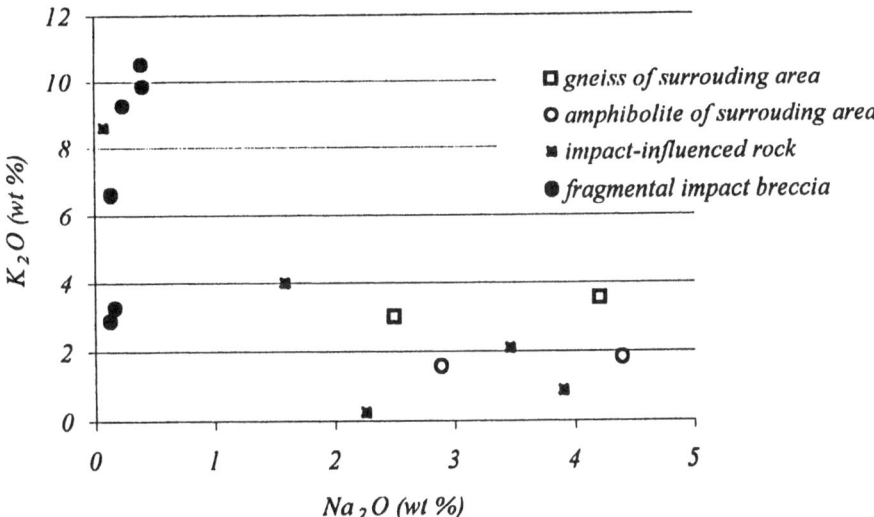

Fig. 15. Na₂O versus K₂O plot of the Precambrian target rocks and impact breccias of the Neugrund impact structure.

In addition to the structure, the major criterion for the recognition and confirmation of an impact structure is the occurrence of shock metamorphic effects (Stöffler 1972; Stöffler 1974; Stöffler and Langenhorst 1996; Grieve et al. 1996). The evidence of shock metamorphism in the impact-derived rocks of the Neugrund crater has been recognized in 20 thin sections of hand specimens collected from clast-supported impact breccias of the rim-wall ejected blocks and erratic boulders. Melt-supported impact breccias, (suevite), as well as massive bodies of clast-supported impact breccias have not been discovered in the Neugrund crater area, probably because the fill, where they possibly could occur, is buried under post-impact sedimentary rocks. All evidence of shock metamorphism is connected with impact breccias formed in metamorphic rocks of the crystalline basement. In the rocks of the Neugrund crater are four types of impact breccias can be distinguished:

1. breccias forming clear-cut dike-like allochthonous bodies in crystalline basement rocks, with a mineralogical composition that differs from the surrounding (host) rocks. Evidence of low stages of shock metamorphism are common (Fig. 4)

2. autochthonous breccias in crystalline basement rocks, mineralogical composition similar to host rocks (Fig. 4). The evidence of shock metamorphism is limited to 0 stage (Stöffler 1972).

3. Autochthonous breccias or brecciated rocks of crystalline basement (Fig. 5). The evidence of shock metamorphism absent;

4. Brecciated sedimentary rocks (clays, silt- and sandstones) are observed on in sections of the drill holes no. 331 and 410.

In general, the evidence of shock metamorphism in the Neugrund crater breccias are the following:

1. widespread fracturing of the target rocks (Fig. 5);
2. conical fracture surfaces (shatter cones);
3. mosaic extinction of quartz and feldspar (Fig. 13c);
4. planar fractures (PF) of quartz corresponding to planes (0001) and {1011};
5. rare and weakly developed planar deformation features (PDF) in quartz (Fig. 13a), 2 sets parallel to {1013} and (0001)
6. kink bands in biotite (Fig. 13b);
7. diaplectic glass by quartz and feldspar;
8. rare occurrence of glassy grains, partly recrystallized to chlorite (Fig. 13 d);
9. occurrence of thin (0.5–2 cm) pseudotachylitic veins of melt.

The small number of PDFs in the clast-supported impact breccias of the Neugrund crater can be explained by the limited exposures that are available for sampling and that microscopic investigation has been limited to impact breccias from the rim-wall and ejected blocks. As shown from experience with Kärdla impact crater (Puura and Suuroja 1992) PDFs in the impact breccias of this part of the crater are rare or absent.

As a result of the impact in the impact-derived crystalline basement target rocks of the Neugrund crater, unusual chemical alterations have taken place (Fig. 14; 15), greatly resembling the chemical alterations observed in the Kärdla impact crater (Puura and Suuroja 1992). The K_2O content of the crystalline basement target rocks (amphibolites, biotite gneisses, and migmatite granites) is 1.6–1.8 wt%; 3.0–3.5 wt% and 4–5 wt% respectively, but the impact breccias formed from them contain 8–10.5 wt% of K_2O (Table 1). At the same time, Na_2O content of the same rocks decreased from 3–4.5 wt%; 2.5–4.2 wt% and 2–2.5 wt% respectively to 0.1–0.4 wt% in clast supported impact breccias (Table 1). The decreased sodium content in these rocks can be explained by weathering and hydrothermal alteration of plagioclase during post-impact processes, which has been observed in thin sections. The reason of enrichment of the rocks with potassium and especially its source, however, is difficult to identify. In some cases the enrichment of impact breccias with potassium has been explained by forming new potassium-rich K-feldspars (orthoclase) on account of plagioclase (Puura et al. 1996).

Due to impact (or post-impact) processes the silica content of brecciated amphibolitic rocks has rapidly decreased (Fig. 7). The SiO_2 content of the unaltered amphibolitic target rocks is 48–52 wt%, but the impact breccias formed from these rocks contain often only 26–28 wt% of SiO_2 (Table 1).

Summary and conclusions

The Neugrund impact crater was discovered began only three years ago in 1996. The present paper provides a preliminary review of these studies. The first samples from the crater structure, verifying the hypothesis of the impact structure locating under the Neugrund Bank and its surrounding area were obtained in 1998. In the course of marine investigations, we used the methods for investigating a submarine impact structure situated in a shallow sea. The continuous seismo-acoustic profiling record, gravimetric and magnetometric mapping, sidescan sonar profiling, as well as diving to study the submarine exposures, are the most suitable methods for investigating such structures. On Neugrund Bank, all the above-cited investigation methods, except gravimetric survey, were used. The data obtained have served as the basis for reconstruction of the morphology of the Neugrund structure. Its major parameters are presented in the tables and figures. The Neugrund impact crater has several common features with other Paleozoic impact structures discovered in Fennoscandia (French et al. 1997; Lindström et al. 1993; Sturkell 1988, Sturkell 1997, Henkel and Pesonen 1992).

In the Neugrund impact structure only the weakly eroded structural uplift of the rim-wall, some ejected megablocks, the upper part of the crater filling and the whole overlying complex are available for direct investigation. This means that although only a small part of the crater is visible, it still provides abundant information on the whole crater. One of the most clearly expressed structural elements of the Neugrund impact structure is the ring-fault with a zone of disturbed target rocks. In the northern part the ring-fault is expressed as up to 60 m high terrace in bedrocks, which is buried under Quaternary deposits. Between the rim-wall and ringfault, the sedimentary target rocks are strongly disturbed. At present, data about the extent of the disturbance down to the rocks of crystalline basement are missing.

The impact origin of the Neugrund structure may be considered verified. In the crater area shock metamorphic effects up to II shock stage or 40 GPa according to the scale of Stöffer and Langenhorst (1994) have been found. PDFs in quartz, the main indicator of an impact origin for crater-like structures, are, in the case of Neugrund, rarely observed. This may be explained by the fact that impact-related rocks, which are the primary hosts of these features (clast- and melt supported impact breccias), are located mainly inner the crater and are not accessible.

Autochthonous impact breccias of crater rim are in most cases low in PDFs (Puura and Suuroja 1992).The problem of the so-called potassium phenomena (elevated K contents and Lower Na contents in the impact breccias compared to in the target rocks), which is expressed in the Kärdla as well in the Neugrund case, stay still unsettled. This phenomena has been found in a variety of impact craters around the world (Ames – Koeberl et al. 1997; Boltysh – Gurov et al. 1986; Brent – Grieve 1978; Ilynets – Gurov et al. 1998; Kärdla – Puura and Suuroja 1992; Newport – Koeberl and Reimold 1995; Rotter Kamm – Reimold 1994 etc.). The mobilization of the alkali elements in a post-impact hydrothermal system (Koeberl et al. 1997) seems to be the most likely explanation for this phenomenon. But a direct connection with the age, composition of target, peculiarities of post-impact history, etc. of the impact crater, and the potassium phenomena has not yet been found. The potassium phenomena seems to be restricted to plagioclase-rich rocks (granitoids, gneisses, amphibolites) and is connected with decomposition of plagioclase and forming of new K-feldspar (Puura et al. 1996) or sericite.

Acknowledgements

The authors are grateful to T. Floden (Stockholm University) and the crew of r/v "Strombus" for help in organization the marine expedition in 1996. We also wish to thank the Estonian Maritime Museum (Director U. Dresen), who kindly allowed using r/v "Mare" for investigations. Our special thanks belong to V. Mäss and A. Eero for assistance in investigation and sampling submarine objects. Funding of this work was provided by Geological Survey of Estonia. S. Peetermann corrected the English of the manuscript. A detailed and helpful review by H. Dypvik (Oslo University) led to an improvement of this manuscript.

References

French B M, Koeberl C, Gilmour I, Shirey S B, Dons J A, Naterstad J (1997) The Gardnos impact structure, Norway: Petrology and geochemistry of target rocks and impactites. Geochim Cosmochim Acta 61: 873–904

Grieve R A F (1978) The petrochemistry of the melt rocks at Brent crater and their implications for the condition of impact. Meteoritics 13: 484-487

Grieve RAF, Langenhorst F, Stöffler D (1996) Shock metamorphism of quartz in nature and experiment: II. Significance in geoscience. Meteoritics 24: 83–88

Gurov E P, Koeberl C, Reimold W U (1998) Petrography and geochemistry of target rocks and impactites from the Ilyinets Crater, Ukraine. Meteoritics Planet Sci 33: 1317-1333

Gurov E P, Kolesov G M, Gurova E P (1986) Composition of impactites of the Boltysh astrobleme. Meteoritika 45: 150-155 (in Russian)

Henkel H, Pesonen L (1992) Impact craters and craterform structures in Fennoscandia. Tectonophysics 216: 31-40

Koeberl C, Reimold (1995) The Newporte impact structure, North Dakota, USA. Geochim Cosmochim Acta 59: 4747-4767

Koeberl C, Anderson R R (1996) Manson and company: Impact structures in the United States. In: Koeberl C, Anderson RR (eds) The Manson impact structure: anatomy of an impact crater. Geological Society of America. Special Paper 302, pp 1–29

Koeberl C, Reimold W U, Brandt D, Dallmeyer, Powell RA (1997). Target rocks and breccias from the Ames impact structure, Oklahoma: Petrology, mineralogy, geochemistry and age. In: The Ames Structure and Similar Features. Oklahoma Geological Survey Circular 100: 169-198

Koppelmaa H, Niin M, Kivisilla J (1996) About the petrography and mineralogy of the crystalline basement rocks in the Kärdla crater area. Bulletin of the Geological Survey of Estonia 6/1: 4–24

Lindström M, Sturkell EFF, Törnberg R, Ormö J (1996) The marine impact crater at Lockne, central Sweden. GFF 118: 193–206

Lutt J, Raukas A (1993) Geology of the Estonian Shelf. Academy of Sciences of Estonia: 1-192

Melosh HJ (1989) Impact Cratering: A Geologic Process. Oxford Univ Press, Oxford New York, 245 pp

Mens K, Pirrus E (1997) Formation of the territory: Vendian – Tremadoc clastogenic sedimentation basins: geology and mineral resources of Estonia. Estonian Academy Publishers, pp 184–191

Mens K, Bergström J, Lendzion K (1990) The Cambrian System on the East European Platform: Correlation Chart and Explanatory Notes. IU Geol Sci Publ 25: 1–73

Mozhayev BN (1973) Recent tectonics of the nortwestern part of the Russian plain. (in Russian)

Öpik A (1927) Die Inseln Odensholm und Rogö. TÜ Geol Inst Toim 9: 1 69

Orviku K (1935) Gneiss-breccia suurte rändrahnude kivimina. Eesti Loodus III/4: 98–99

Puura V (1991) Origin of the Baltic Sea depression. In: Grigalis A (ed.) Geology and Geomorphology of the Baltic Sea, pp 267-290

Puura V, Suuroja K (1992) Ordovician impact crater at Kärdla, Hiiumaa Island, Estonia. Tectonophysics 216: 143–156

Puura V, Floden T (1997) The Baltic sea drainage basin – a model of a Cenozoic morphostructuree rflecting the early Precambrian crustal pattern. In: Proceedings of the Fourth Marine Geological Conference – the Baltic, Uppsala 1995. Sveriges Geologiska Undersökning. Ser Ca 86. 131-137

Puura V, Plado J, Kirsimäe K, Kivisilla J, Niin M, Suuroja K (1996) Impact induced metasomatism of granitic target rocks, Kärdla crater, Estonia. In: The Role of Impact Processes in the Geological and Biological Evolution of Planet Earth. Abstracts. Ljubljana, pp 68-69

Raukas A, Hyvärinen H (1992) Geology of the Gulf of Finland. Estonian Academy of Sciences: 1-421 (In Russian)

Reimold WU, Koeberl C, Bishop J (1994) Roter Kamm impact crater, Namibia: Geochemistry of basement rocks and breccias. Geochim Cosmochim Acta 58: 2689-2710

Stöffler D (1972) Deformation and transformation of rock-forming minerals by natural and experimental shock processes: 1. Behaviour of minerals under shock compression: Fortschritte der Mineralogie 49: 50–113

Stöffler D (1974) Deformation and transformation of rock-forming minerals by natural and experimental processes: 2. Physical properties of shoked minerals: Fortschritte der Mineralogie 51: 256–289

Stöffler D, Grieve RAF (1994) Classification and nomenclature of impact metamorphic rocks: A proposal to the IUGS Subkommission on the systematic of metamorphic rocks: Lunar Planet Sci 25: 1347–1348

Stöffler D, Langenhorst F (1994) Shock metamorphism of quartz in nature and experiment: I Basic observations and theory. Meteoritics 29: 155–181

Sturkell EFF (1988) The origin of the Marine Lockne impact structure, Jämtland. Stockholms Universitetes Insitutions för geologi och geokemi 296: 5–32

Sturkell EFF (1997) Impact related clastic injections in the marine Ordovician Lockne impact structure, central Sweden. Sedimentology 44: 793–804

Suuroja K (1996) The geological mapping as a source of geological discoveries. Geological mapping in Baltic states. Newsletter 2: 19–22

Suuroja K, Saadre T (1995) Loode-Eesti gneissbret_ad senitundmatu impaktstruktuuri tunnistajaina. Bulletin of the Geological Survey of Estonia 5/1: 26–28

Suuroja K, Suuroja S, Puurmann T (1997) Neugrund structure an impact crater. Eesti Geoloogia Seltsi bülletään 2/96: 32–41

Thamm N (1933) Über eine Gneisbrekzia im Glazialgeschiebe der Insel Osmussaar (Odensholm). TÜ Geol Inst Toi 34. 1–14

Torsvik TH, Smethurt MA, Meert JG, Van der Voo R, Kerrow WS, Brasier MD, Sturt BA, Walderhang HJ (1996) Continental break-up and collision in the Neoproterozoic and Paleozoic – A tale of Baltica and Laurentia: Earth-Science Reviews 40: 229-258

Viiding H (1960) Rändrahnud ja kivikülvid. Looduskomitee teatmik. ERK: 1–37

17 Impact-Induced Replacement of Plagioclase by K-feldspar in Granitoids and Amphibolites at the Kärdla Crater, Estonia

V.Puura[1], A.Kärki[2], J.Kirs[1], K.Kirsimäe[1], A.Kleesment[3], M.Konsa[3], M.Niin[4], J.Plado[1], K.Suuroja[4] and S.Suuroja[4]

[1]Institute of Geology, University of Tartu, Vanemuise 46, 51014 Tartu, Estonia (puura@ut.ee)
[2]Department of Geosciences, University of Oulu, P.O. Box 333, FIN-90571 Oulu, Finland
[3]Institute of Geology at Tallinn Technical University, Estonia ave 7, 10143 Tallinn, Estonia
[4]Geological Survey of Estonia, Kadaka tee 80/82, 12618 Tallinn, Estonia

Abstract. The 4-km-wide and 0.5-km-deep Kärdla Crater, presently buried under a thin sequence of Upper Ordovician limestone, was formed in a complex target: the crystalline basement was covered by about 170 m of poorly consolidated Lower to Middle Ordovician and Cambrian sediments. The basement-derived granitoids and amphibolites that were subjected to low shock pressure (less than 8 GPa) and are at present either in allochthonous or autochthonous position, have undergone significant impact related chemical and mineralogical alteration. Altered rocks and minerals are fractured and brecciated at different scales. Optical microscopy has shown that plagioclase is altered ("sericitization" or "saussuritization") and amphibole chloritized. Bulk chemical compositions indicate that sodium and calcium were removed and potassium was brought in. Powder XRD analysis suggests the disappearance of plagioclase and formation of cryptocrystalline orthoclase, in addition to medium- to coarse-grained microcline present in primary rocks. SEM, EDS and EPMA studies reveal that the main secondary minerals are orthoclase, replacing plagioclase in granites and amphibolites, and chlorite, replacing hornblende in amphibolites. The extent of chemical and mineralogical alteration is closely related to microfractures. However, alteration has selectively changed the minerals. Plagioclase, hornblende and biotite have changed dramatically, whereas zircon, quartz and also primary microcline remained almost unaffected. Geological and mineralogical data suggest that this alteration was related to fireball – target rock interaction during the impact process.

Introduction

Potassium-enriched autochthonous and allochthonous breccias have been found in brecciated and impact-melted target rocks at many craters (Reimold et al. 1994; Koeberl 1997; Koeberl et al. 1996; French et al. 1997). At the Kärdla impact site, a geochemical anomaly of K-enrichment and Na and Ca depletion is strongly related to impact-deformed rocks that are derived from the crystalline basement (Puura and Suuroja 1992; Puura et al. 1996). Core sections drilled into the Kärdla crater provide an opportunity to study the position of K-enriched and Na+Ca-depleted rocks in the succession of allochthonous breccias and in brecciated or fractured sub-crater basement, including the central uplift, annular depression of the crater, and uplifted crystalline blocks in its rim wall. Coarse (breccia clasts) and fine (breccia matrix) particles in allochthonous and parautochthonous breccia units, and fractured, occasionally brecciated autochthonous rocks of the sub-crater basement and rim wall have been subjected to this alteration. At the mineral grain level, the chemical and mineralogical alteration of rocks selectively affects some minerals (plagioclase, amphibole), progressing along systems of impact-induced microfractures. The microfractured but more chemically resistant minerals (zircon, quartz, and primary microcline) remain almost unchanged. The present study explains this geochemical anomaly by means of the crater structure and fracturing, brecciation, chemical and mineralogical alteration of target rocks.

We use the term "alteration" in a broad sense, i.e., for any kind of impact-induced changes of rocks and minerals. It includes different kinds of changes of impact-influenced primary rocks and minerals. Petrophysical alteration refers to changes of petrophysical characteristics, such as density, porosity, etc. Mineralogical alteration is a change of major and accessory minerals, such as decomposition of plagioclase and formation of cryptocrystalline K-feldspar. Chemical alteration is a change in bulk chemical composition, especially - in this case - concerning K, Na, and Ca. However, these processes have not completely modified the target rock. According to visual observation and optical microscopy study, those rocks in sub-crater basement and breccia clasts that are large enough have more or less retained their primary structure and texture.

The Kärdla impact structure

The 4-km-wide and 500-m-deep, circular Kärdla impact structure (Figs. 1 and 2) was formed in a shallow, Middle Ordovician epicontinental sea, and buried under clastic and carbonate sediments immediately after impact. The geologic structure of the crater and surroundings has been established by studies of closely spaced boreholes (Puura and Suuroja 1992).

The structure resulted from a 455 Ma impact into a complex target, including a 170-m thick sedimentary cover overlying Precambrian granites, migmatized gneisses, and amphibolites. The sedimentary cover was composed of an about 20 m thick layer of Ordovician carbonate rocks, underlain by an at least 150 m thick layer of mainly soft, water-rich, poorly compacted siliciclastic and clayey Cambrian sediments. During the post-impact geological history, the sedimentary cover has been compacted and thinned to 150 m.

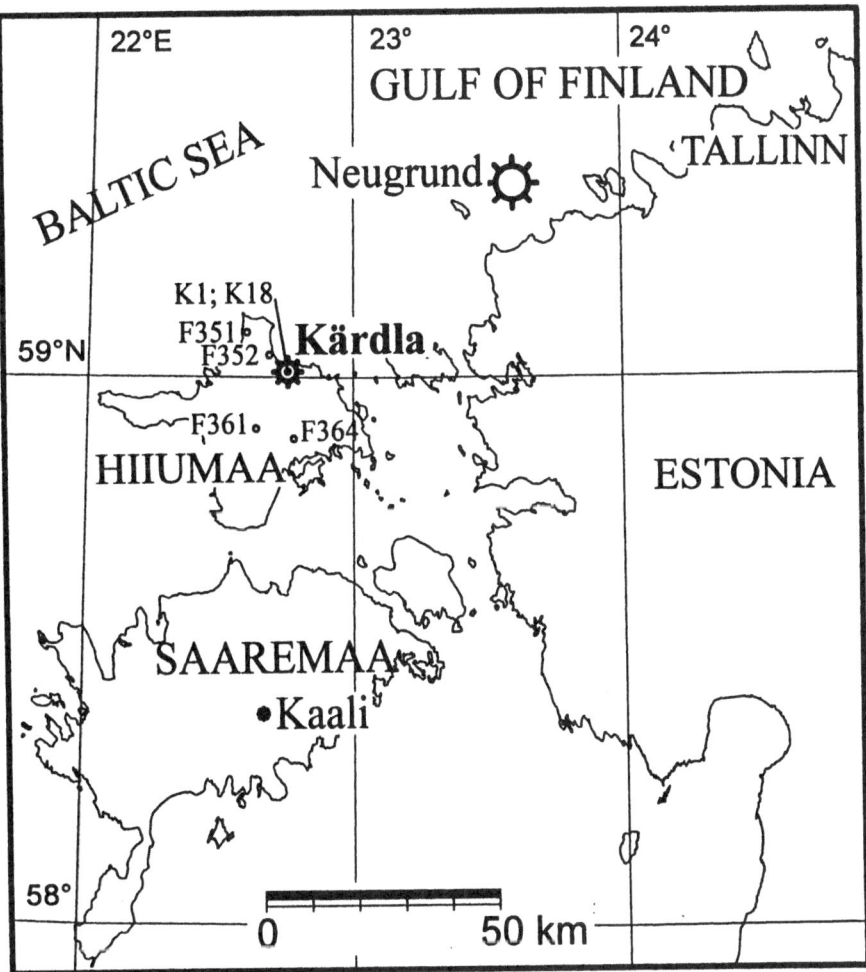

Fig. 1. Location of the Kärdla structure and of boreholes. The locations of the possible Neugrund structure and Kaali crater field are shown.

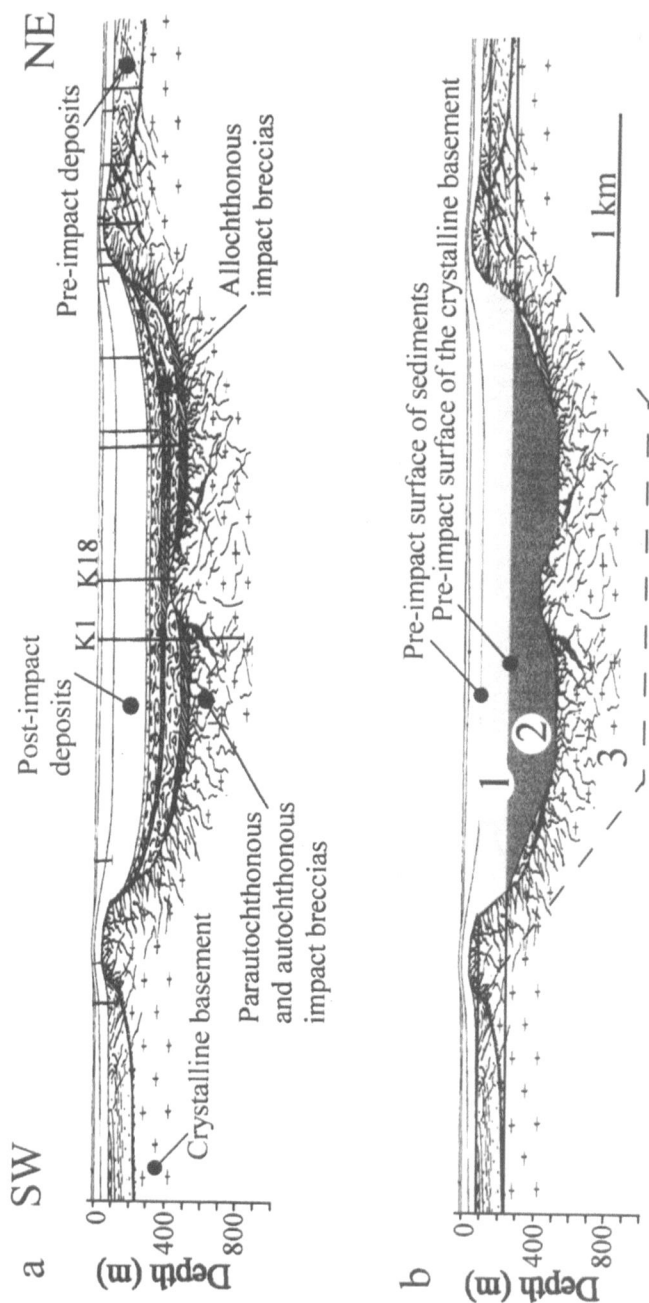

Fig. 2. (a) Cross-section (SW-NE) of the Kärdla impact structure, modified after Puura and Suuroja (1992). (b) Model showing: 1. impact-excavated pre-impact sediments; 2. impact-excavated sub-crater basement rocks; 3. altered sub-crater basement rocks. No vertical exaggeration. K1 etc. deep core drillings.

In core samples, brecciation and fracturing of the target rocks is observable at different scales, from meter-sized blocks and breccia dikes to fine-grained clasts in breccia matrix and microfracturing of mineral grains. In the external part of the crater wall, faults within the crystalline bedrock and along the unconformable contacts of crystalline and sedimentary target rocks were documented (Puura and Suuroja 1992). Breccias of fragments of variable size form allochthonous bodies in the crater. Fragmental breccia dikes are observed at different levels in the crater walls and in the sub-crater basement down to depths of 815 m (more than 250 m below the actual crater floor). Fragmental breccia dikes in crater walls and basement suggests either injection of material or faulting-related brecciation, or both. Basement rocks near faults with breccia filling are more altered than those in monolithic blocks.

In the core K1, the uppermost part of the sub-crater basement is intensely crushed and shattered. Abundant fine-grained to cryptocrystalline matrix supports crystalline clasts of the monomictic breccia, forming the parautochthononous unit between 522.8 and 588.6 m depth in (Fig. 3). From 588.6 m down to the drill-hole bottom at 815.2 m autochthonous, fractured, predominantly granitic basement was penetrated. In some intervals, impact-induced breccia dikes cut massive basement rocks. In the core K18, drilled into the central uplift, the uppermost mainly amphibolitic part of the basement with fine-grained allochthonous breccia dikes is strongly fractured.

Impact lithologies and impact-induced alteration of basement-derived rocks.

Prior to post-impact Middle Ordovician marine sedimentation, the deformed crystalline crater floor was covered and the crater was filled with impact-derived allochthonous, mixed coarse (clasts) to fine (matrix), polymict, clastic deposits of crystalline and sedimentary origin. In the present study (modified after Puura and Suuroja 1992), these deposits are divided into layers of suevitic, slumped and resurge breccia origin (Fig. 3). On the top of the breccias, turbiditic carbonatic, fine-siliciclastic sediments were deposited. Carbonate sedimentation, similar to that found on the surrounding Middle and Late Ordovician shelf, followed. The general synclinal structure of sedimentary layers in the crater (Fig. 2) refers to the gradual post-impact compaction of the crater filling.

The impact produced an about 200 m thick lens of allochthonous impact breccias, which filled the depression and surrounded the central uplift of the crater. Planar deformation features in quartz (up to 3-4 sets) have been documented only in the allochthonous suevitic breccias. Measurements of their crystallographic orientation (Suuroja 1999) indicate shock pressures of roughly 20 GPa (Huffmann and Reimold, 1996). Shock metamorphic signatures have never been found in sub-crater basement rocks or in the parautochthonous breccia.

The breccia units are distinguished using different lithological criteria (Suuroja 1999). The main components of breccias are coarse (> 2 mm) and fine (< 2 mm) particles, forming clasts and matrix, respectively. In both coarse and fine fractions, fragments are derived from both crystalline and sedimentary rocks. However, their relative proportion is highly variable.

Clasts of crystalline rocks subjected to K-enrichment of different intensity are found in every genetic type of breccias. All the crystalline clasts were produced by excavating buried crystalline basement from the central part of the transient cavity.

No evidence of any melt sheet in the crater filling was found. The suevitic breccia units contain rare (<<1% of volume) and small, melted particles (Suuroja 1999). Suevitic breccias also occur as sheet-like bodies or small local lenses (lumps) in the sequence of dominant slump and resurge breccias.

Fig. 3. (Opposite) Log of borehole K1: impact lithologies and distribution of K- and Ca-Na feldspars in granites sampled from core K1. Reference sample of granite is from core F364. Location of boreholes see Fig.1. (Left) Dominating crater filling deposits from the top (depths in meters): ~15-260 m, Upper and Middle Ordovician limestones and marlstones; 260–300.5 m, turbidites - carbonate-siliciclastic fine-grained, in lower part with target rock; 300.5–356 m, clast-supported fragmental (including meter-size blocks) breccia with minor (usually less than 10%, in uppermost 10 m up to 40%) fine-grained (<2 mm) breccia matrix; breccia clasts are composed of sedimentary rocks (60-100%) with variable amount (<10–40%) of crystalline rocks; 356–380 m, mainly matrix-supported suevitic fragmental breccia with the content of < 2mm matrix fragments 10-40%, 2-1000 mm fragments are composed of crystalline rocks (more than 90% of granitoids and less than 10% of amphibolite; 380-471 m, mainly clast-supported fragmental (block) breccia, with <10% fine-grained (<2 mm) matrix, among the fragments blocks >1000 m strongly dominating (70->90%), fragments consist mainly of sandstones (60->90%) and of claystones (<5-30%), with variable, small (less than 5-20%) amount of crystalline granitic rocks; 471-522.8 m, matrix-supported suevitic fragmental breccia, content of fine-grained (<2mm) matrix 40-60 %, among fragments size-fraction 2-100 mm dominates, content of blocks >1000 mm is less than 5-10 %, clasts are mainly (60-80%) of granitoid composition, with rest of amphibolite composition; 522.8–588.6 m, matrix (30-50%) or clast-supported fragmental (dominating size 2-100 mm) breccias composed of crystalline rocks (mostly granitoids); 588.6–815.2 m, fractured and in some intervals brecciated crystalline rocks composed of dominating granitoids and amphibolites in sub-crater crystalline basement. (Right) Powder X-ray diffractograms of whole rock samples showing gradual change of feldspars' contents in granites of the core section. Note, that Ca-Na-feldspar is present only in the lowermost part of the autochthonous breccia sequence in an amount far less than in reference sample from core F364.

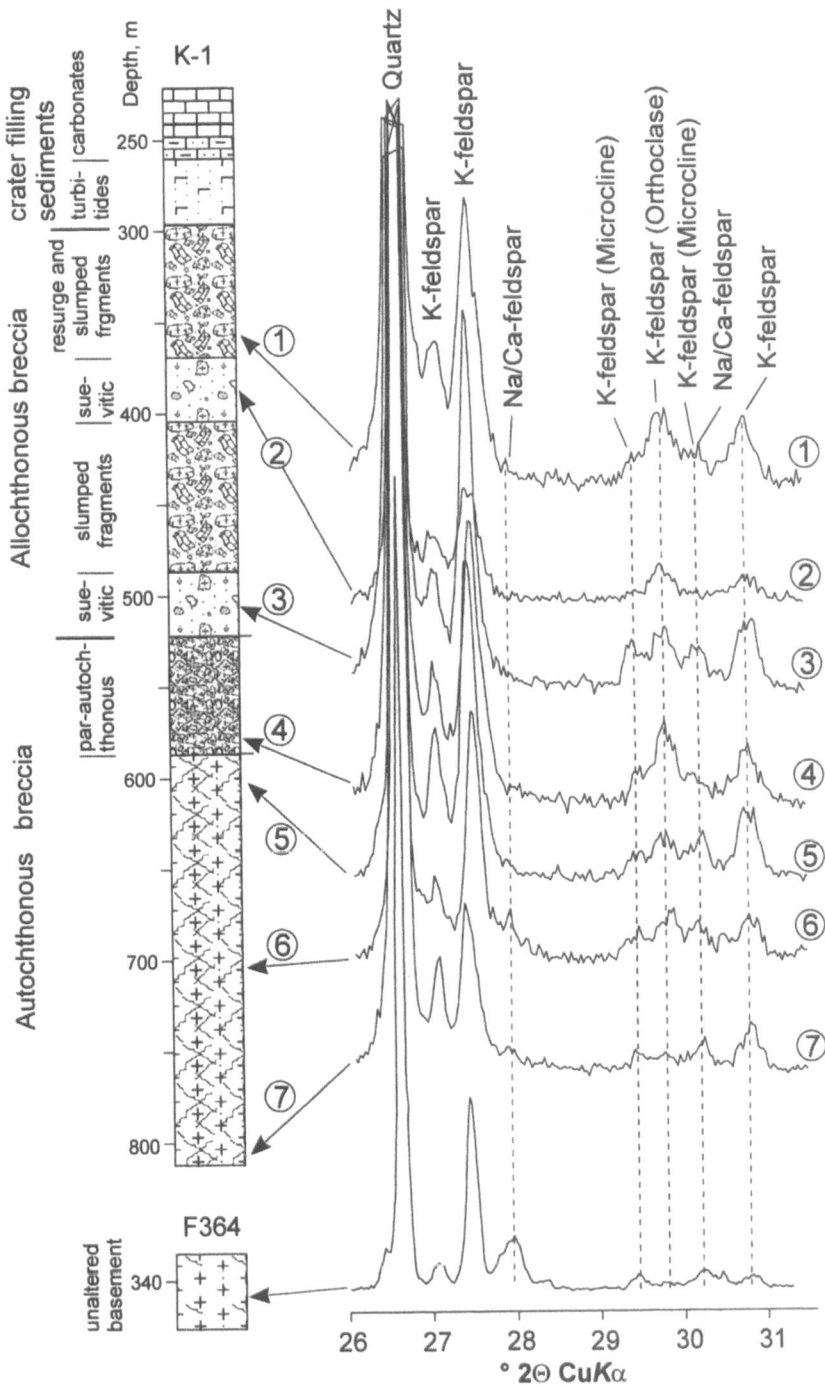

As a result of previous studies, it was concluded that the basement-derived, impact-influenced rock fragments, which presently occur in either autochthonous or allochthonous positions near the center of the crater, have been subjected to maximal chemical and mineralogical alteration. In the crater rim wall, the chemical changes in the rocks are almost half as intense as in the central uplift and allochthonous suevitic breccias. In the sub-crater basement, a downward decrease of potassium enrichment and sodium and calcium depletion of rocks was established (Puura et al. 1996). Therefore, the goal of the present study was to follow the deformation of rocks and alteration of minerals down to the borehole bottom at (K18) and near (K1) the center of the crater. However, even the deepest borehole K1 did not penetrate into the undeformed and unaltered sub-crater basement rocks (Plado et al. 1996). Reference samples of unaltered rocks are available for investigation from neighboring drill cores. At the same time, earlier unpublished geological observations and mineralogical and geochemical studies suggest that no significant impact-related alteration of carbonate or other sedimentary rock-derived, allochthonous or autochthonous breccias occurred. Occurrence of lead and zinc mineralization in both crater rim wall and post-impact, Upper Ordovician sedimentary deposits indicate a post-impact origin. This low-temperature hydrothermal sulfide mineralization is well developed also in zones of tectonic disturbances penetrating both crystalline basement and sedimentary cover in various parts of Estonia (Puura and Sudov 1976). Thus, it is not directly connected with the impact-event.

Methodology and Sampling

An integrated study of macrofracturing and brecciation of impact-influenced rocks and microfracturing and alteration of their minerals was the aim of our investigation. Impact-influenced crystalline rocks of the crater fill, sub-crater basement and ring wall show both open and hidden fracturing and mineralogical and geochemical alteration. In places, core recovery was low and many possible intervals of intense fracturing remained unrepresented. The frequency of hidden, closed fractures in rocks has not been quantitatively measured. We used petrophysical data to characterize the intensity of deformation suggesting that the downward decrease in porosity, and the increase in density of a distinct rock type, such as plagioclase and microcline-bearing granite in core K1, expresses a diminishing degree of rock destruction. Optical microscopy of rocks in thin sections and of separated minerals in oil immersion was used to study the microfracturing and alteration stage of mineral grains down to 0.01 mm size.

Backscattered electron microscopy (SEM) images, together with X-ray energy-dispersive microanalysis (EDS), and X-ray microprobe (EPMA) analyses were used to determine the fracturing and compositional changes of minerals at the micrometer scale. X-ray diffractometry (XRD) was used to determine the bulk mineral composition of primary and altered rocks, especially the relative abundance of plagioclase, K-feldspar, amphibole and chlorite. Traditional methods of optical microscopy, chemical analysis and petrophysics established the spatial trends of alteration intensity.

The results of 47 petrophysical measurements of parautochthonous and fractured autochthonous granitic rocks in core K1 by Plado et al. (1996) and Jõeleht (1995) were combined. The petrophysical data (wet density, grain density and porosity) were measured in the Petrophysical Laboratory of the Geological Survey of Finland, using the instruments and techniques described by Puranen and Sulkanen (1995). For comparison, 15 samples of unfractured granitic rocks from outside the crater (cores F361 and F364; Fig. 1) were also analyzed.

For the present study, around 100 thin sections of granites, amphibolites and breccias from cores K1 and K18 were re-studied with the polarizing microscope at the Geological Survey of Estonia, at the Institute of Geology, University of Tartu, and at the Department of Geosciences, University of Oulu.

Seven granitic and five amphibolitic whole rock unoriented samples were studied using standard XRD techniques on a DRON-3M diffractometer. Three out of seven granitic samples and two out of five amphibolitic samples were taken from the most intensely altered allochthonous breccia. Four other granite and three amphibolite samples came from the autochthonous breccia of the sub-crater basement. In addition, granites and amphibolites were taken from the F364 reference core.

The XRD data were used to estimate the proportions of plagioclase, microcline and orthoclase in variably altered granites, and to classify primary and secondary K-feldspars according to their triclinicity (Wright, 1968). In amphibolite samples, the composition and content of primary hornblende and secondary chlorite was studied in detail.

To characterize altered minerals from the sub-crater basement rocks, six samples from the cores K1 and K18 were studied in oil immersion. From core K18, we also studied a sample of breccia matrix from the boundary between allochthonous and autochthonous breccia on the central uplift. For reference, results of a systematic immersion study of minerals (quartz, feldspars, zircon, garnet, pyroxenes, amphiboles, micas etc.) separated from fresh and weathered Precambrian crystalline and Cambrian sedimentary rocks from all over Estonia were used (Konsa and Puura 1999). During the last three decades, this study was carried out to investigate the provenance of the clastic sedimentary rocks overlying the basement. The aim of the present study was to investigate gradual changes in fracturing of minerals against changing whole-rock physical parameters (Plado et al. 1996) and chemical alteration (Puura et al. 1996) through the fractured sub-crater basement rocks (Fig 4). For morphological studies, quartz, zircon and other accessory mineral grains were handpicked from the 0.1-0.05 mm fraction separated in high-density liquids from crushed rock samples. Using the immersion method, the internal structure (e.g., zonation, fracturing, inclusions and relative transparency of zircon - Fig. 5, right) and quartz grains was studied. The surface structure of zircon grains was studied by scanning electron microscopy at the Tallinn Technical University (Fig. 5, left).

To investigate the character of alteration of plagioclase and amphibole 8 selected samples were studied using SEM and EDS. In four samples of granite and amphibolite, the composition of altered, partially altered and unaltered feldspars and hornblende was studied by electron microprobe techniques (JEOL Super Probe 733, acceleration voltage 15 kV, sample current 15 nA, standard MAC 3056 for silicate analysis).

Results

The Crystalline Basement: Composition of Original and Impact-affected Target Rocks

The basement of the Kärdla crater area belongs to a uniform high-grade amphibolite facies terrain of the 1.9-1.8 Ga Svecofennian crustal domain (Puura and Huhma 1993). These rocks are mainly mafic and intermediate metavolcanics that are everywhere injected with plagioclase-microcline granite veins and larger bodies. In the sub-crater basement (K1 and K18) and in the SW (Tubala) rim wall granites (together with granite gneisses and, occasionally, quartzites) prevail. In contrast, the NE (Paluküla) rim wall contains over 50% amphibolites (Puura and Suuroja 1992; Koppelmaa et al. 1996). In the allochthonous breccias, granites dominate among the basement-derived clasts, but amphibolite clasts are also frequent.

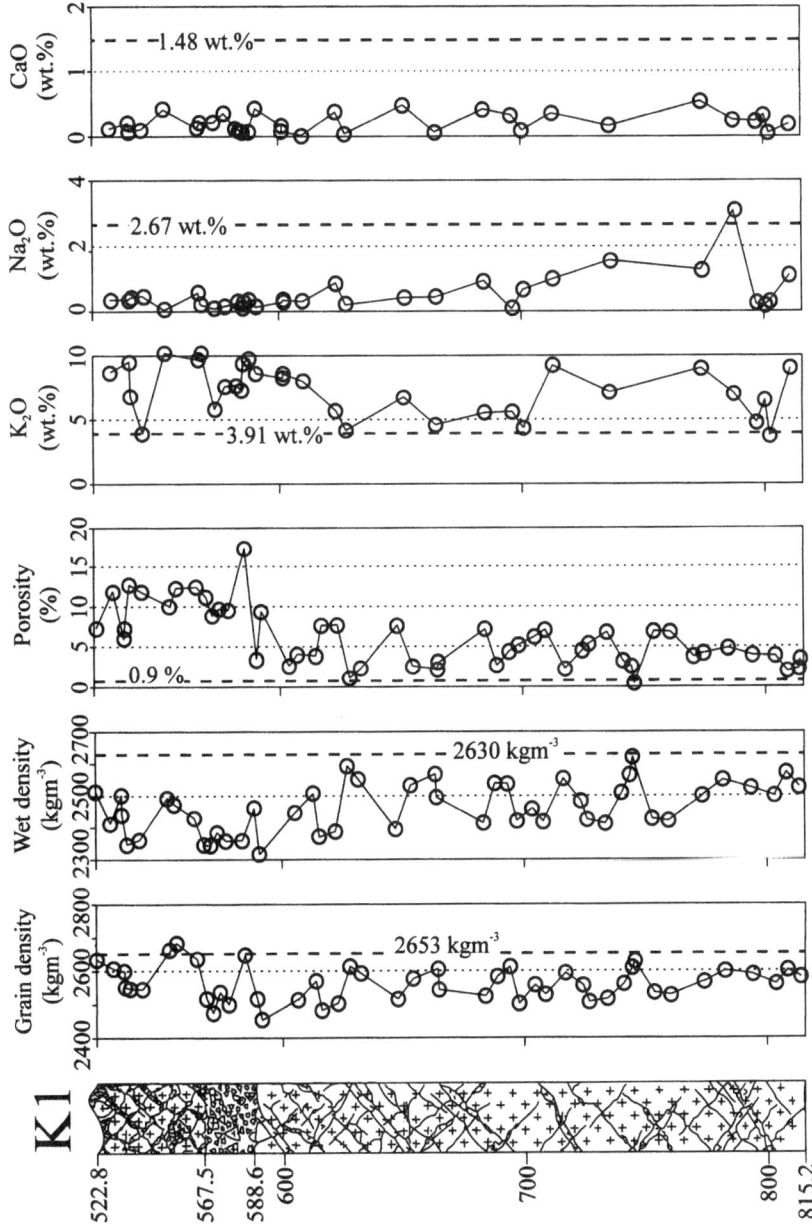

Fig. 4. Petrophysical properties and content of oxides of parautochthonous (522.8-588.6 m) and autochthonous (588.6-815.2 m) granitic rocks from core K1 (for location see Figs. 1 and 2a). From left to right: grain density (δ_g), wet density (δ_w), porosity (Φ), content of K_2O, Na_2O and CaO. Dashed lines indicate the average parameters of the same type of granites outside the Kärdla structure.

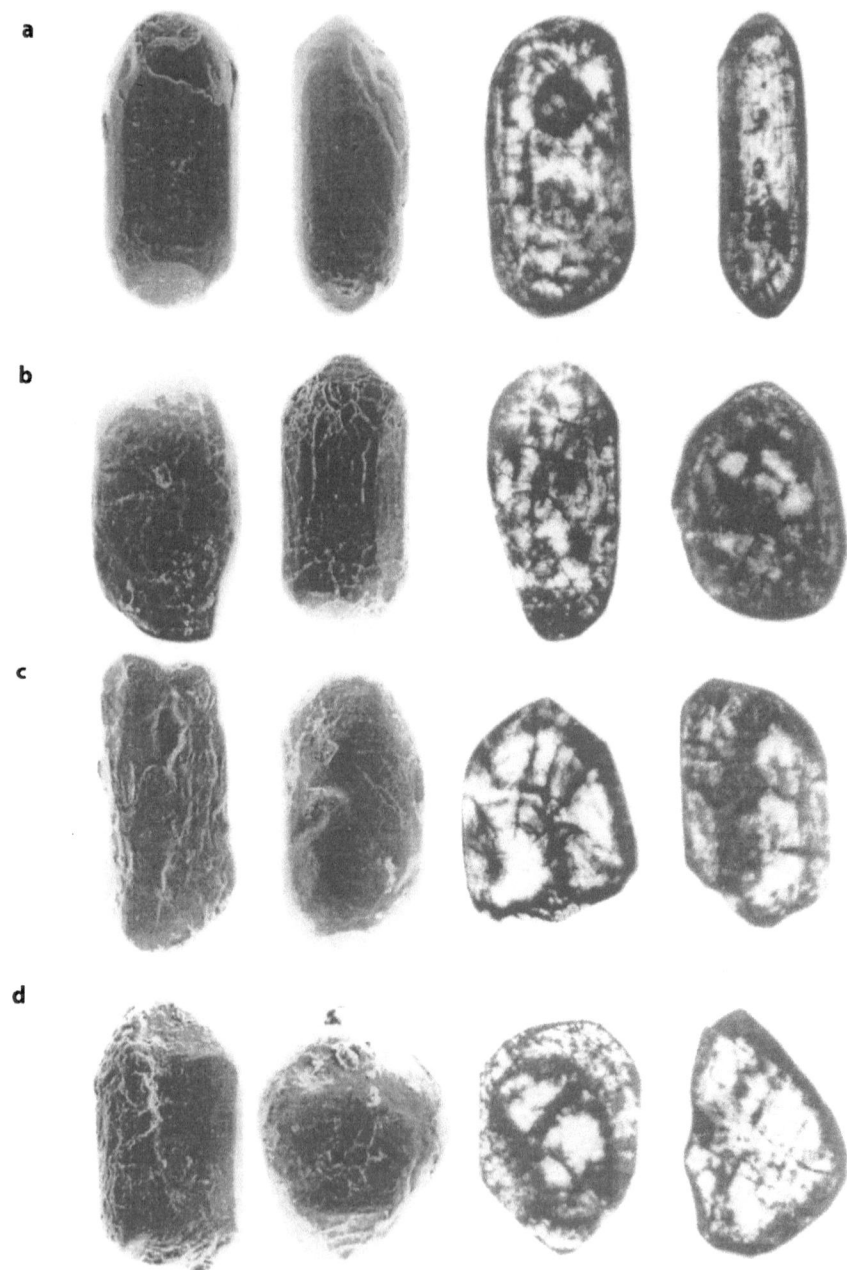

Fig. 5. (above and opposite) SEM micrographs (on the left) and optical immersion (on the right) of zircon grains from (a) unshocked granites outside and (b-e) shocked granites inside the Kärdla structure. Locations: (a) core K364, depth 311.2 m, (b) core K18, depth 383.0 m, (c) core K1, depth 606.0 m, (d) core K1, depth 705.3 m and (e) core K1, depth 806.4 m.

e

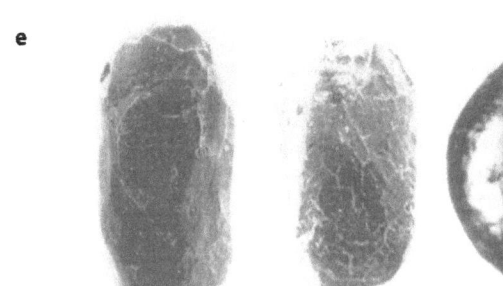

Migmatitic plagioclase-microcline granites in the basement of Hiiumaa Island are pinkish-red, mostly massive, medium- to coarse-grained, felsic rocks with sporadic xenoliths from amphibolites and gneisses. They contain 30-40 vol% xenomorphous quartz, 35-40 vol% cross-hatched microcline, 5-20 vol% albite to oligoclase plagioclase and 5-10 vol% biotite. Sporadically, intergrowth of quartz in plagioclase appears as myrmekitic texture (Koppelmaa et al. 1996). Apatite, zircon and orthite, together with few opaque ore minerals (e.g., magnetite and ilmenite), are common. In macroscopically porous, fractured, reddish granites of the sub-crater basement, quartz has strong undulatory extinction and is often fractured. Deformed biotite flakes are weakly chloritized.

In the surroundings of the Kärdla crater, the average content in CaO, Na_2O and K_2O is 1.32 wt%, 2.71 wt% and 4.74 wt%, in reference granitic rocks (calculated from data by Kivisilla et al. 1999). Granitic clasts and breccia matrix in allochthonous and parautochthonous units of the crater contain 0.42 wt% CaO, 0.24 wt% Na_2O and 8.34 wt% K_2O. In the sub-crater basement, the average content of these components is 0.22 wt%, 0.71 wt% and 7.12 wt%, respectively. In the sub-crater basement, the content of Na and Ca increases downwards, whereas the K_2O content decreases. In the Al_2O_3-K_2O-Na_2O ternary diagram, the composition of impact-influenced rocks shows a continuous change from the reference rocks of the surrounding area to the most altered crater rim and sub-crater granites, and further to allochthonous breccia clasts and breccia matrix (Fig. 6). In the rocks most depleted of Na and Ca (Fig. 4), the XRD study shows disappearance of plagioclase in granites (Fig. 3).

The primary (unaltered) amphibolites of the crystalline basement are medium- or coarse-grained, greenish or dark-grey rocks of linear or massive structure and heterogranoblastic, in thick bodies also of blastogabbroic texture. Mineralogically, they consist of hastingsitic hornblende and andesine- to labrador-type plagioclase (35-45 vol%; Koppelmaa et al. 1996). In the amphibolite from the crater fill and sub-crater basement, hornblende is brecciated and partly or totally replaced by pseudomorphic chloritic aggregates with calcite and ferric hydroxides (see below). Plagioclase is heavily replaced with cryprocrystalline aggregates of secondary minerals, commonly referred to as "saussurite and sericite" (Koppelmaa et al. 1996). However, our studies showed that, among the secondary minerals replacing plagioclase, K-feldspar dominates (see discussion on feldspars below).

Physical properties of sub-crater parautochthonous and autochthonous, brecciated and fractured basement rocks systematically differ from reference basement rocks (Plado et al. 1996). Along the core K1, a trend of gradual changes in physical properties of granites occurs (Fig. 4.) Porosity, having high values (around 10-12 %) in parautochthonous breccias, decreases with depth in autochthonous breccias to less than 5 % near the drill-hole bottom. However, it is still higher than the porosity of surrounding granites. Both, grain and wet densities increase in depth. The trend in grain density between 567.5 and 815.2 m is distinct approximately from 2500 to 2600 kgm^{-3}. Wet density increases in the same depth interval from 2350 to 2520 kgm^{-3}. We suppose that the trend in wet density is partly due to gradual decrease in shock-induced rarefaction wave with depth. The difference between the grain density of surrounding (2653 kgm^{-3}) and sub-crater shocked granites, and the presence of depth-related trend in grain density, shows that the within-grain rarefaction (revealed as closed porosity during the density measurement experiment) occurred.

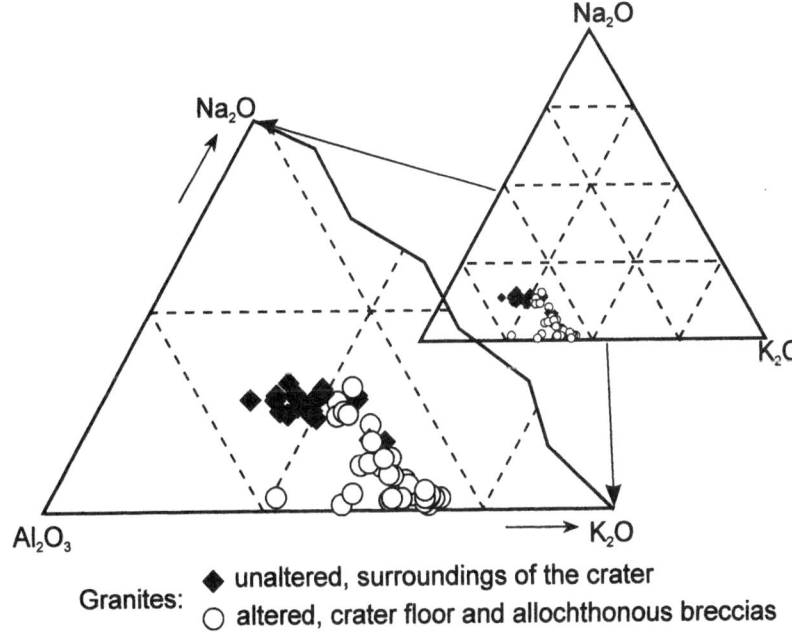

Fig. 6. Chemical analyses of reference granitoids from surrounding lithologies and of impact influenced rocks at Kärdla crater, plotted in an $Al_2O_3 - Na_2O - K_2O$ ternary diagram.

Fracturing and alteration of minerals of sub-crater basement and basement-derived breccia clasts

Zircon

Fracturing is one of the main impact-induced changes observed. During our long-term research into the mineralogy and typology of accessory zircon in unshocked crystalline and sedimentary rocks, the background pattern for this mineral has been well-studied (Konsa and Puura 1999). In impact-influenced rocks, an increase of the number of fractures and widening of pre-existing fractures was established by optical microscopy. Weak planar microdeformation of uncertain nature was rarely observed. The average number of fractured zircon grains in the breccia matrix increases to 75%, compared to 30-40% in reference basement rocks. Inside of intensely fractured grains the proportion of fresh, tight fractures increases. The proportion of turbid (opaque) zircons has also increased to 80% against the background of 40-60%. At the same time, the proportion of zircons with fluid inclusions has decreased to 10-30%, compared to the normal level of 50-60%.

Primarily, zircon grains were often zoned, and the impact-related fracturing depended on this zoning. Irregular fracturing on zircon surfaces is observable on SEM micrographs. Figure 5 shows examples of the most intensely fractured zircon grains characteristic of the surrounding unshocked basement (core F364, depth 311.2 m), and of the most intensely fractured grains from shocked granites of the crater floor (cores K18 and K1). Impact-influenced zircon differs from unshocked grains by fracturing of internal and external zones of crystals. The unshocked zircon grains (Fig. 5a) have a moderately fractured internal structure and basically intact overgrowth. In contrast, zircons from the top of the central uplift (Fig. 5b) have both widened old and tight fresh fractures, showing a maximum degree of destruction level among the studied grains. These samples have also the most strongly fractured overgrowth. In the whole profile of parautochthonous and autochthonous breccias in drill core K1, more or less intensely fractured zircon occurs. It must be mentioned that at the maximum sampling depth of 806.4 m, the fracturing of both internal and external zones of zircon grains is more frequent than at the intermediate depths. Such of increase in chemical, mineralogical and petrophysical anomalies at the depths of 770-810 m (Fig. 4) is in good connection with the frequent distribution of breccia dikes and, consequently, fault zones at these depths.

Quartz

Quartz grains and clasts with up to 3-4 systems of planar deformation features (PDFs) are observed in samples of allochthonous breccia only (Suuroja 1999). The average proportion of grains with PDFs never exceeds 5%. The proportion of quartz with wavy extinction may reach 25%. Crystals with turbid (opaque) outer layers are frequent. SEM observations revealed that fractured quartz crystals are abundant in allochthonous breccia and rare in sub-crater basement rocks. Mosaicism of quartz has been observed in allochthonous breccia only. The extent of irregular fracturing of impact-deformed quartz from the autochthonous breccias is close to that of the reference crystalline basement rocks. Besides fractured quartz and zircon, we found fractured garnet in a sample of parautochthonous breccia in core K1.

Feldspars (Fsp)

In allochthonous breccia matrix and, at a lesser extent throughout the whole profile of autochthonous breccias in core K1, both potassium feldspar (K-Fsp) and plagioclase are brecciated and fractured (Figs. 7-8). In contrast, few observable relict "plagioclase" crystals (actually, pseudomorphs after plagioclase – see below) in breccias are completely turbid (opaque), and their twin structure is either not observable or undulated. Only SEM, EDS, EPMA, and XRD studies could identify authigenic feldspars in the shocked, altered rocks.

Plagioclase (Pl)

Both conventional light microscopy and SEM studies show extensive decomposition of plagioclase in granitic samples from the most intensely altered parts of the crater floor (Figs. 7 and 8). In light microscopy, strongly altered plagioclase crystals are called "saussuritic" and "sericitic." Disappearance of visible twin structure, revealed by immersion studies of altered plagioclase, is probably a result of decomposition of plagioclase. Undulation of twin lamellae is probably an evidence of shock deformation. In SEM images altered plagioclase contains abundant 1-10 µm sized pores and channels, less than 100 µm from each other (Fig. 8b). In amphibolites decomposition of plagioclase is more variable. Partly altered plagioclase grains consist of (1) areas of relict primary plagioclase and (2) zones or spots of secondary minerals replacing the primary plagioclase. In backscattered electron images, the boundaries between the unchanged and changed parts of a grain are usually relatively sharp. In partially changed grains, cryptocrystalline masses typically localize around the pores and channels.

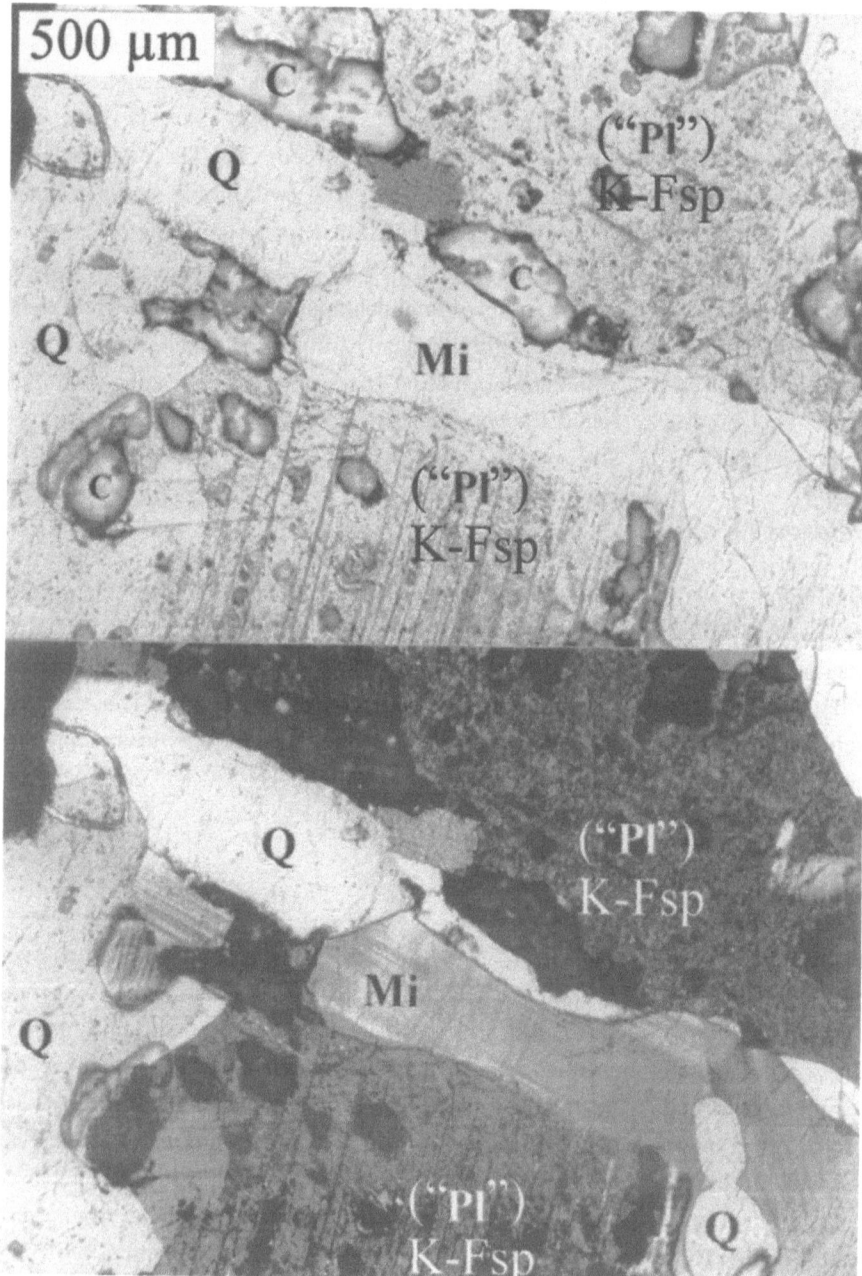

Fig. 7. Selective alteration of feldspars in sub-crater basement granite. Photomicrograph of a sample from core K-1, depth 806.4 m (Top: parallel polarizers, bottom: crossed polarizers, view width 2.2 mm): Mi - fractured but chemically unaltered microcline; Pl – fractured and altered "plagioclase", with twinning pattern locally preserved; Q – quartz; C – cavities in plagioclase, filled with balsam.

Potassium-feldspar (K-Fsp)

The impact-influenced granitoids contain two structural types of K-feldspar. Light microscopy and SEM studies revealed that the primary crosshatched microcline in pre-impact granitoids survived, as a rule, through the impact and related alteration process (Figs. 7 and 8). Except for occasional, local, extensive microbrecciation, the structure and composition of the primary microcline has mostly survived unaffected. It is very remarkable that in perthitic primary microcline, even the albitic intergrowths survived their original composition despite decomposition of independent plagioclase crystals in surrounding altered rock (Table 1). The porosity increase of albitic intergrowth, as observed in SEM images, is negligible. Thus, the surrounding K-feldspar protected the intracrystal lamellae of plagioclase from replacement or dissolution.

However, occasionally in most altered breccia samples, some specific changes in microcline are clearly visible in immersion. In matrix and clasts of allochthonous breccia, and in sub-crater brecciated granites, the crosshatched twin structure of microcline occasionally disappears. In immersion, a large part of microcline grains have lost their transparency: they are turbid (opaque). In both allochthonous and autochthonous breccia, these grains are twice as common than in surrounding reference basement rocks.

The second type of K-feldspar observed replaces altered plagioclase. In thin sections it appears as a secondary cryptocrystalline "sericitic" or "saussuritic" aggregate replacing locally or all the primary plagioclase crystal (Figs. 7 and 8). Backscattered SEM images and EDS microanalysis reveal that both the primary cross-hatched microcline and the cryptocrystalline secondary aggregate inside plagioclase have K-feldspar composition (Fig. 8). The data for partially replaced plagioclase from amphibolites also show that the cryptocrystalline zones and patches inside it are of K-feldspar composition.

The XRD data for reference granites confirmed the low microcline structural state of K-feldspar. In the core intervals of K-enrichment and Na-Ca depletion, XRD peaks (Fig. 3) indicate the presence of abundant orthoclase, and plagioclase disappearance. The X-ray diffractograms of altered granitoids from the monomict and polymict breccias also show a dominating orthoclase structural state together with phases of microcline (Fig. 9). A question rises whether the primary microcline is partially re-crystallized into orthoclase.

An EPMA study of primary microcline and of K-feldspar replacing plagioclase in granitoids, and primary plagioclase and of K-feldspar replacing plagioclase in amphibolites (Tables 1 and 2) revealed that the primary microcline and secondary K-feldspar replacing plagioclase have, however, a small difference in chemical compositions. The primary crosshatched microcline from the altered granitoid has a normative composition of Or_{79-87}, free of Ca (Table 1). The primary plagioclase An_{31-33} from altered amphibolite is K-free. The secondary orthoclase in granitoids has a composition of Or_{96-99}, and the secondary orthoclase partly replacing plagioclase in amphibolite is very pure Or_{99-100} (Table 2).

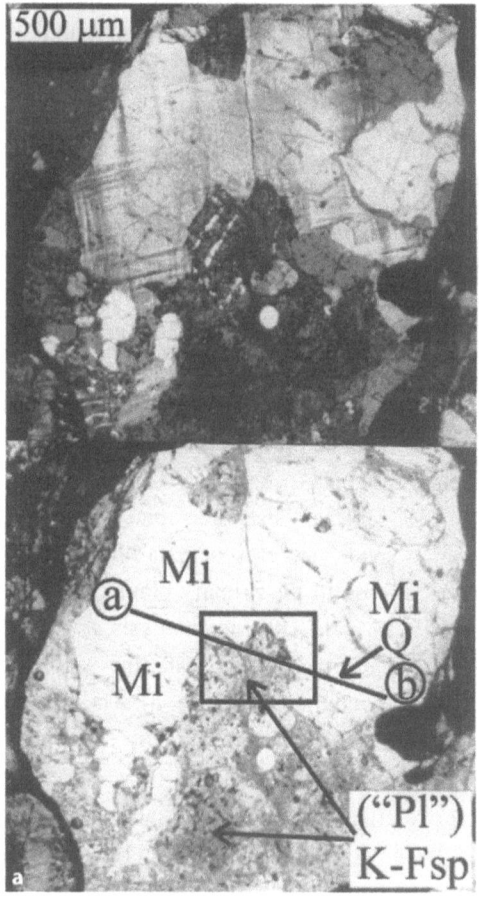

Fig. 8. (Above and opposite) Sub-crater basement. Optical, SEM and EPMA study of a granite sample from autochthonous breccia, core K1, depth 594.5 m. (a) Fabric and composition of feldspars in the granite. Top: parallel polarizers, bottom: crossed polarizers. Mineral indices as in Fig. 7. Insets: SEM image Fig. 8b and EPMA profile 8c. (b) Back-scattered electron image, inset to Fig. 8a: Plagioclase is of high porosity, whereas microcline is comparatively massive. (c) EPMA profile across microcline, altered plagioclase and quartz crystals – see profile line in Figs. 8a and 8b. No significant difference between the composition of primary microcline and altered plagioclase is observed

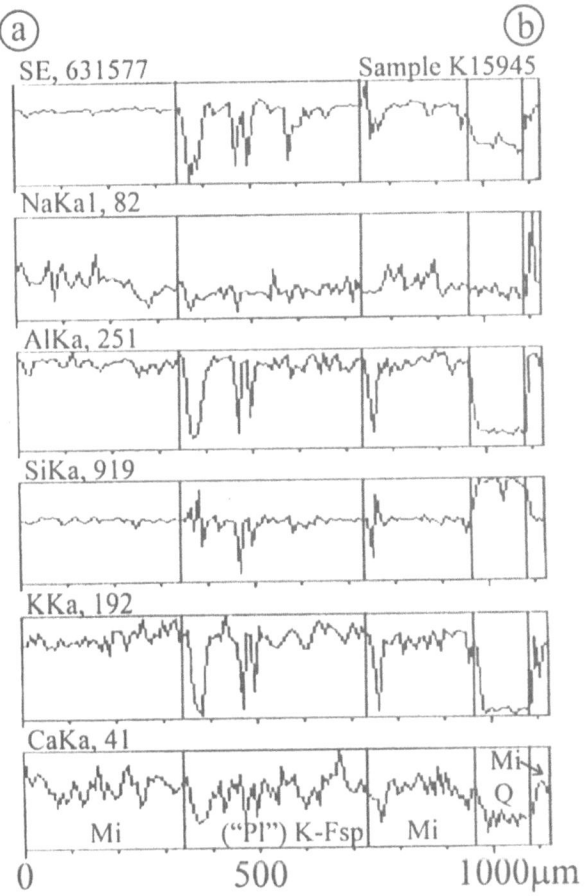

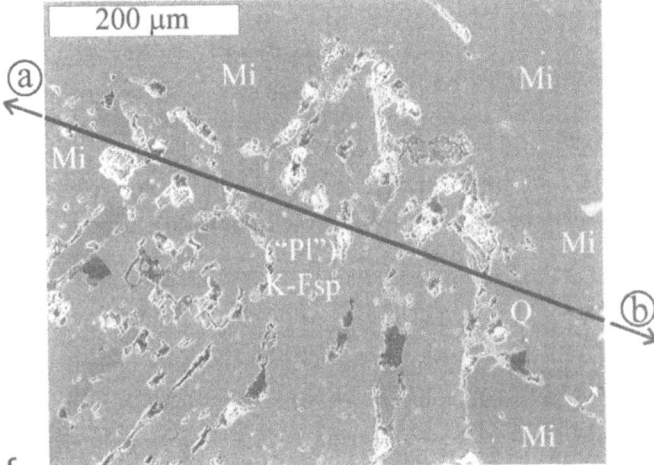

Sample K15945

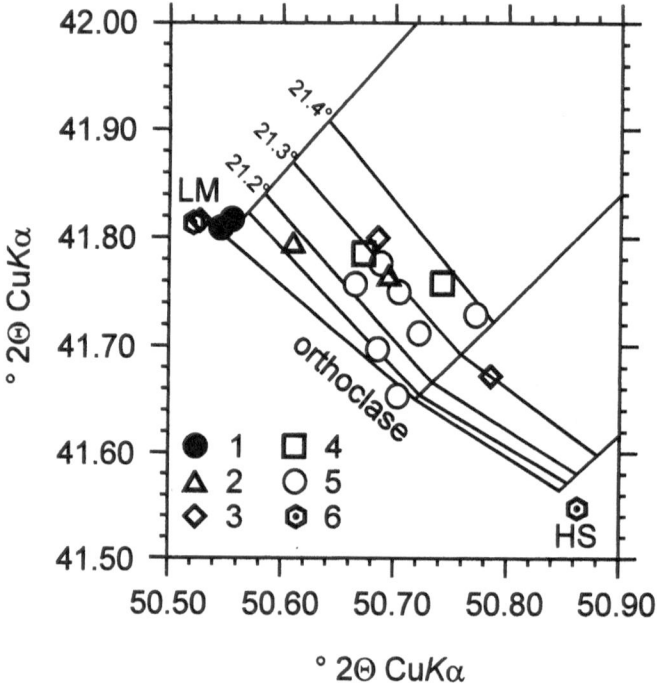

Fig. 9. Prevailing structural state of K-feldspar from granitoids of the Kärdla impact crater plotted on Wright's (1968) diagram. 1 – unchanged granite basement; 2-5 – crater rocks: 2-crater rim, 3 – autochthonous breccia and fractured rocks of the sub-crater basement, 4 – parautochthonous breccia (522.8-588.6 m in core K1), 5 – allochthonous suevitic breccia; 6 – K-feldspar structural series end-members (from Kroll and Ribbe 1987): HS – high sanidine, LM – low microcline.

Mafic rock-forming minerals

As shown by XRD, SEM-EDS and EPMA studies, amphibole and biotite - besides the deformation and brecciation that are observed - are replaced by secondary mixed-layered chlorite-smectite to chlorite-type mineral aggregates.

Amphibole

In shocked amphibolites, the hastingsitic hornblende is brecciated and partly or totally replaced by pseudomorphic chloritic aggregates with secondary quartz, calcite and ferric hydroxides. The main products of hornblende alteration were identified by XRD, SEM and EPMA as trioctahedral chlorite and mixed-layered chlorite-smectite phases mixed with secondary quartz. Amphibolite clasts within allochthonous breccia are completely replaced by chlorite and mixed-layered Fe-rich chlorite phases. The maximum alteration in the sub-crater basement and allochthonous breccia leads to complete replacement of clasts with secondary minerals. However, many clasts have alteration haloes around the fractures, where hornblende was replaced by chlorite and probably chlorite-corrensite type interstratified phases. Cores of amphibolite clasts within the parautochthonous breccia are replaced with trioctahedral mixed-layered chlorite-smectite (corrensite).

Discussion

Shock metamorphic features in minerals and small particles of impact-melted material suggesting high pressure and high temperature of impact processes were found as a minor admixture only in allochthonous suevitic breccia matrix. Consequently, the whole volume of K-enriched and Na+Ca-depleted rocks in the sub-crater basement and almost the whole mass of allochthonous breccia were formed in environments of relatively low impact pressure (far less than 8-10 GPa) as it is observed in distribution of impact-metamorphic signatures in minerals. At present, Na and Ca depletion is well documented for granitic and amphibolitic rocks. In granitoids, decomposition of plagioclase is the main process. No new Ca- and Na-rich phases have been found in altered granitoids. In amphibolites, Na and Ca are removed from plagioclase and from the rock. Hornblende that lost its calcium was replaced by chlorite with intergrowths of secondary quartz. Neither new Na nor Ca silicate minerals have formed in the Kärdla altered amphibolites. Carbonate minerals formed in altered amphibolites are of unknown age.

The main carrier of excess potassium in granites and amphibolites is a pure, cryptocrystalline, orthoclasic K-feldspar that replaces both albite-oligoclase in granitoids and oligoclase-andesine in amphibolites. Excess potassium was found in both granitic and amphibolitic rocks even if the latter occur as a large, massive bodies in the uppermost part of the central uplift.

Table 1. Chemical (wt%) and normative feldspar composition of unaffected primary microcline (1-2), perthitic albite intergrowths in primary microcline (8-9) and plagioclase altered to secondary orthoclase (3-7) from sample of granite K15945 (EPMA results).

	1. Microcline	2. Microcline	3. Orth. Sec.	4. Orth. Sec	5. Orth. Sec.	6. Orth. Sec.	7. Orth. Sec.	8. Albite	9. Albite
SiO$_2$	64.3	67.4	64.0	64.2	64.1	63.6	64.7	74.1	72.3
Al$_2$O$_3$	18.8	18.9	17.9	18.4	18.6	18.3	18.5	21.1	20.5
FeO	0.1	0	0.2	0.2	0.2	0.4	0.2	0	0.2
CaO	0.1	0.1	0	0	0.1	0.1	0	0.1	0
Na$_2$O	1.4	2.4	0.1	0.3	0.3	0.4	0.2	6.6	6.4
K$_2$O	14.7	13.7	17.0	16.9	16.8	16.6	17.1	0.9	1.3
Total	99.3	102.5	99.1	99.9	100.1	99.3	100.6	102.8	100.7
Ab	0.12	0.21	0.01	0.02	0.03	0.04	0.01	0.91	0.88
An	0.01	0.00	0.00	0.00	0.00	0.00	0.00	0.01	0.00
Or	0.87	0.79	0.99	0.98	0.97	0.96	0.99	0.09	0.12

Pl: plagioclase; Orth.Sec: secondary orthoclase; Ab: normative albite; An: normative anorthite; Or: normative orthoclase

Table 2. Chemical (wt%) and normative feldspar composition of unaffected primary plagioclase (1-2, 5, 8) and plagioclase altered to secondary orthoclase (3-4, 6-7, 9) from sample of amphibolite K16030 (EPMA results).

	1. Pl primary	2. Pl primary	3. Orth.Sec	4. Orth.Sec.	5. Pl primary	6. Orth.Sec.	7. Orth.Sec.	8. Pl primary	9. Orth.Sec.
SiO_2	59.4	59.4	68.2	63.8	59.4	65.4	64.6	59.6	65.0
Al_2O	24.8	25.0	18.8	18.0	25.0	17.7	17.8	24.9	17.6
FeO	0.2	0.2	0.2	0.3	0.1	0.1	0.1	0.0	0.0
CaO	6.7	6.7	0.0	0.0	6.9	0.0	0.2	6.5	0.0
Na_2O	7.3	7.3	0.0	0.0	7.4	0.0	0.0	7.6	0.0
K2O	0.4	0.4	15.6	16.4	0.4	16.9	16.5	0.4	16.7
Total	98.8	98.9	102.8	98.5	99.1	100.1	99.1	98.9	99.3
Ab	0.65	0.65	0.00	0.00	0.63	0.00	0.00	0.66	0.00
An	0.33	0.33	0.00	0.00	0.34	0.00	0.01	0.31	0.00
Or	0.02	0.02	1.00	1.00	0.02	1.00	0.99	0.02	1.00

Pl: plagioclase; Orth.Sec: secondary orthoclase; Ab: normative albite; An: normative anorthite; Or: normative orthoclase

No source rocks for potassium have been found at Kärdla. Selective mobility of alkalis is possible, as is shown by geochemical studies of impactites. In some large craters, strongly shock metamorphosed sub-crater rocks have lost their potassium and sodium, e.g., from feldspars (Yakovlev and Parfenova 1982). In the Kärdla case, however, the sub-crater rocks and also rock fragments and fine-grained clastic matrix material of the allochthonous breccias are enriched with potassium, but depleted with sodium and calcium. As far as we know at present, also melt rock droplets at Kärdla are enriched in potassium.

In the granites and amphibolites of the sub-crater basement, the K-enriched and Na-Ca-depleted rocks form a half-spherical body (Figs. 2b and 4). In allochthonous position, the K-enriched and Na-Ca-depleted suevitic breccias form (a) a considerably large lens overlying the parautochthonous basement, (b) a continuous layer in the lower part of the slumped breccia succession, and (c) small bodies (lumps) within the upper part of the slumped and resurge breccias. The slumped and resurge breccias are mainly composed of sedimentary (mainly fine-grained siliciclastic, and carbonate) rocks without any detectable signatures of alteration. These data suggest that the K-enrichment and Ca+Na-depletion proceeded in an environment that did not attack carbonate rocks. Thus, acid fluids as agents of K-metasomatism are unlikely. No essential differences in the structure, texture and composition of K-enriched suevitic breccia samples from different positions were found. There are no data to infer the pathways of hydrothermal systems and a late post-impact origin of potassium enrichment of these breccias. There is no principal difference between the nature of alteration of allochthonous suevitic breccia clasts and sub-crater fractured basement rocks. Therefore, we assume that this process was uniform throughout the altered zone, but less intense in marginal zones of the half-sphere.

On mineral grain level, microfracturing controlled alteration of chemically unstable material. Shock wave-induced microfracturing occurred in all minerals including chemically resistant quartz and zircon. Both K-feldspar and plagioclase are microfractured. However, the chemical reactions that occurred in the two feldspars were completely different. Along the microfractures in plagioclase, the replacement of plagioclase by secondary microcrystalline orthoclasic K-feldspar occurred. In the granites of the most K-enriched zone, the replacement of plagioclase by orthoclase was usually complete. In massive blocks of amphibolite it was only partial. Impact-induced fractures within the crystals have likely promoted secondary porosity and alteration initiation. In primary K-feldspar microcline, no chemical changes are observed along the fractures or elsewhere. Bearing in mind the orthoclase-generating process of plagioclase replacing, temperature conditions over 450°C must have been existed at that time (Brown and Parsons 1994).

Decomposition of hornblende resulted in the formation of Fe-rich chlorite-quartz aggregates, which is a typical process under hydrothermal conditions.

Microfracturing and mineralogical-geochemical alteration of different minerals are the reasons for increased porosity of the impact-influenced rocks, and for the decrease of their grain and wet density (Fig. 4). Substantial decrease of grain density of the samples from the upper part of the autochthonous sub-crater basement and difference between the two kinds of density suggest that there exists a considerable amount of micropores that remained closed during the measurement of the wet density. However, as observed by SEM studies, these pores and microfractures could have been the main pathways for migration of fluids or gases that carried alkaline elements during the alteration process(es). Reopening of pre-impact inclusion-bearing cavities (pores) in minerals was also confirmed by the study of fluid inclusions in granites. The primary high-temperature inclusions were lost in impact-affected rocks (Suuroja 1999).

The half-spherical body of sub-crater basement rocks with decreased density is detectable by the gravity survey. The volume of basement rocks with decreased porosity is about 5.4 km^3, which is at least twice as much as the present crater cavity below the basement surface level. Our data suggest that the volume of rocks subjected to petrophysical, chemical and mineralogical alteration is approximately the same.

Conclusions

The main conclusion from our study is that the impact caused significant alteration of (1) allochthonous crystalline material that was involved in the process forming the suevitic breccias, and of (2) sub-crater rocks that formed the basement of the transient cavity. Shortly after the passing shock wave, these rocks were subjected to a chemical attack of superhigh-pressure and superhigh-temperature of gases and fluids, probably related to the fireball.

Usually, the K-enriched crater-related rocks are considered to have formed during some post-impact alteration process under the influence of relic impact heat. Our data suggest that evidence for post-impact hydrothermal origin is insufficient. A strong K, Na and Ca fractionation is a process not characteristic for near-surface, low-energy geological processes. The most characteristic chemical and structural breakdown of plagioclase, together with the formation of cryptocrystalline, pseudomorphic orthoclase inside it, is not an ordinary mineralogical phenomenon of hydrothermal metasomatism.

We have tentatively considered three possible scenarios responsible for the described rock and mineral alteration: a) low-temperature post-impact hydrothermal systems, b) high-temperature post-impact hydrothermal systems, c) impact-induced high-temperature and super-high-pressure fluid and gas systems. Bearing in mind the general knowledge of conditions of mineral formation, an extensive formation of orthoclase could be evidence for supercritical fluid conditions with temperatures above 450°C. Formation of chlorite, however, could be an evidence of hydrothermal phase of mineralization. A speculative scenario of fireball-induced, early impact, high-temperature and high-pressure processes may fit the evidence of localization, composition and properties of K-enriched rocks. However, the source of potassium still remains an unsolved problem. Data on high-pressure and high-temperature reactivity between the impact-derived fluids, on one hand, and alumosilicate target rocks are critical for solving this problem and for better a understanding of these processes.

Acknowledgements

We appreciate very much the criticism and suggestions made by Philippe Claeys, Bevan French, Hartwig Frimmel and W. Uwe Reimold to our preliminary manuscripts. Philippe Claeys and W. Uwe Reimold reviewed also the re-written second version of our combined manuscript. Ivar Puura improved the text of the manuscript. This study was supported by the Estonian Science Foundation, grants No 2063 and 2191.

References

Brown WL, Parsons I (1994) Feldspars in igneous rocks. In: Parsons I (ed.) Feldspars and their reactions. Kluwer. Dordecht-Boston-London, pp 449-499

French BM, Koeberl C, Gilmour I, Shirey SB, Dons JA, Naterstad J (1997) The Gardnos impact structure, Norway: Petrology and geochemistry of target rocks and impactites. Geochim Cosmochim Acta 61: 873-904

Huffman AR, Reimold WU (1996) Experimental constraints on shock-induced microstructures in naturally deformed silicates. Tectonophysics 256: 165-217

Jõeleht A (1995) Petrophysical properties of Estonian bedrock. Report, Department of Marine Geology and Geophysics, Geological Survey of Estonia, Tallinn, 32 pp (In Estonian)

Kivisilla J, Niin M, Koppelmaa H (1999) Catalogue of chemical analyses of major elements in the rocks of the crystalline basement of Estonia. Geological Survey of Estonia, Tallinn, 97 pp

Koeberl C (1997) Impact cratering: the mineralogical and geochemical evidence. Oklahoma Geological Survey Circular 100: 30-54

Koeberl C, Reimold WU, Kracher A, Träxler B, Vormaier A, Körner W (1996) Mineralogical and petrological, and geochemical studies of drill cores from the Manson impact structure, Iowa. In: Koeberl C, Anderson RR (eds.) The Manson Impact Structure, Iowa: Anatomy of an Impact Crater. Geological Society of America Special Paper 302: 145-219

Konsa M, Puura V (1999) Provenance of zircon of the lowermost sedimentary cover, Estonia, East-European Craton. Bulletin of the Geological Society of Finland 71 (In press)

Koppelmaa H, Niin M, Kivisilla J (1996) About the petrography of the crystalline basement rocks in the Kärdla crater area, Hiiumaa Island, Estonia. Bulletin of the Geological Survey of Estonia 6: 4-24

Kroll H, Ribbe PH (1987) Determining (Al, Si) distribution and strain in alkali feldspars using lattice parameters and diffraction peak positions: A review. American Mineralogist 72: 491-506

Plado J, Pesonen LJ, Elo S, Puura V, Suuroja K (1996) Geophysical research on the Kärdla impact structure, Hiiumaa Island, Estonia. Meteoritics Planet Sci 31: 289-298

Puranen R, Sulkanen K (1995) Technical description of microcomputer-controlled petrophysical laboratory. Open File Report Q15/27/85/1, Department of Geophysics, Geological Survey of Finland, Espoo, 252 pp

Puura V, Huhma H (1993) Palaeoproterozoic age of the East Baltic granulitic crust. Precambrian Research 64: 289-294

Puura V, Kirsimäe K, Kivisilla J, Plado J, Puura I, Suuroja K (1996) Geochemical anomalies of terrestrial compounds in nonmelted impactites at Kärdla, Estonia. Meteoritics Planet Sci 31: A112-A113

Puura V, Sudov B (1976) On the platform tectonic activation zones in the southern slope of the Baltic shield, and their metallogeny. Proceedings of the Academy of Sciences of Estonian SSR. Geology 25: 206-214

Puura V, Suuroja K (1992) Ordovician impact crater at Kärdla, Hiiumaa Island, Estonia. Tectonophysics 216: 143-156

Puura V, Vaher R, Klein V, Koppelmaa H, Niin M, Vanamb V, Kirs J (1983) The Crystalline Basement of Estonian Territory. Nauka, Moscow, 207 pp (In Russian)

Reimold WU, Koeberl C, Bishop J (1994) Roter Kamm impact crater, Namibia: Geochemistry of basement rocks and breccias. Geochim Cosmochim Acta 58: 2689-2710

Suuroja S (1999) Lithologies of within-crater breccias of the Kärdla crater. MSc Thesis, Institute of Geology, University of Tartu, 78 pp (In Estonian)

Wright TL (1968) X-ray and optical study of alkali feldspar: II. An X-ray method for determinig the composition and structural state from measurement of 2θ values for three reflections. American Mineralogist 53: 88-104

Yakovlev OI, Parfenova OB (1982) Petrochemical characteristics of impactites, and consequences of vaporization and condensation in course of meteorite cratering. Meteoritika 40: 113-121 (In Russian)

Lecture Notes in Earth Sciences

For information about Vols. 1–19
please contact your bookseller or Springer-Verlag